Maksym Luz, Mikhail Moklyachuk

Non-Stationary Stochastic Processes Estimation

Also of Interest

Stochastic Dynamics of Economic Cycles
Karmalita, 2020
ISBN 978-3-11-070698-7, e-ISBN (PDF) 978-3-11-070702-1,
e-ISBN (EPUB) 978-3-11-070703-8

Weak Convergence of Stochastic Processes.
With Applications to Statistical Limit Theorems
Mandrekar, 2016
ISBN 978-3-11-047542-5, e-ISBN (PDF) 978-3-11-047631-6,
e-ISBN (EPUB) 978-3-11-047545-6

Brownian Motion.
A Guide to Random Processes and Stochastic Calculus
Schilling, Böttcher, 2021
ISBN 978-3-11-074125-4, e-ISBN (PDF) 978-3-11-074127-8,
e-ISBN (EPUB) 978-3-11-074149-0

Scientific Computing.
For Scientists and Engineers
Heister, Rebholz, 2023
ISBN 978-3-11-099961-7, e-ISBN (PDF) 978-3-11-098845-1,
e-ISBN (EPUB) 978-3-11-098875-8

Maksym Luz, Mikhail Moklyachuk

Non-Stationary Stochastic Processes Estimation

Vector Stationary Increments, Periodically Stationary Multi-Seasonal Increments

DE GRUYTER

Mathematics Subject Classification 2020
60-02, 62-02, 93-02, 60G10, 60G25, 60G35, 62M20, 62P20, 93E10, 93E11

Authors
Maksym Luz
BNP Paribas Cardif in Ukraine
Illinska st., 8
Kyiv 04070
Ukraine

Mikhail Moklyachuk
Taras Shevchenko
National University of Kyiv
Department of Probability Theory,
Statistics and Actuarial Mathematics
Volodymyrska str., 60
Kyiv 01601
Ukraine

ISBN 978-3-11-132533-0
e-ISBN (PDF) 978-3-11-132562-0
e-ISBN (EPUB) 978-3-11-132625-2

Library of Congress Control Number: 2024930853

Bibliographic information published by the Deutsche Nationalbibliothek
The Deutsche Nationalbibliothek lists this publication in the Deutsche Nationalbibliografie; detailed bibliographic data are available on the Internet at http://dnb.dnb.de.

Cover image: Peach iStock / iStock / Getty Images Plus
Typesetting: VTeX UAB, Lithuania
Printing and binding: CPI books GmbH, Leck

www.degruyter.com

Introduction

This book is dedicated to the problem of estimation of various types of non-stationary stochastic sequences and processes, namely, discrete time sequences with periodically correlated multi-seasonal increments, multidimensional (vector-valued) sequences with stationary fractional multi-seasonal increments, cointegrated vector sequences, and continuous time processes with periodically stationary (periodically correlated) increments. The authors consider the classical mean square optimal and minimax (robust) approaches to estimation of unobserved values of the linear functionals of stochastic sequences and processes.

Our investigation of the sequences with periodically stationary long memory multi-seasonal increments was motivated by an interest in models which combine different types of non-stationarity and multiple seasonal patterns. Such interest is illustrated, for example, by the articles by Dudek [46], Gould et al. [61], and Reisen et al. [178], who considered models with multiple seasonal patterns for inference and forecasting, and Hurd and Piparas [75], who introduced two models of periodic autoregressive time series with multiple periodic coefficients. This book develops our previous results with respect to the estimation of stochastic sequences and processes with stationary increments [115]. The functional estimation problem for periodically correlated increment sequences was initially introduced by Kozak and Moklyachuk [95]. The periodically correlated sequences with stationary multi-seasonal, or generalized multiple (GM), increments have been introduced in the authors' article [117].

The estimation (extrapolation, interpolation, and filtering) of stochastic processes plays a key role in various areas of applied research. Forecasting of economic and physical processes, restoration of lost information, and cleaning signals or removing noise from other data flows are essential in our information-laden world. Therefore, the development of estimation methods, which are adjusted for the wide range of time series and processes models, is considered as one of the main tasks of the modern research field of stochastic processes.

Stochastic process estimation involves consideration of two fundamental factors. The first one is the selection of a model which reflects the properties and behavior of the process. Filtering (smoothing), interpolation (estimating missing values), and extrapolation (forecasting) problems are widely investigated for stationary processes and sequences and are viewed as representative models of actual processes. The second critical factor guiding the choice of estimation method is the level of available information about the structure of the studied process. In the context of stationary processes, theoretical outcomes are derived based on complete information about the spectral structure of these processes. When working with real-world data, researchers often obtain incomplete information about the process characteristics, specifically, parametric or non-parametric estimates of the spectral function or spectral density. Thus, at a certain confidence level, the available information can be interpreted as a set of spectral functions within the confidence intervals of the estimates. Hence, when developing estimation

https://doi.org/10.1515/9783111325620-201

methods for stochastic processes, it is essential to consider both the theoretical model of the process and the fact that complete information about the investigated process may not be fully accessible.

Nowadays, the theory of (wide sense) stationary stochastic processes is well developed. For a process $x(t), t \in \mathbb{Z}$ or $t \in \mathbb{R}$, its key characteristics – the mean value $a = \mathsf{E}x(t)$ and the covariance function $\gamma(h) = \operatorname{Cov}(x(t), x(t+h))$ – are finite and do not depend on t. The main properties of stationary sequences were obtained based on the results by Cramér [36] and Stone [195]. Using the results by Wold [208, 209], Kolmogorov proposed a decomposition of stationary sequences into a sum of regular and singular stationary sequences. He also introduced a concept of minimality and was the first who proved the minimality condition $\int_{-\pi}^{\pi} f^{-1}(\lambda)d\lambda < \infty$ for the spectral density $f(\lambda)$. These properties allowed Wiener, Kolmogorov, and Yaglom to develop methods of solving the interpolation, extrapolation, and filtering problems for such processes (see selected works by Kolmogorov [90] and books by Wiener [204] and Yaglom [215, 216]). For a particular problem, they found an estimate $\widetilde{x}(t)$ constructed from a linear combination of available observations that minimizes the mean square error (MSE) $\Delta(\widetilde{x}(t), f) = \mathsf{E}|x(t) - \widetilde{x}(t)|^2$ for the fixed spectral density $f(\lambda)$ of the stationary process $x(t)$. Such estimates are called the optimal linear estimate in this book.

An extension of the theory to multidimensional processes was made in the works by Zasuhin [218], where a criterion of regularity of multidimensional stationary processes of maximal rank and a solution to the extrapolation problem were proposed, and Wiener and Masani [205, 206], Rozanov [182, 183], Hennan [68], and Yaglom [214], where the one-dimensional theory of stationary processes and estimation problems was generalized to the multidimensional case. All these studies rely on Hilbert space theory [3, 83, 84, 179].

Having developed the theory of stationary processes, researchers intended to expand it to non-stationary processes. Yaglom and Pinsker [168, 217] introduced a concept of stationarity for the nth increments of the stochastic processes $x(t)$. The generated increment process $x^{(n)}(t, \tau) = \Delta_\tau^n x(t)$ is stationary by the variable t, namely, the mathematical expectations $c^{(n)}(\tau) = \mathsf{E}x^{(n)}(t, \tau)$ and $D^{(n)}(s, \tau_1, \tau_2) = \mathsf{E}x^{(n)}(t+s, \tau_1)x^{(n)}(t, \tau_2)$ do not depend on t. Yaglom and Pinsker described the spectral representation of the increment $\Delta_\tau^{(n)} x(t)$ and the spectral density canonical factorization. Yaglom [213] solved the extrapolation and filtering problems and illustrated the solutions by examples with the rational spectral density and the spectral density $f(\lambda) = C\frac{|\lambda|^{2(n-\alpha)}}{(1+\lambda^n)^2}$. Similar results for these processes were obtained by Pinsker [167], who also introduced a self-similar structural function

$$D^{(n)}(kt, k\tau_1, k\tau_2) = a(k)D^{(n)}(t, \tau_1, \tau_2)$$

for processes with stationary increments. For more results, we refer to the book by Yaglom [215].

Another way to investigate the random fields with stationary nth increments is to apply the concept of generalized stochastic processes [212]

$$\Xi(\varphi) = \int_{\mathbb{R}^n} \xi(\mathbf{x})\varphi(\mathbf{x})d\mathbf{x}, \quad d\mathbf{x} = dx_1 \dots dx_n.$$

A wide range of works are dedicated to the wavelet transform of non-stationary processes, particularly the works by Mallat [128] and Masray [129], who solved some problems of processing of non-stationary signals, Van Bellegem [199], who solved the forecasting problem for non-stationary processes, and Krim and Pesquet [97] and Albeverio and Kawasaki [4, 5], who studied processes with stationary increments with the help of wavelets. For other applications of the wavelet approach, we refer to Nguyen and Milanfar [163].

For further insights into the spectral theory of random fields, we refer to Yadrenko's monograph [211]. For Kalman filtering theory and practice, we refer to the book by Grewal and Andrews [66]. We also would like to mention some books on the statistics of stochastic processes: Basawa and Prakasa Rao [13], Kallianpur [82], Kutoyants [98–100], Aggoun and Elliott [1], Prakasa Rao [172, 173], Rajarshi [176], Cohen and Elliott [35], and Rozovsky and Lototsky [184].

A wide range of time series models rely on the concept of stationary increment sequences, including seasonal time series, autoregressive integrated moving average (ARIMA), and seasonal autoregressive integrated moving average (SARIMA). The article by White [203] was one of the first works dedicated to non-stationary time series. It was followed by the works by Brelsford and Jones [23] and Harvey and Stock [70]. The classical book by Box and Jenkins [21] gives a description of different time series models. In the field of statistical inference for seasonal long memory sequences, recent research has yielded several notable results. One such contribution is the work by Tsai, Rachinger, and Lin [198], who developed methods for estimating model parameters when measurement errors are present. In another study, Baillie, Kongcharoen, and Kapetanios [10] compared two commonly used estimation procedures for prediction problems based on ARFIMA models. Specifically, they compared the performance of maximum likelihood estimation (MLE) to that of a two-step local Whittle estimator. Through a simulation study, they found that the MLE estimator outperformed the two-step local Whittle estimator. In addition, Hassler and Pohle [71] evaluated the predictive performance of various forecasting methods for inflation and return volatility time series. Their analysis provided compelling evidence in support of models with a fractional integration component. Recent studies have been conducted, for example, by Bell [15], Bell and Martin [16], Findley and Martin [50], Ahlgren and Nyblom [2], Seong et al. [186], Chambers and Thornton [31], Spangl and Dutter [194], Zhong [219], Chambers [30], Arvanitis and Magdalinos [7], Proietti et al. [174], and Roueff and Sanchez-Perez [181].

The next step in time series studies was the concept of cointegration, introduced and explored by Engle and Granger [49, 62]. It appears when a vector sequence with stationary increments allows decreasing an order of integration by taking some linear combination of its entries. The idea of cointegration became popular in financial and econometrics studies. It was used by Campbell and Schiller [24] to describe the present value of stocks, by Juselius and MacDonald [81] to describe relationships between exchange rates, bond prices, and consumer prices, and by Stresing, Lindenberger, and Kummel [196] to describe relationships between output, capital, labor, and energy. The cointegrated sequences, representation models, and forecasting have been studied in the papers by Phillips [166], Christoffersen and Diebold [33], Clements and Hendry [34], Avarucci and Marinucci [8], Chigira and Yamamoto [32], Johansen [79, 80], Gregoir [64], and Lazaridis [101]. The main results and further references on cointegrated sequences can be found in the book by Banerjee et al. [11].

An important problem related to sequences with stationary increments is signal extraction in the presence of noise, studied by Hannan [67] and Sobel [188–190]. Bell [14] and Sobel [189] investigated the model

$$y(t) = s(t) + n(t), \quad t \in \mathbb{Z},$$

with a non-stationary signal $s(t)$ and a stationary noise sequence $n(t)$ uncorrelated with $s(t)$. The author derived an asymptotic formula for an estimate $\widehat{s}(T)$ as $T \to \infty$ based on an infinite set of observations. Bell [14] considered the case of non-stationary noise $n(t)$ and obtained an exact formula for estimating the sequence $s(t)$ from observations of the sequence $y(t)$, making specific assumptions on initial values of the sequences. The author also considered the cases of a finite observation set and multidimensional sequences. For further insights into signal extraction, we refer to Koopman and Harvey [91], McElroy and Sutcliffe [133], Wildi [207], and McElroy [132]. Koroliouk [92] investigated the difference stochastic equation $\Delta\alpha_{t+1} = -V_0\alpha_t + \sigma_0\Delta W_{t+1}$, $t \geq 0$, which determines the sequence α_t, $t \geq 0$. Koroliouk and Koroliuk [93] have studied the problem of filtration of stationary Gaussian statistical experiments for the solution α_t, $t \geq 0$.

The problem of prediction, focusing on observations within a finite segment, was introduced by Krein [96], with the original work published in 1954. Subsequent developments were made in the article by Dym [47] and the book coauthored by Dym and McKean [48], who found solutions of the extrapolation of Gaussian stationary processes by determining an orthogonal projection in the space $L_2(\Delta)$ onto a subspace $Z^{[-T,0]}(\Delta) = \overline{\mathrm{span}}\{e_t(\lambda)\xi; -T \leq t \leq 0, \xi \in \mathbb{C}\}$, where $e_t(\lambda) = e^{i\lambda t}$ for the stationary process case and $e_t(\lambda) = (e^{i\lambda t} - 1)(i\lambda)^{-1}$ for processes with stationary increments. In this context, Dym and McKean considered the space $Z^{[-T,0]}(\Delta)$ as a de Branges space of entire functions [37] and determined projections using reproducing kernels. The generalization of these problems to p-dimensional stationary processes with stationary increments was investigated by Arov and Dym [6]. Bondon [19, 20] explored the prediction problem based on incomplete past information, while Soltani and Mohammadpour [192] studied multivariate stationary processes.

Another type of non-stationarity is described by periodically correlated (periodically stationary), or cyclostationary, processes introduced by Gladyshev [58]. The correlation function $K(t,s) = \mathsf{E}x(t)\overline{x(s)}$ of such processes is a T-periodic function, $K(t,s) = K(t+T, s+T)$, which implies a time-dependent spectrum. These processes are widely used in signal processing and communications, which is described in the books by Gardner [54], Hurd and Miamee [76], and Napolitano [159, 162]. For further works on cyclostationarity and its applications, see the reviews by Napolitano [160, 161], Gardner et al. [55], and Serpedin et al. [187], as well as contributions to the Workshops on Cyclostationary Systems and Their Applications, Grodek, Poland [26–29]. Periodic time series can be considered as an extension of the SARIMA model [9, 12, 103, 164] and are suitable for forecasting stream flows with quarterly, monthly, or weekly cycles; see Osborn [164]. Particularly, Lund [103] proposed a test assessing if a PARMA model is preferable to a SARMA one. The model's performance improves when it includes a fractional integration. A variety of articles investigated fractional patterns. As an example, see Porter-Hudak [171] for the investigation of the seasonal ARFIMA model with application to the monetary aggregates used by the US Federal Reserve.

Baek, Davis, and Pipiras [9] introduced a periodic dynamic factor model (PDFM) with periodic vector autoregressive (PVAR) factors, in contrast to seasonal VARIMA factors.

In the previously mentioned works, the authors assumed that the covariations or spectral densities of the processes are exactly known. The methods used for parameter estimation and filtering of time series data often fail to account for real-world challenges such as outliers, measurement errors, and incomplete information about the spectral structure. As a result, interest in robust estimation methods that can effectively handle such issues is growing. For example, Reisen et al. [177] proposed a semi-parametric robust estimator for the fractional parameters in the SARFIMA model and illustrated its application in forecasting of SO_2 concentrations. Solci et al. [191] proposed robust estimates of a periodic autoregressive (PAR) model.

Earlier, Vastola and Poor [200] demonstrated that, for certain classes of spectral densities, estimates of unknown values are sensitive to even minor changes in the spectral structure of the process. This sensitivity can lead to a substantial increase in estimation error. Thus, estimation methods need robust adaptation for practical use. One potential approach to address this challenge is to derive the minimax estimates. The method consists in choosing a certain class of admissible spectral densities and finding the spectral density from this class which results in the highest estimation error.

The worst sequence for forecasting of a single value of a stationary sequence is a sequence of uncorrelated random variables (white noise). When the value of a linear functional of unobserved values is estimated, the white noise sequence does not appear as the worst sequence. For the first time, this result was shown by Grenander [65] for a linear functional $Ax = \int_0^1 a(t)x(t)dt$ of a continuous time stationary stochastic process $\{x(t), t \in \mathbb{R}\}$. He considered a game of two players, X and P; player X chooses a stationary process $x(t)$ with a fixed variance (a spectral density f) and player P builds a forecast $\widehat{A}x$

of the functional Ax based on the observations at points $t < 0$. Player X tries to maximize the value $\Delta(f, \widehat{A}x) = \mathsf{E}|Ax - \widehat{A}x|^2$ and player P tries to minimize it. This game has a saddle point under proper conditions:

$$\max_f \min_{\widehat{A}x} \Delta(f, \widehat{A}x) = \min_{\widehat{A}x} \max_f \Delta(f, \widehat{A}x) = \nu.$$

The function $a(t)$, $t \in (0, 1)$, determines the game solution. Similar two-person games for stationary sequences and processes were considered by Moklyachuk [136, 140] (see also the book [145] and the review article [146]).

Uncertainty of the spectral structure may occur in different ways. For example, Huber [73] studied the normal distribution with ε-mixture and solved the estimation problem with the help of an M-estimator. Later on, he found the robustness conditions (insensitivity to small deviations) of the censored likelihood ratio test and proposed how to check if a spectral density is the least favorable in a given class [74]. Kassam and Poor [88], Breyman [22], Hosoya [72], Taniguchi [197], and Liu et al. [102] extended these results. Kassam [86, 87] summarized the results for the interpolation problem of stationary sequences. By applying the statistical hypothesis testing approach, he found the least favorable spectral densities.

For the stationary sequence observed with noise, Franke [51, 52] found solutions of the minimax extrapolation and interpolation problems in the case of one missing value using the convex optimization method for a given set of admissible spectral densities. Poor [169, 170], Franke and Poor [53], and Vastola and Poor [201] studied the minimax filtering and related problems. We should also mention the work by Verdu and Poor [202] dedicated to robust methods of estimation.

The main results of robust estimation techniques and its applications for stationary sequences published up to 1985 were summarized in a review article by Kassam and Poor [88]. Further results on minimax extrapolation, interpolation, and filtering problems for linear functionals have been derived by Moklyachuk [136–146], who investigated the estimation problems for stationary sequences and processes and found the classical mean square optimal and minimax estimates based on observations with and without noise. He illustrated these results with the help of some specific sets of admissible spectral densities.

Robust estimation approaches are widely applied in studies of stochastic processes and their applications. See, for example, the works by Hansen and Sargent [69], Ishihara et al. [77], Gholipour et al. [56], Bertsimas et al. [17], Delage and Ye [38], Spangl and Dutter [194], Ruckdeschel et al. [185], Carrizosaa et al. [25], Jiang et al. [78], Molloy and Ford [158], Song [193], and Bhattacharya and Thomakos [18]. For a deeper understanding of robust methods in statistical forecasting, we refer to the book by Kharin [89].

Robust estimation relies on optimization techniques which are described in the books [70, 134, 175, 180].

This book relies on previous works of Moklyachuk and his students. The book by Moklyachuk and Masyutka [151] summarizes results of minimax estimation of multidimensional stationary sequences and processes. The book by Moklyachuk and Golichenko (Dubovets'ka) [147] describes results for the same problems for periodically correlated sequences and processes. In the book by Luz and Moklyachuk [115], solutions of extrapolation, interpolation, and filtering problems for linear functionals depending on the unobserved values of stochastic sequences and processes with stationary *n*th increments are presented. References to the particular articles are given directly in the corresponding chapters of this book. More references to Moklyachuk's, Masyutka's, and Golichenko's articles are given in the Conclusion sections of the corresponding chapters of the book.

We also refer to the book [135] and the articles [130, 131] by Moklyachuk, Masyutka, and Sidei for the results of the estimation problem for the sequences and processes from observations with missing values and to the articles by Moklyachuk and Ostapenko [155–157], who studied minimax estimation of harmonizable stochastic sequences and processes.

In this book, we describe the results of our investigation of the extrapolation, interpolation, and filtering problems for the linear functionals of stochastic sequences and processes with periodically stationary increments. In the discrete time case, the multi-seasonal increments are in scope. The case of fractional increments is discussed as well. For the continuous time processes, a periodically stationary increment without seasonal patterns is considered, which illustrates the simplest type of the studied increment.

To deal with a periodical correlation, the sequences and processes are presented as multidimensional, finite and infinite, increment sequences.

Formulas for calculating the MSEs and the spectral characteristics of the optimal estimates of the functionals are derived in the case of spectral certainty, where the spectral structures of the involved sequences and processes are exactly known. The Hilbert space projection technique is applied after proper transformations of the functionals in scope. In the case where spectral densities of the sequences and processes are not known exactly, while some sets of admissible spectral densities are given, we apply the minimax-robust method of estimation. This method allows us to find estimates that minimize the maximal values of the MSEs of the estimates for all spectral densities from a given class of admissible spectral densities. Formulas that determine the least favorable spectral densities and the minimax spectral characteristics are proposed for some special classes of admissible densities.

The book is organized as follows.

In Chapter 1, we give a definition of the sequences with periodically correlated multiple seasonal increments. They are represented as multidimensional, or vector-valued, stochastic sequences with stationary GM increments. The chapter presents the spectral theory of stationary GM increment sequences. It also discusses the case of fractional multi-seasonal increments.

In Chapters 2–5, we describe our results on the classical mean square optimal and minimax estimation of the linear functionals of the sequences with periodically correlated GM increments. The estimates are based on the Fourier transformation coefficients of the functions involving spectral densities of the studied stochastic sequences. For the extrapolation and filtering problems, we formulate lemmas which allow us to derive the corresponding formulas in terms of the coefficients of canonical factorizations of the spectral densities.

The extrapolation problem without noise is presented in Chapter 2. It also contains various examples of forecasting, illustrating applications of the proposed estimation technique.

Extrapolation, interpolation, and filtering problems with periodically correlated noise are described in Chapters 3–5. Chapter 3 contains an example of applying the obtained results to solving the extrapolation problem for a special class of cointegrated sequences. Chapter 4 also contains the solution of the interpolation problem without noise.

In the next chapters, we describe our results for continuous time processes with periodically correlated increments.

In Chapter 6, we give a definition of the processes with periodically correlated increments and represent them as infinite-dimensional stochastic sequences with stationary increments. We consider the simplest case of increments which describe only integrated patterns without seasonal behavior. The chapter presents the elements of spectral theory of infinite-dimensional stationary increment sequences.

As in the discrete case, in Chapters 7–11, we present our results on classical and minimax estimation of the linear functionals of the processes with periodically correlated increments.

Chapter 7 contains results of the extrapolation problem without noise, Chapters 8–10 contain results of extrapolation, interpolation, and filtering problems with periodically correlated noise, and Chapter 11 contains results of the filtering problem when both signal and noise processes have periodically correlated increments. The filtering problem in Chapter 11 is solved under additional conditions on the function $a(t)$, $t \geq 0$, compared to the case of Chapter 10 when the noise process is periodically correlated. An illustrative example of the minimax approach is given in Chapter 9 for interpolation without noise.

A list of problems covering the discussed topics is proposed to the readers.

Contents

Notations and abbreviations

$\mathbb{N}$	the set of positive integer numbers
$\mathbb{N}_0$	$= \mathbb{N} \cup \{0\}$
$\mathbb{Z}$	the set of integer numbers
$\mathbb{R}$	the set of real numbers
$\mathbb{C}$	the set of complex numbers
$\binom{n}{l}$	$= \frac{n!}{l!(n-l)!}$, binomial coefficient
$\{x\}_{p=1}^{T}$	a T-dimensional vector $\vec{x} = \{x_p\}_{p=1}^{T} = (x_1, x_2, \dots, x_T)^\top$
$\{x\}_{p=1}^{\infty}$	an infinite-dimensional vector $\vec{x} = \{x_p\}_{p=1}^{\infty} = (x_1, x_2, \dots)^\top$
$\{f_{ij}(\lambda)\}_{i,j=1}^{T}$	a $T \times T$ matrix function $f(\lambda) = \{f_{ij}(\lambda)\}_{i,j=1}^{T}$
$\{f_{ij}(\lambda)\}_{i,j=1}^{\infty}$	an infinite-dimensional matrix function $f(\lambda) = \{f_{ij}(\lambda)\}_{i,j=1}^{\infty}$
$\{\varphi_{ij}\}_{i=\overline{1,T}}^{j=\overline{1,q}}$	a $T \times q$ matrix $\varphi = \{\varphi_{ij}\}_{i=\overline{1,T}}^{j=\overline{1,q}}$
δ_{lp}	Kronecker symbols
$\vec{\delta}_l$	a vector $\vec{\delta}_l = \{\delta_{lp}\}_{p=1}^{T}$, $l = 1, \dots, T$
$\vec{0}$	a T-dimensional or infinite-dimensional vector $\vec{0} = (0, \dots, 0)^\top$ or $\vec{0} = (0, \dots, 0, \dots)^\top$
$\mathcal{O}$	a finite- or infinite-dimensional matrix with zero entries
E_q, E_M	identity $q \times q$ and $M \times M$ matrices
$\mathbb{I}\{\cdot\}$	an indicator function
$a \mod b$	the remainder after dividing an integer a by an integer b
$x \vee y$	the maximum of two numbers x and y
$x \wedge y$	the minimum of two numbers x and y
$[x]$	the integer part of x
$\lceil x \rceil$	the least integer greater than or equal to x
$(\Omega, \mathcal{F}, \mathrm{P})$	the probability space
$\mathrm{E}x$	the mathematical expectation of x
$\langle \cdot \, ; \cdot \rangle$	inner product
$L_2(\Omega, \mathcal{F}, \mathrm{P})$	the Hilbert space of random variables with zero first moment and finite second moment endowed with the inner product $\langle \zeta; \eta \rangle = \mathrm{E}\zeta\overline{\eta}$
$L_2(f)$	the Hilbert space of the integrable vector-valued functions endowed with the inner product $\langle g_1; g_2 \rangle = \int_{-\pi}^{\pi} (g_1(\lambda))^\top f(\lambda) \overline{g_2(\lambda)} d\lambda$
B_μ	a backward shift operator with the step μ, $B_\mu \xi(t) = \xi(t - \mu)$
B_μ^s	a backward shift operator $B_\mu^s = B_\mu B_\mu \cdot \dots \cdot B_\mu$
$\overline{d}$	a multi-increment step $\overline{d} = (d_1, d_2, \dots, d_r)$
$\overline{s}$	a multi-increment season $\overline{s} = (s_1, s_2, \dots, s_r)$
$\chi_{\overline{\mu},\overline{s}}^{(d)}(B)$	an incremental operator $\chi_{\overline{\mu},\overline{s}}^{(d)}(B) = (1 - B_{\mu_1}^{s_1})^{d_1} (1 - B_{\mu_2}^{s_2})^{d_2} \cdot \dots \cdot (1 - B_{\mu_r}^{s_r})^{d_r}$
$\Delta\xi_m$	an increment of the sequence ξ_m, $\Delta\xi_m = \xi_m - \xi_{m-1}$
$\xi^{(d)}(m, \mu)$	stochastic dth increment sequence (process) with step μ
$\chi_{\overline{\mu},\overline{s}}^{(d)}(\xi(m))$	stochastic generalized multiple (GM) increment sequence of order d with a seasonal vector $\overline{s}$ and a step $\overline{\mu}$
$\mathrm{diag}_T(x)$	a $T \times T$ diagonal matrix with entry x on its diagonal
f, g	matrix-valued spectral densities of the vector increment sequences (processes)
F, G	matrix-valued spectral functions of the vector increment sequences (processes)
$A\xi$	a linear functional of a stochastic sequence or process ξ
$\widehat{A}\xi$	an estimate of a functional $A\xi$
$\Delta(f; \widehat{A}\xi)$	the mean square error of an estimate $\widehat{A}\xi$, where f is a spectral density of ξ
$\vec{d}$, F	a finite- or infinite-dimensional vector or matrix

https://doi.org/10.1515/9783111325620-202

d, **F**	a vector or a matrix with the vector $\vec{d}_j$ or the matrix $F_{k,j}$ entries
$\vec{d}^{\top}, F^{\top}$	the transpose of a vector $\vec{d}$ or a matrix F
$\overline{\vec{d}}, \overline{F}$	the conjugate of a vector $\vec{d}$ or a matrix F
$(\mathbf{d})_k$	the kth entry of a vector **d**
F^*	the conjugate transpose of a matrix F
$\det(F)$	the determinant of a matrix F
$\overline{\text{span}}\{\cdot\}$	a closed linear subspace generated by a set of elements
$\text{Proj}_S \xi$	an orthogonal projection of $\xi \in H$ on a closed linear subspace S of a Hilbert space H
h	a spectral characteristic of an estimate
$\mathcal{D}$	a set of admissible spectral densities
$\delta(\cdot \mid \mathcal{D})$	an indicator function of a set $\mathcal{D}$
$\partial\Delta_{\mathcal{D}}(f^0)$	a subdifferential of a functional $\Delta_{\mathcal{D}}(f)$ at point $f = f^0$
$\text{Tr}[f(\lambda)]$	the trace of a spectral density matrix $f(\lambda)$
i. i. d.	independent and identically distributed
GM increment	generalized multiple increment
FM increment	fractional multiple increment
MSE	mean square error
ARMA	autoregressive moving average
ARIMA	integrated ARMA
ARFIMA	fractional integrated ARMA
SARMA	seasonal ARMA
SARFIMA	seasonal fractional integrated ARMA
GARMA	Gegenbauer ARMA
SAR	seasonal autoregressive
VAR	vector autoregressive
VARIMA	vector ARIMA
PAR	periodic autoregressive
PARMA	periodic ARMA
PVAR	periodic vector autoregressive

1 Periodically stationary multi-seasonal increments of stochastic sequences

In this chapter, we describe properties of the periodically stationary long memory multiplicative seasonal increments of discrete time stochastic processes defined and studied in [117]. We give a definition of the increment sequence and introduce stochastic sequences with periodically stationary (periodically correlated, cyclostationary) long memory multiplicative seasonal increments. These non-stationary stochastic sequences combine the periodic structure of covariation functions of sequences as well as the integrating pattern. A short review of the spectral theory of these increment sequences is presented.

1.1 Stochastic sequences with periodically stationary generalized multiple increments

1.1.1 Generalized multiple increment sequence

Consider a stochastic sequence $\xi = \{\xi(m), m \in \mathbb{Z}\}$. Let B_μ denote a backward shift operator with step $\mu \in \mathbb{Z}$, such that $B_\mu \xi(m) = \xi(m-\mu)$, $B := B_1$. Then $B^s_\mu = B_\mu B_\mu \cdot \dots \cdot B_\mu$. Define a *multiple increment operator*

$$\chi^{(d)}_{\overline{\mu},\overline{s}}(B) = (1-B^{s_1}_{\mu_1})^{d_1}(1-B^{s_2}_{\mu_2})^{d_2} \cdot \dots \cdot (1-B^{s_r}_{\mu_r})^{d_r},$$

where $d := d_1 + d_2 + \dots + d_r$, $\overline{d} = (d_1, d_2, \dots, d_r) \in (\mathbb{N})^r$, $\overline{s} = (s_1, s_2, \dots, s_r) \in \mathbb{N}^r$, and $\overline{\mu} = (\mu_1, \mu_2, \dots, \mu_r) \in \mathbb{N}^r$ or $\in (\mathbb{Z} \setminus \mathbb{N}_0)^r$. Here $\mathbb{N}_0 = \mathbb{N} \cup \{0\}$ and $\binom{n}{l} = \frac{n!}{l!(n-l)!}$.

Definition 1.1.1. For a stochastic sequence $\xi(m)$, $m \in \mathbb{Z}$, the sequence

$$\begin{aligned}
\chi^{(d)}_{\overline{\mu},\overline{s}}(\xi(m)) &:= \chi^{(d)}_{\overline{\mu},\overline{s}}(B)\xi(m) = (1-B^{s_1}_{\mu_1})^{d_1}(1-B^{s_2}_{\mu_2})^{d_2} \cdot \dots \cdot (1-B^{s_r}_{\mu_r})^{d_r}\xi(m) \\
&= \sum_{l_1=0}^{d_1} \dots \sum_{l_r=0}^{d_r} (-1)^{l_1+\dots+l_r} \binom{d_1}{l_1} \cdot \dots \cdot \binom{d_r}{l_r} \\
&\quad \times \xi(m - \mu_1 s_1 l_1 - \dots - \mu_r s_r l_r), \quad m \in \mathbb{Z},
\end{aligned} \tag{1.1}$$

is called a stochastic generalized multiple (GM) increment sequence of differentiation order d with a fixed seasonal vector $\overline{s} \in \mathbb{N}^r$ and a varying increment step $\overline{\mu} \in \mathbb{N}^r$ or $\overline{\mu} \in (\mathbb{Z} \setminus \mathbb{N}_0)^r$.

Let γ denote a triple $(\overline{\mu}, \overline{s}, \overline{d})$. For $i = 1, 2, \dots, r$, $j \in \mathbb{Z}$ define coefficients $M^j_i := [\frac{j}{\mu_i s_i}]$ and $I^j_i := \mathbb{I}\{j \mod \mu_i s_i = 0\}$, where $\mathbb{I}\{\cdot\}$ is the indicator function, and notations $n_i :=$

https://doi.org/10.1515/9783111325620-001

$\mu_i s_i d_i$, $\langle \overline{s}, \overline{\mu}, \overline{d}\rangle_k := \sum_{i=1}^{k} \mu_i s_i d_i = \sum_{i=1}^{k} n_i$, $d(\gamma) := \langle \overline{s}, \overline{\mu}, \overline{d}\rangle_r$. Let $x \vee y$ denote the maximum of two numbers x, y and let $x \wedge y$ denote the minimum of two numbers x, y.

Lemma 1.1.1. *The multiple increment operator $\chi^{(d)}_{\overline{\mu},\overline{s}}(B)$ admits a representation*

$$\chi^{(d)}_{\overline{\mu},\overline{s}}(B) = \prod_{i=1}^{r}(1 - B^{s_i}_{\mu_i})^{d_i} = \sum_{k=0}^{d(\gamma)} e_\gamma(k)B^k,$$

where

$$e_\gamma(k) = \sum_{k_{r-1}=0\vee k_r - n_r}^{\langle \overline{s},\overline{\mu},\overline{d}\rangle_{r-1}\wedge k_r} \sum_{k_{r-2}=0\vee k_{r-1}-n_{r-1}}^{\langle \overline{s},\overline{\mu},\overline{d}\rangle_{r-2}\wedge k_{r-1}} \cdots \sum_{k_1=0\vee k_2-n_2}^{\langle \overline{s},\overline{\mu},\overline{d}\rangle_{1}\wedge k_2} e_\gamma(k_1, \dots, k_r)$$

and

$$e_\gamma(k_1, \dots, k_r) = (-1)^{\sum_{i=1}^{r} M_i^{k_i - k_{i-1}}} \prod_{i=1}^{r} I_i^{k_i - k_{i-1}} \prod_{i=1}^{r} \binom{d_i}{M_i^{k_i - k_{i-1}}},$$

$k_0 := 0$, $k_r := k$.

Proof. We have

$$\begin{aligned}
\prod_{i=1}^{r}(1 - B^{s_i}_{\mu_i})^{d_i} &= \prod_{i=1}^{r}\left(\sum_{j_i=0}^{d_i}(-1)^{j_i}\binom{d_i}{j_i}B^{j_i\mu_i s_i}\right)\\
&= \prod_{i=1}^{r}\left(\sum_{j_i=0}^{d_i\mu_i s_i}(-1)^{[j_i/\mu_i s_i]}\binom{d_i}{[j_i/\mu_i s_i]}\mathbb{I}\{j_i \mod \mu_i s_i = 0\}B^{j_i}\right)\\
&= \sum_{j_1=0}^{n_1}\cdots\sum_{j_r=0}^{n_r}\left((-1)^{\sum_{i=1}^{r} M_i^{j_i}}\prod_{i=1}^{r} I_i^{j_i}\prod_{i=1}^{r}\binom{d_i}{M_i^{j_i}}\right)B^{\sum_{i=1}^{r} j_i}.
\end{aligned}$$

By replacing $j_1 \to k_1$, $k_1 + j_2 \to k_2$, $k_2 + j_3 \to k_3, \dots, k_{r-1} + j_r \to k_r := k$, we obtain

$$\begin{aligned}
\prod_{i=1}^{r}(1 - B^{s_i}_{\mu_i})^{d_i} &= \sum_{k_2=0}^{n_1+n_2}\sum_{k_1=0\vee k_2-n_2}^{n_1\wedge k_2}\sum_{j_3=0}^{n_3}\cdots\sum_{j_r=0}^{n_r}(-1)^{M_2^{k_2-k_1}+M_1^{k_1}} I_2^{k_2-k_1} I_1^{k_1}\\
&\quad\times\binom{d_2}{M_2^{k_2-k_1}}\binom{d_1}{M_1^{k_1}}\left((-1)^{\sum_{i=3}^{r} M_i^{j_i}}\prod_{i=3}^{r} I_i^{j_i}\prod_{i=3}^{r}\binom{d_i}{M_i^{j_i}}\right)B^{k_2+\sum_{i=3}^{r} j_i}\\
&= \sum_{k_3=0}^{n_1+n_2+n_3}\sum_{k_2=0\vee k_3-n_3}^{n_1+n_2\wedge k_3}\sum_{k_1=0\vee k_2-n_2}^{n_1\wedge k_2}\sum_{j_4=0}^{n-4}\cdots\sum_{j_r=0}^{n_r}(-1)^{\sum_{i=1}^{3} M_i^{k_i-k_{i-1}}}\prod_{i=1}^{3} I_i^{k_i-k_{i-1}}\\
&\quad\times\prod_{i=1}^{3}\binom{d_i}{M_i^{k_i-k_{i-1}}}\left((-1)^{\sum_{i=4}^{r} M_i^{j_i}}\prod_{i=3}^{r} I_i^{j_i}\prod_{i=3}^{r}\binom{d_i}{M_i^{j_i}}\right)B^{k_3+\sum_{i=4}^{r} j_i}
\end{aligned}$$

$$= \sum_{k=0}^{d(\gamma)} e_{\gamma}(k) B^k,$$

where $e_{\gamma}(k)$ are coefficients from the lemma statement. □

The theory of (wide sense) stationary stochastic sequences describes second order random variables $\eta(m)$, $m \in \mathbb{Z}$, such that the mean value $a = \mathsf{E}\eta(m_0)$ and the covariance function $\gamma(h) = \mathrm{Cov}(\eta(m_0), \eta(m_0 + h))$ are finite and do not depend on m_0. The following definition describes the stationarity of the increment sequence $\chi^{(d)}_{\overline{\mu},\overline{s}}(\xi(m))$, $m \in \mathbb{Z}$, with a generalization of the mean value and covariance function, arising from the presence of the increment step $\overline{\mu}$.

Definition 1.1.2. A stochastic GM increment sequence $\chi^{(d)}_{\overline{\mu},\overline{s}}(\xi(m))$, $m \in \mathbb{Z}$, is called wide sense stationary if the mathematical expectations

$$\mathsf{E}\chi^{(d)}_{\overline{\mu},\overline{s}}(\xi(m_0)) = c^{(d)}_{\overline{s}}(\overline{\mu}),$$

$$\mathsf{E}\chi^{(d)}_{\overline{\mu}_1,\overline{s}}(\xi(m_0 + m))\chi^{(d)}_{\overline{\mu}_2,\overline{s}}(\xi(m_0)) = D^{(d)}_{\overline{s}}(m; \overline{\mu}_1, \overline{\mu}_2)$$

exist for all $m_0, m, \overline{\mu}, \overline{\mu}_1, \overline{\mu}_2$ and do not depend on m_0. The function $c^{(d)}_{\overline{s}}(\overline{\mu})$ is called a mean value and the function $D^{(d)}_{\overline{s}}(m; \overline{\mu}_1, \overline{\mu}_2)$ is called a structural function of the stationary GM increment sequence (of a stochastic sequence with stationary GM increments).

The stochastic sequence $\xi(m)$, $m \in \mathbb{Z}$, determining the stationary GM increment sequence $\chi^{(d)}_{\overline{\mu},\overline{s}}(\xi(m))$ by (1.1) is called a stochastic sequence with stationary GM increments (or GM increment sequence of order d).

Remark 1.1.1. For spectral properties of the one-pattern stochastic increment sequence $\chi^{(d)}_{\mu,1}(\xi(m)) := \xi^{(d)}(m, \mu) = (1 - B_{\mu})^d \xi(m)$, see, e. g., [115, 215]. The corresponding results for the continuous time stochastic increment process $\xi^{(d)}(t, \tau) = (1 - B_{\tau})^d \xi(t)$ are described in [213, 217].

1.1.2 Definition and spectral representation of periodically stationary GM increments

In this section, we present a definition, a justification, and a brief review of the spectral theory of stochastic sequences with periodically stationary multiple seasonal increments.

Definition 1.1.3. A stochastic sequence $\xi(m)$, $m \in \mathbb{Z}$, is called a stochastic sequence with periodically stationary (periodically correlated) GM increments with period T if the mathematical expectations exist and satisfy the relations

$$\mathsf{E}\chi^{(d)}_{\overline{\mu},T\overline{s}}(\xi(m + T)) = \mathsf{E}\chi^{(d)}_{\overline{\mu},T\overline{s}}(\xi(m)) = c^{(d)}_{T\overline{s}}(m, \overline{\mu}),$$

$$\mathsf{E}\chi^{(d)}_{\overline{\mu}_1,T\overline{s}}(\xi(m+T))\chi^{(d)}_{\overline{\mu}_2,T\overline{s}}(\xi(k+T)) = D^{(d)}_{T\overline{s}}(m+T,k+T;\overline{\mu}_1,\overline{\mu}_2)$$
$$= D^{(d)}_{T\overline{s}}(m,k;\overline{\mu}_1,\overline{\mu}_2)$$

for every $m, k, \overline{\mu}_1, \overline{\mu}_2$ and $T > 0$ is the least integer for which these equalities hold.

It follows from Definition 1.1.3 that the sequence

$$\xi_p(m) = \xi(mT + p - 1), \quad p = 1, 2, \dots, T, \quad m \in \mathbb{Z}, \tag{1.2}$$

forms a multidimensional sequence $\vec{\xi}(m) = \{\xi_p(m)\}_{p=1}^{T}$, $m \in \mathbb{Z}$, with stationary GM increments as follows:

$$\begin{aligned}\chi^{(d)}_{\overline{\mu},\overline{s}}(\xi_p(m)) &= \sum_{l_1=0}^{d_1}\cdots\sum_{l_r=0}^{d_r}(-1)^{l_1+\dots+l_r}\binom{d_1}{l_1}\cdot\dots\cdot\binom{d_r}{l_r}\\ &\quad\times\xi_p(m-\mu_1 s_1 l_1-\dots-\mu_r s_r l_r)\\ &= \sum_{l_1=0}^{d_1}\cdots\sum_{l_r=0}^{d_r}(-1)^{l_1+\dots+l_r}\binom{d_1}{l_1}\cdot\dots\cdot\binom{d_r}{l_r}\\ &\quad\times\xi((m-\mu_1 s_1 l_1-\dots-\mu_r s_r l_r)T+p-1)\\ &= \chi^{(d)}_{\overline{\mu},T\overline{s}}(\xi(mT+p-1)), \quad p = 1, 2, \dots, T,\end{aligned}$$

where $\chi^{(d)}_{\overline{\mu},\overline{s}}(\xi_p(m))$ is the GM increment of the pth component of the vector-valued sequence $\vec{\xi}(m)$.

Example 1.1.1. Define a T-periodic seasonal autoregressive integrated moving average model (PSARIMA) $X = \{X_m, m \in \mathbb{Z}\}$ with multiple seasonal patterns by the relation

$$\phi_m(B)(1-B^T)^d\prod_{i=1}^{r}\Phi_{i,m}(B)(1-B^{Ts_i})^{d_i}X_m = \theta_m(B)\prod_{i=1}^{r}\Theta_{i,m}(B)\varepsilon_m,$$

where $\varepsilon = \{\varepsilon_t, t \in \mathbb{Z}\}$ is a sequence of zero mean i. i. d. random variables, and where all polynomials $\phi_m(z)$, $\theta_m(z)$, $\Phi_{i,m}(z)$, and $\Theta_{i,m}(z)$ are T-periodic by parameter m functions, $1 < s_1 < \dots < s_r$. Define

$$\Phi_m(z) := \phi_m(z)\prod_{i=1}^{r}\Phi_{i,m}(z) = \sum_{k=0}^{q_1}\Phi_m(k)z^k,$$
$$\Theta_m(z) := \theta_m(z)\prod_{i=1}^{r}\Theta_{i,m}(z) = \sum_{k=0}^{q_2}\Theta_m(k)z^k$$

and put $\Phi_m(k) = 0$ for $k > q_1$ and $\Theta_m(k) = 0$ for $k > q_2$. Then the increment sequence

$$Y_m = (1-B^T)^d\prod_{i=1}^{r}(1-B^{Ts_i})^{d_i}X_m$$

is periodically stationary and allows a stationary vector representation

$$\mathbf{Y}_m = (1-B)^d \prod_{i=1}^{r} (1-B^{s_i})^{d_i} \mathbf{X}_m$$

with

$$\mathbf{Y}_m = (Y_{mT}, Y_{mT+1}, \dots, Y_{mT+T-1})^\top,$$
$$\mathbf{X}_m = (X_{mT}, X_{mT+1}, \dots, X_{mT+T-1})^\top,$$

$\boldsymbol{\varepsilon}_m = (\varepsilon_{mT}, \varepsilon_{mT+1}, \dots, \varepsilon_{mT+T-1})^\top$. We can write the relation

$$\mathbf{\Pi Y}_m + \sum_{l=1}^{q_1^*} \mathbf{\Pi}_l \mathbf{Y}_{m-l} = \Xi \boldsymbol{\varepsilon}_m + \sum_{l=1}^{q_2^*} \Xi_l \boldsymbol{\varepsilon}_{m-l},$$

where $\mathbf{\Pi}(k,j) = \Phi_k(k-j), \Xi(k,j) = \Theta_k(k-j)$ for $k \ge j$ and $\mathbf{\Pi}(k,j) = 0, \Xi(k,j) = 0$ otherwise. We have $\mathbf{\Pi}_l(k,j) = \Phi_k(lT+k-j), \Xi_l(k,j) = \Theta_k(lT+k-j)$ [45], provided $\det(\mathbf{\Pi}+\sum_{l=1}^{q_1^*} \mathbf{\Pi}_l z^l) \neq 0$ for $|z| \le 1$ [68]. A GM increment sequence is defined as

$$\chi_{\overline{\mu},\overline{s}}^{(d)}(\mathbf{X}_m) = (1-B^{\mu_0})^d \prod_{i=1}^{r} (1-B^{s_i \mu_i})^{d_i} \mathbf{X}_m, \quad m \in \mathbb{Z}.$$

The following theorem describes the spectral structure of the GM increment [85, 117].

Theorem 1.1.1. 1. *The mean value and the structural function of the vector-valued stationary stochastic GM increment sequence $\chi_{\overline{\mu},\overline{s}}^{(d)}(\vec{\xi}(m))$ can be represented in the form*

$$c_{\overline{s}}^{(d)}(\overline{\mu}) = \vec{c} \prod_{i=1}^{r} \mu_i^{d_i}, \tag{1.3}$$

$$D_{\overline{s}}^{(d)}(m; \overline{\mu}_1, \overline{\mu}_2) = \int_{-\pi}^{\pi} e^{i\lambda m} \chi_{\overline{\mu}_1}^{(d)}(e^{-i\lambda}) \chi_{\overline{\mu}_2}^{(d)}(e^{i\lambda}) \frac{1}{|\beta^{(d)}(i\lambda)|^2} dF(\lambda), \tag{1.4}$$

where

$$\chi_{\overline{\mu}}^{(d)}(e^{-i\lambda}) = \prod_{j=1}^{r} (1-e^{-i\lambda \mu_j s_j})^{d_j},$$

$$\beta^{(d)}(i\lambda) = \prod_{j=1}^{r} \prod_{k_j=-[s_j/2]}^{[s_j/2]} (i\lambda - 2\pi i k_j / s_j)^{d_j},$$

$\vec{c}$ is a vector, and $F(\lambda)$ is the matrix-valued spectral function of the stationary stochastic sequence $\chi^{(d)}_{\overline{\mu},\overline{s}}(\vec{\xi}(m))$. The vector c and the matrix-valued function $F(\lambda)$ are determined uniquely by the GM increment sequence $\chi^{(d)}_{\overline{\mu},\overline{s}}(\vec{\xi}(m))$.

2. *The stationary GM increment sequence $\chi^{(d)}_{\overline{\mu},\overline{s}}(\vec{\xi}(m))$ admits*

$$\chi^{(d)}_{\overline{\mu},\overline{s}}(\vec{\xi}(m)) = \int_{-\pi}^{\pi} e^{im\lambda}\chi^{(d)}_{\overline{\mu}}(e^{-i\lambda})\frac{1}{\beta^{(d)}(i\lambda)}d\vec{Z}_{\xi^{(d)}}(\lambda), \tag{1.5}$$

where $\vec{Z}_{\xi^{(d)}}(\lambda) = \{Z_p(\lambda)\}_{p=1}^{T}$ is a (multidimensional) stochastic process with uncorrelated increments on $[-\pi,\pi)$ connected with the spectral function $F(\lambda)$ by the relation

$$\mathsf{E}(Z_p(\lambda_2) - Z_p(\lambda_1))(\overline{Z_q(\lambda_2) - Z_q(\lambda_1)}) = F_{pq}(\lambda_2) - F_{pq}(\lambda_1),$$
$$-\pi \le \lambda_1 < \lambda_2 < \pi, \quad p,q = 1,2,\dots,T.$$

Proof. We follow the idea from [213] for continuous time stationary increments. Consider a one-dimensional GM increment sequence with one seasonal factor $\chi^{(d)}_{\mu,s}(\xi(m)) = \xi^{(d)}_s(m,\mu) = (1 - B^s_\mu)^d\xi(m)$. The equality

$$\xi^{(d)}_s(m,\mu) = \sum_{l=0}^{(\mu-1)d} A_l\xi^{(d)}(m - ls, 1), \quad \mu > 0, \tag{1.6}$$

where $\{A_l, l = 0,1,2,\dots,(\mu-1)d\}$ are coefficients from the representation

$$(1 + x + \dots + x^{\mu-1})^d = \sum_{l=0}^{(\mu-1)d} A_l x^l,$$

implies

$$c^{(d)}_s(\mu) = \mathsf{E}\xi^{(d)}_s(m,\mu) = (A_0 + A_1 + \dots + A_{(\mu-1)d})c^{(d)}_s(1) = \mu^d c^{(d)}_s(1) = c\mu^d,$$

where $c = c^{(d)}_s(1)$ does not depend on μ.

Since $D^{(d)}_s(m;\mu,\mu)$ is a positive-definite function with respect to variable m, one can define a function $F_{\mu,s}(\lambda)$ depending on the parameter μ, which is a real bounded non-decreasing left-continuous with respect to a $\lambda \in [-\pi,\pi)$ function, such that

$$D^{(d)}_s(m;\mu,\mu) = \int_{-\pi}^{\pi} e^{i\lambda m}dF_{\mu,s}(\lambda). \tag{1.7}$$

Again, equality (1.6) implies

$$\mathsf{E}\xi^{(d)}_s(m_1 + m,\mu)\overline{\xi^{(d)}_s(m,\mu)} = \sum_{p=0}^{(\mu-1)d}\sum_{q=0}^{(\mu-1)d} A_pA_q\mathsf{E}\xi^{(d)}_s(m_1 + m - ps, 1)\overline{\xi^{(d)}_s(m_2 - qs, 1)}$$

$$= \int_{-\pi}^{\pi} \sum_{p=0}^{(\mu-1)d} \sum_{q=0}^{(\mu-1)d} A_p A_q e^{i\lambda(m-(p-q)s)} dF_{1,s}(\lambda)$$

$$= \int_{-\pi}^{\pi} e^{i\lambda m} \sum_{p=0}^{(k-1)d} A_p e^{-ips\lambda} \sum_{q=0}^{(\mu-1)d} A_q e^{iqs\lambda} dF_{1,s}(\lambda)$$

$$= \int_{-\pi}^{\pi} e^{im\lambda} \frac{(1-e^{-i\mu s\lambda})^d}{(1-e^{-is\lambda})^d} \frac{(1-e^{i\mu s\lambda})^d}{(1-e^{is\lambda})^d} dF_{1,s}(\lambda).$$

Thus,

$$\int_{-\pi}^{\pi} e^{im\lambda} dF_{\mu,s}(\lambda) = \int_{-\pi}^{\pi} e^{im\lambda} \frac{(1-\cos \mu s\lambda)^d}{(1-\cos s\lambda)^d} dF_{1,s}(\lambda). \tag{1.8}$$

The latter equality implies

$$F_{\mu,s}(\lambda) = \int_{-\pi}^{\lambda} \frac{(1-\cos \mu s u)^d}{(1-\cos s u)^d} F_{1,s}(u)$$

and

$$\int_{-\pi}^{\lambda} \frac{|\beta^{(d)}(iu)|^2}{(1-\cos \mu s u)^d} dF_{\mu,s}(u) = \int_{-\pi}^{\lambda} \frac{|\beta^{(d)}(iu)|^2}{(1-\cos s u)^d} dF_{1,s}(u), \tag{1.9}$$

where the function $\beta^{(d)}(iu)$ is to be chosen in the way that the integrals are defined for $\lambda \in [-\pi, \pi)$ and converge at the neighborhoods of the points $\cos su = 1$, $u \in [-\pi, \pi]$, namely, the points $u = 2\pi k/s$ for $|k| \le s/2$, $k \in \mathbb{Z}$. Thus, we choose the function

$$\beta^{(d)}(iu) = \prod_{k=-[s/2]}^{[s/2]} (iu - 2\pi i k/s)^d.$$

The right side of equality (1.9) does not depend on k; thus, we put

$$F(\lambda) = \frac{1}{2^d} \int_{-\pi}^{\lambda} \frac{|\beta^{(d)}(iu)|^2}{(1-\cos \mu s u)^d} dF_{\mu,s}(u). \tag{1.10}$$

The function $F(\lambda)$ is a real-valued non-decreasing bounded function defined on $[-\pi, \pi)$ such that $F(-\pi) = 0$. Consider the function $F(\lambda)$ as left-continuous. This function is the one stated in the theorem.

Relation (1.4) for $r = 1$ is obtained by considering the following equality for positive μ_1, μ_2:

$$D_s^{(d)}(m,\mu_1,\mu_2) = \int_{-\pi}^{\pi} e^{im\lambda} \frac{(1-e^{-i\mu_1 s\lambda})^d}{(1-e^{-is\lambda})^d} \frac{(1-e^{i\mu_2 s\lambda})^d}{(1-e^{is\lambda})^d} dF_{1,s}(\lambda)$$

$$= \int_{-\pi}^{\pi} e^{im\lambda}\left(1-e^{-i\mu_1 s\lambda}\right)^d\left(1-e^{i\mu_2 s\lambda}\right)^d \frac{1}{|\beta^{(d)}(i\lambda)|^2} dF(\lambda).$$

For negative μ_1, μ_2, an equality $\xi_s^{(d)}(m,-\mu) = (-1)^d \xi_s^{(d)}(m+d\mu s,\mu)$ is applied. By generalizing for $r > 1$ the given reasonings, we obtain relations (1.3) and (1.4).

The spectral representation of the increment sequence follows from representation (1.4) and the Karhunen theorem [85]. □

Consider another multidimensional stochastic sequence with the stationary GM increments $\vec{\zeta}(m) = \vec{\xi}(m) + \vec{\eta}(m)$, where $\vec{\eta}(m)$ is a multidimensional stationary stochastic sequence, uncorrelated with $\vec{\xi}(m)$, with a spectral representation

$$\vec{\eta}(m) = \int_{-\pi}^{\pi} e^{i\lambda m} d\vec{Z}_\eta(\lambda),$$

where $\vec{Z}_\eta(\lambda) = \{Z_{\eta,p}(\lambda)\}_{p=1}^T$, $\lambda \in [-\pi,\pi)$, is a stochastic process with uncorrelated increments that corresponds to the spectral function $G(\lambda)$ [68]. The stochastic stationary GM increment $\chi_{\overline{\mu},\overline{s}}^{(d)}(\vec{\zeta}(m))$ allows the spectral representation

$$\chi_{\overline{\mu},\overline{s}}^{(d)}(\vec{\zeta}(m)) = \int_{-\pi}^{\pi} e^{i\lambda m} \frac{\chi_{\overline{\mu}}^{(d)}(e^{-i\lambda})}{\beta^{(d)}(i\lambda)} d\vec{Z}_{\xi^{(d)}}(\lambda) + \int_{-\pi}^{\pi} e^{i\lambda m} \chi_{\overline{\mu}}^{(d)}(e^{-i\lambda}) d\vec{Z}_\eta(\lambda),$$

while $d\vec{Z}_\eta(\lambda) = (\beta^{(d)}(i\lambda))^{-1} d\vec{Z}_{\eta^{(d)}}(\lambda)$, $\lambda \in [-\pi,\pi)$. Therefore, in the case where the spectral functions $F(\lambda)$ and $G(\lambda)$ have the spectral densities $f(\lambda)$ and $g(\lambda)$, the spectral density matrix $p(\lambda) = \{p_{ij}(\lambda)\}_{i,j=1}^T$ of the stochastic sequence $\vec{\zeta}(m)$ is determined by the formula

$$p(\lambda) = f(\lambda) + \left|\beta^{(d)}(i\lambda)\right|^2 g(\lambda).$$

1.1.3 Moving average representation of vector stationary GM increment

Denote by $H = L_2(\Omega,\mathcal{F},\mathsf{P})$ the Hilbert space of random variables ζ with zero first moment, $\mathsf{E}\zeta = 0$, and finite second moment, $\mathsf{E}|\zeta|^2 < \infty$, endowed with the inner product $\langle\zeta,\eta\rangle = \mathsf{E}\zeta\overline{\eta}$. In the space H, consider a closed linear subspace

$$H(\vec{\xi}^{(d)}) = \overline{\text{span}}\{\chi_{\overline{\mu},\overline{s}}^{(d)}(\xi_p(m)), p = 1,\dots,T;\ m \in \mathbb{Z}\}$$

generated by the components of the stationary stochastic GM increment sequence $\vec{\xi}^{(d)} = \{\{\chi^{(d)}_{\overline{\mu},\overline{s}}(\xi_p(m))\}_{p=1}^{T}, m \in \mathbb{Z}\}$ with the steps $\overline{\mu} > \overline{0}$. For $q \in \mathbb{Z}$, consider also a closed linear subspace

$$H^q(\vec{\xi}^{(d)}) = \overline{\operatorname{span}}\{\chi^{(d)}_{\overline{\mu},\overline{s}}(\xi_p(m)),\ p = 1, \dots, T;\ m \leqslant q\}.$$

Define a subspace

$$S(\vec{\xi}^{(d)}) = \bigcap_{q\in\mathbb{Z}} H^q(\vec{\xi}^{(d)})$$

of the Hilbert space $H(\vec{\xi}^{(d)})$. The space $H(\vec{\xi}^{(d)})$ admits a decomposition $H(\vec{\xi}^{(d)}) = S(\vec{\xi}^{(d)}) \oplus R(\vec{\xi}^{(d)})$, where $R(\vec{\xi}^{(d)})$ is the orthogonal complement of the subspace $S(\vec{\xi}^{(d)})$ in the space $H(\vec{\xi}^{(d)})$.

Definition 1.1.4. A stationary (wide sense) stochastic vector GM increment sequence $\chi^{(d)}_{\overline{\mu},\overline{s}}(\vec{\xi}(m)) = \{\chi^{(d)}_{\overline{\mu},\overline{s}}(\xi_p(m))\}_{p=1}^{T}$ is called regular if $H(\vec{\xi}^{(d)}) = R(\vec{\xi}^{(d)})$, and it is called singular if $H(\vec{\xi}^{(d)}) = S(\vec{\xi}^{(d)})$.

Theorem 1.1.2. *A stationary stochastic vector GM increment sequence $\chi^{(d)}_{\overline{\mu},\overline{s}}(\vec{\xi}(m)) = \{\chi^{(d)}_{\overline{\mu},\overline{s}}(\xi_p(m))\}_{p=1}^{T}$ is uniquely represented in the form*

$$\chi^{(d)}_{\overline{\mu},\overline{s}}(\xi_p(m)) = \chi^{(d)}_{\overline{\mu},\overline{s}}(\xi_{S,p}(m)) + \chi^{(d)}_{\overline{\mu},\overline{s}}(\xi_{R,p}(m)), \tag{1.11}$$

where $\chi^{(d)}_{\overline{\mu},\overline{s}}(\xi_{R,p}(m))$, $p = 1, \dots, T$, is a regular stationary GM increment sequence and $\chi^{(d)}_{\overline{\mu},\overline{s}}(\xi_{S,p}(m))$, $p = 1, \dots, T$, is a singular stationary GM increment sequence. The GM increment sequences $\chi^{(d)}_{\overline{\mu},\overline{s}}(\xi_{R,p}(m))$, $p = 1, \dots, T$, and $\chi^{(d)}_{\overline{\mu},\overline{s}}(\xi_{S,p}(m))$, $p = 1, \dots, T$, are orthogonal for all $m, k \in \mathbb{Z}$. They are defined by the formulas

$$\chi^{(d)}_{\overline{\mu},\overline{s}}(\xi_{S,p}(m)) = \mathsf{E}[\chi^{(d)}_{\overline{\mu},\overline{s}}(\xi_p(m))|S(\vec{\xi}^{(d)})],$$
$$\chi^{(d)}_{\overline{\mu},\overline{s}}(\xi_{R,p}(m)) = \chi^{(d)}_{\overline{\mu},\overline{s}}(\xi_p(m)) - \chi^{(d)}_{\overline{\mu},\overline{s}}(\xi_{S,p}(m)), \quad p = 1, \dots, T.$$

Consider a vector innovation sequence $\vec{\varepsilon}(u) = \{\varepsilon_k(u)\}_{k=1}^{q}$, $u \in \mathbb{Z}$, for a regular stationary GM increment, namely, a sequence of uncorrelated random variables such that $\mathsf{E}\varepsilon_k(u)\overline{\varepsilon}_j(v) = \delta_{kj}\delta_{uv}$, $\mathsf{E}|\varepsilon_k(u)|^2 = 1$, $k, j = 1, \dots, q$, $u \in \mathbb{Z}$, and $H^r(\vec{\xi}^{(d)}) = H^r(\vec{\varepsilon})$ holds for all $r \in \mathbb{Z}$, where $H^r(\vec{\varepsilon})$ is the Hilbert space generated by elements $\{\varepsilon_k(u) : k = 1, \dots, q; u \leq r\}$; δ_{kj} and δ_{uv} are Kronecker symbols.

Theorem 1.1.3. *A stationary vector GM increment sequence $\chi^{(d)}_{\overline{\mu},\overline{s}}(\vec{\xi}(m))$ is regular if and only if there exists an innovation sequence $\vec{\varepsilon}(u)$, $u \in \mathbb{Z}$, and a sequence of matrix-valued functions $\varphi^{(d)}(k, \overline{\mu}) = \{\varphi^{(d)}_{ij}(k, \overline{\mu})\}_{i=\overline{1,T}}^{j=\overline{1,q}}$, $k \geq 0$, such that*

$$\sum_{k=0}^{\infty}\sum_{i=1}^{T}\sum_{j=1}^{q}|\varphi_{ij}^{(d)}(k,\overline{\mu})|^2 < \infty, \quad \chi_{\overline{\mu},\overline{s}}^{(d)}(\vec{\xi}(m)) = \sum_{k=0}^{\infty}\varphi^{(d)}(k,\overline{\mu})\vec{\varepsilon}(m-k). \tag{1.12}$$

Representation (1.12) is called the canonical moving average representation of the stochastic stationary GM increment sequence $\chi_{\overline{\mu},\overline{s}}^{(d)}(\vec{\xi}(m))$.

The spectral function $F(\lambda)$ of a stationary GM increment sequence $\chi_{\overline{\mu},\overline{s}}^{(d)}(\vec{\xi}(m))$ which admits the canonical representation (1.12) has the spectral density $f(\lambda) = \{f_{ij}(\lambda)\}_{i,j=1}^{T}$ admitting the canonical factorization

$$f(\lambda) = \varphi(e^{-i\lambda})\varphi^*(e^{-i\lambda}), \tag{1.13}$$

where the function $\varphi(z) = \sum_{k=0}^{\infty}\varphi(k)z^k$ has analytic in the unit circle $\{z : |z| \leq 1\}$ components $\varphi_{ij}(z) = \sum_{k=0}^{\infty}\varphi_{ij}(k)z^k$, $i = 1,\dots,T, j = 1,\dots,q$. Based on the moving average representation (1.12), define

$$\varphi_{\overline{\mu}}(z) = \sum_{k=0}^{\infty}\varphi^{(d)}(k,\overline{\mu})z^k = \sum_{k=0}^{\infty}\varphi_{\overline{\mu}}(k)z^k.$$

Then the following relation holds:

$$\varphi_{\overline{\mu}}(e^{-i\lambda})\varphi_{\overline{\mu}}^*(e^{-i\lambda}) = \frac{|\chi_{\overline{\mu}}^{(d)}(e^{-i\lambda})|^2}{|\beta^{(d)}(i\lambda)|^2}f(\lambda) = \prod_{j=1}^{r}\frac{|1-e^{-i\lambda\mu_j s_j}|^{2d_j}}{\prod_{k_j=-[s_j/2]}^{[s_j/2]}|\lambda-2\pi k_j/s_j|^{2d_j}}f(\lambda). \tag{1.14}$$

1.2 Stochastic sequences with fractional multiple increments

In the previous subsection, we described the GM increment sequence $\chi_{\overline{\mu},\overline{s}}^{(d)}(\vec{\xi}(m))$ of the positive integer orders $(d_1,\dots,d_r)$. Here we consider the case of possibly fractional increment orders d_i.

In this subsection, we use the step $\overline{\mu} = (1,1,\dots,1)$. We present the increment operator $\chi_{\overline{s}}^{(d)}(B)$ in the form

$$\chi_{\overline{s}}^{(R+D)}(B) = (1-B)^{R_0+D_0}\prod_{j=1}^{r}(1-B^{s_j})^{R_j+D_j}, \tag{1.15}$$

where $(1-B)^{R_0+D_0}$ is an integration component, $R_j, j = 0,1,\dots,r$, are non-negative integer numbers, and $1 < s_1 < \dots < s_r$. Below we describe a representation $d_j = R_j + D_j$, $j = 0,1,\dots,r$, of the increment orders d_j by stating conditions on the fractional parts D_j, such that the increment sequence $\vec{y}(m) := (1-B)^{R_0}\prod_{j=1}^{r}(1-B^{s_j})^{R_i}\vec{\xi}(m)$ is a stationary fractionally integrated seasonal stochastic sequence. For example, in case of the single increment pattern $(1-B^{s^*})^{R^*+D^*}$, this condition is $|D^*| < 1/2$.

We will call a sequence $\chi_{\overline{s}}^{(R+D)}(\vec{\xi}(m))$ *a fractional multiple (FM) increment sequence.*

Lemma 1.2.1. *The fractional increment operator* $\chi_{\bar{s}}^{(D)}(B) := (1-B)^{D_0}\prod_{j=1}^{r}(1-B^{s_j})^{D_j}$ *admits a representation*

$$\chi_{\bar{s}}^{(D)}(B) = \prod_{j=0}^{r}\prod_{k_j=0}^{[s_j/2]}(1-2\cos\nu_{k_j}B+B^2)^{D_{k_j}}$$
$$= (1-B)^{D_0+D_1+\dots+D_r}\prod_{j=1}^{r}\prod_{k_j=1}^{[s_j/2]}(1-2\cos\nu_{k_j}B+B^2)^{D_{k_j}},$$

where $s_0 = 1$, $\nu_{k_j} = 2\pi k_j/s_j$, $k_j = 0,1,\dots,[s_j/2]$, $D_{k_j} = D_0/2$ *for* $k_j = 0$, $D_{k_j} = D_j$ *for* $k_j = 1,2,\dots,[s_j/2]-1$, $D_{[s_j/2]} = D_j$ *for odd* s_j, *and* $D_{[s_j/2]} = D_j/2$ *for even* s_j.

Note that Lemma 1.2.1 follows from the representation

$$(1-B^{s_j})^{D_r} = \prod_{k_j=0}^{[s_j/2]}(z_{k_j}-B)^{D_{k_j}}(z_{-k_j}-B)^{D_{k_j}},$$

where $z_{k_j} = \exp(\nu_{k_j}i)$, $k_j = 0,1,\dots,s_j-1$, are solutions of the equation $1-B^{s_j} = 0$. Lemma 1.2.1 implies the following statement.

Lemma 1.2.2. *Define the sets* $\mathcal{M}_j = \{\nu_{k_j} = 2\pi k_j/s_j : k_j = 0,1,\dots,[s_j/2]\}$, $j = 0,1,\dots,r$, *and the set* $\mathcal{M} = \bigcup_{j=0}^{r}\mathcal{M}_j$. *Then*

$$\chi_{\bar{s}}^{(D)}(B) = \prod_{\nu\in\mathcal{M}}(1-2\cos\nu B+B^2)^{\tilde{D}_\nu}$$
$$= (1-B)^{D_0+D_1+\dots+D_r}(1+B)^{D_\pi}\prod_{\nu\in\mathcal{M}\setminus\{0,\pi\}}(1-2\cos\nu B+B^2)^{D_\nu},$$

where $D_\nu = \sum_{j=0}^{r}D_j\mathbb{I}\{\nu\in\mathcal{M}_j\}$, $\tilde{D}_\nu = D_\nu$ *for* $\nu\in\mathcal{M}\setminus\{0,\pi\}$, *and* $\tilde{D}_\nu = D_\nu/2$ *for* $\nu = 0$ *or* $\nu = \pi$.

Lemma 1.2.2 shows that a multi-seasonal fractional increment sequence can be represented as a k-factor Gegenbauer process:

$$\prod_{i=1}^{k}(1-2u_iB+B^2)^{d_i}x(m) = \xi(m). \tag{1.16}$$

In the case where $\xi(m)$ is an ARMA(p,q) sequence, the model $x(m)$ defined by (1.16) is called a k-factor GARMA(p,d_i,u_i,q) sequence. It is stationary and invertible if $|d_i| < 1/2$ for $|u_i| < 1$ and $|d_i| < 1/4$ for $|u_i| = 1$. If additionally $d_i > 0$, the model exhibits long memory behavior [210]. The function $(1-2u_iB+B^2)^{-d_i}$ is a generating function of the Gegenbauer polynomial:

$$(1-2uB+B^2)^{-d}=\sum_{n=0}^{\infty}C_n^{(d)}(u)B^n,$$

where

$$C_n^{(d)}(u)=\sum_{k=0}^{[n/2]}\frac{(-1)^k(2u)^{n-2k}\Gamma(d-k+n)}{k!(n-2k)!\Gamma(d)}.$$

Thus, denoting by $k^*=|\mathcal{M}|$ the number of elements of the set $\mathcal{M}$, we obtain

$$\begin{aligned}(\chi_{\overline{s}}^{(D)}(B))^{-1}&=\prod_{\nu\in\mathcal{M}}(1-2\cos\nu B+B^2)^{-\widetilde{D}_\nu}\\&=\sum_{m=0}^{\infty}G_{k^*}^{+}(m)B^m=\left(\sum_{m=0}^{\infty}G_{k^*}^{-}(m)B^m\right)^{-1},\end{aligned}$$

where

$$G_{k^*}^{+}(m)=\sum_{0\le n_1,\dots,n_{k^*}\le m,n_1+\dots+n_{k^*}=m}\;\prod_{\nu\in\mathcal{M}}C_{n_\nu}^{(\widetilde{D}_\nu)}(\cos\nu),\tag{1.17}$$

$$G_{k^*}^{-}(m)=\sum_{0\le n_1,\dots,n_{k^*}\le m,n_1+\dots+n_{k^*}=m}\;\prod_{\nu\in\mathcal{M}}C_{n_\nu}^{(-\widetilde{D}_\nu)}(\cos\nu).\tag{1.18}$$

The derived representations of the fractional increment operator $\chi_{\overline{s}}^{(D)}(B)$ imply the following theorem.

Theorem 1.2.1. *Assume that for a stochastic vector sequence $\vec{\xi}(m)$ and fractional differentiation orders $d_j=R_j+D_j$, $j=0,1,\dots,r$, the FM increment sequence $\chi_{\overline{1,s}}^{(R+D)}(\vec{\xi}(m))$ generated by increment operator (1.15) is a stationary sequence with a bounded from zero and from infinity spectral density $\widetilde{f}_{\overline{1}}(\lambda)$. Then for the non-negative integer numbers R_j, $j=0,1,\dots,r$, the GM increment sequence $\chi_{\overline{1,s}}^{(R)}(\vec{\xi}(m))$ is stationary if $-1/2<D_\nu<1/2$ for all $\nu\in\mathcal{M}$, where D_ν are defined by real numbers D_j, $j=0,1,\dots,r$, in Lemma 1.2.2, and it is long memory if $0<D_\nu<1/2$ for at least one $\nu\in\mathcal{M}$ and invertible if $-1/2<D_\nu<0$. The spectral density $f(\lambda)$ of the stationary GM increment sequence $\chi_{\overline{1,s}}^{(R)}(\vec{\xi}(m))$ admits a representation*

$$\begin{aligned}f(\lambda)&=|\beta^{(R)}(i\lambda)|^2|\chi_{\overline{1}}^{(R)}(e^{-i\lambda})|^{-2}|\chi_{\overline{1}}^{(D)}(e^{-i\lambda})|^{-2}\widetilde{f}_{\overline{1}}(\lambda)\\&=:|\chi_{\overline{1}}^{(D)}(e^{-i\lambda})|^{-2}\widetilde{f}(\lambda),\end{aligned}$$

where

$$|\chi_{\overline{1}}^{(D)}(e^{-i\lambda})|^{-2}=\left|\sum_{m=0}^{\infty}G_{k^*}^{+}(m)e^{-i\lambda m}\right|^2=\left|\sum_{m=0}^{\infty}G_{k^*}^{-}(m)e^{-i\lambda m}\right|^{-2}$$

$$= \prod_{\nu\in\mathcal{M}} |(e^{-i\nu} - e^{i\lambda})(e^{i\nu} - e^{i\lambda})|^{-2\widetilde{D}_\nu},$$

where coefficients $G^{+}_{k^*}(m)$ and $G^{-}_{k^*}(m)$ are defined by (1.17) *and* (1.18), *respectively.*

The spectral density $f(\lambda)$ and the structural function $D^{(R)}_{\bar{s}}(m,\bar{1},\bar{1})$ of a stationary GM increment sequence $\chi^{(R)}_{\bar{1},\bar{s}}(\vec{\xi}(m))$ exhibit the following behavior for the constant matrices C and K:

- $|\beta^{(R)}(i\lambda)|^{-2}|\chi^{(R)}_{\bar{1}}(e^{-i\lambda})|^2 f(\lambda) \sim C|\nu-\lambda|^{-2\widetilde{D}_\nu}$ when $\lambda\to\nu$, $\nu\in\mathcal{M}$ (for properties of eigenvalues of generalized fractional process, we refer to Palma and Bondon [165]),
- $D^{(R)}_{\bar{s}}(m,\bar{1},\bar{1}) \sim K\sum_{\nu\in\mathcal{M}:\widetilde{D}_\nu>0} |m|^{2\widetilde{D}_\nu-1}\cos(m\nu)$, as $m\to\infty$ (see Giraitis and Leipus [57]).

Let the conditions of Theorem 1.2.1 for the stochastic sequence $\vec{\xi}(m)$ be fulfilled. Consider a stationary stochastic sequence $\vec{\eta}(m)$, uncorrelated with $\vec{\xi}(m)$, having the spectral density $g(\lambda)$. The stationary GM increment sequence $\chi^{(R)}_{\bar{1},\bar{s}}(\vec{\xi}(m)+\vec{\eta}(m))$ has the spectral density

$$\begin{aligned} p(\lambda) &= |\beta^{(R)}(i\lambda)|^2|\chi^{(R)}_{\bar{1}}(e^{-i\lambda})|^{-2}|\chi^{(D)}_{\bar{1}}(e^{-i\lambda})|^{-2}\widetilde{f}_{\bar{1}}(\lambda) + |\beta^{(R)}(i\lambda)|^2 g(\lambda) \\ &= |\chi^{(D)}_{\bar{1}}(e^{-i\lambda})|^{-2}\widetilde{f}(\lambda) + |\beta^{(R)}(i\lambda)|^2 g(\lambda). \end{aligned}$$

Example 1.2.1. 1. For the increment operator $(1-B)^{R_0+D_0}(1-B^2)^{R_1+D_1}$, $\mathcal{M}_0=\{0\}$, $\mathcal{M}_1=\{0,\pi\}$, $\mathcal{M}=\{0,\pi\}$, the Gegenbauer representation is

$$(1-B)^{D_0+D_1}(1+B)^{D_1}.$$

Stationarity conditions: $|D|=|D_0+D_1|<1/2$, $|D_\pi|=|D_1|<1/2$.

2. For the increment operator $(1-B^2)^{R_1+D_1}(1-B^3)^{R_2+D_2}$, $\mathcal{M}_0=\{0,\pi\}$, $\mathcal{M}_1=\{0,2\pi/3\}$, $\mathcal{M}=\{0,2\pi/3,\pi\}$, the Gegenbauer representation is

$$(1-B)^{D_1+D_2}(1-2\cos(2\pi/3)B+B^2)^{D_2}(1+B)^{D_1}.$$

Stationarity conditions: $|D|=|D_1+D_2|<1/2$, $|D_{2\pi/3}|=|D_2|<1/2$, $|D_\pi|=|D_1|<1/2$.

3. For the increment operator $(1-B^2)^{R_1+D_1}(1-B^4)^{R_2+D_2}$, $\mathcal{M}_0=\{0,\pi\}$, $\mathcal{M}_1=\{0,\pi/2,\pi\}$, $\mathcal{M}=\{0,\pi/2,\pi\}$, the Gegenbauer representation is

$$(1-B)^{D_1+D_2}(1+B^2)^{D_2}(1+B)^{D_1+D_2}.$$

Stationarity conditions: $|D|=|D_\pi|=|D_1+D_2|<1/2$, $|D_{\pi/2}|=|D_2|<1/2$.

Conclusions

In this chapter, we propose a description of properties of the periodically stationary long memory multiplicative seasonal increments of discrete time stochastic processes defined and studied in [117]. We introduce stochastic sequences with periodically stationary (periodically correlated, cyclostationary) long memory multiplicative seasonal increments. These non-stationary stochastic sequences combine the periodic structure of covariation functions of sequences as well as the integrating pattern. Spectral properties of such stochastic sequences are described. Stochastic sequences with fractional multiple increments are introduced. Their properties are described. In particular, it is shown that the multi-seasonal fractional increment sequence can be represented as a k-factor Gegenbauer process.

2 Extrapolation of sequences with periodically stationary increments

In this chapter, we present results of the investigation of the extrapolation problem for stochastic sequences with periodically stationary long memory multi-seasonal increments.

The extrapolation problem consists in the mean square optimal linear estimation of some functionals which depend on the future unobservable values of a stochastic sequence with periodically stationary long memory multi-seasonal increments. Estimates are based on observations of the sequence in the past.

2.1 Hilbert space projection method of extrapolation

2.1.1 Extrapolation of multidimensional stationary GM increment

Consider a multidimensional stochastic sequence with stationary GM increments $\vec{\xi}(m)$ constructed from the sequence $\zeta(m)$ with the help of transformation (1.2). Let the stationary GM increment sequence $\chi_{\overline{\mu},\overline{s}}^{(d)}(\vec{\xi}(m)) = \{\chi_{\overline{\mu},\overline{s}}^{(d)}(\xi_p(m))\}_{p=1}^{T}$ have an absolutely continuous spectral function $F(\lambda)$ and the spectral density $f(\lambda) = \{f_{ij}(\lambda)\}_{i,j=1}^{T}$. Without loss of generality we will assume that $\mathsf{E}\chi_{\overline{\mu},\overline{s}}^{(d)}(\vec{\xi}(m)) = 0$ and $\overline{\mu} > \overline{0}$.

Extrapolation problem

Consider the problem of mean square optimal linear estimation of the functionals

$$A\vec{\xi} = \sum_{k=0}^{\infty} (\vec{a}(k))^{\top} \vec{\xi}(k), \quad A_N\vec{\xi} = \sum_{k=0}^{N} (\vec{a}(k))^{\top} \vec{\xi}(k), \tag{2.1}$$

which depend on the unobserved values of a stochastic sequence $\vec{\xi}(k) = \{\xi_p(k)\}_{p=1}^{T}$ with stationary GM increments. Estimates are based on observations of the sequence $\vec{\xi}(k)$ at points $k = -1, -2, \ldots$. As a *mean square optimal linear estimate* (optimal estimate) of the functional $A\vec{\xi}$ we understand the estimate $\widehat{A}\xi$ which minimizes the mean square error (MSE) $\Delta(f, \widetilde{A}\vec{\xi}) = \mathsf{E}|A\vec{\xi} - \widetilde{A}\vec{\xi}|^2$ for the fixed spectral density $f(\lambda)$.

We suppose that the following conditions are satisfied:

– conditions on coefficients $\vec{a}(k) = \{a_p(k)\}_{p=1}^{T}$, $k \geq 0$, and a linear transformation $D^{\overline{\mu}}$ to be defined in Lemma 2.1.1,

$$\sum_{k=0}^{\infty} \|\vec{a}(k)\| < \infty, \quad \sum_{k=0}^{\infty} (k+1)\|\vec{a}(k)\|^2 < \infty, \tag{2.2}$$

$$\sum_{k=0}^{\infty} \|(D^{\overline{\mu}}\mathbf{a})_k\| < \infty, \quad \sum_{k=0}^{\infty} (k+1)\|(D^{\overline{\mu}}\mathbf{a})_k\|^2 < \infty, \tag{2.3}$$

https://doi.org/10.1515/9783111325620-002

– the *minimality condition* on the spectral density $f(\lambda)$

$$\int_{-\pi}^{\pi} \mathrm{Tr}\left[\frac{|\beta^{(d)}(i\lambda)|^2}{|\chi_{\overline{\mu}}^{(d)}(e^{-i\lambda})|^2} f^{-1}(\lambda)\right] d\lambda < \infty. \tag{2.4}$$

The latter is the necessary and sufficient condition under which the MSEs of estimates of functionals $A\vec{\xi}$ and $A_N\vec{\xi}$ are not equal to 0.

The Hilbert space estimation technique can be described as a three-stage procedure: (i) define a target element of the space $H = L_2(\Omega, \mathcal{F}, \mathsf{P})$ to be estimated, (ii) define a subspace of H generated by observations, and (iii) find an estimate of the target element as an orthogonal projection of the target element on the defined subspace.

Stage (i)

Neither functional $A\vec{\xi}$ nor $A_N\vec{\xi}$ belongs to the space H. The following lemma describes a representation of the functional $A\vec{\xi}$ as sums of a functional with finite second moment belonging to H and a functional depending on the observed values of the sequence $\vec{\xi}(k)$. The latter is an *initial value* functional.

Lemma 2.1.1. *The functional $A\vec{\xi}$ admits a representation*

$$A\vec{\xi} = B\chi\vec{\xi} - V\vec{\xi},$$

where

$$B\chi\vec{\xi} = \sum_{k=0}^{\infty} (\vec{b}_{\overline{\mu}}(k))^{\top} \chi_{\overline{\mu},\overline{s}}^{(d)}(\vec{\xi}(k)), \quad V\vec{\xi} = \sum_{k=-d(\gamma)}^{-1} (\vec{v}_{\overline{\mu}}(k))^{\top} \vec{\xi}(k),$$

$$\vec{v}_{\overline{\mu}}(k) = \sum_{l=0}^{k+d(\gamma)} \mathrm{diag}_T(e_\nu(l-k))\vec{b}_{\overline{\mu}}(l), \quad k = -1, -2, \dots, -d(\gamma),$$

$$\vec{b}_{\overline{\mu}}(k) = \sum_{m=k}^{\infty} \mathrm{diag}_T(d_{\overline{\mu}}(m-k))\vec{a}(m) = (D^{\overline{\mu}}\mathbf{a})_k, \quad k = 0, 1, 2, \dots,$$

$\vec{b}_{\overline{\mu}}(k) = \{b_{\overline{\mu},p}(k)\}_{p=1}^{T}$ and $\vec{v}_{\overline{\mu}}(k) = \{v_{\overline{\mu},p}(k)\}_{p=1}^{T}$ are T-dimensional vectors,

$$\mathbf{a} = ((\vec{a}(0))^{\top}, (\vec{a}(1))^{\top}, (\vec{a}(2))^{\top}, \dots)^{\top},$$

and $D^{\overline{\mu}}$ is a linear transformation determined by a matrix with $T \times T$ entries $D^{\overline{\mu}}(k,j)$, $k,j = 0,1,2,\dots$, such that $D^{\overline{\mu}}(k,j) = \mathrm{diag}_T(d_{\overline{\mu}}(j-k))$ if $0 \le k \le j$ and $D^{\overline{\mu}}(k,j) =$ for $0 \le j < k$; $\mathrm{diag}_T(x)$ denotes a $T \times T$ diagonal matrix with the entry x on its diagonal, $\mathcal{O}$ denotes a $T{\times}T$ matrix with zero entries, and the coefficients $\{d_{\overline{\mu}}(k) : k \ge 0\}$ are determined *by the relationship*

$$\sum_{k=0}^{\infty} d_{\overline{\mu}}(k)x^k = \prod_{i=1}^{r}\left(\sum_{j_i=0}^{\infty} x^{\mu_i s_i j_i}\right)^{d_i}.$$

Proof. Using Definition 1.1.1 we obtain the formal equality

$$\xi_p(k) = \frac{1}{\chi_{\overline{\mu}}^{(d)}(e^{-i\lambda})}\chi_{\overline{\mu},\overline{s}}^{(d)}(\xi_p(k)) = \sum_{j=-\infty}^{k} d_{\overline{\mu}}(k-j)\chi_{\overline{\mu},\overline{s}}^{(d)}(\xi_p(j)),$$

which implies

$$\sum_{k=0}^{\infty} a_p(k)\xi_p(k) = -\sum_{i=-d(\gamma)}^{-1} v_p^{\overline{\mu}}(i)\xi_p(i) + \sum_{i=0}^{\infty}\left(\sum_{k=i}^{\infty} a_p(k)d_{\overline{\mu}}(k-i)\right)\chi_{\overline{\mu},\overline{s}}^{(d)}(\xi_p(i)) \tag{2.5}$$

and

$$\sum_{i=0}^{\infty} b_p^{\overline{\mu}}(i)\chi_{\overline{\mu},\overline{s}}^{(d)}(\xi_p(i)) = \sum_{k=-d(\gamma)}^{-1} \xi_p(k)\sum_{l=0}^{k+d(\gamma)} e_\nu(l-k)b_p^{\overline{\mu}}(k) + \sum_{k=0}^{\infty} \xi_p(k)\sum_{l=k}^{k+d(\gamma)} e_\nu(l-k)b_p^{\overline{\mu}}(k), \tag{2.6}$$

where coefficients $e_\nu(k)$ are defined in Lemma 1.1.1. Relations (2.5) and (2.6) imply a representation $A\vec{\xi} = B\vec{\xi} - V\vec{\xi}$ and relations which prove the lemma:

$$v_p^{\overline{\mu}}(k) = \sum_{l=0}^{k+d(\gamma)} e_\nu(l-k)b_p^{\overline{\mu}}(l), \quad k=-1,-2,\dots,-d(\gamma), \quad p=1,2,\dots,T,$$

$$b_p^{\overline{\mu}}(k) = \sum_{m=k}^{\infty} d_{\overline{\mu}}(m-k)a_p(m), \quad k=0,1,2,\dots, \quad p=1,2,\dots,T. \qquad \square$$

Lemma 2.1.1 provides a representation of the functional $A\vec{\xi}$ as the sum of an element $B\chi\vec{\xi}$ from the space $H = L_2(\Omega, \mathcal{F}, \mathsf{P})$ under conditions (2.2)–(2.3) and a linear combination $V\vec{\xi}$ of a finite number of initial values $\vec{\xi}(k)$, $k = -1,-2,\dots,-d(\gamma)$, which are observed. Thus, the following equality holds:

$$\widehat{A}\vec{\xi} = \widehat{B}\chi\vec{\xi} - V\vec{\xi}. \tag{2.7}$$

Denote by $\Delta(f, \widehat{A}\vec{\xi}) := \mathsf{E}|A\vec{\xi} - \widehat{A}\vec{\xi}|^2$ the MSE of the optimal estimate $\widehat{A}\vec{\xi}$ of the functional $A\vec{\xi}$ and let $\Delta(f, \widehat{B}\chi\vec{\xi}) := \mathsf{E}|B\chi\vec{\xi} - \widehat{B}\chi\vec{\xi}|^2$ denote the MSE of the optimal estimate $\widehat{B}\chi\vec{\xi}$ of the functional $B\chi\vec{\xi}$. Then

$$\Delta(f; \widehat{A}\vec{\xi}) = \mathsf{E}|A\vec{\xi} - \widehat{A}\vec{\xi}|^2 = \mathsf{E}|B\chi\vec{\xi} - V\vec{\xi} - \widehat{B}\chi\vec{\xi} + V\vec{\xi}|^2$$

$$= \mathsf{E}|B\chi\vec{\xi} - \widehat{B}\chi\vec{\xi}|^2 = \Delta(f; \widehat{B}\chi\vec{\xi}).$$

Thus, we have defined the functional $B\chi\vec{\xi}$ to be used in finding the optimal linear estimate of the functional $A\vec{\xi}$.

Stage (ii)

In the Hilbert space $H = L_2(\Omega, \mathcal{F}, \mathsf{P})$, define a subspace

$$H^{-1}(\vec{\xi}^{(d)}) = \overline{\text{span}}\{\chi^{(d)}_{\overline{\mu},\overline{s}}(\xi_p(k)), p = 1, \dots, T;\ k \leq -1\}$$

generated by the observations. Define a closed linear subspace

$$L_2^{-1}(f) = \overline{\text{span}}\{e^{i\lambda k}\chi^{(d)}_{\overline{\mu}}(e^{-i\lambda})(\beta^{(d)}(i\lambda))^{-1}\vec{\delta}_p,\ p = 1, \dots, T; k \leq -1\}$$

of the Hilbert space $L_2(f)$ of vector-valued functions endowed with the inner product $\langle g_1; g_2\rangle = \int_{-\pi}^{\pi}(g_1(\lambda))^\top f(\lambda)\overline{g_2(\lambda)}d\lambda$, where $\vec{\delta}_p = \{\delta_{lp}\}_{l=1}^T$, where δ_{lp} are Kronecker symbols.

Remark 2.1.1. The relation

$$\chi^{(d)}_{\overline{\mu},\overline{s}}(\xi_p(m)) = \int_{-\pi}^{\pi} e^{i\lambda m}\chi^{(d)}_{\overline{\mu}}(e^{-i\lambda})\frac{1}{\beta^{(d)}(i\lambda)}dZ_p(\lambda), \quad p = 1, 2, \dots, T, \tag{2.8}$$

implies a map between elements $\chi^{(d)}_{\overline{\mu},\overline{s}}(\xi_p(m))$ of the space $H(\vec{\xi}^{(d)})$ and elements $e^{i\lambda m}\chi^{(d)}_{\overline{\mu}}(e^{-i\lambda})(\beta^{(d)}(i\lambda))^{-1}\vec{\delta}_p$ of the space $L_2(f)$.

Stage (iii)

Consider the spectral representation of the functional $B\chi\vec{\xi}$ is of the form

$$B\chi\vec{\xi} = \int_{-\pi}^{\pi}(\vec{B}_{\overline{\mu}}(e^{i\lambda}))^\top\frac{\chi^{(d)}_{\overline{\mu}}(e^{-i\lambda})}{\beta^{(d)}(i\lambda)}d\vec{Z}_{\xi^{(d)}}(\lambda),$$

where

$$\vec{B}_{\overline{\mu}}(e^{i\lambda}) = \sum_{k=0}^{\infty}\vec{b}_{\overline{\mu}}(k)e^{i\lambda k} = \sum_{k=0}^{\infty}(D^{\overline{\mu}}\mathbf{a})_k e^{i\lambda k}.$$

Remark 2.1.1 implies that the estimation problem is equivalent to finding an orthogonal projection of the element $\vec{B}_{\overline{\mu}}(e^{i\lambda})\frac{\chi^{(d)}_{\overline{\mu}}(e^{-i\lambda})}{\beta^{(d)}(i\lambda)}$ of the Hilbert space $L_2(f)$ on the subspace $L_2^{0-}(f)$:

$$\vec{h}_{\overline{\mu}}(\lambda) := \text{Proj}_{L_2^{0-}(f)}(\vec{B}_{\overline{\mu}}(e^{i\lambda})\chi^{(d)}_{\overline{\mu}}(e^{-i\lambda})(\beta^{(d)}(i\lambda))^{-1}).$$

Then relation (2.7) implies that every linear estimate $\widehat{A}\vec{\xi}$ of the functional $A\vec{\xi}$ can be written in the form

$$\widehat{A}\vec{\xi} = \int_{-\pi}^{\pi} (\vec{h}_{\overline{\mu}}(\lambda))^{\top} d\vec{Z}_{\xi^{(d)}}(\lambda) - \sum_{k=-d(\gamma)}^{-1} (\vec{v}_{\overline{\mu}}(k))^{\top} \vec{\xi}(k), \tag{2.9}$$

where $\vec{h}_{\overline{\mu}}(\lambda) = \{h_p(\lambda)\}_{p=1}^{T}$ is the spectral characteristic of the estimate $\widehat{B}\vec{\xi}$.

Define the matrix-valued Fourier coefficients

$$F^{\overline{\mu}}(j,k) = \frac{1}{2\pi} \int_{-\pi}^{\pi} e^{-i\lambda(j-k)} \frac{|\beta^{(d)}(i\lambda)|^2}{|\chi_{\overline{\mu}}^{(d)}(e^{-i\lambda})|^2} (f^{-1}(\lambda))^{\top} d\lambda, \quad j,k \geq 0, \tag{2.10}$$

and a linear operator $\mathbf{F}_{\overline{\mu}}$ in the space ℓ_2 determined by a matrix with the $T \times T$ matrix entries $\mathbf{F}_{\overline{\mu}}(j,k) = F^{\overline{\mu}}(j,k), j,k \geq 0$.

Define a vector

$$\mathbf{a} = ((\vec{a}(0))^{\top}, (\vec{a}(1))^{\top}, (\vec{a}(2))^{\top}, \ldots)^{\top}.$$

Theorem 2.1.1. *Let a vector-valued stochastic sequence $\{\vec{\xi}(m), m \in \mathbb{Z}\}$ determine a stationary stochastic GM increment sequence $\chi_{\overline{\mu},\overline{s}}^{(d)}(\vec{\xi}(m))$ with the spectral density matrix $f(\lambda) = \{f_{ij}(\lambda)\}_{i,j=1}^{T}$ which satisfies the minimality condition* (2.4). *Let the coefficients $\vec{a}(j)$, $j \geqslant 0$ satisfy conditions* (2.2)–(2.3). *Then the optimal linear estimate $\widehat{A}\vec{\xi}$ of the functional $A\vec{\xi}$ based on observations of the sequence $\vec{\xi}(m)$ at points $m = -1, -2, \ldots$ is calculated by formula* (2.9) *with the spectral characteristic*

$$(\vec{h}_{\overline{\mu}}(\lambda))^{\top} = (\vec{B}_{\overline{\mu}}(e^{i\lambda}))^{\top} \frac{\chi_{\overline{\mu}}^{(d)}(e^{-i\lambda})}{\beta^{(d)}(i\lambda)} - \frac{\overline{\beta^{(d)}(i\lambda)}}{\chi_{\overline{\mu}}^{(d)}(e^{i\lambda})} \left(\sum_{k=0}^{\infty} (\mathbf{F}_{\overline{\mu}}^{-1} D^{\overline{\mu}} \mathbf{a})_k e^{i\lambda k} \right)^{\top} f^{-1}(\lambda). \tag{2.11}$$

The MSE of the estimate $\widehat{A}\vec{\xi}$ is calculated by the formula

$$\begin{aligned}
\Delta(f; \widehat{A}\vec{\xi}) &= \Delta(f; \widehat{B}\chi\vec{\xi}) = \mathsf{E}|B\chi\vec{\xi} - \widehat{B}\chi\vec{\xi}|^2 \\
&= \frac{1}{2\pi} \int_{-\pi}^{\pi} \frac{\overline{\beta^{(d)}(i\lambda)}}{\chi_{\overline{\mu}}^{(d)}(e^{i\lambda})} \left(\sum_{k=0}^{\infty} (\mathbf{F}_{\overline{\mu}}^{-1} D^{\overline{\mu}} \mathbf{a})_k e^{i\lambda k} \right)^{\top} f(\lambda) \\
&\quad \times \left(\overline{\sum_{k=0}^{\infty} (\mathbf{F}_{\overline{\mu}}^{-1} D^{\overline{\mu}} \mathbf{a})_k e^{i\lambda k}} \right) \frac{\beta^{(d)}(i\lambda)}{\chi_{\overline{\mu}}^{(d)}(e^{-i\lambda})} d\lambda \\
&= \langle D^{\overline{\mu}} \mathbf{a}, \mathbf{F}_{\overline{\mu}}^{-1} D^{\overline{\mu}} \mathbf{a} \rangle.
\end{aligned} \tag{2.12}$$

Proof. The projection $\vec{h}_{\overline{\mu}}(\lambda)$ is characterized by the following conditions:

$$\vec{h}_{\overline{\mu}}(\lambda) \in L_2^{-1}(f), \tag{2.13}$$

$$\left(\vec{B}_{\overline{\mu}}(e^{i\lambda})\frac{\chi_{\overline{\mu}}^{(d)}(e^{-i\lambda})}{\beta^{(d)}(i\lambda)} - \vec{h}_{\overline{\mu}}(\lambda)\right) \perp L_2^{-1}(f). \tag{2.14}$$

Condition (2.14) implies the following relation holding for all $k \leq -1$ and for $l = 1, \ldots, T$:

$$\int_{-\pi}^{\pi} \left(\vec{B}_{\overline{\mu}}(e^{i\lambda})\frac{\chi_{\overline{\mu}}^{(d)}(e^{-i\lambda})}{\beta^{(d)}(i\lambda)} - \vec{h}_{\overline{\mu}}(\lambda)\right)^{\top} f(\lambda)\vec{\delta}_l e^{-i\lambda k}\frac{\overline{\chi_{\overline{\mu}}^{(d)}(e^{i\lambda})}}{\overline{\beta^{(d)}(i\lambda)}} d\lambda = 0. \tag{2.15}$$

Thus, the spectral characteristic $\vec{h}_{\overline{\mu}}(\lambda)$ can be represented in the form

$$(\vec{h}_{\overline{\mu}}(\lambda))^{\top} = (\vec{B}_{\overline{\mu}}(e^{i\lambda}))^{\top}\frac{\chi_{\overline{\mu}}^{(d)}(e^{-i\lambda})}{\beta^{(d)}(i\lambda)} - \frac{\overline{\beta^{(d)}(i\lambda)}}{\chi_{\overline{\mu}}^{(d)}(e^{i\lambda})}(\vec{C}_{\overline{\mu}}(e^{i\lambda}))^{\top} f^{-1}(\lambda),$$

where

$$\vec{C}_{\overline{\mu}}(e^{i\lambda}) = \sum_{k=0}^{\infty} \vec{c}_{\overline{\mu}}(k)e^{i\lambda k},$$

where $\vec{c}(k) = \{c_p(k)\}_{p=1}^{T}$, $k \geqslant 0$, are unknown coefficients to be found.

Condition (2.13) implies the following representation of the spectral characteristic $\vec{h}_{\overline{\mu}}(\lambda)$:

$$\vec{h}_{\overline{\mu}}(\lambda) = \vec{h}(\lambda)\chi_{\overline{\mu}}^{(d)}(e^{-i\lambda})\frac{1}{\beta^{(d)}(i\lambda)}, \quad \vec{h}(\lambda) = \sum_{k=1}^{\infty} \vec{s}(k)e^{-i\lambda k},$$

which allows us to write the equations

$$\int_{-\pi}^{\pi} \left((\vec{B}_{\overline{\mu}}(e^{i\lambda}))^{\top} - \frac{|\beta^{(d)}(i\lambda)|^2}{\chi_{\overline{\mu}}^{(d)}(e^{-i\lambda})\chi_{\overline{\mu}}^{(d)}(e^{i\lambda})}(\vec{C}_{\overline{\mu}}(e^{i\lambda}))^{\top} f^{-1}(\lambda)\right)e^{-i\lambda j} d\lambda = 0, \quad j \geqslant 0. \tag{2.16}$$

Making use of Fourier coefficients (2.10), we rewrite equations (2.16) as a system of linear vector equations

$$\vec{b}_{\overline{\mu}}(j) = \sum_{k=0}^{\infty} F^{\overline{\mu}}(j,k)\,\vec{c}_{\overline{\mu}}(k), \quad j \geq 0,$$

determining the unknown coefficients $\vec{c}_{\overline{\mu}}(k)$, $k \geq 0$. This system is presented in matrix form as

$$D^{\overline{\mu}}\mathbf{a} = \mathbf{F}_{\overline{\mu}}\mathbf{c}_{\overline{\mu}}, \tag{2.17}$$

where

$$\mathbf{c}_{\overline{\mu}} = \left((\vec{c}_{\overline{\mu}}(0))^{\top}, (\vec{c}_{\overline{\mu}}(1))^{\top}, (\vec{c}_{\overline{\mu}}(2))^{\top}, \dots\right)^{\top}.$$

Therefore, coefficients $\vec{c}_{\overline{\mu}}(k)$, $k \geq 0$, which determine the spectral characteristic $\vec{h}_{\overline{\mu}}(\lambda)$, are calculated as

$$\vec{c}_{\mu}(k) = (\mathbf{F}_{\overline{\mu}}^{-1} D^{\overline{\mu}} \mathbf{a})_k, \quad k \geq 0, \tag{2.18}$$

where $(\mathbf{F}_{\overline{\mu}}^{-1} D^{\overline{\mu}} \mathbf{a})_k$, $k \geq 0$, is the kth T-dimensional vector entry of the vector $\mathbf{F}_{\overline{\mu}}^{-1} D^{\overline{\mu}} \mathbf{a}$.

The derived expressions justify the use of formulas (2.11) and (2.12) for calculating the spectral characteristic $\vec{h}_{\overline{\mu}}(\lambda)$ and the MSE $\Delta(f; \widehat{A}\xi)$ of the estimate $\widehat{A}\xi$. □

Next consider the case where the GM increment sequence of the stochastic sequence $\vec{\xi}(m)$ admits the moving average representation (1.12) and its spectral density $f(\lambda) = \{f_{ij}(\lambda)\}_{i,j=1}^{T}$ admits canonical factorizations (1.13)–(1.14), namely,

$$f(\lambda) = \varphi(e^{-i\lambda})\varphi^{*}(e^{-i\lambda}), \qquad \frac{|\chi_{\overline{\mu}}^{(d)}(e^{-i\lambda})|^2}{|\beta^{(d)}(i\lambda)|^2} f(\lambda) = \varphi_{\overline{\mu}}(e^{-i\lambda})\varphi_{\overline{\mu}}^{*}(e^{-i\lambda}), \tag{2.19}$$

where

$$\varphi(e^{-i\lambda}) = \sum_{k=0}^{\infty} \varphi(k) e^{-i\lambda k}, \quad \varphi_{\overline{\mu}}(e^{-i\lambda}) = \sum_{k=0}^{\infty} \varphi_{\overline{\mu}}(k) e^{-i\lambda k},$$

and $\varphi_{\overline{\mu}}(k) = \{\varphi_{\overline{\mu},ij}(k)\}_{i=\overline{1,T}}^{j=\overline{1,q}}$, $k = 0, 1, 2, \dots$. Let E_q denote the identity $q \times q$ matrix. Define the matrix-valued function $\psi_{\overline{\mu}}(e^{-i\lambda}) = \{\psi_{\overline{\mu},ij}(e^{-i\lambda})\}_{i=\overline{1,q}}^{j=\overline{1,T}}$ by the equation

$$\psi_{\overline{\mu}}(e^{-i\lambda})\varphi_{\overline{\mu}}(e^{-i\lambda}) = E_q.$$

Formulas for calculating the spectral characteristic $\vec{h}_{\overline{\mu}}(\lambda)$ and the MSE $\Delta(f; \widehat{A}\vec{\xi})$ of the estimate $\widehat{A}\vec{\xi}$ can be presented in terms of the function $\psi_{\overline{\mu}}(e^{-i\lambda})$ and the factorization coefficients $\varphi_{\overline{\mu}}(k)$, $k = 0, 1, 2, \dots$.

Define a linear symmetric operator $\mathbf{A}$ determined by a matrix with the vector entries $\mathbf{A}(k, j) = \vec{a}(k + j)$, $k, j \geq 0$. The operators $D^{\overline{\mu}}\mathbf{A}$ and $\mathbf{A}$ are compact under conditions (2.2)–(2.3).

Theorem 2.1.2. *Let a vector-valued stochastic sequence $\{\vec{\xi}(m), m \in \mathbb{Z}\}$ with stationary GM increments and coefficients $\vec{a}(j)$, $j \geqslant 0$ satisfy the conditions of Theorem 2.1.1. Let the spectral density $f(\lambda)$ satisfy the minimality condition* (2.4) *and admit canonical factorization* (2.19). *Then the optimal linear estimate $\widehat{A}\vec{\xi}$ of the functional $A\vec{\xi}$ based on observations of the sequence $\vec{\xi}(m)$ at points $m = -1, -2, \dots$ is calculated by formula* (2.9) *with the spectral characteristic*

$$\vec{h}_{\overline{\mu}}(\lambda) = \frac{\chi_{\overline{\mu}}^{(d)}(e^{-i\lambda})}{\beta^{(d)}(i\lambda)}(\vec{B}_{\overline{\mu}}(e^{i\lambda}) - (\psi_{\overline{\mu}}(e^{-i\lambda}))^{\top}\vec{r}_{\overline{\mu}}(e^{i\lambda})), \tag{2.20}$$

where

$$\vec{r}_{\overline{\mu}}(e^{i\lambda}) = \sum_{k=0}^{\infty}(D^{\overline{\mu}}\mathbf{A}\varphi_{\overline{\mu}})_k e^{i\lambda k},$$

$$(D^{\overline{\mu}}\mathbf{A}\varphi_{\overline{\mu}})_k = \sum_{m=0}^{\infty}\sum_{l=m}^{\infty}(\varphi_{\overline{\mu}}(m))^{\top}D^{\overline{\mu}}(m,l)\vec{a}(l+k)$$
$$= \sum_{m=0}^{\infty}\sum_{l=k}^{\infty}(\varphi_{\overline{\mu}}(m))^{\top}\vec{a}(m+l)d_{\overline{\mu}}(l-k).$$

The MSE of the estimate $\widehat{A}\xi$ can be calculated by the formula

$$\Delta(f;\widehat{A}\vec{\xi}) = \frac{1}{2\pi}\int_{-\pi}^{\pi}\left(\sum_{k=0}^{\infty}(D^{\overline{\mu}}\mathbf{A}\varphi_{\overline{\mu}})_k e^{i\lambda k}\right)^{\top}\overline{\left(\sum_{k=0}^{\infty}(D^{\overline{\mu}}\mathbf{A}\varphi_{\overline{\mu}})_k e^{i\lambda k}\right)}d\lambda$$
$$= \frac{1}{2\pi}\int_{-\pi}^{\pi}\|\vec{r}_{\overline{\mu}}(e^{i\lambda})\|^2 d\lambda = \|D^{\overline{\mu}}\mathbf{A}\varphi_{\overline{\mu}}\|^2. \tag{2.21}$$

Proof. One can directly check that conditions (2.13) and (2.14) are satisfied by function (2.20). □

2.1.2 Estimates of the functional $A_N\vec{\xi}$ and the value $\xi_p(N)$

The following corollary from Lemma 2.1.1 holds.

Corollary 2.1.1. *The functional $A_N\vec{\xi}$ allows a representation*

$$A_N\vec{\xi} = B_N\chi\vec{\xi} - V_N\vec{\xi},$$

where

$$B_N\chi\vec{\xi} = \sum_{k=0}^{N}(\vec{b}_{\overline{\mu},N}(k))^{\top}\chi_{\overline{\mu},\overline{s}}^{(d)}(\vec{\xi}(k)), \quad V_N\vec{\xi} = \sum_{k=-d(\gamma)}^{-1}(\vec{v}_{\overline{\mu},N}(k))^{\top}\vec{\xi}(k),$$

the vectors

$$\vec{b}_{\overline{\mu},N}(k) = \{b_{\overline{\mu},N,p}(k)\}_{p=1}^{T}, \quad k = 0,1,\dots,N,$$
$$\vec{v}_{\overline{\mu},N}(k) = \{v_{\overline{\mu},N,p}(k)\}_{p=1}^{T}, \quad k = -1,-2,\dots,-d(\gamma),$$

are calculated by the formulas

$$\vec{v}_{\overline{\mu},N}(k) = \sum_{l=0}^{\min\{N,k+d(\gamma)\}} \mathrm{diag}_T(e_\nu(l-k))\vec{b}_{\overline{\mu},N}(l), \quad k=-1,-2,\dots,-d(\gamma), \tag{2.22}$$

$$\vec{b}_{\overline{\mu},N}(k) = \sum_{m=k}^{N} \mathrm{diag}_T(d_{\overline{\mu}}(m-k))\vec{a}(m) = (D_N^{\overline{\mu}}\mathbf{a}_N)_k, \quad k=0,1,\dots,N, \tag{2.23}$$

and $D_N^{\overline{\mu}}$ is the linear transformation determined by an infinite matrix with the entries $(D_N^{\overline{\mu}})(k,j) = \mathrm{diag}_T(d_{\overline{\mu}}(j-k))$ if $0 \le k \le j \le N$ and $(D_N^{\overline{\mu}})(k,j) = \mathcal{O}$ if $j < k$ or $j,k > N$;

$$\mathbf{a}_N = ((\vec{a}(0))^\top, (\vec{a}(1))^\top, \dots, (\vec{a}(N))^\top, \vec{0}^\top, \dots)^\top.$$

Theorem 2.1.1 and Corollary 2.1.1 allow to find an optimal estimate $\widehat{A}_N\vec{\xi}$ of the functional $A_N\vec{\xi}$ which depends on the unobserved values $\vec{\xi}(m)$, $m=0,1,2,\dots,N$, based on observations of the sequence $\vec{\xi}(m)$ at points $m=-1,-2,\dots$. Set $\vec{a}(k)=\vec{0}$ for $k > N$.

Define

$$B_{\overline{\mu},N}(e^{i\lambda}) = \sum_{k=0}^{N} (D_N^{\overline{\mu}}\mathbf{a}_N)_k e^{i\lambda k}$$

and $\varphi_{\overline{\mu},N} = (\varphi_{\overline{\mu}}(0), \varphi_{\overline{\mu}}(1), \dots, \varphi_{\overline{\mu}}(N))$. Define a linear operator $\mathbf{A}_N$ by the vectors $\vec{a}(k)$, $k=0,1,\dots,N$, as follows: $(\mathbf{A}_N)(k,j) = \vec{a}(k+j)$, $0 \le k+j \le N$, $(\mathbf{A}_N)(k,j) = \vec{0}$, $k+j > N$, $0 \le k,j \le N$. Define a linear transformation $\widetilde{D}_N^{\overline{\mu}}$ by the matrix of dimension $(N+1)\times(N+1)$ with the $T\times T$ entries $\widetilde{D}_N^{\overline{\mu}}(k,j) = \mathrm{diag}_T(d_{\overline{\mu}}(j-k))$ if $0 \le k \le j \le N$ and $\widetilde{D}_N^{\overline{\mu}}(k,j) = \mathcal{O}$ if $0 \le j < k \le N$.

The following theorem holds.

Theorem 2.1.3. *Let $\{\vec{\xi}(m), m \in \mathbb{Z}\}$ be a stochastic sequence which determines a stationary GM increment sequence $\chi_{\overline{\mu},\overline{s}}^{(d)}(\vec{\xi}(m))$ with the spectral density matrix $f(\lambda)$ satisfying minimality condition (2.4). The optimal linear estimate $\widehat{A}_N\vec{\xi}$ of the functional $A_N\vec{\xi}$ based on observations of the sequence $\vec{\xi}(m)$ at points $m=-1,-2,\dots$ is calculated by the formula*

$$\widehat{A}_N\vec{\xi} = \int_{-\pi}^{\pi} (\vec{h}_{\overline{\mu},N}(\lambda))^\top d\vec{Z}_{\xi^{(d)}}(\lambda) - \sum_{k=-d(\gamma)}^{-1} (\vec{v}_{\overline{\mu},N}(k))^\top \vec{\xi}(k). \tag{2.24}$$

The spectral characteristic $\vec{h}_{\overline{\mu},N}(\lambda) = \{h_{\mu,N,p}(\lambda)\}_{p=1}^T$ and the MSE $\Delta(f;\widehat{A}_N\vec{\xi})$ are calculated by the formulas

$$(\vec{h}_{\overline{\mu},N}(\lambda))^\top = (\vec{B}_{\overline{\mu},N}(e^{i\lambda}))^\top \frac{\chi_{\overline{\mu}}^{(d)}(e^{-i\lambda})}{\beta^{(d)}(i\lambda)} - \frac{\overline{\beta^{(d)}(i\lambda)}}{\chi_{\overline{\mu}}^{(d)}(e^{i\lambda})} \left(\sum_{k=0}^{\infty} (\mathbf{F}_{\overline{\mu}}^{-1} D_N^{\overline{\mu}}\mathbf{a}_N)_k e^{i\lambda k}\right)^\top f^{-1}(\lambda) \tag{2.25}$$

and

$$\Delta(f,\widehat{A}_N\vec{\xi}) = \Delta(f,\widehat{B}_N\chi\vec{\xi}) = \mathsf{E}|B_N\chi\vec{\xi} - \widehat{B}_N\chi\vec{\xi}|^2$$

$$
\begin{aligned}
&= \frac{1}{2\pi}\int_{-\pi}^{\pi} \overline{\frac{\beta^{(d)}(i\lambda)}{\chi_{\overline{\mu}}^{(d)}(e^{i\lambda})}} \left(\sum_{k=0}^{\infty} (\mathbf{F}_{\overline{\mu}}^{-1} D_N^{\overline{\mu}} \mathbf{a}_N)_k e^{i\lambda k} \right)^{\top} f(\lambda) \\
&\quad \times \overline{\left(\sum_{k=0}^{\infty} (\mathbf{F}_{\overline{\mu}}^{-1} D_N^{\overline{\mu}} \mathbf{a}_N)_k e^{i\lambda k} \right)} \frac{\beta^{(d)}(i\lambda)}{\chi_{\overline{\mu}}^{(d)}(e^{-i\lambda})} d\lambda \\
&= \langle D_N^{\overline{\mu}} \mathbf{a}_N, \mathbf{F}_{\overline{\mu}}^{-1} D_N^{\overline{\mu}} \mathbf{a}_N \rangle,
\end{aligned} \tag{2.26}
$$

respectively. If the spectral density $f(\lambda)$ admits canonical factorization (2.19), *the spectral characteristic $\vec{h}_{\overline{\mu},N}(\lambda)$ and the MSE of the estimate $\widehat{A}_N\vec{\xi}$ can be calculated by the formulas*

$$
\vec{h}_{\overline{\mu},N}(\lambda) = \frac{\chi_{\overline{\mu}}^{(d)}(e^{-i\lambda})}{\beta^{(d)}(i\lambda)} (\vec{B}_{\overline{\mu},N}(e^{i\lambda}) - (\psi_{\overline{\mu}}(e^{-i\lambda}))^{\top} \vec{r}_{\overline{\mu},N}(e^{i\lambda})), \tag{2.27}
$$

where

$$
\begin{aligned}
\vec{r}_{\overline{\mu},N}(e^{i\lambda}) &= \sum_{k=0}^{N} (\widetilde{D}_N^{\overline{\mu}} \mathbf{A}_N \varphi_{\overline{\mu},N})_k e^{i\lambda k}, \\
(\widetilde{D}_N^{\overline{\mu}} \mathbf{A}_N \varphi_{\overline{\mu},N})_k &= \sum_{m=0}^{N} \sum_{l=k}^{N} (\varphi_{\overline{\mu}}(m))^{\top} \vec{a}(m+l) d_{\overline{\mu}}(l-k),
\end{aligned}
$$

and

$$
\begin{aligned}
\Delta(f;\widehat{A}_N\vec{\xi}) &= \frac{1}{2\pi}\int_{-\pi}^{\pi} \left(\sum_{k=0}^{N} (\widetilde{D}_N^{\overline{\mu}} \mathbf{A}_N \varphi_{\overline{\mu},N})_k e^{i\lambda k} \right)^{\top} \overline{\left(\sum_{k=0}^{N} (\widetilde{D}_N^{\overline{\mu}} \mathbf{A}_N \varphi_{\overline{\mu},N})_k e^{i\lambda k} \right)} d\lambda \\
&= \frac{1}{2\pi}\int_{-\pi}^{\pi} \|\vec{r}_{\overline{\mu},N}(e^{i\lambda})\|^2 d\lambda = \|\widetilde{D}_N^{\overline{\mu}} \mathbf{A}_N \varphi_{\overline{\mu},N}\|^2,
\end{aligned} \tag{2.28}
$$

respectively.

The mean square optimal estimate of the unobserved value $A_{N,p}\vec{\xi} = \xi_p(N) = \vec{\xi}(N)\vec{\delta}_p$, $p = 1, 2, \ldots, T$, $N \geq 0$, of the stochastic sequence $\vec{\xi}(m)$ with GM stationary increments based on its observations at points $m = -1, -2, \ldots$ is obtained as a corollary from Theorem 2.1.3.

Corollary 2.1.2. *The optimal linear estimate $\widehat{\xi}_p(N)$ of the value $\xi_p(N)$, $p = 1, 2, \ldots, T$, $N \geq 0$, of the stochastic sequence $\xi(m)$ with GM stationary increments from observations of the sequence $\vec{\xi}(m)$ at points $m = -1, -2, \ldots$ is calculated by the formula*

$$
\widehat{\xi}_p(N) = \int_{-\pi}^{\pi} (\vec{h}_{\overline{\mu},N,p}(\lambda))^{\top} d\vec{Z}_{\xi^{(d)}}(\lambda) - \sum_{k=-d(\gamma)}^{-1} (\vec{v}_{\overline{\mu},N,p}(k))^{\top} \vec{\xi}(k). \tag{2.29}
$$

The spectral characteristic $\vec{h}_{\overline{\mu},N,p}(\lambda)$ of the estimate $\widehat{\xi}_p(N)$ is calculated by the formula

$$\begin{aligned}(\vec{h}_{\overline{\mu},N,p}(\lambda))^{\top} &= \frac{\chi_{\overline{\mu}}^{(d)}(e^{-i\lambda})}{\beta^{(d)}(i\lambda)}\left(\vec{\delta}_p\sum_{k=0}^{N} d_{\overline{\mu}}(N-k)e^{i\lambda k}\right)^{\top} \\ &\quad - \frac{\overline{\beta^{(d)}(i\lambda)}}{\chi_{\overline{\mu}}^{(d)}(e^{i\lambda})}\left(\sum_{k=0}^{\infty}(\mathbf{F}_{\overline{\mu}}^{-1}\mathbf{d}_{\overline{\mu},N,p})_k e^{i\lambda k}\right)^{\top} f^{-1}(\lambda),\end{aligned} \tag{2.30}$$

where $\mathbf{d}_{\overline{\mu},N,p} = (d_{\overline{\mu}}(N)\vec{\delta}_p^{\top}, d_{\overline{\mu}}(N-1)\vec{\delta}_p^{\top}, \dots, d_{\overline{\mu}}(0)\vec{\delta}_p^{\top}, \vec{0}^{\top}, \dots)^{\top}$. *The MSE of the estimate $\widehat{\xi}_p(N)$ is calculated by the formula*

$$\begin{aligned}\Delta(f;\widehat{\xi}_p(N)) &= \Delta(f;\chi_{\overline{\mu},\overline{s}}^{(d)}(\widehat{\xi}_p(m))) = \mathsf{E}|\chi_{\overline{\mu},\overline{s}}^{(d)}(\xi_p(m)) - \chi_{\overline{\mu},\overline{s}}^{(d)}(\widehat{\xi}_p(m))|^2 \\ &= \frac{1}{2\pi}\int_{-\pi}^{\pi} \frac{\overline{\beta^{(d)}(i\lambda)}}{\chi_{\overline{\mu}}^{(d)}(e^{i\lambda})}\left(\sum_{k=0}^{\infty}(\mathbf{F}_{\overline{\mu}}^{-1}\mathbf{d}_{\overline{\mu},N,p})_k e^{i\lambda k}\right)^{\top} f(\lambda) \\ &\quad \times \overline{\left(\sum_{k=0}^{\infty}(\mathbf{F}_{\overline{\mu}}^{-1}\mathbf{d}_{\overline{\mu},N,p})_k e^{i\lambda k}\right)} \frac{\beta^{(d)}(i\lambda)}{\chi_{\overline{\mu}}^{(d)}(e^{-i\lambda})} d\lambda \\ &= \langle \mathbf{d}_{\overline{\mu},N,p}, \mathbf{F}_{\overline{\mu}}^{-1}\mathbf{d}_{\overline{\mu},N,p}\rangle.\end{aligned} \tag{2.31}$$

If the spectral density $f(\lambda)$ admits canonical factorization (2.19) *and the condition* $\min_{i=\overline{1,r}} \mu_i s_i > N$ *is satisfied, the spectral characteristic and the MSE of the estimate $\widehat{\xi}_p(N)$ can be calculated by the formulas*

$$\vec{h}_{\overline{\mu},N,p}(\lambda) = \frac{\chi_{\overline{\mu}}^{(d)}(e^{-i\lambda})}{\beta^{(d)}(i\lambda)} e^{iN\lambda}\left(\vec{\delta}_p - (\psi_{\overline{\mu}}(e^{-i\lambda}))^{\top}\left(\sum_{k=0}^{N}\varphi_{\overline{\mu}}(k)e^{-i\lambda k}\right)^{\top}\vec{\delta}_p\right) \tag{2.32}$$

and

$$\begin{aligned}\Delta(f;\widehat{\xi}_p(N)) &= \frac{1}{2\pi}\int_{-\pi}^{\pi}\left(\vec{\delta}_p^{\top}\sum_{k=0}^{N}\varphi_{\overline{\mu}}(k)e^{-i\lambda k}\right)\left(\vec{\delta}_p^{\top}\sum_{k=0}^{N}\varphi_{\overline{\mu}}(k)e^{-i\lambda k}\right)^{*} d\lambda \\ &= \sum_{k=0}^{N}\sum_{j=1}^{q}|\varphi_{\mu,p,j}(k)|^2,\end{aligned} \tag{2.33}$$

respectively.

Remark 2.1.2. Since for all $\overline{d} \geq \overline{1}$ and $\overline{\mu} \geq \overline{1}$ the condition

$$\int_{-\pi}^{\pi}\left|\ln\frac{|\chi_{\overline{\mu}}^{(d)}(e^{-i\lambda})|^2}{|\beta^{(d)}(i\lambda)|^2}\right| d\lambda < \infty$$

holds, there exists a function

$$w_{\overline{\mu}}(z) = \sum_{k=0}^{\infty} w_{\overline{\mu}}(k) z^k, \quad \sum_{k=0}^{\infty} |w_{\overline{\mu}}(k)|^2 < \infty$$

such that [68]

$$|w_{\overline{\mu}}(e^{-i\lambda})|^2 = \frac{|\chi_{\overline{\mu}}^{(d)}(e^{-i\lambda})|^2}{|\beta^{(d)}(i\lambda)|^2},$$

which can be calculated by the formula

$$w_{\overline{\mu}}(z) = \exp\left\{\frac{1}{4\pi}\int_{-\pi}^{\pi} \frac{e^{i\lambda}+z}{e^{i\lambda}-z} \ln \frac{|\chi_{\overline{\mu}}^{(d)}(e^{-i\lambda})|^2}{|\beta^{(d)}(i\lambda)|^2} d\lambda\right\}. \tag{2.34}$$

Then the following relation holds:

$$\varphi_{\overline{\mu}}(e^{-i\lambda}) = w_{\overline{\mu}}(e^{-i\lambda})\varphi(e^{-i\lambda}), \tag{2.35}$$

which implies

$$\varphi_{\overline{\mu}}(k) = \sum_{m=0}^{k} w_{\overline{\mu}}(k-m)\varphi(m), \quad k = 0, 1, \ldots,$$

that is,

$$\varphi_{\mu,ij}(k) = \sum_{m=0}^{k} w_{\overline{\mu}}(k-m)\varphi_{ij}(m), \quad i = 1, \ldots, T,\ j = 1, \ldots, q,\ k = 0, 1, \ldots.$$

This relation can be represented in the form

$$\boldsymbol{\varphi}_{\overline{\mu}} = \mathbf{W}^{\overline{\mu}}\boldsymbol{\varphi}, \tag{2.36}$$

where

$$\boldsymbol{\varphi}_{\overline{\mu}} = ((\varphi_{\overline{\mu}}(0))^{\top}, (\varphi_{\overline{\mu}}(1))^{\top}, (\varphi_{\overline{\mu}}(2))^{\top}, \ldots)^{\top},$$
$$\boldsymbol{\varphi} = ((\varphi(0))^{\top}, (\varphi(1))^{\top}, (\varphi(2))^{\top}, \ldots)^{\top}$$

are vectors composed of matrices $\varphi_{\overline{\mu}}(k) = \{\varphi_{\mu,ij}(k)\}_{i=\overline{1,T}}^{j=\overline{1,q}}$, $k = 0, 1, 2, \ldots$, and $\varphi(k) = \{\varphi_{ij}(k)\}_{i=\overline{1,T}}^{j=\overline{1,q}}$, $k = 0, 1, 2, \ldots$, and where $\mathbf{W}^{\overline{\mu}}$ is a linear operator with entries $(\mathbf{W}^{\overline{\mu}})_{j,k} = w_{\overline{\mu}}(j-k)$ if $0 \le k \le j$ and $(\mathbf{W}^{\overline{\mu}})_{j,k} = \mathcal{O}$ if $0 \le j < k$.

2.1.3 Extrapolation of sequences with periodically stationary GM increment

Consider the problem of mean square optimal linear estimation of the functionals

$$A\zeta = \sum_{k=0}^{\infty} a^{(\zeta)}(k)\zeta(k), \quad A_M\zeta = \sum_{k=0}^{N} a^{(\zeta)}(k)\zeta(k), \tag{2.37}$$

which depend on unobserved values of the stochastic sequence $\zeta(m)$ with periodically stationary increments. Estimates are based on observations of the sequence $\zeta(m)$ at points $m = -1, -2, \ldots$.

The functional $A\zeta$ can be represented in the form

$$\begin{aligned} A\zeta &= \sum_{k=0}^{\infty} a^{(\zeta)}(k)\zeta(k) = \sum_{m=0}^{\infty}\sum_{p=1}^{T} a^{(\zeta)}(mT+p-1)\zeta(mT+p-1) \\ &= \sum_{m=0}^{\infty}\sum_{p=1}^{T} a_p(m)\xi_p(m) = \sum_{m=0}^{\infty} (\vec{a}(m))^{\top}\vec{\xi}(m) = A\vec{\xi}, \end{aligned}$$

where

$$\vec{\xi}(m) = (\xi_1(m), \xi_2(m), \ldots, \xi_T(m))^{\top}, \quad \xi_p(m) = \zeta(mT+p-1),\ p = 1, 2, \ldots, T, \tag{2.38}$$

$$\vec{a}(m) = (a_1(m), a_2(m), \ldots, a_T(m))^{\top}, \quad a_p(m) = a^{(\zeta)}(mT+p-1),\ p = 1, 2, \ldots, T. \tag{2.39}$$

Making use of the introduced notations and statements of Theorem 2.1.1, we can claim that the following theorem holds.

Theorem 2.1.4. *Let a stochastic sequence $\zeta(m)$ with periodically stationary increments generate by formula* (2.38) *a multidimensional stochastic sequence $\vec{\xi}(m)$ which determines a stationary stochastic GM increment sequence $\chi^{(d)}_{\overline{\mu},\overline{s}}(\vec{\xi}(m))$ with the spectral density matrix $f(\lambda) = \{f_{ij}(\lambda)\}_{i,j=1}^{T}$ satisfying minimality condition* (2.4). *Let the coefficients $\vec{a}(k)$, $k \geq 0$, determined by formula* (2.39) *satisfy conditions* (2.2)–(2.3). *Then the optimal linear estimate $\widehat{A}\zeta$ of the functional $A\zeta$ based on observations of the sequence $\zeta(m)$ at points $m = -1, -2, \ldots$ is calculated by formula* (2.9). *The spectral characteristic $\vec{h}_{\overline{\mu}}(\lambda)$ and the MSE $\Delta(f; \widehat{A}\zeta)$ of the estimate $\widehat{A}\zeta$ are calculated by formulas* (2.11) *and* (2.12), *respectively.*

In the case where the spectral density matrix $f(\lambda)$ admits the canonical factorization (2.19), *the spectral characteristic and the MSE of the estimate $\widehat{A}\xi$ can be calculated by formulas* (2.20) *and* (2.21), *respectively.*

The functional $A_M\zeta$ can be represented in the form

$$A_M\zeta = \sum_{k=0}^{M} a^{(\zeta)}(k)\zeta(k) = \sum_{m=0}^{N}\sum_{p=1}^{T} a^{(\zeta)}(mT+p-1)\zeta(mT+p-1)$$

$$= \sum_{m=0}^{N} \sum_{p=1}^{T} a_p(m)\xi_p(m) = \sum_{m=0}^{N} (\vec{a}(m))^{\top} \vec{\xi}(m) = A_N \vec{\xi},$$

where $N = [\frac{M}{T}]$, the sequence $\vec{\xi}(m)$ is determined by formula (2.38), and

$$\begin{aligned} \vec{a}(m) &= (a_1(m), a_2(m), \dots, a_T(m))^{\top}, \\ a_p(m) &= a^{\zeta}(mT + p - 1), \quad 0 \le m \le N,\ 1 \le p \le T,\ mT + p - 1 \le M, \\ a_p(N) &= 0, \quad M + 1 \le NT + p - 1 \le (N+1)T - 1,\ 1 \le p \le T. \end{aligned} \tag{2.40}$$

Making use of the introduced notations and statements of Theorem 2.1.3, we can claim that the following corollary holds.

Corollary 2.1.3. *Suppose that the conditions of Theorem* 2.1.4 *are satisfied. Let coefficients* $\vec{a}(k)$, $k = \overline{1, N}$, *be determined by formula* (2.40). *The optimal linear estimate* $\widehat{A}_M\zeta$ *of the functional* $A_M\zeta = A_N\vec{\xi}$ *based on observations of the sequence* $\zeta(m)$ *at points* $m = -1, -2, \dots$ *is calculated by formula* (2.24). *The spectral characteristic* $\vec{h}_{\overline{\mu},N}(\lambda)$ *and the MSE* $\Delta(f; \widehat{A}_M\zeta)$ *are calculated by formulas* (2.25) *and* (2.26), *respectively. If the spectral density matrix* $f(\lambda)$ *admits canonical factorization* (2.19), *the spectral characteristic* $\vec{h}_{\overline{\mu},N}(\lambda)$ *and the MSE of the estimate* $\widehat{A}_M\zeta$ *can be calculated by formulas* (2.27) *and* (2.28), *respectively.*

As a corollary from the proposed theorem, one can obtain the mean square optimal estimate of the unobserved value $\zeta(M)$, $M \ge 0$, of a stochastic sequence $\zeta(m)$ with periodically stationary GM increments based on observations of the sequence $\zeta(m)$ at points $m = -1, -2, \dots$ Making use of the notations $\zeta(M) = \xi_p(N) = (\vec{\xi}(N))^{\top}\vec{\delta}_p$, $N = [\frac{M}{T}]$, $p = M + 1 - NT$, and the obtained results we can conclude that the following corollary holds.

Corollary 2.1.4. *Suppose that the conditions of Theorem* 2.1.4 *are satisfied. The optimal linear estimate* $\widehat{\zeta}(M)$ *of the unobserved value* $\zeta(M)$, $M \ge 0$, *based on observations of the sequence* $\zeta(m)$ *at points* $m = -1, -2, \dots$ *is calculated by formula* (2.29). *The spectral characteristic* $\vec{h}_{\overline{\mu},N,p}(\lambda)$ *of the estimate is calculated by formula* (2.30). *The MSE of the estimate* $\widehat{\zeta}(M)$ *is calculated by formula* (2.31). *If the spectral density* $f(\lambda)$ *admits canonical factorization* (2.19), *the spectral characteristic and the MSE of the estimate* $\widehat{\zeta}(M)$ *can be calculated by formulas* (2.32) *and* (2.33), *respectively.*

2.1.4 Extrapolation of sequences with FM increments

In the previous subsections, we solved the extrapolation problem for the increment sequence $\chi_{\overline{\mu},\overline{s}}^{(d)}(\vec{\xi}(m))$ of the positive integer orders $(d_1, \dots, d_r)$. Here we consider a forecasting problem in the case of possibly fractional increment orders d_i.

Let $d_j = R_j + D_j, j = 0, 1, \dots, r$, denote the fractional differentiation order and let R_j, $j = 0, 1, \dots, r$, be non-negative integer numbers. Set the increment step $\overline{\mu} = (1, 1, \dots, 1)$.

Let $1 < s_1 < \cdots < s_r$. Consider a stochastic vector sequence $\vec{\xi}(m)$ with stationary FM increments $\chi^{(R+D)}_{\overline{1},\overline{s}}(\vec{\xi}(m))$ having a bounded from zero and from infinity spectral density $\widetilde{f}_{\overline{1}}(\lambda)$. Assume that the numbers $R_j, D_j, j = 0, 1, \dots, r$, can be chosen in such way that $-1/2 < D_v < 1/2$ for all $v \in \mathcal{M}$, where the numbers D_v and the set are defined in Lemma 1.2.2. Then Theorem 1.2.1 implies that the GM increment sequence $\chi^{(R)}_{\overline{1},\overline{s}}(\vec{\xi}(m))$ is stationary. The following remarks describe how to apply the theorems from the previous subsection to solve the extrapolation problem without noise.

Remark 2.1.3. Theorem 1.2.1 implies that the Fourier coefficients (2.10) of the function

$$|\beta^{(R)}(i\lambda)|^2 |\chi^{(R)}_{\overline{1}}(e^{-i\lambda})|^{-2} f^{-1}(\lambda)$$

are calculated by the formula

$$F^{\overline{1}}(k,j) = \frac{1}{2\pi}\int_{-\pi}^{\pi} e^{i\lambda(j-k)} |\chi^{(D)}_{\overline{1}}(e^{-i\lambda})|^2 (\widetilde{f}^{-1}_{\overline{1}}(\lambda))^{\top} d\lambda, \quad k,j \geq 0.$$

Remark 2.1.4. Assume that the spectral density $\widetilde{f}_{\overline{1}}(\lambda)$ admits a factorization

$$\widetilde{f}_{\overline{1}}(\lambda) = \widetilde{\Phi}_{\overline{1}}(e^{-i\lambda})\widetilde{\Phi}^*_{\overline{1}}(e^{-i\lambda}) = \left(\sum_{k=0}^{\infty} \widetilde{\varphi}_{\overline{1}}(k)(e^{-i\lambda})\right)\left(\sum_{k=0}^{\infty} \widetilde{\varphi}_{\overline{1}}(k)(e^{-i\lambda})\right)^*,$$

where $\widetilde{\varphi}_{\overline{1}}(k) = \{\varphi_{\overline{1},ij}(k)\}_{i=\overline{1,T}}^{j=\overline{1,q}}$, $k = 0, 1, 2, \dots$. Then coefficients $\{\varphi_{\overline{1},ij}\}_{i=\overline{1,T}}^{j=\overline{1,q}}$, $k = 0, 1, 2, \dots$, from factorization (2.19) are calculated by the formula

$$\varphi_{\overline{1},ij}(k) = \sum_{m=0}^{k} G^{+}_{k^*}(k-m)\widetilde{\varphi}_{\overline{1},ij}(m) = (G^{+}_{k^*} * \widetilde{\varphi}_{\overline{1},ij})(k).$$

Remark 2.1.5. Define a matrix-valued function $\widetilde{\Psi}_{\overline{1}}(e^{-i\lambda}) = \{\widetilde{\Psi}_{\overline{1},ij}(e^{-i\lambda})\}_{i=\overline{1,q}}^{j=\overline{1,T}}$ by the equation $\widetilde{\Psi}_{\overline{1}}(e^{-i\lambda})\widetilde{\Phi}_{\overline{1}}(e^{-i\lambda}) = E_q$, where E_q is the identity $q \times q$ matrix. Then $\Psi_{\overline{1}}(e^{-i\lambda}) = \chi^{(D)}_{\overline{1}}(e^{-i\lambda})\widetilde{\Psi}_{\overline{1}}(e^{-i\lambda})$.

2.1.5 Examples of forecasting for some special models

Example 2.1.1. Basawa et al. [12] consider a so-called first order seasonal periodic autoregressive process (SPAR(1, 1)) defined by the difference equation

$$X_{nT+v} = \phi(v)X_{mT+v-1} + \alpha(v)X_{(m-1)T+v} - \phi(v)\alpha(v)X_{(m-1)T+v-1} + \varepsilon_{mT+v}. \tag{2.41}$$

where $\varepsilon_{mT+\nu}$ is an uncorrelated periodic white noise process with $\mathsf{E}\varepsilon_{mT+\nu} = 0$ and $\mathsf{E}|\varepsilon_{mT+\nu}|^2 = \sigma^2(\nu)$, $1 \le \nu \le T$ (we follow notations from [12]). Depending on coefficients $\phi(\nu)$, $\alpha(\nu)$, model (2.41) has the following properties:

- if $\phi(\nu) \equiv \phi$, $\alpha(\nu) \equiv \alpha$, $\sigma^2(\nu) \equiv \sigma^2$ for $1 \le \nu \le T$, then model (2.41) reduces to the Box–Jenkins SAR(1, 1) model,
- if $\alpha(\nu) \equiv 0$ for $1 \le \nu \le T$, then model (2.41) reduces to the PAR(1) model,
- if $\prod_{\nu=1}^{T} |\phi(\nu)| < 1$ and $|\alpha(\nu)| < 1$ for all ν, then model (2.41) admits a causal and stationary T-dimensional VAR representation

  $$\mathbf{\Phi}_0\mathbf{X}_m = \mathbf{\Phi}_1\mathbf{X}_{m-1} + \mathbf{\Phi}_2\mathbf{X}_{m-2} + \boldsymbol{\varepsilon}_n,$$

  where

  $$\mathbf{X}_m = (X_{mT+1}, X_{mT+2}, \dots, X_{mT+T})^\top,$$
  $$\boldsymbol{\varepsilon}_m = (\varepsilon_{mT+1}, \varepsilon_{mT+2}, \dots, \varepsilon_{mT+T})^\top.$$

Here we consider another case where $\prod_{\nu=1}^{T} |\phi(\nu)| < 1$ and $\alpha(\nu) \equiv 1$ for all ν. Without loss of generality, we assume that $\sigma^2(\nu) \equiv 1$ for all ν. Then, taking into account $B^T X_{mT+\nu} = X_{(m-1)T+\nu}$, model (2.41) reduces to the integrated PAR, or PARIMA, model

$$(1 - B^T)(X_{mT+\nu} - \phi(\nu)X_{mT+\nu-1}) = \varepsilon_{mT+\nu},$$

which admits a VARIMA(1,1,0) representation

$$\mathbf{\Psi}_0\Delta\mathbf{X}_m + \mathbf{\Psi}_1\Delta\mathbf{X}_{m-1} = \boldsymbol{\varepsilon}_m,$$

where $\mathbf{\Psi}_0(r, s) = 1$ for $r = s$, $\mathbf{\Psi}_0(r, s) = -\phi(s)$ for $r = s + 1$, and $\mathbf{\Psi}_0(r, s) = 0$ otherwise; $\mathbf{\Psi}_1(1, T) = -\phi(T)$ and $\Psi_1(r, s) = 0$ otherwise, $1 \le r, s \le T$. Note that $\Delta\mathbf{X}_m = (1 - B)\mathbf{X}_m = \chi^{(1)}_{1,1}(\mathbf{X}_m)$ in terms of GM increments. The spectral density of the one-step increment sequence $\Delta\mathbf{X}_m$ is the following:

$$f(\lambda) = \frac{\lambda^2}{|1 - e^{-i\lambda}|^2} |\Psi_0 + \Psi_1 e^{-i\lambda}|^{-2}.$$

Consider the following estimation problem. Let us assume that we observe a time series $X_{mT+\nu}$ at points $m \le -1$, $1 \le \nu \le T$. It is necessary to find an estimate $\widehat{A}X$ of the functional which depends on future values of $X_{mT+\nu}$, $m \ge 0$, $1 \le \nu \le T$, with a discount factor ρ, $0 < \rho < 1$:

$$AX = \sum_{m=0}^{\infty}\sum_{\nu=1}^{T} \rho^{mT+\nu} X_{mT+\nu} = \sum_{m=0}^{\infty} \mathbf{a}_m^\top \mathbf{X}_m =: A\mathbf{X},$$

where $\mathbf{a}_m = \rho^{mT}(\rho, \rho^2, \dots, \rho^T)^\top = \rho^{mT}\mathbf{a}$, $\mathbf{a} = (\rho, \rho^2, \dots, \rho^T)^\top$. Coefficients $\mathbf{b}_m$, $m \ge 0$, and $\mathbf{v}_{-1}$ from the representation $A\mathbf{X} = B\Delta\mathbf{X} - V\mathbf{X}$ are the following: $\mathbf{b}_m = \frac{\rho^{mT}}{1-\rho^T}\mathbf{a}$, $\mathbf{v}_{-1} =$

$\mathbf{b}_0 = -\frac{1}{1-\rho^T}\mathbf{a}$. Since $\|\mathbf{a}_m\|^2 = c_1(T)\rho^{mT}$, $\|\mathbf{b}_m\|^2 = c_2(T)\rho^{mT}$, conditions (2.2) and (2.3) are satisfied. Thus, we apply Theorem 2.1.4 to find the spectral characteristic of the estimate $\widehat{A}X$. We have

$$\Phi_{(1)}(e^{-i\lambda}) = \Psi_0^{-1}\sum_{k=0}^{\infty}(-1)^k(\Psi_1\Psi_0^{-1})^k e^{-i\lambda k}, \quad \Psi_{(1)}(e^{-i\lambda}) = \Psi_0 + \Psi_1 e^{-i\lambda}$$

and

$$\vec{r}_{(1)}(e^{i\lambda}) = \frac{1}{1-\rho^T}\Theta^\top \mathbf{a}\sum_{k=0}^{\infty}\rho^{kT}e^{i\lambda k},$$

where $\Theta := (\Psi_0 + \rho^T\Psi_1)^{-1}$,

$$(\Psi_{(1)}(e^{-i\lambda}))^\top \vec{r}_{(1)}(e^{i\lambda}) = \frac{1}{1-\rho^T}\Psi_1^\top\Theta^\top \mathbf{a}e^{-i\lambda} + \frac{1}{1-\rho^T}\mathbf{a}\sum_{k=0}^{\infty}\rho^{kT}e^{i\lambda k}.$$

Then the optimal estimate of the value of the functional AX is calculated by the formula

$$\begin{aligned}\widehat{A}X &= -\frac{1}{1-\rho^T}\mathbf{a}^\top\Theta\Psi_1\Delta\mathbf{X}_{-1} + \frac{1}{1-\rho^T}\mathbf{a}^\top\mathbf{X}_{-1}\\ &= \frac{1}{1-\rho^T}\mathbf{a}^\top((E_T - \Theta\Psi_1)\mathbf{X}_{-1} + \Theta\Psi_1\mathbf{X}_{-2}).\end{aligned}$$

The MSE of the estimate is calculated by the formula

$$\Delta(f;\widehat{A}X) = \frac{1}{(1-\rho^T)^3(1+\rho^T)}\|\Theta^\top\mathbf{a}\|^2.$$

Example 2.1.2. To illustrate the extrapolation technique developed in Sections 2.1.3 and 2.1.4, we consider a seasonal time series $\xi(t)$, $t \in \mathbb{Z}$, exhibiting two fractional seasonal patterns and a periodic covariance behavior:

$$(1-B^s)^{d_0}(1-B^{us})^{d_1}\xi(t) = \varepsilon(t) - a_0\varepsilon(t-1) - a_{i(t)}\varepsilon(t-s), \quad i(t) = (t \mod s) + 1,$$

where $d_0 = 1 + D_0$, $d_1 = 1 + D_1$, and $\varepsilon(t)$, $t \in \mathbb{Z}$, are i. i. d. random variables with $\mathsf{E}\varepsilon(t) = 0$, $\mathsf{E}|\varepsilon(t)|^2 = 1$. The first cycle s may refer to seven days within a week, and this pattern shows a different correlation structure for each “season,” namely, day of the week. The second seasonal pattern us may describe a year period, assuming that $u = 52$ corresponds to the number of weeks within a year. Under the conditions stated below, an increment $w(t) = (1-B^s)(1-B^{us})\xi(t)$ is cyclostationary since coefficients $a_{i(t)}$ are periodic with period $T = s$.

Define the vector-valued sequences $\vec{\xi}(m) = (\xi_1(m), \xi_2(m), \dots, \xi_s(m))^\top$, where $\xi_p(m) = \xi(sm+p-1)$, and $\vec{\varepsilon}(m) = (\varepsilon_1(m), \varepsilon_2(m), \dots, \varepsilon_s(m))^\top$, where $\varepsilon_p(m) = \varepsilon(sm+p-1)$.

Consider an increment function $\chi^{(2)}_{(1,1),(1,u)}(B) = (1-B)(1-B^u)$ with step $\overline{\mu} = (1,1)$. The GM increment sequence $\chi^{(2)}_{(1,1),(1,u)}(\vec{\xi}(m))$ admits a representation

$$(1-B)^{D_0}(1-B^u)^{D_1}\chi^{(2)}_{(1,1),(1,u)}(\vec{\xi}(m)) = \Phi_0\vec{\varepsilon}(m) + \Phi_1\vec{\varepsilon}(m-1),$$

where

$$\Phi_0 = \begin{pmatrix} 1 & 0 & 0 & \dots & 0 \\ -a_0 & 1 & 0 & \dots & 0 \\ 0 & -a_0 & 1 & \ddots & 0 \\ \vdots & \vdots & \ddots & \ddots & \vdots \\ 0 & 0 & 0 & \dots & 1 \end{pmatrix}$$

and

$$\Phi_1 = \begin{pmatrix} -a_1 & 0 & 0 & \dots & -a_0 \\ 0 & -a_2 & 0 & \dots & 0 \\ 0 & 0 & -a_3 & \ddots & 0 \\ \vdots & \vdots & \ddots & \ddots & \vdots \\ 0 & 0 & 0 & \dots & -a_s \end{pmatrix}.$$

It is stationary under conditions $|D_0 + D_1| < 1/2$, $|D_1| < 1/2$. For instance, if $d_0 = 0.7$ and $d_1 = 1.2$, then $D_0 = -0.3$, $D_1 = 0.2$, and the process exhibits long memory behavior.

The spectral density $f(\lambda)$ of the GM increment sequence $\chi^{(2)}_{(1,1),(1,d)}\vec{\xi}(m)$ is

$$\begin{aligned} f(\lambda) &= \frac{|\beta^{(2)}(i\lambda)|^2}{|\chi^{(2)}_{(1,1)}(e^{-i\lambda})|^2|\chi^{(D)}_{(1,1)}(e^{-i\lambda})|^2}|\Phi_{(1,1)}(e^{-i\lambda})|^2 \\ &= \frac{\lambda^2\prod_{k=-[u/2]}^{[u/2]}(\lambda-2\pi k/u)^2}{|1-e^{-i\lambda}|^2|1-e^{-iu\lambda}|^2|1-e^{-i\lambda}|^{2D_0}|1-e^{-iu\lambda}|^{2D_1}}|\Phi_0+\Phi_1e^{-i\lambda}|^2, \end{aligned}$$

where

$$\begin{aligned} (\chi^{(D)}_{(1,1)}(e^{-i\lambda}))^{-1} &= (1-e^{-i\lambda})^{-D_0}(1-e^{-iu\lambda})^{-D_1} \\ &= (1-e^{-i\lambda})^{-D_0-D_1}\prod_{k=1}^{[u/2]}(1-2\cos(2\pi k/u)e^{-i\lambda}+e^{-2i\lambda})^{-\widetilde{D}_k} \\ &= \sum_{k=0}^{\infty}G^{+}_{k^*}(k)e^{-i\lambda k} = \left(\sum_{k=0}^{\infty}G^{-}_{k^*}(k)e^{-i\lambda k}\right)^{-1}, \end{aligned}$$

$k^* = [u/2] + 1$, $G^+_{k^*}(m)$ and $G^-_{k^*}(m)$, $m \geq 0$, are defined by (1.17) and (1.18), respectively, $\widetilde{D}_k = D_1$ for $1 \leq k \leq [u/2] - 1$, $\widetilde{D}_{[u/2]} = D_1$ for odd u, and $\widetilde{D}_{[u/2]} = D_1/2$ for even u. Note that $G^+_{k^*}(0) = G^-_{k^*}(0) = 1$.

Let us find an estimate of a weighted sum of two average weekly values of the time series $\xi(t)$,

$$A_{2s}\xi = \alpha\left(\frac{1}{s}\sum_{k=0}^{s-1}\xi(k)\right) + (1-\alpha)\left(\frac{1}{s}\sum_{k=s}^{2s-1}\xi(k)\right),$$

based on observations of $\xi(t)$ at points $t = -1, -2, \ldots$.

In terms of the sequence $\vec{\xi}(m)$, the functional $A_{2s}\xi$ is rewritten as

$$A_2\vec{\xi} = (\vec{a}(0))^\top\vec{\xi}(0) + (\vec{a}(1))^\top\vec{\xi}(1),$$

where

$$\vec{a}(0) = (\alpha s^{-1}, \alpha s^{-1}, \ldots, \alpha s^{-1})^\top,$$
$$\vec{a}(1) = ((1-\alpha)s^{-1}, (1-\alpha)s^{-1}, \ldots, (1-\alpha)s^{-1})^\top,$$

and admits a representation

$$\begin{aligned} A_2\vec{\xi} &= B_2\chi\vec{\xi} - V_2\vec{\xi} \\ &= (\vec{b}_{(1,1),2}(0))^\top\chi(\vec{\xi}(0)) + (\vec{b}_{(1,1),2}(1))^\top\chi(\vec{\xi}(1)) - \sum_{k=-u-1}^{-1}(\vec{v}_{(1,1),2}(k))^\top\vec{\xi}(k). \end{aligned}$$

Here

$$\vec{b}_{(1,1),2}(0) = \vec{a}(0) + \vec{a}(1) = (s^{-1}, s^{-1}, \ldots, s^{-1})^\top = s^{-1}\vec{\mathbf{1}},$$
$$\vec{b}_{(1,1),2}(1) = \vec{a}(1) = ((1-\alpha)s^{-1}, (1-\alpha)s^{-1}, \ldots, (1-\alpha)s^{-1})^\top = (1-\alpha)s^{-1}\vec{\mathbf{1}};$$

and further

$$\begin{aligned} \vec{v}_{(1,1),2}(k) &= \vec{0}, \quad k = -1, -2, \ldots, -u+2, \\ \vec{v}_{(1,1),2}(-u+1) &= -\vec{b}_{(1,1),2}(1) = -(1-\alpha)s^{-1}\vec{\mathbf{1}}, \\ \vec{v}_{(1,1),2}(-u) &= -\vec{b}_{(1,1),2}(0) + \vec{b}_{(1,1),2}(1) = -\alpha s^{-1}\vec{\mathbf{1}}, \\ \vec{v}_{(1,1),2}(-u-1) &= \vec{b}_{(1,1),2}(0) = s^{-1}\vec{\mathbf{1}}. \end{aligned}$$

By $\vec{\mathbf{1}}$ we denote a vector $(1, 1, \ldots, 1)^\top$ of dimension s. Note that

$$(1-B)^{-1}(1-B^u)^{-1} = \sum_{k=0}^{\infty} d(k)B^k = \sum_{k=0}^{\infty}\left(1 + \left[\frac{k}{u}\right]\right)B^k.$$

We find the spectral characteristic $\vec{h}_{(1,1),2}(\lambda)$ of the estimate $\widehat{A}_2\vec{\xi}$ using Theorem 2.1.3 as well as remarks to Theorem 1.2.1. First we obtain

$$\begin{aligned}\Phi_{(1,1)}(e^{-i\lambda}) &= \sum_{k=0}^{\infty}(G^{+}_{k^*} * \Phi)(k)e^{-i\lambda k}\\ &= G^{+}_{k^*}(0)\Phi_0 + \sum_{k=1}^{\infty}(G^{+}_{k^*}(k)\Phi_0 + G^{+}_{k^*}(k-1)\Phi_1)e^{-i\lambda k},\\ \Psi_{(1,1)}(e^{-i\lambda}) &= \sum_{k=0}^{\infty}(G^{-}_{k^*} * \Psi)(k)e^{-i\lambda k} = \Phi_0^{-1}\sum_{k=0}^{\infty}(G^{-}_{k^*} * \widetilde{\Psi})(k)e^{-i\lambda k},\end{aligned}$$

where $(G^{-}_{k^*} * \Psi)(k)$, $k \geq 0$, is a convolution of two sequences $G^{-}_{k^*}(k)$ and Ψ_k, $k \geq 0$, $\Psi_k = (-1)^k\Phi_0^{-1}(\Phi_1\Phi_0^{-1})^k$, $\widetilde{\Psi}_k = \Phi_0^{-1}\Psi_k = (-1)^k(\Phi_1\Phi_0^{-1})^k$, and

$$\Phi_1\Phi_0^{-1} = \begin{pmatrix} -a_1 - a_0^s & -a_0^{s-1} & -a_0^{s-2} & \dots & -a_0\\ -a_2a_0 & -a_2 & 0 & \dots & 0\\ -a_3a_0^2 & -a_3a_0 & -a_3 & \dots & 0\\ \vdots & \vdots & \vdots & \ddots & \vdots\\ -a_sa_0^{s-1} & -a_sa_0^{s-2} & -a_sa_0^{s-3} & \dots & -a_s \end{pmatrix}.$$

Then

$$\begin{aligned}&(\Psi_{(1,1)}(e^{-i\lambda}))^{\top}\vec{r}_{(1,1),2}(e^{i\lambda})\\ &= \vec{b}_{(1,1),2}(1)e^{i\lambda} + \vec{b}_{(1,1),2}(0)\\ &\quad + \sum_{k=1}^{\infty}((G^{-}_{k^*} * \widetilde{\Psi}^{\top})(k)(\vec{b}_{(1,1),2}(0) + G^{+}_{k^*}(1)\vec{b}_{(1,1),2}(1))\\ &\quad + G^{-}_{k^*}(k+1)\vec{b}_{(1,1),2}(1))e^{-i\lambda k}\\ &= \chi^{(D)}_{(1,u)}(e^{-i\lambda})((\widetilde{\Psi}(e^{-i\lambda}))^{\top}(\vec{b}_{(1,1),2}(0) + D_0\vec{b}_{(1,1),2}(1)) + \vec{b}_{(1,1),2}(1)e^{i\lambda})\end{aligned}$$

and

$$\begin{aligned}\vec{h}_{(1,1),2}(\lambda) &= -\frac{\chi^{(2)}_{(1,1)}(e^{-i\lambda})}{\beta^{(2)}(i\lambda)}\sum_{k=1}^{\infty}s^{-1}((1 + (1-\alpha)D_0)(G^{-}_{k^*} * \widetilde{\Psi}^{\top})(k)\\ &\quad + (1-\alpha)G^{-}_{k^*}(k+1)E_T)\vec{\mathbf{1}}e^{-i\lambda k}\\ &= \frac{\chi^{(2)}_{(1,1)}(e^{-i\lambda})}{\beta^{(2)}(i\lambda)}\sum_{k=1}^{\infty}\vec{h}(k).\end{aligned}$$

The optimal estimate of the functional $A_{2s}\xi$ is calculated by the formula

$$\begin{aligned}\widehat{A}_2\vec{\xi} &= -s^{-1}(\vec{\mathbf{1}})^{\top}\vec{\xi}(-u-1) + \alpha s^{-1}(\vec{\mathbf{1}})^{\top}\vec{\xi}(-u) + (1-\alpha)s^{-1}(\vec{\mathbf{1}})^{\top}\vec{\xi}(-u+1)\\ &\quad + \sum_{k=1}^{\infty}(h(k))^{\top}(\vec{\xi}(-k) - \vec{\xi}(-k-1) - \vec{\xi}(-k-u) + \vec{\xi}(-k-u-1)).\end{aligned}$$

The MSE of the estimate is calculated by the formula

$$\Delta(f;\hat{A}_2\vec{\xi}) = s^{-2}\|(\Phi_0 + (1-\alpha)(D_0\Phi_0 + \Phi_1))^{\top}\bar{\mathbf{1}}\|^2 + (1-\alpha)^2 s^{-2}\|\Phi_0^{\top}\bar{\mathbf{1}}\|^2.$$

In the particular case where $d_0 = 1, d_1 = 1, D_0 = 0$, and $D_1 = 0$, we have $\chi_{(1,L)}^{(D)}(e^{-i\lambda}) \equiv 1$ and $G_{k^*}^{\pm}(k) = 0$ for $k \geq 1$. In this case the estimate of the functional $A_{2s}\xi$ and the MSE are calculated by the formulas

$$\begin{aligned}\hat{A}_2\vec{\xi} &= -s^{-1}(\bar{\mathbf{1}})^{\top}\vec{\xi}(-u-1) + \alpha s^{-1}(\bar{\mathbf{1}})^{\top}\vec{\xi}(-u) + (1-\alpha)s^{-1}(\bar{\mathbf{1}})^{\top}\vec{\xi}(-u+1)\\ &\quad + s^{-1}\sum_{k=1}^{\infty}(-1)^{k+1}(\bar{\mathbf{1}})^{\top}(\Phi_1\Phi_0^{-1})^k(\vec{\xi}(-k) - \vec{\xi}(-k-1) - \vec{\xi}(-k-u) + \vec{\xi}(-k-u-1))\end{aligned}$$

and

$$\Delta(\hat{A}_2\vec{\xi};f) = s^{-2}\left(\sum_{k=1}^{s}(1-(1-\alpha)^{\delta_{ks}}a_0 - (1-\alpha)a_k)^2 + (1-\alpha)^2(s-1)(1-a_0)^2 + (1-\alpha)^2\right),$$

respectively.

2.2 Minimax (robust) method of extrapolation

2.2.1 Least favorable spectral density and minimax-robust spectral characteristic

Values of the MSEs and spectral characteristics of the optimal estimates of functionals $A\vec{\xi}$ and $A_N\vec{\xi}$ constructed from unobserved values of stochastic sequence $\vec{\xi}(m)$ determining a stationary GM increment sequence $\chi_{\overline{\mu},\overline{s}}^{(d)}(\vec{\xi}(m))$ based on observations $\vec{\xi}(m)$ at points $m = -1, -2, \dots$ can be calculated by formulas (2.12) and (2.11) and formulas (2.26) and (2.25), respectively, in the case where the spectral density matrix $f(\lambda)$ of the increment sequence $\chi_{\overline{\mu},\overline{s}}^{(d)}(\vec{\xi}(m))$ is exactly known. When $f(\lambda)$ admits canonical factorization (2.19), formulas (2.48) and (2.20) and formulas (2.28) and (2.27) are derived for calculating the MSEs and the spectral characteristics, respectively.

When complete information about the spectral density matrix is not available, while a set $\mathcal{D}$ of admissible spectral densities can be defined, the minimax method of estimation is suitable. This method consists in finding an estimate that minimizes the maximal values of the MSEs for all spectral densities from a given class $\mathcal{D}$ simultaneously. The following two definitions formalize the minimax method of estimation.

Definition 2.2.1. For a given class of spectral densities $\mathcal{D}$, a spectral density $f^0(\lambda) \in \mathcal{D}$ is called the least favorable in $\mathcal{D}$ for the optimal linear estimation of the functional $A\vec{\xi}$ if the following relation holds:

$$\Delta(f^0) = \Delta(h_{\overline{\mu}}(f^0);f^0) = \max_{f\in\mathcal{D}}\Delta(h_{\overline{\mu}}(f);f).$$

Definition 2.2.2. For a given class of spectral densities $\mathcal{D}$, a spectral characteristic $h^0(\lambda)$ of the optimal linear estimate of the functional $A\xi$ is called minimax-robust if the following conditions are satisfied:

$$h^0(\lambda) \in H_{\mathcal{D}} = \bigcap_{f \in \mathcal{D}} L_2^{-1}(f), \quad \min_{h \in H_{\mathcal{D}}} \max_{f \in \mathcal{D}} \Delta(h;f) = \sup_{f \in \mathcal{D}} \Delta(h^0;f).$$

Taking into account the introduced definitions and the relations derived in the previous sections we can verify that the following lemmas hold.

Lemma 2.2.1. *A spectral density $f^0(\lambda) \in \mathcal{D}$ satisfying the minimality condition* (2.4) *is the least favorable density in the class $\mathcal{D}$ for the optimal linear estimation of the functional $A\vec{\xi}$ based on observations of the sequence $\vec{\xi}(m)$ at points $m = -1, -2, \dots$ if the operator $\mathbf{F}^0_{\overline{\mu}}$ defined by the Fourier coefficients of the function $|\beta^{(d)}(i\lambda)|^2 |\chi^{(d)}_{\overline{\mu}}(e^{-i\lambda})|^{-2}(f^0(\lambda))^{-1}$ determines a solution of the constrained optimization problem*

$$\max_{f \in \mathcal{D}}(\langle D^{\overline{\mu}}\mathbf{a}, \mathbf{F}^{-1}_{\overline{\mu}} D^{\overline{\mu}}\mathbf{a}\rangle) = \langle D^{\overline{\mu}}\mathbf{a}, (\mathbf{F}^0_{\overline{\mu}})^{-1} D^{\overline{\mu}}\mathbf{a}\rangle. \tag{2.42}$$

The minimax spectral characteristic $h^0 = \vec{h}_{\overline{\mu}}(f^0)$ is calculated by formula (2.11) *if $\vec{h}_{\overline{\mu}}(f^0) \in H_{\mathcal{D}}$.*

Lemma 2.2.2. *A spectral density $f^0(\lambda) \in \mathcal{D}$ which admits the canonical factorization* (2.19) *is the least favorable density in the class $\mathcal{D}$ for the optimal linear estimation of the functional $A\vec{\xi}$ based on observations of the sequence $\vec{\xi}(m)$ at points $m = -1, -2, \dots$ if coefficients $\{\varphi^0(k) : k \ge 0\}$ of the canonical factorization*

$$f^0(\lambda) = \left(\sum_{k=0}^{\infty} \varphi^0(k) e^{-i\lambda k}\right)\left(\sum_{k=0}^{\infty} \varphi^0(k) e^{-i\lambda k}\right)^* \tag{2.43}$$

of the spectral density $f^0(\lambda)$ determine a solution to the constrained optimization problem

$$\|D^{\overline{\mu}}\mathbf{A}\varphi_{\overline{\mu}}\|^2 \to \max, \quad f(\lambda) = \left(\sum_{k=0}^{\infty} \varphi(k) e^{-i\lambda k}\right)\left(\sum_{k=0}^{\infty} \varphi(k) e^{-i\lambda k}\right)^* \in \mathcal{D}. \tag{2.44}$$

The minimax spectral characteristic $h^0 = \vec{h}_{\overline{\mu}}(f^0)$ is calculated by formula (2.20) *if $\vec{h}_{\overline{\mu}}(f^0) \in H_{\mathcal{D}}$.*

Lemma 2.2.3. *A spectral density $f^0(\lambda) \in \mathcal{D}$ which admits the canonical factorization* (2.19) *is the least favorable density in the class $\mathcal{D}$ for the optimal linear extrapolation of the functional $A_N\vec{\xi}$ based on observations of the sequence $\vec{\xi}(m)$ at points $m = -1, -2, \dots$ if coefficients $\{\varphi^0(k) : k = 0, 1, \dots, N\}$ from the canonical factorization*

$$f^0(\lambda) = \left(\sum_{k=0}^{N} \varphi^0(k) e^{-i\lambda k}\right)\left(\sum_{k=0}^{N} \varphi^0(k) e^{-i\lambda k}\right)^* \tag{2.45}$$

of the spectral density $f^0(\lambda)$ determine a solution to the constrained optimization problem

$$\|D^{\overline{\mu}}_N \mathbf{A}_N \varphi_{\overline{\mu},N}\|^2 \to \max, \quad f(\lambda) = \left(\sum_{k=0}^{N} \varphi(k)e^{-i\lambda k}\right)\left(\sum_{k=0}^{N} \varphi(k)e^{-i\lambda k}\right)^* \in \mathcal{D}. \tag{2.46}$$

The minimax spectral characteristic $h^0 = \vec{h}_{\overline{\mu},N}(f^0)$ is calculated by formula (2.27) *if $\vec{h}_{\overline{\mu},N}(f^0) \in H_{\mathcal{D}}$.*

For more detailed analysis, we conclude that the minimax spectral characteristic h^0 and the least favorable spectral density f^0 form a saddle point of the function $\Delta(h;f)$ on the set $H_{\mathcal{D}} \times \mathcal{D}$. The saddle point inequalities

$$\Delta(h;f^0) \geq \Delta(h^0;f^0) \geq \Delta(h^0;f) \quad \forall f \in \mathcal{D}, \forall h \in H_{\mathcal{D}}$$

hold if $h^0 = \vec{h}_{\overline{\mu}}(f^0)$, $\vec{h}_{\overline{\mu}}(f^0) \in H_{\mathcal{D}}$, and f^0 is a solution of the constrained optimization problem

$$\widetilde{\Delta}(f) = -\Delta(\vec{h}_{\overline{\mu}}(f^0);f) \to \inf, \quad f \in \mathcal{D}, \tag{2.47}$$

where the functional $\Delta(\vec{h}_{\overline{\mu}}(f^0);f)$ is calculated by the formula

$$\Delta(\vec{h}_{\overline{\mu}}(f^0);f) = \frac{1}{2\pi}\int_{-\pi}^{\pi} \frac{\overline{\beta^{(d)}(i\lambda)}}{\chi^{(d)}_{\overline{\mu}}(e^{-i\lambda})}\left(\sum_{k=0}^{\infty}((\mathbf{F}^0_{\overline{\mu}})^{-1}D^{\overline{\mu}}\mathbf{a})_k e^{i\lambda k}\right)^{\top}(f^0(\lambda))^{-1}$$
$$\times f(\lambda)(f^0(\lambda))^{-1}\overline{\left(\sum_{k=0}^{\infty}((\mathbf{F}^0_{\overline{\mu}})^{-1}D^{\overline{\mu}}\mathbf{a})_k e^{i\lambda k}\right)}\frac{\beta^{(d)}(i\lambda)}{\chi^{(d)}_{\overline{\mu}}(e^{-i\lambda})}d\lambda$$

or by the formula

$$\Delta(\vec{h}_{\overline{\mu}}(f^0);f) = \frac{1}{2\pi}\int_{-\pi}^{\pi} \frac{\chi^{(d)}_{\overline{\mu}}(e^{-i\lambda})}{\beta^{(d)}(i\lambda)}\left(\sum_{k=0}^{\infty}(D^{\overline{\mu}}\mathbf{A}\varphi^0_{\overline{\mu}})_k e^{i\lambda k}\right)^{\top}(\Psi^0_{\overline{\mu}}(e^{-i\lambda}))$$
$$\times f(\lambda)(\Psi^0_{\overline{\mu}}(e^{-i\lambda}))^*\overline{\left(\sum_{k=0}^{\infty}(D^{\overline{\mu}}\mathbf{A}\varphi^0_{\overline{\mu}})_k e^{i\lambda k}\right)}\frac{\overline{\chi^{(d)}_{\overline{\mu}}(e^{-i\lambda})}}{\overline{\beta^{(d)}(i\lambda)}}d\lambda \tag{2.48}$$

in the case where the spectral density admits the canonical factorization (2.19).

The constrained optimization problem (2.47) is equivalent to the unconstrained optimization problem

$$\Delta_{\mathcal{D}}(f) = \widetilde{\Delta}(f) + \delta(f|\mathcal{D}) \to \inf,$$

where $\delta(f|\mathcal{D})$ is the indicator function of the set $\mathcal{D}$, namely, $\delta(f|\mathcal{D}) = 0$ if $f \in \mathcal{D}$ and $\delta(f|\mathcal{D}) = +\infty$ if $f \notin \mathcal{D}$. A solution f^0 of this unconstrained optimization problem is characterized by the condition $0 \in \partial\Delta_{\mathcal{D}}(f^0)$, which is the necessary and sufficient condition under which a point f^0 belongs to the set of minimums of the convex functional $\Delta_{\mathcal{D}}(f)$ [52, 145, 151, 180]. This condition makes it possible to find the least favorable spectral densities in some special classes of spectral densities D. The form of the functional $\widetilde{\Delta}(f)$ allows us to apply the Lagrange method of indefinite multipliers for investigating the constrained optimization problem (2.47).

2.2.2 Least favorable spectral densities in classes $\mathcal{D}_0$

Consider the extrapolation problem for the functional $A\vec{\xi}$ which depends on unobserved values of a sequence $\vec{\xi}(m)$ with stationary GM increments based on observations of the sequence at points $m = -1, -2, \dots$ under the condition that sets of admissible spectral densities $\mathcal{D}_0^k$, $k = 1, 2, 3, 4$, are defined as follows:

$$\mathcal{D}_0^1 = \left\{ f(\lambda) \left| \frac{1}{2\pi} \int_{-\pi}^{\pi} \frac{|\chi_{\overline{\mu}}^{(d)}(e^{-i\lambda})|^2}{|\beta^{(d)}(i\lambda)|^2} f(\lambda) d\lambda = P \right. \right\},$$

$$\mathcal{D}_0^2 = \left\{ f(\lambda) \left| \frac{1}{2\pi} \int_{-\pi}^{\pi} \frac{|\chi_{\overline{\mu}}^{(d)}(e^{-i\lambda})|^2}{|\beta^{(d)}(i\lambda)|^2} \operatorname{Tr}[f(\lambda)] d\lambda = p \right. \right\},$$

$$\mathcal{D}_0^3 = \left\{ f(\lambda) \left| \frac{1}{2\pi} \int_{-\pi}^{\pi} \frac{|\chi_{\overline{\mu}}^{(d)}(e^{-i\lambda})|^2}{|\beta^{(d)}(i\lambda)|^2} f_{kk}(\lambda) d\lambda = p_k, k = \overline{1, T} \right. \right\},$$

$$\mathcal{D}_0^4 = \left\{ f(\lambda) \left| \frac{1}{2\pi} \int_{-\pi}^{\pi} \frac{|\chi_{\overline{\mu}}^{(d)}(e^{-i\lambda})|^2}{|\beta^{(d)}(i\lambda)|^2} \langle B_1, f(\lambda) \rangle d\lambda = p \right. \right\},$$

where $p, p_k, k = \overline{1, T}$, are given numbers and P, B_1 are given positive-definite Hermitian matrices.

From the condition $0 \in \partial\Delta_{\mathcal{D}}(f^0)$ we find the following equations, which determine the least favorable spectral densities for these given sets of admissible spectral densities.

For the first set of admissible spectral densities $\mathcal{D}_0^1$, we have

$$\begin{aligned} &\left(\sum_{k=0}^{\infty} ((\mathbf{F}_{\overline{\mu}}^0)^{-1} D^{\overline{\mu}} \mathbf{a})_k e^{i\lambda k} \right) \left(\sum_{k=0}^{\infty} ((\mathbf{F}_{\overline{\mu}}^0)^{-1} D^{\overline{\mu}} \mathbf{a})_k e^{i\lambda k} \right)^* \\ &\quad = \left(\frac{|\chi_{\overline{\mu}}^{(d)}(e^{-i\lambda})|^2}{|\beta^{(d)}(i\lambda)|^2} f^0(\lambda) \right) \vec{\alpha} \cdot \vec{\alpha}^* \left(\frac{|\chi_{\overline{\mu}}^{(d)}(e^{-i\lambda})|^2}{|\beta^{(d)}(i\lambda)|^2} f^0(\lambda) \right), \end{aligned} \tag{2.49}$$

where $\vec{\alpha}$ is a vector of Lagrange multipliers.

For the second set of admissible spectral densities $\mathcal{D}_0^2$, we have

$$\left(\sum_{k=0}^{\infty}((\mathbf{F}_{\overline{\mu}}^0)^{-1}D^{\overline{\mu}}\mathbf{a})_k e^{i\lambda k}\right)\left(\sum_{k=0}^{\infty}((\mathbf{F}_{\overline{\mu}}^0)^{-1}D^{\overline{\mu}}\mathbf{a})_k e^{i\lambda k}\right)^* = \alpha^2\left(\frac{|\chi_{\overline{\mu}}^{(d)}(e^{-i\lambda})|^2}{|\beta^{(d)}(i\lambda)|^2}f^0(\lambda)\right)^2 \tag{2.50}$$

where α^2 is a Lagrange multiplier.

For the third set of admissible spectral densities $\mathcal{D}_0^3$, we have

$$\begin{aligned}&\left(\sum_{k=0}^{\infty}((\mathbf{F}_{\overline{\mu}}^0)^{-1}D^{\overline{\mu}}\mathbf{a})_k e^{i\lambda k}\right)\left(\sum_{k=0}^{\infty}((\mathbf{F}_{\overline{\mu}}^0)^{-1}D^{\overline{\mu}}\mathbf{a})_k e^{i\lambda k}\right)^* \\ &= \left(\frac{|\chi_{\overline{\mu}}^{(d)}(e^{-i\lambda})|^2}{|\beta^{(d)}(i\lambda)|^2}f^0(\lambda)\right)\{\alpha_k^2\delta_{kl}\}_{k,l=1}^T\left(\frac{|\chi_{\overline{\mu}}^{(d)}(e^{-i\lambda})|^2}{|\beta^{(d)}(i\lambda)|^2}f^0(\lambda)\right),\end{aligned} \tag{2.51}$$

where α_k^2 are Lagrange multipliers and δ_{kl} are Kronecker symbols.

For the fourth set of admissible spectral densities $\mathcal{D}_0^4$, we have

$$\begin{aligned}&\left(\sum_{k=0}^{\infty}((\mathbf{F}_{\overline{\mu}}^0)^{-1}D^{\overline{\mu}}\mathbf{a})_k e^{i\lambda k}\right)\left(\sum_{k=0}^{\infty}((\mathbf{F}_{\overline{\mu}}^0)^{-1}D^{\overline{\mu}}\mathbf{a})_k e^{i\lambda k}\right)^* \\ &= \alpha^2\left(\frac{|\chi_{\overline{\mu}}^{(d)}(e^{-i\lambda})|^2}{|\beta^{(d)}(i\lambda)|^2}f^0(\lambda)\right)B_1^{\top}\left(\frac{|\chi_{\overline{\mu}}^{(d)}(e^{-i\lambda})|^2}{|\beta^{(d)}(i\lambda)|^2}f^0(\lambda)\right),\end{aligned} \tag{2.52}$$

where α^2 is a Lagrange multiplier.

In the case where the spectral density admits the canonical factorization (2.19), we have the following equations, respectively:

$$\left(\sum_{k=0}^{\infty}(D^{\overline{\mu}}\mathbf{A}\varphi_{\overline{\mu}}^0)_k e^{i\lambda k}\right)\left(\sum_{k=0}^{\infty}(D^{\overline{\mu}}\mathbf{A}\varphi_{\overline{\mu}}^0)_k e^{i\lambda k}\right)^* = (\Phi_{\overline{\mu}}^0(e^{-i\lambda}))^{\top}\vec{\alpha}\cdot\vec{\alpha}^*\overline{\Phi_{\overline{\mu}}^0(e^{-i\lambda})}, \tag{2.53}$$

$$\left(\sum_{k=0}^{\infty}(D^{\overline{\mu}}\mathbf{A}\varphi_{\overline{\mu}}^0)_k e^{i\lambda k}\right)\left(\sum_{k=0}^{\infty}(D^{\overline{\mu}}\mathbf{A}\varphi_{\overline{\mu}}^0)_k e^{i\lambda k}\right)^* = \alpha^2(\Phi_{\overline{\mu}}^0(e^{-i\lambda}))^{\top}\overline{\Phi_{\overline{\mu}}^0(e^{-i\lambda})}, \tag{2.54}$$

$$\left(\sum_{k=0}^{\infty}(D^{\overline{\mu}}\mathbf{A}\varphi_{\overline{\mu}}^0)_k e^{i\lambda k}\right)\left(\sum_{k=0}^{\infty}(D^{\overline{\mu}}\mathbf{A}\varphi_{\overline{\mu}}^0)_k e^{i\lambda k}\right)^* = (\Phi_{\overline{\mu}}^0(e^{-i\lambda}))^{\top}\{\alpha_k^2\delta_{kl}\}_{k,l=1}^T\overline{\Phi_{\overline{\mu}}^0(e^{-i\lambda})}, \tag{2.55}$$

$$\left(\sum_{k=0}^{\infty}(D^{\overline{\mu}}\mathbf{A}\varphi_{\overline{\mu}}^0)_k e^{i\lambda k}\right)\left(\sum_{k=0}^{\infty}(D^{\overline{\mu}}\mathbf{A}\varphi_{\overline{\mu}}^0)_k e^{i\lambda k}\right)^* = \alpha^2(\Phi_{\overline{\mu}}^0(e^{-i\lambda}))^{\top}B_1^{\top}\overline{\Phi_{\overline{\mu}}^0(e^{-i\lambda})}. \tag{2.56}$$

The following theorem holds.

Theorem 2.2.1. *The least favorable spectral densities $f^0(\lambda)$ in the classes $\mathcal{D}_0^k$, $k = 1, 2, 3, 4$, for the optimal linear extrapolation of the functional $A\vec{\xi}$ from observations of the sequence $\vec{\xi}(m)$ at points $m = -1, -2, \ldots$ are determined by equations* (2.49), (2.50), (2.51), *and* (2.52) *(or equations* (2.53), (2.54), (2.55), *and* (2.56) *in the case where the spectral densities admit the canonical factorization* (2.19), *respectively), the minimality condition* (2.4), *the constrained optimization problem* (2.42), *and restrictions on densities from the corresponding classes* $\mathcal{D}_0^k$, $k = 1, 2, 3, 4$. *The minimax-robust spectral characteristic of the optimal estimate of the functional* $A\vec{\xi}$ *is determined by formula* (2.11).

2.2.3 Least favorable spectral densities in classes $\mathcal{D}_V^U$

Consider the extrapolation problem for the functional $A\vec{\xi}$ which depends on unobserved values of a sequence $\vec{\xi}(m)$ with stationary GM increments based on observations of the sequence at points $m = -1, -2, \ldots$ under the condition that sets of admissible spectral densities $\mathcal{D}_V^{Uk}$, $k = 1, 2, 3, 4$, are defined as follows:

$$\mathcal{D}_V^{U1} = \left\{ f(\lambda) \middle| V(\lambda) \le f(\lambda) \le U(\lambda), \frac{1}{2\pi}\int_{-\pi}^{\pi} \frac{|\chi_{\overline{\mu}}^{(d)}(e^{-i\lambda})|^2}{|\beta^{(d)}(i\lambda)|^2} f(\lambda)d\lambda = Q \right\},$$

$$\mathcal{D}_V^{U2} = \left\{ f(\lambda) \middle| \operatorname{Tr}[V(\lambda)] \le \operatorname{Tr}[f(\lambda)] \le \operatorname{Tr}[U(\lambda)], \frac{1}{2\pi}\int_{-\pi}^{\pi} \frac{|\chi_{\overline{\mu}}^{(d)}(e^{-i\lambda})|^2}{|\beta^{(d)}(i\lambda)|^2} \operatorname{Tr}[f(\lambda)]d\lambda = q \right\},$$

$$\mathcal{D}_V^{U3} = \left\{ f(\lambda) \middle| v_{kk}(\lambda) \le f_{kk}(\lambda) \le u_{kk}(\lambda), \frac{1}{2\pi}\int_{-\pi}^{\pi} \frac{|\chi_{\overline{\mu}}^{(d)}(e^{-i\lambda})|^2}{|\beta^{(d)}(i\lambda)|^2} f_{kk}(\lambda)d\lambda = q_k, k = \overline{1,T} \right\},$$

$$\mathcal{D}_V^{U4} = \left\{ f(\lambda) \middle| \langle B_2, V(\lambda)\rangle \le \langle B_2, f(\lambda)\rangle \le \langle B_2, U(\lambda)\rangle, \frac{1}{2\pi}\int_{-\pi}^{\pi} \frac{|\chi_{\overline{\mu}}^{(d)}(e^{-i\lambda})|^2}{|\beta^{(d)}(i\lambda)|^2} \langle B_2, f(\lambda)\rangle d\lambda = q \right\}.$$

Here the spectral densities $V(\lambda)$ and $U(\lambda)$ are known and fixed, $q, q_k, k = \overline{1,T}$, are given numbers, and Q, B_2 are given positive-definite Hermitian matrices.

From the condition $0 \in \partial\Delta_{\mathcal{D}}(f^0)$ we find the following equations, which determine the least favorable spectral densities for these given sets of admissible spectral densities.

For the first set of admissible spectral densities $\mathcal{D}_V^{U1}$, we have

$$\begin{gathered}\left(\sum_{k=0}^{\infty}((\mathbf{F}_{\overline{\mu}}^0)^{-1}D^{\overline{\mu}}\mathbf{a})_k e^{i\lambda k}\right)\left(\sum_{k=0}^{\infty}((\mathbf{F}_{\overline{\mu}}^0)^{-1}D^{\overline{\mu}}\mathbf{a})_k e^{i\lambda k}\right)^* \\ = \left(\frac{|\chi_{\overline{\mu}}^{(d)}(e^{-i\lambda})|^2}{|\beta^{(d)}(i\lambda)|^2} f^0(\lambda)\right)(\vec{\beta}\cdot\vec{\beta}^* + \Gamma_1(\lambda) + \Gamma_2(\lambda))\left(\frac{|\chi_{\overline{\mu}}^{(d)}(e^{-i\lambda})|^2}{|\beta^{(d)}(i\lambda)|^2} f^0(\lambda)\right),\end{gathered} \tag{2.57}$$

where $\vec{\beta}$ is a vector of Lagrange multipliers, $\Gamma_1(\lambda) \le 0$ and $\Gamma_1(\lambda) = 0$ if $f^0(\lambda) > V(\lambda)$, and $\Gamma_2(\lambda) \ge 0$ and $\Gamma_2(\lambda) = 0$ if $f^0(\lambda) < U(\lambda)$.

For the second set of admissible spectral densities $\mathcal{D}_V^{U^2}$, we have

$$\left(\sum_{k=0}^{\infty}((\mathbf{F}_{\overline{\mu}}^0)^{-1}D^{\overline{\mu}}\mathbf{a})_k e^{i\lambda k}\right)\left(\sum_{k=0}^{\infty}((\mathbf{F}_{\overline{\mu}}^0)^{-1}D^{\overline{\mu}}\mathbf{a})_k e^{i\lambda k}\right)^* \\ = (\beta^2+\gamma_1(\lambda)+\gamma_2(\lambda))\left(\frac{|\chi_{\overline{\mu}}^{(d)}(e^{-i\lambda})|^2}{|\beta^{(d)}(i\lambda)|^2}f^0(\lambda)\right)^2, \tag{2.58}$$

where β^2 is a Lagrange multiplier, $\gamma_1(\lambda)\leq 0$ and $\gamma_1(\lambda)=0$ if $\mathrm{Tr}[f^0(\lambda)]>\mathrm{Tr}[V(\lambda)]$, and $\gamma_2(\lambda)\geq 0$ and $\gamma_2(\lambda)=0$ if $\mathrm{Tr}[f^0(\lambda)]<\mathrm{Tr}[U(\lambda)]$.

For the third set of admissible spectral densities $\mathcal{D}_V^{U^3}$, we have

$$\left(\sum_{k=0}^{\infty}((\mathbf{F}_{\overline{\mu}}^0)^{-1}D^{\overline{\mu}}\mathbf{a})_k e^{i\lambda k}\right)\left(\sum_{k=0}^{\infty}((\mathbf{F}_{\overline{\mu}}^0)^{-1}D^{\overline{\mu}}\mathbf{a})_k e^{i\lambda k}\right)^* \\ = \left(\frac{|\chi_{\overline{\mu}}^{(d)}(e^{-i\lambda})|^2}{|\beta^{(d)}(i\lambda)|^2}f^0(\lambda)\right)\{(\beta_k^2+\gamma_{1k}(\lambda)+\gamma_{2k}(\lambda))\delta_{kl}\}_{k,l=1}^T\left(\frac{|\chi_{\overline{\mu}}^{(d)}(e^{-i\lambda})|^2}{|\beta^{(d)}(i\lambda)|^2}f^0(\lambda)\right) \tag{2.59}$$

where β_k^2 are Lagrange multipliers, δ_{kl} are Kronecker symbols, $\gamma_{1k}(\lambda)\leq 0$ and $\gamma_{1k}(\lambda)=0$ if $f_{kk}^0(\lambda)>v_{kk}(\lambda)$, and $\gamma_{2k}(\lambda)\geq 0$ and $\gamma_{2k}(\lambda)=0$ if $f_{kk}^0(\lambda)<u_{kk}(\lambda)$.

For the fourth set of admissible spectral densities $\mathcal{D}_V^{U^4}$, we have

$$\left(\sum_{k=0}^{\infty}((\mathbf{F}_{\overline{\mu}}^0)^{-1}D^{\overline{\mu}}\mathbf{a})_k e^{i\lambda k}\right)\left(\sum_{k=0}^{\infty}((\mathbf{F}_{\overline{\mu}}^0)^{-1}D^{\overline{\mu}}\mathbf{a})_k e^{i\lambda k}\right)^* \\ = (\beta^2+\gamma_1'(\lambda)+\gamma_2'(\lambda))\left(\frac{|\chi_{\overline{\mu}}^{(d)}(e^{-i\lambda})|^2}{|\beta^{(d)}(i\lambda)|^2}f^0(\lambda)\right)B_2^{\top}\left(\frac{|\chi_{\overline{\mu}}^{(d)}(e^{-i\lambda})|^2}{|\beta^{(d)}(i\lambda)|^2}f^0(\lambda)\right), \tag{2.60}$$

where β^2 is a Lagrange multiplier, $\gamma_1'(\lambda)\leq 0$ and $\gamma_1'(\lambda)=0$ if $\langle B_2,f^0(\lambda)\rangle>\langle B_2,V(\lambda)\rangle$, and $\gamma_2'(\lambda)\geq 0$ and $\gamma_2'(\lambda)=0$ if $\langle B_2,f^0(\lambda)\rangle<\langle B_2,U(\lambda)\rangle$.

In the case where the spectral density admits the canonical factorization (2.19), we have the following equations:

$$\left(\sum_{k=0}^{\infty}(D^{\overline{\mu}}\mathbf{A}\varphi_{\overline{\mu}}^0)_k e^{i\lambda k}\right)\left(\sum_{k=0}^{\infty}(D^{\overline{\mu}}\mathbf{A}\varphi_{\overline{\mu}}^0)_k e^{i\lambda k}\right)^* \\ = (\Phi_{\overline{\mu}}^0(e^{-i\lambda}))^{\top}(\vec{\beta}\cdot\vec{\beta}^*+\Gamma_1(\lambda)+\Gamma_2(\lambda))\,\overline{\Phi_{\overline{\mu}}^0(e^{-i\lambda})}, \tag{2.61}$$

$$\left(\sum_{k=0}^{\infty}(D^{\overline{\mu}}\mathbf{A}\varphi_{\overline{\mu}}^0)_k e^{i\lambda k}\right)\left(\sum_{k=0}^{\infty}(D^{\overline{\mu}}\mathbf{A}\varphi_{\overline{\mu}}^0)_k e^{i\lambda k}\right)^* \\ = (\beta^2+\gamma_1(\lambda)+\gamma_2(\lambda))(\Phi_{\overline{\mu}}^0(e^{-i\lambda}))^{\top}\,\overline{\Phi_{\overline{\mu}}^0(e^{-i\lambda})}, \tag{2.62}$$

$$\left(\sum_{k=0}^{\infty}(D^{\overline{\mu}}\mathbf{A}\varphi_{\overline{\mu}}^{0})_k e^{i\lambda k}\right)\left(\sum_{k=0}^{\infty}(D^{\overline{\mu}}\mathbf{A}\varphi_{\overline{\mu}}^{0})_k e^{i\lambda k}\right)^{*}$$
$$= (\Phi_{\overline{\mu}}^{0}(e^{-i\lambda}))^{\top}\{(\beta_k^2+\gamma_{1k}(\lambda)+\gamma_{2k}(\lambda))\delta_{kl}\}_{k,l=1}^{T}\overline{\Phi_{\overline{\mu}}^{0}(e^{-i\lambda})}, \tag{2.63}$$

$$\left(\sum_{k=0}^{\infty}(D^{\overline{\mu}}\mathbf{A}\varphi_{\overline{\mu}}^{0})_k e^{i\lambda k}\right)\left(\sum_{k=0}^{\infty}(D^{\overline{\mu}}\mathbf{A}\varphi_{\overline{\mu}}^{0})_k e^{i\lambda k}\right)^{*}$$
$$= (\beta^2+\gamma_1'(\lambda)+\gamma_2'(\lambda))(\Phi_{\overline{\mu}}^{0}(e^{-i\lambda}))^{\top}B_2^{\top}\,\overline{\Phi_{\overline{\mu}}^{0}(e^{-i\lambda})}. \tag{2.64}$$

Theorem 2.2.2. *The least favorable spectral densities $f^0(\lambda)$ in the classes $\mathcal{D}_V^{U^k}$, $k = 1,2,3,4$, for the optimal linear extrapolation of the functional $A\vec{\xi}$ from observations of the sequence $\vec{\xi}(m)$ at points $m = -1,-2,\dots$ are determined by equations* (2.57), (2.58), (2.59), *and* (2.60) *(or equations* (2.61), (2.62), (2.63), *and* (2.64) *in the case where the spectral densities admit canonical factorization* (2.19), *respectively), the minimality condition* (2.4), *the constrained optimization problem* (2.42), *and restrictions on densities from the corresponding classes $\mathcal{D}_V^{U^k}$, $k = 1,2,3,4$. The minimax-robust spectral characteristic of the optimal estimate of the functional $A\vec{\xi}$ is determined by formula* (2.11).

2.2.4 Least favorable spectral densities in classes $\mathcal{D}_\varepsilon$

Consider the extrapolation problem for the functional $A\vec{\xi}$ which depends on unobserved values of a sequence $\vec{\xi}(m)$ with stationary GM increments based on observations of the sequence at points $m = -1,-2,\dots$ under the condition that sets of admissible spectral densities $\mathcal{D}_\varepsilon^k$, $k = 1,2,3,4$, are defined as follows:

$$\mathcal{D}_\varepsilon^1 = \left\{f(\lambda)\,\middle|\,\operatorname{Tr}[f(\lambda)] = (1-\varepsilon)\operatorname{Tr}[f_1(\lambda)] + \varepsilon\operatorname{Tr}[W(\lambda)],\right.$$
$$\left.\frac{1}{2\pi}\int_{-\pi}^{\pi}\frac{|\chi_{\overline{\mu}}^{(d)}(e^{-i\lambda})|^2}{|\beta^{(d)}(i\lambda)|^2}\operatorname{Tr}[f(\lambda)]d\lambda = p\right\},$$

$$\mathcal{D}_\varepsilon^2 = \left\{f(\lambda)\,\middle|\,f_{kk}(\lambda) = (1-\varepsilon)f_{kk}^1(\lambda) + \varepsilon w_{kk}(\lambda),\right.$$
$$\left.\frac{1}{2\pi}\int_{-\pi}^{\pi}\frac{|\chi_{\overline{\mu}}^{(d)}(e^{-i\lambda})|^2}{|\beta^{(d)}(i\lambda)|^2}f_{kk}(\lambda)d\lambda = p_k, k=\overline{1,T}\right\},$$

$$\mathcal{D}_\varepsilon^3 = \left\{f(\lambda)\,\middle|\,\langle B_1, f(\lambda)\rangle = (1-\varepsilon)\langle B_1, f_1(\lambda)\rangle + \varepsilon\langle B_1, W(\lambda)\rangle,\right.$$
$$\left.\frac{1}{2\pi}\int_{-\pi}^{\pi}\frac{|\chi_{\overline{\mu}}^{(d)}(e^{-i\lambda})|^2}{|\beta^{(d)}(i\lambda)|^2}\langle B_1, f(\lambda)\rangle d\lambda = p\right\},$$

$$\mathcal{D}_\varepsilon^4 = \left\{ f(\lambda) \,\middle|\, f(\lambda) = (1-\varepsilon) f_1(\lambda) + \varepsilon W(\lambda), \frac{1}{2\pi} \int_{-\pi}^{\pi} \frac{|\chi_{\overline{\mu}}^{(d)}(e^{-i\lambda})|^2}{|\beta^{(d)}(i\lambda)|^2} f(\lambda) d\lambda = P \right\}.$$

Here $f_1(\lambda)$ is a fixed spectral density, $W(\lambda)$ is an unknown spectral density, $p, p_k, k = \overline{1,T}$, are given numbers, and P is a given positive-definite Hermitian matrix.

From the condition $0 \in \partial\Delta_{\mathcal{D}}(f^0)$ we find the following equations, which determine the least favorable spectral densities for these given sets of admissible spectral densities.

For the first set of admissible spectral densities $\mathcal{D}_\varepsilon^1$, we have

$$\left(\sum_{k=0}^{\infty} ((\mathbf{F}_{\overline{\mu}}^0)^{-1} D^{\overline{\mu}} \mathbf{a})_k e^{i\lambda k} \right) \left(\sum_{k=0}^{\infty} ((\mathbf{F}_{\overline{\mu}}^0)^{-1} D^{\overline{\mu}} \mathbf{a})_k e^{i\lambda k} \right)^* = (\alpha^2 + \gamma_1(\lambda)) \left(\frac{|\chi_{\overline{\mu}}^{(d)}(e^{-i\lambda})|^2}{|\beta^{(d)}(i\lambda)|^2} f^0(\lambda) \right)^2, \tag{2.65}$$

where α^2 is a Lagrange multiplier, $\gamma_1(\lambda) \le 0$, and $\gamma_1(\lambda) = 0$ if $\mathrm{Tr}[f^0(\lambda)] > (1-\varepsilon)\mathrm{Tr}[f_1(\lambda)]$.

For the second set of admissible spectral densities $\mathcal{D}_\varepsilon^2$, we have

$$\left(\sum_{k=0}^{\infty} ((\mathbf{F}_{\overline{\mu}}^0)^{-1} D^{\overline{\mu}} \mathbf{a})_k e^{i\lambda k} \right) \left(\sum_{k=0}^{\infty} ((\mathbf{F}_{\overline{\mu}}^0)^{-1} D^{\overline{\mu}} \mathbf{a})_k e^{i\lambda k} \right)^*$$
$$= \left(\frac{|\chi_{\overline{\mu}}^{(d)}(e^{-i\lambda})|^2}{|\beta^{(d)}(i\lambda)|^2} f^0(\lambda) \right) \{(\alpha_k^2 + \gamma_k^1(\lambda))\delta_{kl}\}_{k,l=1}^T \left(\frac{|\chi_{\overline{\mu}}^{(d)}(e^{-i\lambda})|^2}{|\beta^{(d)}(i\lambda)|^2} f^0(\lambda) \right), \tag{2.66}$$

where α_k^2 are Lagrange multipliers, $\gamma_k^1(\lambda) \le 0$, and $\gamma_k^1(\lambda) = 0$ if $f_{kk}^0(\lambda) > (1-\varepsilon) f_{kk}^1(\lambda)$.

For the third set of admissible spectral densities $\mathcal{D}_\varepsilon^3$, we have

$$\left(\sum_{k=0}^{\infty} ((\mathbf{F}_{\overline{\mu}}^0)^{-1} D^{\overline{\mu}} \mathbf{a})_k e^{i\lambda k} \right) \left(\sum_{k=0}^{\infty} ((\mathbf{F}_{\overline{\mu}}^0)^{-1} D^{\overline{\mu}} \mathbf{a})_k e^{i\lambda k} \right)^*$$
$$= (\alpha^2 + \gamma_1'(\lambda)) \left(\frac{|\chi_{\overline{\mu}}^{(d)}(e^{-i\lambda})|^2}{|\beta^{(d)}(i\lambda)|^2} f^0(\lambda) \right) B_1^{\top} \left(\frac{|\chi_{\overline{\mu}}^{(d)}(e^{-i\lambda})|^2}{|\beta^{(d)}(i\lambda)|^2} f^0(\lambda) \right), \tag{2.67}$$

where α^2 is a Lagrange multiplier, $\gamma_1'(\lambda) \le 0$, and $\gamma_1'(\lambda) = 0$ if $\langle B_1, f^0(\lambda) \rangle > (1-\varepsilon)\langle B_1, f_1(\lambda) \rangle$.

For the fourth set of admissible spectral densities $\mathcal{D}_\varepsilon^4$, we have

$$\left(\sum_{k=0}^{\infty} ((\mathbf{F}_{\overline{\mu}}^0)^{-1} D^{\overline{\mu}} \mathbf{a})_k e^{i\lambda k} \right) \left(\sum_{k=0}^{\infty} ((\mathbf{F}_{\overline{\mu}}^0)^{-1} D^{\overline{\mu}} \mathbf{a})_k e^{i\lambda k} \right)^*$$
$$= \left(\frac{|\chi_{\overline{\mu}}^{(d)}(e^{-i\lambda})|^2}{|\beta^{(d)}(i\lambda)|^2} f^0(\lambda) \right) (\vec{\alpha} \cdot \vec{\alpha}^* + \Gamma(\lambda)) \left(\frac{|\chi_{\overline{\mu}}^{(d)}(e^{-i\lambda})|^2}{|\beta^{(d)}(i\lambda)|^2} f^0(\lambda) \right), \tag{2.68}$$

where $\vec{\alpha}$ is a vector of Lagrange multipliers, $\Gamma(\lambda) \le 0$, and $\Gamma(\lambda) = 0$ if $f^0(\lambda) > (1-\varepsilon) f_1(\lambda)$.

In the case where the spectral density admits the canonical factorization (2.19), we have the following equations:

$$\left(\sum_{k=0}^{\infty}(D^{\overline{\mu}}\mathbf{A}\varphi^0_{\overline{\mu}})_k e^{i\lambda k}\right)\left(\sum_{k=0}^{\infty}(D^{\overline{\mu}}\mathbf{A}\varphi^0_{\overline{\mu}})_k e^{i\lambda k}\right)^* = (\alpha^2+\gamma_1(\lambda))(\Phi^0_{\overline{\mu}}(e^{-i\lambda}))^{\top}\overline{\Phi^0_{\overline{\mu}}(e^{-i\lambda})}, \tag{2.69}$$

$$\left(\sum_{k=0}^{\infty}(D^{\overline{\mu}}\mathbf{A}\varphi^0_{\overline{\mu}})_k e^{i\lambda k}\right)\left(\sum_{k=0}^{\infty}(D^{\overline{\mu}}\mathbf{A}\varphi^0_{\overline{\mu}})_k e^{i\lambda k}\right)^* = (\Phi^0_{\overline{\mu}}(e^{-i\lambda}))^{\top}\{(\alpha_k^2+\gamma_k^1(\lambda))\delta_{kl}\}_{k,l=1}^{T}\overline{\Phi^0_{\overline{\mu}}(e^{-i\lambda})}, \tag{2.70}$$

$$\left(\sum_{k=0}^{\infty}(D^{\overline{\mu}}\mathbf{A}\varphi^0_{\overline{\mu}})_k e^{i\lambda k}\right)\left(\sum_{k=0}^{\infty}(D^{\overline{\mu}}\mathbf{A}\varphi^0_{\overline{\mu}})_k e^{i\lambda k}\right)^* = (\alpha^2+\gamma_1'(\lambda))(\Phi^0_{\overline{\mu}}(e^{-i\lambda}))^{\top}B_1^{\top}\overline{\Phi^0_{\overline{\mu}}(e^{-i\lambda})}, \tag{2.71}$$

$$\left(\sum_{k=0}^{\infty}(D^{\overline{\mu}}\mathbf{A}\varphi^0_{\overline{\mu}})_k e^{i\lambda k}\right)\left(\sum_{k=0}^{\infty}(D^{\overline{\mu}}\mathbf{A}\varphi^0_{\overline{\mu}})_k e^{i\lambda k}\right)^* = (\Phi^0_{\overline{\mu}}(e^{-i\lambda}))^{\top}(\vec{\alpha}\cdot\vec{\alpha}^*+\Gamma(\lambda))\overline{\Phi^0_{\overline{\mu}}(e^{-i\lambda})}. \tag{2.72}$$

The following theorem holds.

Theorem 2.2.3. *The least favorable spectral densities $f^0(\lambda)$ in the classes $\mathcal{D}_{\varepsilon}^k$, $k=1,2,3,4$, for the optimal linear extrapolation of the functional $A\vec{\xi}$ from observations of the sequence $\vec{\xi}(m)$ at points $m=-1,-2,\ldots$ are determined by equations* (2.65), (2.66), (2.67), *and* (2.68) *(or equations* (2.69), (2.70), (2.71), *and* (2.72) *in the case where the spectral densities admit the canonical factorization* (2.19), *respectively), the minimality condition* (2.4), *the constrained optimization problem* (2.42), *and restrictions on densities from the corresponding classes $\mathcal{D}_{\varepsilon}^k$, $k=1,2,3,4$. The minimax-robust spectral characteristic of the optimal estimate of the functional $A\vec{\xi}$ is determined by formula* (2.11).

2.2.5 Least favorable spectral densities in classes $\mathcal{D}_{1\delta}$

Consider the extrapolation problem for the functional $A\vec{\xi}$ which depends on unobserved values of a sequence $\vec{\xi}(m)$ with stationary GM increments based on observations of the sequence at points $m=-1,-2,\ldots$ under the condition that the sets of admissible spectral densities $\mathcal{D}_{1\delta}^k$, $k=1,2,3,4$, are defined as follows:

$$\mathcal{D}_{1\delta}^1=\left\{f(\lambda)\left|\frac{1}{2\pi}\int_{-\pi}^{\pi}\frac{|\chi_{\overline{\mu}}^{(d)}(e^{-i\lambda})|^2}{|\beta^{(d)}(i\lambda)|^2}|\mathrm{Tr}(f(\lambda)-f_1(\lambda))|d\lambda\le\delta\right.\right\},$$

$$\mathcal{D}_{1\delta}^2=\left\{f(\lambda)\left|\frac{1}{2\pi}\int_{-\pi}^{\pi}\frac{|\chi_{\overline{\mu}}^{(d)}(e^{-i\lambda})|^2}{|\beta^{(d)}(i\lambda)|^2}|f_{kk}(\lambda)-f_{kk}^1(\lambda)|d\lambda\le\delta_k,\ k=\overline{1,T}\right.\right\},$$

$$\mathcal{D}_{1\delta}^3=\left\{f(\lambda)\left|\frac{1}{2\pi}\int_{-\pi}^{\pi}\frac{|\chi_{\overline{\mu}}^{(d)}(e^{-i\lambda})|^2}{|\beta^{(d)}(i\lambda)|^2}|\langle B_2,f(\lambda)-f_1(\lambda)\rangle|d\lambda\le\delta\right.\right\},$$

$$\mathcal{D}_{1\delta}^4=\left\{f(\lambda)\left|\frac{1}{2\pi}\int_{-\pi}^{\pi}\frac{|\chi_{\overline{\mu}}^{(d)}(e^{-i\lambda})|^2}{|\beta^{(d)}(i\lambda)|^2}|f_{ij}(\lambda)-f_{ij}^1(\lambda)|d\lambda\le\delta_i^j,\ i,j=\overline{1,T}\right.\right\}.$$

Here $f_1(\lambda)$ is a fixed spectral density and δ, δ_k, $k = \overline{1,T}$, and δ_i^j, $i,j = \overline{1,T}$ are given numbers. The condition $0 \in \partial\Delta_{\mathcal{D}}(f^0)$ implies the following equations determining the least favorable spectral densities for these sets.

For the first set of admissible spectral densities $\mathcal{D}_{1\delta}^1$, we have

$$\left(\sum_{k=0}^{\infty}((\mathbf{F}_{\overline{\mu}}^0)^{-1}D^{\overline{\mu}}\mathbf{a})_k e^{i\lambda k}\right)\left(\sum_{k=0}^{\infty}((\mathbf{F}_{\overline{\mu}}^0)^{-1}D^{\overline{\mu}}\mathbf{a})_k e^{i\lambda k}\right)^* = \beta^2\gamma_2(\lambda)\left(\frac{|\chi_{\overline{\mu}}^{(d)}(e^{-i\lambda})|^2}{|\beta^{(d)}(i\lambda)|^2}f^0(\lambda)\right)^2, \tag{2.73}$$

$$\frac{1}{2\pi}\int_{-\pi}^{\pi}\frac{|\chi_{\overline{\mu}}^{(d)}(e^{-i\lambda})|^2}{|\beta^{(d)}(i\lambda)|^2}|\mathrm{Tr}(f^0(\lambda)-f_1(\lambda))|d\lambda = \delta, \tag{2.74}$$

where β^2 is a Lagrange multiplier, $|\gamma_2(\lambda)| \le 1$, and

$$\gamma_2(\lambda) = \mathrm{sign}(\mathrm{Tr}(f^0(\lambda)-f_1(\lambda))) : \mathrm{Tr}(f^0(\lambda)-f_1(\lambda)) \ne 0.$$

For the second set of admissible spectral densities $\mathcal{D}_{1\delta}^2$, we have

$$\left(\sum_{k=0}^{\infty}((\mathbf{F}_{\overline{\mu}}^0)^{-1}D^{\overline{\mu}}\mathbf{a})_k e^{i\lambda k}\right)\left(\sum_{k=0}^{\infty}((\mathbf{F}_{\overline{\mu}}^0)^{-1}D^{\overline{\mu}}\mathbf{a})_k e^{i\lambda k}\right)^*$$
$$= \left(\frac{|\chi_{\overline{\mu}}^{(d)}(e^{-i\lambda})|^2}{|\beta^{(d)}(i\lambda)|^2}f^0(\lambda)\right)\{\beta_k^2\gamma_k^2(\lambda)\delta_{kl}\}_{k,l=1}^T\left(\frac{|\chi_{\overline{\mu}}^{(d)}(e^{-i\lambda})|^2}{|\beta^{(d)}(i\lambda)|^2}f^0(\lambda)\right), \tag{2.75}$$

$$\frac{1}{2\pi}\int_{-\pi}^{\pi}\frac{|\chi_{\overline{\mu}}^{(d)}(e^{-i\lambda})|^2}{|\beta^{(d)}(i\lambda)|^2}|f_{kk}^0(\lambda)-f_{kk}^1(\lambda)|d\lambda = \delta_k, \tag{2.76}$$

where β_k^2 are Lagrange multipliers, $|\gamma_k^2(\lambda)| \le 1$, and

$$\gamma_k^2(\lambda) = \mathrm{sign}(f_{kk}^0(\lambda)-f_{kk}^1(\lambda)) : f_{kk}^0(\lambda)-f_{kk}^1(\lambda) \ne 0,\ k=\overline{1,T}.$$

For the third set of admissible spectral densities $\mathcal{D}_{1\delta}^3$, we have

$$\left(\sum_{k=0}^{\infty}((\mathbf{F}_{\overline{\mu}}^0)^{-1}D^{\overline{\mu}}\mathbf{a})_k e^{i\lambda k}\right)\left(\sum_{k=0}^{\infty}((\mathbf{F}_{\overline{\mu}}^0)^{-1}D^{\overline{\mu}}\mathbf{a})_k e^{i\lambda k}\right)^*$$
$$= \beta^2\gamma_2'(\lambda)\left(\frac{|\chi_{\overline{\mu}}^{(d)}(e^{-i\lambda})|^2}{|\beta^{(d)}(i\lambda)|^2}f^0(\lambda)\right)B_2^{\top}\left(\frac{|\chi_{\overline{\mu}}^{(d)}(e^{-i\lambda})|^2}{|\beta^{(d)}(i\lambda)|^2}f^0(\lambda)\right), \tag{2.77}$$

$$\frac{1}{2\pi}\int_{-\pi}^{\pi}\frac{|\chi_{\overline{\mu}}^{(d)}(e^{-i\lambda})|^2}{|\beta^{(d)}(i\lambda)|^2}|\langle B_2, f^0(\lambda)-f_1(\lambda)\rangle|d\lambda = \delta, \tag{2.78}$$

where β^2 is a Lagrange multiplier, $|\gamma_2'(\lambda)| \le 1$, and

$$\gamma_2'(\lambda) = \mathrm{sign}\langle B_2, f^0(\lambda)-f_1(\lambda)\rangle : \langle B_2, f^0(\lambda)-f_1(\lambda)\rangle \ne 0.$$

For the fourth set of admissible spectral densities $\mathcal{D}_{1\delta}^4$, we have

$$\left(\sum_{k=0}^{\infty}((\mathbf{F}_{\overline{\mu}}^0)^{-1}D^{\overline{\mu}}\mathbf{a})_k e^{i\lambda k}\right)\left(\sum_{k=0}^{\infty}((\mathbf{F}_{\overline{\mu}}^0)^{-1}D^{\overline{\mu}}\mathbf{a})_k e^{i\lambda k}\right)^* \\ = \left(\frac{|\chi_{\overline{\mu}}^{(d)}(e^{-i\lambda})|^2}{|\beta^{(d)}(i\lambda)|^2}f^0(\lambda)\right)\{\beta_{ij}(\lambda)\gamma_{ij}(\lambda)\}_{i,j=1}^T\left(\frac{|\chi_{\overline{\mu}}^{(d)}(e^{-i\lambda})|^2}{|\beta^{(d)}(i\lambda)|^2}f^0(\lambda)\right), \tag{2.79}$$

$$\frac{1}{2\pi}\int_{-\pi}^{\pi}\frac{|\chi_{\overline{\mu}}^{(d)}(e^{-i\lambda})|^2}{|\beta^{(d)}(i\lambda)|^2}|f_{ij}^0(\lambda)-f_{ij}^1(\lambda)|d\lambda = \delta_i^j, \tag{2.80}$$

where β_{ij} are Lagrange multipliers, $|\gamma_{ij}(\lambda)| \leq 1$, and

$$\gamma_{ij}(\lambda) = \frac{f_{ij}^0(\lambda)-f_{ij}^1(\lambda)}{|f_{ij}^0(\lambda)-f_{ij}^1(\lambda)|} : f_{ij}^0(\lambda)-f_{ij}^1(\lambda) \neq 0, \; i,j=\overline{1,T}.$$

In the case where the spectral density admits the canonical factorization (2.19), we have the following equations:

$$\left(\sum_{k=0}^{\infty}(D^{\overline{\mu}}\mathbf{A}\varphi_{\overline{\mu}}^0)_k e^{i\lambda k}\right)\left(\sum_{k=0}^{\infty}(D^{\overline{\mu}}\mathbf{A}\varphi_{\overline{\mu}}^0)_k e^{i\lambda k}\right)^* = \beta^2\gamma_2(\lambda)(\Phi_{\overline{\mu}}^0(e^{-i\lambda}))^\top\overline{\Phi_{\overline{\mu}}^0(e^{-i\lambda})}, \tag{2.81}$$

$$\left(\sum_{k=0}^{\infty}(D^{\overline{\mu}}\mathbf{A}\varphi_{\overline{\mu}}^0)_k e^{i\lambda k}\right)\left(\sum_{k=0}^{\infty}(D^{\overline{\mu}}\mathbf{A}\varphi_{\overline{\mu}}^0)_k e^{i\lambda k}\right)^* = (\Phi_{\overline{\mu}}^0(e^{-i\lambda}))^\top\{\beta_k^2\gamma_k^2(\lambda)\delta_{kl}\}_{k,l=1}^T\overline{\Phi_{\overline{\mu}}^0(e^{-i\lambda})}, \tag{2.82}$$

$$\left(\sum_{k=0}^{\infty}(D^{\overline{\mu}}\mathbf{A}\varphi_{\overline{\mu}}^0)_k e^{i\lambda k}\right)\left(\sum_{k=0}^{\infty}(D^{\overline{\mu}}\mathbf{A}\varphi_{\overline{\mu}}^0)_k e^{i\lambda k}\right)^* = \beta^2\gamma_2'(\lambda)(\Phi_{\overline{\mu}}^0(e^{-i\lambda}))^\top B_2^\top\overline{\Phi_{\overline{\mu}}^0(e^{-i\lambda})}, \tag{2.83}$$

$$\left(\sum_{k=0}^{\infty}(D^{\overline{\mu}}\mathbf{A}\varphi_{\overline{\mu}}^0)_k e^{i\lambda k}\right)\left(\sum_{k=0}^{\infty}(D^{\overline{\mu}}\mathbf{A}\varphi_{\overline{\mu}}^0)_k e^{i\lambda k}\right)^* = (\Phi_{\overline{\mu}}^0(e^{-i\lambda}))^\top\{\beta_{ij}(\lambda)\gamma_{ij}(\lambda)\}_{i,j=1}^T\overline{\Phi_{\overline{\mu}}^0(e^{-i\lambda})}. \tag{2.84}$$

The following theorem holds.

Theorem 2.2.4. *The least favorable spectral densities $f^0(\lambda)$ in the classes $\mathcal{D}_{1\delta}^k$, $k=1,2,3,4$, for the optimal linear extrapolation of the functional $A\vec{\xi}$ from observations of the sequence $\vec{\xi}(m)$ at points $m=-1,-2,\dots$ are determined by equations* (2.73)–(2.74), (2.75)–(2.76), (2.77)–(2.78), *and* (2.79)–(2.80) *(or equations* (2.81), (2.82), (2.83), *and* (2.84) *in the case where the spectral densities admit canonical factorization* (2.19), *respectively), the minimality condition* (2.4), *the constrained optimization problem* (2.42), *and restrictions on densities from the corresponding classes $\mathcal{D}_{1\delta}^k$, $k=1,2,3,4$. The minimax-robust spectral characteristic of the optimal estimate of the functional $A\vec{\xi}$ is determined by formula* (2.11).

Conclusions

In this chapter, we describe methods of solution of the extrapolation problem for functionals from unobserved values of stochastic sequences with periodically stationary long memory multi-seasonal increments. Estimates are based on observations of the considered sequence without noise. The problem is investigated using a representation of the sequence under investigation as a multidimensional sequence with stationary GM increments. We first apply the Hilbert space projection method to derive formulas for calculating the spectral characteristics and the MSEs of the optimal estimates of the functionals in the case of spectral certainty, where the spectral density of the considered sequence is exactly known. The obtained results are illustrated with some examples. In the case of spectral uncertainty, where the spectral density of the considered sequence is not exactly known while, instead, some sets of admissible spectral densities are specified, the minimax-robust method is applied. This method consists in finding an estimate that minimizes the maximal values of the MSEs for all spectral densities from a given class $\mathcal{D}$ of admissible spectral densities simultaneously. We propose a representation of the MSE in the form of a linear functional in L_1 space with respect to spectral densities, which allows us to solve the corresponding constrained optimization problem and describe the minimax (robust) estimates of the functionals. Application of this method is illustrated with the help of concrete sets of admissible spectral densities: $\mathcal{D}_0$, $\mathcal{D}_V^U$, $\mathcal{D}_\varepsilon$, and $\mathcal{D}_{1\delta}$. Here D_0 are classes of spectral densities with the moment restrictions, D_V^U are classes describing the “strip” models of spectral densities, D_ε are classes describing the “ε-contaminated” models of a fixed bounded spectral density, and $D_{1\delta}$ are classes describing the “δ-neighborhood” models in the space L_1 of a fixed bounded spectral density.

The main results of this chapter have been published in [117]. The corresponding results for minimax extrapolation of sequences with periodically stationary increments (a particular case of GM increments) observed without noise are described in the paper by Kozak, Luz, and Moklyachuk [94].

The results presented in this chapter rely on and develop prior studies of the minimax extrapolation method by Moklyachuk and Masyutka [153] for multidimensional stationary sequences, Moklyachuk and Golichenko [60] for periodically correlated sequences, and Luz and Moklyachuk [107] for sequences with stationary increments. For more related results, see the papers by Hosoya [72], Taniguchi [197], and Liu et al. [102].

3 Extrapolation of sequences with periodically stationary increments observed with noise

In this chapter, we deal with the extrapolation problem for stochastic sequences with periodically stationary long memory multi-seasonal increments based on observations with periodically stationary noise.

The extrapolation problem consists in the mean square optimal linear estimation of some functionals which depend on the future unobservable values of a stochastic sequence with periodically stationary long memory multi-seasonal increments. Estimates are based on observations of the sequence in the past with a periodically stationary noise sequence.

3.1 Hilbert space projection method of extrapolation

3.1.1 Extrapolation of multidimensional sequences with stationary GM increments

Consider a vector-valued stochastic sequence $\vec{\xi}(m)$ with stationary GM increments constructed from the sequence $\xi(m) = \{\xi_p(m)\}_{p=1}^T$ with the help of transformation (1.2). Let the stationary GM increment sequence $\chi_{\overline{\mu},\overline{s}}^{(d)}(\vec{\xi}(m)) = \{\chi_{\overline{\mu},\overline{s}}^{(d)}(\xi_p(m))\}_{p=1}^T$ have an absolutely continuous spectral function $F(\lambda)$ and the spectral density $f(\lambda) = \{f_{ij}(\lambda)\}_{i,j=1}^T$.

Let $\vec{\eta}(m) = \{\eta_p(m)\}_{p=1}^T$ be an uncorrelated with the sequence $\vec{\xi}(m)$ stationary stochastic sequence with absolutely continuous spectral function $G(\lambda)$ and spectral density $g(\lambda) = \{g_{ij}(\lambda)\}_{i,j=1}^T$.

Without loss of generality assume that the mean values of the increment sequence $\chi_{\overline{\mu},\overline{s}}^{(d)}(\vec{\xi}(m))$ and the stationary sequence $\vec{\eta}(m)$ equal $\vec{0}$. We will also consider the increment step $\overline{\mu} > \overline{0}$.

Extrapolation (forecasting) problem

Consider the problem of mean square optimal linear estimation of the functionals

$$A\vec{\xi} = \sum_{k=0}^{\infty} (\vec{a}(k))^{\top} \vec{\xi}(k), \quad A_N\vec{\xi} = \sum_{k=0}^{N} (\vec{a}(k))^{\top} \vec{\xi}(k), \tag{3.1}$$

which depend on unobserved values of the stochastic sequence $\vec{\xi}(k)$ with stationary GM increments. Estimates are based on observations of the sequence $\vec{\zeta}(m) = \vec{\xi}(m) + \vec{\eta}(m)$ at points $m = -1, -2, \ldots$. As a mean square optimal linear estimate of the functional $A\vec{\xi}$ we understand the one which minimizes the MSE $\Delta(f, g; \widehat{A}\vec{\xi}) = \mathsf{E}|A\vec{\xi} - \widehat{A}\vec{\xi}|^2$ for a fixed spectral density $f(\lambda)$.

https://doi.org/10.1515/9783111325620-003

First of all we indicate some conditions which are necessary for solving the considered problem. Assume that coefficients $\vec{a}(k) = \{a_p(k)\}_{p=1}^T$, $k \geq 0$, and the linear transformation D^μ which is defined in the following part of the section satisfy the conditions

$$\sum_{k=0}^{\infty} \|\vec{a}(k)\| < \infty, \quad \sum_{k=0}^{\infty} (k+1)\|\vec{a}(k)\|^2 < \infty, \tag{3.2}$$

$$\sum_{k=0}^{\infty} \|(D^\mu \mathbf{a})_k\| < \infty, \quad \sum_{k=0}^{\infty} (k+1)\|(D^\mu \mathbf{a})_k\|^2 < \infty. \tag{3.3}$$

Assume also that spectral densities $f(\lambda)$ and $g(\lambda)$ satisfy the minimality condition

$$\int_{-\pi}^{\pi} \mathrm{Tr}\left[\frac{|\beta^{(d)}(i\lambda)|^2}{|\chi_{\overline{\mu}}^{(d)}(e^{-i\lambda})|^2}(f(\lambda) + |\beta^{(d)}(i\lambda)|^2 g(\lambda))^{-1}\right] d\lambda < \infty. \tag{3.4}$$

This is the necessary and sufficient condition under which the MSEs of the optimal estimates of the functionals $A\vec{\xi}$ and $A_N\vec{\xi}$ are not equal to 0.

We apply the Hilbert space projection technique proposed by Kolmogorov [90], which can be described as a three-stage procedure: (i) define a target element (to be estimated) of the space $H = L_2(\Omega, \mathcal{F}, \mathsf{P})$ of random variables γ which have zero mean values and finite variances, $\mathsf{E}\gamma = 0$, $\mathsf{E}|\gamma|^2 < \infty$, endowed with the inner product $\langle \gamma_1; \gamma_2 \rangle = \mathsf{E}\gamma_1\overline{\gamma_2}$, (ii) define a subspace of H generated by observations, and (iii) find an estimate of the target element as an orthogonal projection of the target element on the defined subspace.

Stage (i)

The functional $A\vec{\xi}$ does not belong to the space $H = L_2(\Omega, \mathcal{F}, \mathsf{P})$. With the help of the following lemma we describe representations of the functional as the sum of a functional with finite second moments belonging to H and a functional depending on the observed values of the sequence $\vec{\zeta}(k)$ ("initial values") (for more details see [115, 117]).

Lemma 3.1.1. *The functional $A\vec{\xi}$ admits the representation*

$$A\vec{\xi} = A\vec{\zeta} - A\vec{\eta} = H\vec{\xi} - V\vec{\zeta}, \tag{3.5}$$

where

$$H\vec{\xi} := B\chi\vec{\zeta} - A\vec{\eta},$$

$$A\vec{\zeta} = \sum_{k=0}^{\infty} (\vec{a}(k))^\top \vec{\zeta}(k), \quad A\vec{\eta} = \sum_{k=0}^{\infty} (\vec{a}(k))^\top \vec{\eta}(k),$$

$$B\chi\vec{\zeta} = \sum_{k=0}^{\infty} (\vec{b}_{\overline{\mu}}(k))^\top \chi_{\overline{\mu},\overline{s}}^{(d)}(\vec{\zeta}(k)), \quad V\vec{\zeta} = \sum_{k=-d(\gamma)}^{-1} (\vec{v}_{\overline{\mu}}(k))^\top \vec{\zeta}(k),$$

where the coefficients $\vec{b}_{\overline{\mu}}(k) = \{b_{\overline{\mu},p}(k)\}_{p=1}^{T}$, $k = 0, 1, \ldots$, *and* $\vec{v}_{\overline{\mu}}(k) = \{v_{\overline{\mu},p}(k)\}_{p=1}^{T}$, $k = -1, -2, \ldots, -d(\gamma)$, *are calculated by the formulas*

$$\vec{v}_{\overline{\mu}}(k) = \sum_{l=0}^{k+d(\gamma)} \mathrm{diag}_T(e_\nu(l-k))\vec{b}_{\overline{\mu}}(l), \quad k = -1, -2, \ldots, -d(\gamma), \tag{3.6}$$

$$\vec{b}_{\overline{\mu}}(k) = \sum_{m=k}^{\infty} \mathrm{diag}_T(d_{\overline{\mu}}(m-k))\vec{a}(m) = (D^{\overline{\mu}}\mathbf{a})_k, \quad k = 0, 1, \ldots, \tag{3.7}$$

where $D^{\overline{\mu}}$ *is the linear transformation determined by a matrix with the entries* $(D^{\overline{\mu}})(k,j) = \mathrm{diag}_T(d_{\overline{\mu}}(j-k))$ *if* $0 \le k \le j$ *and* $(D^{\overline{\mu}})(k,j) = \mathcal{O}$ *if* $0 \le j < k$, $\mathrm{diag}_T(x)$ *denotes a* $T\times T$ *diagonal matrix with the entry* x *on its diagonal,* $\mathbf{a} = ((\vec{a}(0))^\top, (\vec{a}(1))^\top, \ldots)^\top$, *and coefficients* $\{d_{\overline{\mu}}(k) : k \ge 0\}$ *are determined by the relationship*

$$\sum_{k=0}^{\infty} d_{\overline{\mu}}(k)x^k = \prod_{i=1}^{r}\left(\sum_{j_i=0}^{\infty} x^{\mu_i s_i j_i}\right)^{d_i}.$$

Proof. The formulas for the coefficients $\vec{b}_{\overline{\mu}}(k)$ and $\vec{v}_{\overline{\mu}}(k)$ are derived in Lemma 2.1.1. □

The functional $H\vec{\xi}$ from representation (3.5) has finite variance and the functional $V\vec{\zeta}$ depends on the known observations of the stochastic sequence $\vec{\zeta}(k)$ at points $k = -d(\gamma), -d(\gamma)+1, \ldots, -1$. Therefore, estimates $\widehat{A}\vec{\xi}$ and $\widehat{H}\vec{\xi}$ of the functionals $A\vec{\xi}$ and $H\vec{\xi}$ and the MSEs $\Delta(f,g;\widehat{A}\vec{\xi}) = \mathsf{E}|A\vec{\xi} - \widehat{A}\vec{\xi}|^2$, $\Delta(f,g;\widehat{H}\vec{\xi}) = \mathsf{E}|H\vec{\xi} - \widehat{H}\vec{\xi}|^2$ of the estimates $\widehat{A}\vec{\xi}$ and $\widehat{H}\vec{\xi}$ satisfy the following relations:

$$\widehat{A}\vec{\xi} = \widehat{H}\vec{\xi} - V\vec{\zeta}, \tag{3.8}$$

$$\Delta(f,g;\widehat{A}\vec{\xi}) = \mathsf{E}|A\vec{\xi} - \widehat{A}\vec{\xi}|^2 = \mathsf{E}|H\vec{\xi} - \widehat{H}\vec{\xi}|^2 = \Delta(f,g;\widehat{H}\vec{\xi}). \tag{3.9}$$

Therefore, the estimation problem for the functional $A\vec{\xi}$ is equivalent to the one for the functional $H\vec{\xi}$. This problem can be solved by applying the Hilbert space projection method.

The functional $H\vec{\xi}$ admits the spectral representation

$$H\vec{\xi} = \int_{-\pi}^{\pi} (\vec{B}_{\overline{\mu}}(e^{i\lambda}))^\top \frac{\chi_{\overline{\mu}}^{(d)}(e^{-i\lambda})}{\beta^{(d)}(i\lambda)} d\vec{Z}_{\xi^{(d)}+\eta^{(d)}}(\lambda) - \int_{-\pi}^{\pi} (\vec{A}(e^{i\lambda}))^\top d\vec{Z}_\eta(\lambda),$$

where

$$\vec{B}_{\overline{\mu}}(e^{i\lambda}) = \sum_{k=0}^{\infty} \vec{b}_{\overline{\mu}}(k)e^{i\lambda k} = \sum_{k=0}^{\infty} (D^{\overline{\mu}}\mathbf{a})_k e^{i\lambda k}, \quad \vec{A}(e^{i\lambda}) = \sum_{k=0}^{\infty} \vec{a}(k)e^{i\lambda k}.$$

Stage (ii)
Define the closed linear subspace

$$H^{-1}(\vec{\xi}^{(d)}+\vec{\eta}^{(d)})=\overline{\operatorname{span}}\{\chi^{(d)}_{\overline{\mu},\overline{s}}(\xi_p(k))+\chi^{(d)}_{\overline{\mu},\overline{s}}(\eta_p(k)) : k\le -1,\ p=1,\dots,T\}$$

in the Hilbert space $H=L_2(\Omega,\mathcal{F},\mathsf{P})$ of random variables γ with zero mean value, $\mathsf{E}\gamma=0$, finite variance, $\mathsf{E}|\gamma|^2<\infty$, and the inner product $(\gamma_1;\gamma_2)=\mathsf{E}\gamma_1\overline{\gamma_2}$. Define the closed linear subspace

$$L_2^{-1}(f(\lambda)+|\beta^{(d)}(i\lambda)|^2g(\lambda))=\overline{\operatorname{span}}\left\{e^{i\lambda k}\chi^{(d)}_{\overline{\mu}}(e^{-i\lambda})\frac{1}{\beta^{(d)}(i\lambda)}\vec{\delta}_l,\, l=1,\dots,T,\ k\le -1\right\}$$

in the Hilbert space $L_2(f(\lambda)+|\beta^{(d)}(i\lambda)|^2g(\lambda))$ of vector-valued functions endowed with the inner product

$$\langle g_1,g_2\rangle=\int_{-\pi}^{\pi}(g_1(\lambda))^{\top}(f(\lambda)+|\beta^{(d)}(i\lambda)|^2g(\lambda))\overline{g_2(\lambda)}d\lambda.$$

Here $\vec{\delta}_l=\{\delta_{lp}\}_{p=1}^T$, $l=1,\dots,T$; δ_{lp} are Kronecker symbols.

Remark 3.1.1. The representation

$$\chi^{(d)}_{\overline{\mu},\overline{s}}(\vec{\xi}(k))+\chi^{(d)}_{\overline{\mu},\overline{s}}(\vec{\eta}(k))=\int_{-\pi}^{\pi}e^{i\lambda k}\chi^{(d)}_{\overline{\mu}}(e^{-i\lambda})\frac{1}{\beta^{(d)}(i\lambda)}d\vec{Z}_{\xi^{(d)}+\eta^{(d)}}(\lambda)$$

yields a map between elements $e^{i\lambda k}\chi^{(d)}_{\overline{\mu}}(e^{-i\lambda})\frac{1}{\beta^{(d)}(i\lambda)}\vec{\delta}_l$ from the space $L_2^{-1}(f(\lambda)+|\beta^{(d)}(i\lambda)|^2g(\lambda))$ and elements $\chi^{(d)}_{\overline{\mu},\overline{s}}(\vec{\xi}_l(k))+\chi^{(d)}_{\overline{\mu},\overline{s}}(\vec{\eta}_l(k))$ from the space $H^{-1}(\vec{\xi}^{(d)}+\vec{\eta}^{(d)})$.

Stage (iii)
At this stage, we find the mean square optimal estimate $\widehat{H}\vec{\xi}$ as an orthogonal projection of the element $H\vec{\xi}\in H$ on the subspace $H^{-1}(\vec{\xi}^{(d)}+\vec{\eta}^{(d)})$:

$$\widehat{H}\vec{\xi}=\operatorname{Proj}_{H^{-1}(\vec{\xi}^{(d)}+\vec{\eta}^{(d)})}H\vec{\xi}.$$

Relation (3.8) implies that an optimal linear estimate $\widehat{A}\vec{\xi}$ of the functional $A\vec{\xi}$ can be represented in the form

$$\widehat{A}\vec{\xi}=\operatorname{Proj}_{H^{-1}(\vec{\xi}^{(d)}+\vec{\eta}^{(d)})}H\vec{\xi}-V\vec{\zeta}$$

or in the form

$$\widehat{A}\vec{\xi} = \int_{-\pi}^{\pi} (\vec{h}_{\overline{\mu}}(\lambda))^{\top} d\vec{Z}_{\xi^{(d)}+\eta^{(d)}}(\lambda) - \sum_{k=-d(\gamma)}^{-1} (\vec{v}_{\overline{\mu}}(k))^{\top} (\vec{\xi}(k) + \vec{\eta}(k)), \tag{3.10}$$

where $\vec{h}_{\overline{\mu}}(\lambda) = \{h_{\overline{\mu},p}(\lambda)\}_{p=1}^{T}$ is the spectral characteristic of the projection $\widehat{H}\vec{\xi}$.

Define for $k, j \geq 0$ the following Fourier coefficients:

$$T_{k,j}^{\overline{\mu}} = \frac{1}{2\pi} \int_{-\pi}^{\pi} e^{-i\lambda(k-j)} \frac{|\beta^{(d)}(i\lambda)|^2}{|\chi_{\overline{\mu}}^{(d)}(e^{-i\lambda})|^2} (g(\lambda)(f(\lambda) + |\beta^{(d)}(i\lambda)|^2 g(\lambda))^{-1})^{\top} d\lambda,$$

$$P_{k,j}^{\overline{\mu}} = \frac{1}{2\pi} \int_{-\pi}^{\pi} e^{-i\lambda(k-j)} \frac{|\beta^{(d)}(i\lambda)|^2}{|\chi_{\overline{\mu}}^{(d)}(e^{-i\lambda})|^2} ((f(\lambda) + |\beta^{(d)}(i\lambda)|^2 g(\lambda))^{-1})^{\top} d\lambda,$$

$$Q_{k,j} = \frac{1}{2\pi} \int_{-\pi}^{\pi} e^{-i\lambda(k-j)} (f(\lambda)(f(\lambda) + |\beta^{(d)}(i\lambda)|^2 g(\lambda))^{-1} g(\lambda))^{\top} d\lambda,$$

and corresponding linear operators $\mathbf{P}_{\overline{\mu}}$ and $\mathbf{T}_{\overline{\mu}}$ in the space ℓ_2 determined by matrices with the $T \times T$ matrix entries $(\mathbf{P}_{\overline{\mu}})_{l,k} = P_{l,k}^{\overline{\mu}}$, $l, k \geq 0$, and $(\mathbf{T}_{\overline{\mu}})_{l,k} = T_{l,k}^{\overline{\mu}}$, $l, k \geq 0$, and a linear operator in the space ℓ_2 defined by the matrix with the $T \times T$ matrix entries $(\mathbf{Q})_{l,k} = Q_{l,k}$, $l, k \geq 0$.

By the inner product $\langle \cdot, \cdot \rangle$ for vectors $\mathbf{x} = ((\vec{x}(0))^{\top}, (\vec{x}(1))^{\top}, (\vec{x}(2))^{\top}, \ldots)^{\top}$, $\mathbf{y} = ((\vec{y}(0))^{\top}, (\vec{y}(1))^{\top}, (\vec{y}(2))^{\top}, \ldots)^{\top}$ we understand $\langle \mathbf{x}, \mathbf{y} \rangle = \sum_{k=0}^{\infty} (\vec{x}(k))^{\top} \vec{y}(k)$.

Define a functional

$$\vec{A}_{\overline{\mu}}(e^{i\lambda}) = \vec{A}(e^{i\lambda}) \overline{\chi_{\overline{\mu}}^{(d)}(e^{-i\lambda})} = \sum_{k=0}^{\infty} \vec{a}_{\overline{\mu}}(k) e^{ik\lambda},$$

with coefficients

$$\vec{a}_{\overline{\mu}}(m) = \sum_{l=\max\{m-d(\gamma),0\}}^{m} e_{\gamma}(m-l) \vec{a}(l), \quad m \geq 0, \tag{3.11}$$

and a vector

$$\mathbf{a}_{\overline{\mu}} = ((\vec{a}_{\overline{\mu}}(0))^{\top}, (\vec{a}_{\overline{\mu}}(1))^{\top}, (\vec{a}_{\overline{\mu}}(2))^{\top}, \ldots)^{\top}.$$

Theorem 3.1.1. *Let $\vec{\xi}(m)$, $m \in \mathbb{Z}$, be a multidimensional stochastic sequence which defines a stationary GM increment sequence $\chi_{\overline{\mu},\overline{s}}^{(d)}(\vec{\xi}(m))$ with absolutely continuous spectral function $F(\lambda)$ which has spectral density $f(\lambda)$. Let $\vec{\eta}(m)$, $m \in \mathbb{Z}$, be an uncorrelated with the sequence $\vec{\xi}(m)$ multidimensional stationary stochastic sequence with absolutely continuous spectral function $G(\lambda)$ which has spectral density $g(\lambda)$. Let minimality condition* (3.4) *be satisfied. Let coefficients $\vec{a}(k)$, $k \geq 0$, satisfy conditions* (3.2)–(3.3). *The optimal linear estimate $\widehat{A}\vec{\xi}$ of the functional $A\vec{\xi}$ based on observations of the sequence $\vec{\xi}(m) + \vec{\eta}(m)$ at* *points* $m = -1, -2, \ldots$ *is calculated by formula* (3.10). *The spectral characteristic* $\vec{h}_{\overline{\mu}}(\lambda)$ *of the estimate* $\widehat{A}\vec{\xi}$ *is calculated by the formula*

$$\begin{aligned}(\vec{h}_{\overline{\mu}}(\lambda))^{\top} &= (\vec{B}_{\overline{\mu}}(e^{i\lambda}))^{\top}\frac{\chi_{\overline{\mu}}^{(d)}(e^{-i\lambda})}{\beta^{(d)}(i\lambda)}\\ &- (\vec{A}_{\overline{\mu}}(e^{i\lambda}))^{\top}g(\lambda)\frac{\overline{\beta^{(d)}(i\lambda)}}{\chi_{\overline{\mu}}^{(d)}(e^{-i\lambda})}(f(\lambda)+|\beta^{(d)}(i\lambda)|^2g(\lambda))^{-1}\\ &- (\vec{C}_{\overline{\mu}}(e^{i\lambda}))^{\top}\frac{\overline{\beta^{(d)}(i\lambda)}}{\chi_{\overline{\mu}}^{(d)}(e^{-i\lambda})}(f(\lambda)+|\beta^{(d)}(i\lambda)|^2g(\lambda))^{-1}, \end{aligned}\tag{3.12}$$

where

$$\vec{C}_{\overline{\mu}}(e^{i\lambda}) = \sum_{k=0}^{\infty}(\mathbf{P}_{\overline{\mu}}^{-1}D^{\overline{\mu}}\mathbf{a} - \mathbf{P}_{\overline{\mu}}^{-1}\mathbf{T}_{\overline{\mu}}\mathbf{a}_{\overline{\mu}})_k e^{ik\lambda}.$$

The MSE $\Delta(f,g;\widehat{A}\vec{\xi})$ *is calculated by the formula*

$$\begin{aligned}\Delta(f,g;\widehat{A}\vec{\xi}) &= \Delta(f,g;\widehat{H}\vec{\xi}) = \mathsf{E}|H\vec{\xi}-\widehat{H}\vec{\xi}|^2\\ &= \frac{1}{2\pi}\int_{-\pi}^{\pi}\frac{|\beta^{(d)}(i\lambda)|^2}{|\chi_{\overline{\mu}}^{(d)}(e^{-i\lambda})|^2}(\mathbf{C}_{\overline{\mu}}^{f}(e^{i\lambda}))^{\top}(f(\lambda)+|\beta^{(d)}(i\lambda)|^2g(\lambda))^{-1}\\ &\times f(\lambda)\,(f(\lambda)+|\beta^{(d)}(i\lambda)|^2g(\lambda))^{-1}\overline{\mathbf{C}_{\overline{\mu}}^{f}(e^{i\lambda})}d\lambda\\ &+\frac{1}{2\pi}\int_{-\pi}^{\pi}\left(\frac{|\beta^{(d)}(i\lambda)|^2}{|\chi_{\overline{\mu}}^{(d)}(e^{-i\lambda})|^2}\right)^2(\mathbf{C}_{\overline{\mu}}^{g}(e^{i\lambda}))^{\top}(f(\lambda)+|\beta^{(d)}(i\lambda)|^2g(\lambda))^{-1}\\ &\times g(\lambda)\,(f(\lambda)+|\beta^{(d)}(i\lambda)|^2g(\lambda))^{-1}\overline{\mathbf{C}_{\overline{\mu}}^{g}(e^{i\lambda})}d\lambda\\ &= \langle D^{\overline{\mu}}\mathbf{a}-\mathbf{T}_{\overline{\mu}}\mathbf{a}_{\overline{\mu}}, \mathbf{P}_{\overline{\mu}}^{-1}D^{\overline{\mu}}\mathbf{a}-\mathbf{P}_{\overline{\mu}}^{-1}\mathbf{T}_{\overline{\mu}}\mathbf{a}_{\overline{\mu}}\rangle + \langle\mathbf{Q}\mathbf{a},\mathbf{a}\rangle, \end{aligned}\tag{3.13}$$

where

$$\begin{aligned}\mathbf{C}_{\overline{\mu}}^{f}(e^{i\lambda}) &= \overline{g(\lambda)}\vec{A}_{\overline{\mu}}(e^{i\lambda}) + \vec{C}_{\overline{\mu}}(e^{i\lambda}),\\ \mathbf{C}_{\overline{\mu}}^{g}(e^{i\lambda}) &= \frac{|\chi_{\overline{\mu}}^{(d)}(e^{-i\lambda})|^2}{|\beta^{(d)}(i\lambda)|^2}\overline{f(\lambda)}\vec{A}(e^{i\lambda}) - \chi_{\overline{\mu}}^{(d)}(e^{-i\lambda})\vec{C}_{\overline{\mu}}(e^{i\lambda}).\end{aligned}$$

Proof. The estimate $\widehat{H}\vec{\xi}$ is determined by two conditions:

(1) $\widehat{H}\vec{\xi} \in H^{-1}(\vec{\xi}^{(d)} + \vec{\eta}^{(d)})$;

(2) $(H\vec{\xi} - \widehat{H}\vec{\xi}) \perp H^{-1}(\vec{\xi}^{(d)} + \vec{\eta}^{(d)})$.

The second condition implies the following relation, which holds for all $k \leq -1$ and for $l = 1, \dots, T$:

$$\int_{-\pi}^{\pi} \left(\left(\vec{B}_{\overline{\mu}}(e^{i\lambda})^{\top} \frac{\chi_{\overline{\mu}}^{(d)}(e^{-i\lambda})}{\beta^{(d)}(i\lambda)} - \vec{h}_{\overline{\mu}}(\lambda) \right)^{\top} (f(\lambda) + |\beta^{(d)}(i\lambda)|^2 g(\lambda)) \right.$$
$$\left. - (\vec{A}(e^{i\lambda}))^{\top} g(\lambda) \overline{\beta^{(d)}(i\lambda)} \right) \overline{\frac{\chi_{\overline{\mu}}^{(d)}(e^{-i\lambda})}{\beta^{(d)}(i\lambda)}} \vec{\delta}_l e^{-i\lambda k} d\lambda = 0.$$

This relation justifies the form (3.12) of the spectral characteristic $\vec{h}_{\overline{\mu}}(\lambda)$ of the estimate $\widehat{H}\vec{\xi}$, where $\vec{C}_{\overline{\mu}}(e^{i\lambda}) = \sum_{k=0}^{\infty} \vec{c}_{\overline{\mu}}(k) e^{ik\lambda}$, with coefficients $\vec{c}_{\overline{\mu}}(k) = \{c_{\overline{\mu},p}(k)\}_{p=1}^{T}$, $k = 0, 1, \dots$, which are unknown and have to be found.

It follows from condition (1) that the following equations should be satisfied for $j \geq 0$:

$$\int_{-\pi}^{\pi} \left((\vec{B}_{\overline{\mu}}(e^{i\lambda}))^{\top} - (\vec{A}_{\overline{\mu}}(e^{i\lambda}))^{\top} \frac{|\beta^{(d)}(i\lambda)|^2}{|\chi_{\overline{\mu}}^{(d)}(e^{-i\lambda})|^2} g(\lambda)(f(\lambda) + |\beta^{(d)}(i\lambda)|^2 g(\lambda))^{-1} \right.$$
$$\left. - (\vec{C}_{\overline{\mu}}(e^{i\lambda}))^{\top} \frac{|\beta^{(d)}(i\lambda)|^2}{|\chi_{\overline{\mu}}^{(d)}(e^{-i\lambda})|^2} (f(\lambda) + |\beta^{(d)}(i\lambda)|^2 g(\lambda))^{-1} \right) e^{-i\lambda j} d\lambda = 0. \tag{3.14}$$

Relation (3.14) can be presented in term of the Fourier coefficients $T_{k,j}^{\overline{\mu}}$ and $P_{k,j}^{\overline{\mu}}$ as a system of linear equations,

$$\vec{b}_{\overline{\mu}}(j) - \sum_{m=0}^{\infty} T_{j,m}^{\overline{\mu}} \vec{a}_{\overline{\mu}}(m) = \sum_{k=0}^{\infty} P_{j,k}^{\overline{\mu}} \vec{c}_{\overline{\mu}}(k), \quad j \geq 0, \tag{3.15}$$

determining the unknown coefficients $\vec{c}_{\overline{\mu}}(k)$, $k \geq 0$. This system of equations can be written in matrix form as

$$D^{\overline{\mu}} \mathbf{a} - \mathbf{T}_{\overline{\mu}} \mathbf{a}_{\overline{\mu}} = \mathbf{P}_{\overline{\mu}} \mathbf{c}_{\overline{\mu}},$$

where $\mathbf{c}_{\overline{\mu}} = ((\vec{c}_{\overline{\mu}}(0))^{\top}, (\vec{c}_{\overline{\mu}}(1))^{\top}, (\vec{c}_{\overline{\mu}}(2))^{\top}, \dots)^{\top}$ and the linear transformation $D^{\overline{\mu}}$ is defined in Lemma 3.1.1. Consequently, the unknown coefficients $\vec{c}_{\overline{\mu}}(k)$, $k \geq 0$, are calculated by the formula

$$\vec{c}_{\overline{\mu}}(k) = (\mathbf{P}_{\overline{\mu}}^{-1} D^{\overline{\mu}} \mathbf{a} - \mathbf{P}_{\overline{\mu}}^{-1} \mathbf{T}_{\overline{\mu}} \mathbf{a}_{\overline{\mu}})_k, \quad k \geq 0, \tag{3.16}$$

where $(\mathbf{P}_{\overline{\mu}}^{-1} D^{\overline{\mu}} \mathbf{a} - \mathbf{P}_{\overline{\mu}}^{-1} \mathbf{T}_{\overline{\mu}} \mathbf{a}_{\overline{\mu}})_k$, $k \geq 0$, is the kth T-dimensional vector entry of the vector $\mathbf{P}_{\overline{\mu}}^{-1} D^{\overline{\mu}} \mathbf{a} - \mathbf{P}_{\overline{\mu}}^{-1} \mathbf{T}_{\overline{\mu}} \mathbf{a}_{\overline{\mu}}$.

The derived expressions justify formulas (3.12) and (3.13) for calculating the spectral characteristic $\vec{h}_{\overline{\mu}}(\lambda)$ and the value of the MSE $\Delta(f, g; \widehat{A}\xi)$ of the estimate $\widehat{A}\xi$ of the functional $A\xi$. □

Corollary 3.1.1. *The spectral characteristic* $\vec{h}_{\overline{\mu}}(\lambda)$ *admits the representation* $\vec{h}_{\overline{\mu}}(\lambda) = \vec{h}^1_{\overline{\mu}}(\lambda) - \vec{h}^2_{\overline{\mu}}(\lambda)$, *where*

$$(\vec{h}^1_{\overline{\mu}}(\lambda))^\top = (\vec{B}_{\overline{\mu}}(e^{i\lambda}))^\top \frac{\chi^{(d)}_{\overline{\mu}}(e^{-i\lambda})}{\beta^{(d)}(i\lambda)} - \frac{\overline{\beta^{(d)}(i\lambda)}}{\chi^{(d)}_{\overline{\mu}}(e^{-i\lambda})}\left(\sum_{k=0}^{\infty}(\mathbf{P}^{-1}_{\overline{\mu}}D^{\overline{\mu}}\mathbf{a})_k e^{ik\lambda}\right)^\top (f(\lambda) + |\beta^{(d)}(i\lambda)|^2 g(\lambda))^{-1}, \tag{3.17}$$

$$(\vec{h}^2_{\overline{\mu}}(\lambda))^\top = \frac{\overline{\beta^{(d)}(i\lambda)}}{\chi^{(d)}_{\overline{\mu}}(e^{-i\lambda})}(\vec{A}_{\overline{\mu}}(e^{i\lambda}))^\top g(\lambda)(f(\lambda) + |\beta^{(d)}(i\lambda)|^2 g(\lambda))^{-1} - \frac{\overline{\beta^{(d)}(i\lambda)}}{\chi^{(d)}_{\overline{\mu}}(e^{-i\lambda})}\left(\sum_{k=0}^{\infty}(\mathbf{P}^{-1}_{\overline{\mu}}\mathbf{T}_{\overline{\mu}}\mathbf{a}_{\overline{\mu}})_k e^{ik\lambda}\right)^\top (f(\lambda) + |\beta^{(d)}(i\lambda)|^2 g(\lambda))^{-1}. \tag{3.18}$$

Here $\vec{h}^1_{\overline{\mu}}(\lambda)$ *and* $\vec{h}^2_{\overline{\mu}}(\lambda)$ *are spectral characteristics of the optimal estimates* $\widehat{B}\vec{\zeta}$ *and* $\widehat{A}\vec{\eta}$ *of the functionals* $B\vec{\zeta}$ *and* $A\vec{\eta}$, *respectively, based on observations* $\vec{\xi}(m) + \vec{\eta}(m)$ *at points* $m = -1, -2, \ldots$.

Making use of Theorem 3.1.1 we can obtain the optimal estimate $\widehat{A}_N\vec{\xi}$ of the functional $A_N\vec{\xi}$ of the unknown values of elements $\vec{\xi}(m)$, $m = 0, 1, 2, \ldots, N$, based on observations of the sequence $\vec{\xi}(m) + \vec{\eta}(m)$ at points $m = -1, -2, \ldots$. Let $\vec{a}(k) = 0$, $k > N$.

From Lemma 3.1.1 we obtain the following corollary.

Corollary 3.1.2. *The functional* $A_N\vec{\xi}$ *admits the representation*

$$A_N\vec{\xi} = A_N\vec{\zeta} - A_N\vec{\eta} = H_N\vec{\xi} - V_N\vec{\zeta}, \tag{3.19}$$

where

$$H_N\vec{\xi} := B_N\chi\vec{\zeta} - A_N\vec{\eta},$$

$$A_N\vec{\zeta} = \sum_{k=0}^{N}(\vec{a}(k))^\top\vec{\zeta}(k), \quad A_N\vec{\eta} = \sum_{k=0}^{N}(\vec{a}(k))^\top\vec{\eta}(k),$$

$$B_N\chi\vec{\zeta} = \sum_{k=0}^{N}(\vec{b}_{\overline{\mu},N}(k))^\top\chi^{(d)}_{\overline{\mu},\overline{s}}(\vec{\zeta}(k)), \quad V_N\vec{\zeta} = \sum_{k=-d(\gamma)}^{-1}(\vec{v}_{\overline{\mu},N}(k))^\top\vec{\zeta}(k),$$

where the coefficients

$$\vec{b}_{\overline{\mu},N}(k) = \{b_{\overline{\mu},N,p}(k)\}_{p=1}^{T}, \quad k = 0, 1, \ldots, N,$$

and

$$\vec{v}_{\overline{\mu},N}(k) = \{v_{\overline{\mu},N,p}(k)\}_{p=1}^{T}, \quad k = -1, -2, \ldots, -d(\gamma),$$

are calculated by the formulas

$$\vec{v}_{\overline{\mu},N}(k)=\sum_{l=0}^{\min\{N,k+d(\gamma)\}}\operatorname{diag}_T(e_\gamma(l-k))\vec{b}_{\overline{\mu},N}(l),\quad k=-1,-2,\dots,-d(\gamma),\tag{3.20}$$

$$\vec{b}_{\overline{\mu},N}(k)=\sum_{m=k}^{N}\operatorname{diag}_T(d_{\overline{\mu}}(m-k))\vec{a}(m)=(D_N^{\overline{\mu}}\mathbf{a}_N)_k,\quad k=0,1,\dots,N,\tag{3.21}$$

where $D_N^{\overline{\mu}}$ is the linear transformation determined by an infinite matrix with the entries $(D_N^{\overline{\mu}})(k,j)=\operatorname{diag}_T(d_{\overline{\mu}}(j-k))$ if $0\le k\le j\le N$ and $(D_N^{\overline{\mu}})(k,j)=\mathcal{O}$ if $j<k$ or $j,k>N$ and

$$\mathbf{a}_N=((\vec{a}(0))^\top,(\vec{a}(1))^\top,\dots,(\vec{a}(N))^\top,\vec{0}^\top,\dots)^\top.$$

Define

$$\vec{B}_{\overline{\mu},N}(e^{i\lambda})=\sum_{k=0}^{N}(D_N^{\overline{\mu}}\mathbf{a}_N)_k e^{i\lambda k},$$

$$\vec{A}_N(e^{i\lambda})=\sum_{k=0}^{N}\vec{a}(k)e^{i\lambda k},\quad \vec{A}_{\overline{\mu},N}(e^{i\lambda})=\sum_{k=0}^{N}\vec{a}_{\overline{\mu},N}(k)e^{i\lambda k},$$

where we have the vectors

$$\mathbf{a}_N=((\vec{a}(0))^\top,(\vec{a}(1))^\top,\dots,(\vec{a}(N))^\top,\vec{0}^\top,\dots)^\top,$$

$$\mathbf{a}_{\overline{\mu},N}=((\vec{a}_{\overline{\mu},N}(0))^\top,(\vec{a}_{\overline{\mu},N}(1))^\top,\dots,(\vec{a}_{\overline{\mu},N}(N+d(\gamma)))^\top,\vec{0}^\top,\dots)^\top$$

and the coefficients

$$\vec{a}_{\overline{\mu},N}(m)=\sum_{l=\max\{m-d(\gamma),0\}}^{\min\{m,N\}}e_\gamma(m-l)\vec{a}(l),\quad 0\le m\le N+d(\gamma).\tag{3.22}$$

Define a linear operator $\mathbf{T}_{\overline{\mu},N}$ in the space ℓ_2 by the matrix with the $T\times T$ matrix entries $(\mathbf{T}_{\overline{\mu},N})_{l,m}=T^{\overline{\mu}}_{l,m}$, $l\ge 0$, $0\le m\le N+d(\gamma)$, and $(\mathbf{T}_{\overline{\mu},N})_{l,m}=\mathcal{O}$, $l\ge 0$, $m>N+d(\gamma)$, and a linear operator $\mathbf{Q}_N$ in the space ℓ_2 by the matrix with the $T\times T$ matrix entries $(\mathbf{Q}_N)_{l,k}=Q_{l,k}$, $0\le l,k\le N$, and $(\mathbf{Q}_N)_{l,k}=\mathcal{O}$ otherwise.

The following theorem holds.

Theorem 3.1.2. *Let a stochastic vector sequence $\vec{\xi}(m)$, $m\in\mathbb{Z}$, with stationary GM increments and a stochastic stationary vector sequence $\vec{\eta}(m)$, $m\in\mathbb{Z}$, satisfy the conditions of Theorem* 3.1.1*, and let their spectral densities $f(\lambda)$ and $g(\lambda)$ satisfy minimality condition* (3.4)*. The optimal linear estimate $\widehat{A}_N\vec{\xi}$ of the functional $A_N\vec{\xi}$ which depends on the unknown values $\vec{\xi}(k)$, $k=0,1,2,\dots,N$, from observations of the sequence $\vec{\xi}(m)+\vec{\eta}(m)$ at points $m=-1,-2,\dots$ is calculated by the formula*

$$\widehat{A}_N\vec{\xi} = \int_{-\pi}^{\pi} (\vec{h}_{\overline{\mu},N}(\lambda))^\top d\vec{Z}_{\xi^{(d)}+\eta^{(d)}}(\lambda) - \sum_{k=-d(\gamma)}^{-1} (\vec{v}_{\overline{\mu},N}(k))^\top (\vec{\xi}(k) + \vec{\eta}(k)). \tag{3.23}$$

The spectral characteristic $\vec{h}_{\overline{\mu},N}(\lambda)$ of the estimate $\widehat{A}_N\vec{\xi}$ is calculated by the formula

$$\begin{aligned}(\vec{h}_{\overline{\mu},N}(\lambda))^\top &= (\vec{B}_{\overline{\mu},N}(e^{i\lambda}))^\top \frac{\chi_{\overline{\mu}}^{(d)}(e^{-i\lambda})}{\beta^{(d)}(i\lambda)} \\ &- \frac{\overline{\beta^{(d)}(i\lambda)}}{\chi_{\overline{\mu}}^{(d)}(e^{-i\lambda})} (g(\lambda)\overline{\vec{A}_{\overline{\mu},N}(e^{i\lambda})} + \vec{C}_{\overline{\mu},N}(e^{i\lambda}))^\top (f(\lambda) + |\beta^{(d)}(i\lambda)|^2 g(\lambda))^{-1},\end{aligned} \tag{3.24}$$

where

$$\vec{C}_{\overline{\mu},N}(e^{i\lambda}) = \sum_{k=0}^{\infty} (\mathbf{P}_{\overline{\mu}}^{-1} D_N^{\overline{\mu}} \mathbf{a}_N - \mathbf{P}_{\overline{\mu}}^{-1} \mathbf{T}_{\overline{\mu},N} \mathbf{a}_{\overline{\mu},N})_k e^{i\lambda k}.$$

The MSE of the estimate $\widehat{A}_N\vec{\xi}$ is calculated by the formula

$$\begin{aligned}\Delta(f,g;\widehat{A}_N\vec{\xi}) &= \Delta(f,g;\widehat{H}_N\vec{\xi}) = \mathsf{E}|H_N\vec{\xi} - \widehat{H}_N\vec{\xi}|^2 \\ &= \frac{1}{2\pi}\int_{-\pi}^{\pi} \frac{|\beta^{(d)}(i\lambda)|^2}{|\chi_{\overline{\mu}}^{(d)}(e^{-i\lambda})|^2} ((\vec{A}_{\overline{\mu},N}(e^{i\lambda}))^\top g(\lambda) + (\vec{C}_{\overline{\mu},N}(e^{i\lambda}))^\top) \\ &\times (f(\lambda) + |\beta^{(d)}(i\lambda)|^2 g(\lambda))^{-1} f(\lambda)\,(f(\lambda) + |\beta^{(d)}(i\lambda)|^2 g(\lambda))^{-1} \\ &\times (g(\lambda)\overline{\vec{A}_{\overline{\mu},N}(e^{i\lambda})} + \overline{\vec{C}_{\overline{\mu},N}(e^{i\lambda})})d\lambda \\ &+ \frac{1}{2\pi}\int_{-\pi}^{\pi} \frac{1}{|\chi_{\overline{\mu}}^{(d)}(e^{-i\lambda})|^2} ((\vec{A}_{\overline{\mu},N}(e^{i\lambda}))^\top f(\lambda) - |\beta^{(d)}(i\lambda)|^2 (\vec{C}_{\overline{\mu},N}(e^{i\lambda}))^\top) \\ &\times (f(\lambda) + |\beta^{(d)}(i\lambda)|^2 g(\lambda))^{-1} g(\lambda)\,(f(\lambda) + |\beta^{(d)}(i\lambda)|^2 g(\lambda))^{-1} \\ &\times (f(\lambda)\overline{\vec{A}_{\overline{\mu},N}(e^{i\lambda})} - |\beta^{(d)}(i\lambda)|^2 \overline{\vec{C}_{\overline{\mu},N}(e^{i\lambda})})d\lambda \\ &= \langle D_N^{\overline{\mu}} \mathbf{a}_N - \mathbf{T}_{\overline{\mu},N}\mathbf{a}_{\overline{\mu},N}, \mathbf{P}_{\overline{\mu}}^{-1} D_N^{\overline{\mu}} \mathbf{a}_N - \mathbf{P}_{\overline{\mu}}^{-1}\mathbf{T}_{\overline{\mu},N}\mathbf{a}_{\overline{\mu},N}\rangle + \langle \mathbf{Q}_N \mathbf{a}_N, \mathbf{a}_N\rangle\end{aligned} \tag{3.25}$$

As a corollary from Theorem 3.1.2, one can obtain the mean square optimal estimate of the unobserved value $A_{N,p}\vec{\xi} = \xi_p(N) = (\vec{\xi}(N))^\top \vec{\delta}_p$, $p = 1,2,\dots,T$, $N \geq 0$, of the stochastic sequence with GM stationary increments based on observations of the sequence $\vec{\xi}(m) + \vec{\eta}(m)$ at points $m = -1, -2, \dots$.

Corollary 3.1.3. *The optimal linear estimate $\widehat{\xi}_p(N)$ of the unobserved value $\xi_p(N)$, $p = 1,2,\dots,T$, $N \geq 0$, of the stochastic sequence with stationary GM increments from observations of the sequence $\vec{\xi}(m) + \vec{\eta}(m)$ at points $m = -1,-2,\dots$ is calculated by the formula*

$$\widehat{\xi}_p(N) = \int_{-\pi}^{\pi} (\vec{h}_{\overline{\mu},N,p}(\lambda))^{\top} d\vec{Z}_{\xi^{(d)}+\eta^{(d)}}(\lambda) - \sum_{k=-d(\gamma)}^{-1} (\vec{v}_{\overline{\mu},N,p}(k))^{\top}(\vec{\xi}(k) + \vec{\eta}(k)). \tag{3.26}$$

The spectral characteristic $\vec{h}_{\overline{\mu},N,p}(\lambda)$ of the estimate is calculated by the formula

$$\begin{aligned}
(\vec{h}_{\overline{\mu},N,p}(\lambda))^{\top} = {} & \frac{\chi_{\overline{\mu}}^{(d)}(e^{-i\lambda})}{\beta^{(d)}(i\lambda)}\left(\vec{\delta}_p \sum_{k=0}^{N} d_{\overline{\mu}}(N-k)e^{i\lambda k}\right)^{\top} \\
& - e^{i\lambda N}(\vec{\delta}_p)^{\top} g(\lambda)\overline{\beta^{(d)}(i\lambda)}(f(\lambda) + |\beta^{(d)}(i\lambda)|^2 g(\lambda))^{-1} \\
& - \left(\sum_{k=0}^{\infty}(\mathbf{P}_{\overline{\mu}}^{-1}\mathbf{d}_{\overline{\mu},N,p} - \mathbf{P}_{\overline{\mu}}^{-1}\mathbf{T}_{\overline{\mu},N}\vec{\mathbf{a}}_{\overline{\mu},N,p})_k e^{i\lambda k}\right)^{\top} \\
& \times \frac{\overline{\beta^{(d)}(i\lambda)}}{\chi_{\overline{\mu}}^{(d)}(e^{-i\lambda})}(f(\lambda) + |\beta^{(d)}(i\lambda)|^2 g(\lambda))^{-1},
\end{aligned} \tag{3.27}$$

where

$$\begin{aligned}
\mathbf{d}_{\overline{\mu},N,p} &= ((d_{\overline{\mu}}(N)\vec{\delta}_p)^{\top}, (d_{\overline{\mu}}(N-1)\vec{\delta}_p)^{\top}, (d_{\overline{\mu}}(N-2)\vec{\delta}_p)^{\top}, \dots, (d_{\overline{\mu}}(0)\vec{\delta}_p)^{\top}, \vec{0}^{\top}, \dots)^{\top}, \\
\vec{\mathbf{a}}_{\overline{\mu},N,p} &= (\vec{0}^{\top}, \dots, \vec{0}^{\top}, (\vec{a}_{\overline{\mu},N,p}(N))^{\top}, (\vec{a}_{\overline{\mu},N,p}(N+1))^{\top}, \dots, (\vec{a}_{\overline{\mu},N,p}(N+d(\gamma)))^{\top}, \vec{0}^{\top}, \dots)^{\top},
\end{aligned}$$

and

$$\vec{a}_{\overline{\mu},N,p}(m) = e_{\gamma}(m-N)\vec{\delta}_p, \quad N \le m \le N + d(\gamma).$$

The MSE of the estimate $\widehat{\xi}_p(N)$ is calculated by the formula

$$\begin{aligned}
\Delta(f,g;\widehat{\xi}_p(N)) &= \mathsf{E}|\xi_p(N) - \widehat{\xi}_p(N)|^2 \\
&= \frac{1}{2\pi}\int_{-\pi}^{\pi} \frac{|\beta^{(d)}(i\lambda)|^2}{|\chi_{\overline{\mu}}^{(d)}(e^{-i\lambda})|^2}(\overline{\chi_{\overline{\mu}}^{(d)}(e^{-i\lambda})e^{i\lambda N}(\vec{\delta}_p)^{\top} g(\lambda) + (\vec{C}_{\overline{\mu},N,p}(e^{i\lambda}))^{\top}}) \\
&\quad \times (f(\lambda) + |\beta^{(d)}(i\lambda)|^2 g(\lambda))^{-1} f(\lambda)\,(f(\lambda) + |\beta^{(d)}(i\lambda)|^2 g(\lambda))^{-1} \\
&\quad \times (\chi_{\overline{\mu}}^{(d)}(e^{-i\lambda})e^{-i\lambda N} g(\lambda)\vec{\delta}_p + \overline{\vec{C}_{\overline{\mu},N,p}(e^{i\lambda})})d\lambda \\
&\quad + \frac{1}{2\pi}\int_{-\pi}^{\pi} \frac{1}{|\chi_{\overline{\mu}}^{(d)}(e^{-i\lambda})|^2}(\overline{\chi_{\overline{\mu}}^{(d)}(e^{-i\lambda})e^{i\lambda N}(\vec{\delta}_p)^{\top} f(\lambda) - |\beta^{(d)}(i\lambda)|^2(\vec{C}_{\overline{\mu},N,p}(e^{i\lambda}))^{\top}}) \\
&\quad \times (f(\lambda) + |\beta^{(d)}(i\lambda)|^2 g(\lambda))^{-1} g(\lambda)\,(f(\lambda) + |\beta^{(d)}(i\lambda)|^2 g(\lambda))^{-1} \\
&\quad \times (\chi_{\overline{\mu}}^{(d)}(e^{-i\lambda})e^{-i\lambda N} f(\lambda)\vec{\delta}_p - |\beta^{(d)}(i\lambda)|^2\overline{\vec{C}_{\overline{\mu},N,p}(e^{i\lambda})})d\lambda \\
&= \langle \mathbf{d}_{\overline{\mu},N,p} - \mathbf{T}_{\overline{\mu},N}\vec{\mathbf{a}}_{\overline{\mu},N,p}, \mathbf{P}_{\overline{\mu}}^{-1}\mathbf{d}_{\overline{\mu},N,p} - \mathbf{P}_{\overline{\mu}}^{-1}\mathbf{T}_{\overline{\mu},N}\vec{\mathbf{a}}_{\overline{\mu},N,p}\rangle + \langle \mathbf{Q}_{0,0}\vec{\delta}_p, \vec{\delta}_p\rangle,
\end{aligned} \tag{3.28}$$

where

$$\vec{C}_{\overline{\mu},N,p}(e^{i\lambda}) = \sum_{k=0}^{\infty} (\mathbf{P}_{\overline{\mu}}^{-1}\mathbf{d}_{\overline{\mu},N,p} - \mathbf{P}_{\overline{\mu}}^{-1}\mathbf{T}_{\overline{\mu},N}\vec{\mathbf{a}}_{\overline{\mu},N,p})_{\kappa} e^{i\lambda k}.$$

3.1.2 Extrapolation of sequences with FM increments

Let $d_j = R_j + D_j$, $j = 0,1,\dots,r$, denote the fractional differentiation order and let R_j, $j = 0,1,\dots,r$, be non-negative integer numbers. Set the increment step $\overline{\mu} = (1,1,\dots,1)$. Let $1 < s_1 < \cdots < s_r$. Consider a stochastic vector sequence $\vec{\xi}(m)$ with stationary FM increments $\chi_{\overline{1},\overline{s}}^{(R+D)}(\vec{\xi}(m))$ having a bounded from zero and from infinity spectral density $\widetilde{f}_{\overline{1}}(\lambda)$. Assume that the numbers $R_j, D_j, j = 0,1,\dots,r$, can be chosen in such way that $-1/2 < D_\nu < 1/2$ for all $\nu \in \mathcal{M}$, where the numbers D_ν and the set are defined in Lemma 1.2.2. Then Theorem 1.2.1 implies that the GM increment sequence $\chi_{\overline{1},\overline{s}}^{(R)}(\vec{\xi}(m))$ is stationary.

Corollary 3.1.4. *A solution of the extrapolation problem for the multidimensional sequence $\vec{\xi}(m)$ with stationary FM increments based on its observations with stationary noise can be obtained from Theorem* 3.1.1, *Theorem* 3.1.2, *or Corollary* 3.1.3 *for the GM increment sequence $\chi_{\overline{1},\overline{s}}^{(R)}(\vec{\xi}(m))$ making use of the Fourier coefficients*

$$T_{k,j}^{\overline{1}} = \frac{1}{2\pi}\int_{-\pi}^{\pi} e^{-i\lambda(k-j)} \frac{|\beta^{(R)}(i\lambda)|^2}{|\chi_{\overline{1}}^{(R)}(e^{-i\lambda})|^2} (g(\lambda)p^{-1}(\lambda))^{\top} d\lambda, \quad k,j \geq 0,$$

$$P_{k,j}^{\overline{1}} = \frac{1}{2\pi}\int_{-\pi}^{\pi} e^{-i\lambda(k-j)} \frac{|\beta^{(R)}(i\lambda)|^2}{|\chi_{\overline{1}}^{(R)}(e^{-i\lambda})|^2} (p^{-1}(\lambda))^{\top} d\lambda, \quad k,j \geq 0,$$

$$Q_{k,j} = \frac{1}{2\pi}\int_{-\pi}^{\pi} e^{-i\lambda(k-j)} (|\chi_{\overline{1}}^{(D)}(e^{-i\lambda})|^{-2}\widetilde{f}(\lambda)p^{-1}(\lambda)g(\lambda))^{\top} d\lambda, \quad k,j \geq 0,$$

where

$$p(\lambda) = |\chi_{\overline{1}}^{(D)}(e^{-i\lambda})|^{-2}\widetilde{f}(\lambda) + |\beta^{(R)}(i\lambda)|^2 g(\lambda),$$
$$\widetilde{f}(\lambda) = |\beta^{(R)}(i\lambda)|^2 |\chi_{\overline{1}}^{(R)}(e^{-i\lambda})|^{-2}\widetilde{f}_{\overline{1}}(\lambda),$$

under the conditions of Theorem 1.2.1.

3.1.3 Extrapolation based on factorizations of spectral densities

In Theorem 3.1.1, Theorem 3.1.2, and Corollary 3.1.3, formulas for finding estimates of the linear functionals $A\xi$, $A_N\xi$ and the value $\xi(p)$, $p \geq 0$, are derived by using the Fourier coefficients of the functions

$$\frac{|\beta^{(d)}(i\lambda)|^2}{|\chi_{\overline{\mu}}^{(d)}(e^{-i\lambda})|^2}(g(\lambda)(f(\lambda)+|\beta^{(d)}(i\lambda)|^2g(\lambda))^{-1})^{\top},$$
$$\frac{|\beta^{(d)}(i\lambda)|^2}{|\chi_{\overline{\mu}}^{(d)}(e^{-i\lambda})|^2}((f(\lambda)+|\beta^{(d)}(i\lambda)|^2g(\lambda))^{-1})^{\top}.$$

In this section, we propose formulas for finding estimates based on factorizations of the spectral densities. Assume that the canonical factorizations

$$\frac{|\chi_{\overline{\mu}}^{(d)}(e^{-i\lambda})|^2}{|\beta^{(d)}(i\lambda)|^2}(f(\lambda)+|\beta^{(d)}(i\lambda)|^2g(\lambda))=\Theta_{\overline{\mu}}(e^{-i\lambda})\Theta_{\overline{\mu}}^*(e^{-i\lambda}),\quad \Theta_{\overline{\mu}}(e^{-i\lambda})=\sum_{k=0}^{\infty}\theta_{\overline{\mu}}(k)e^{-i\lambda k},\tag{3.29}$$

and

$$g(\lambda)=\sum_{k=-\infty}^{\infty}g(k)e^{i\lambda k}=\Phi(e^{-i\lambda})\Phi^*(e^{-i\lambda}),\quad \Phi(e^{-i\lambda})=\sum_{k=0}^{\infty}\phi(k)e^{-i\lambda k}.\tag{3.30}$$

take place. Determine the matrix-valued function $\Psi_{\overline{\mu}}(e^{-i\lambda})=\{\Psi_{ij}(e^{-i\lambda})\}_{i=\overline{1,q}}^{j=\overline{1,T}}$ by the equation

$$\Psi_{\overline{\mu}}(e^{-i\lambda})\Theta_{\overline{\mu}}(e^{-i\lambda})=E_q,$$

where E_q is an identity $q\times q$ matrix. One can check that the following factorization takes place:

$$\frac{|\beta^{(d)}(i\lambda)|^2}{|\chi_{\overline{\mu}}^{(d)}(e^{-i\lambda})|^2}(f(\lambda)+|\beta^{(d)}(i\lambda)|^2g(\lambda))^{-1}=\Psi_{\overline{\mu}}^*(e^{-i\lambda})\Psi_{\overline{\mu}}(e^{-i\lambda}),\quad \Psi_{\overline{\mu}}(e^{-i\lambda})=\sum_{k=0}^{\infty}\psi_{\overline{\mu}}(k)e^{-i\lambda k}.\tag{3.31}$$

Remark 3.1.2. Any spectral density matrix $f(\lambda)$ is self-adjoint: $f(\lambda)=f^*(\lambda)$. Thus, $(f(\lambda))^{\top}=\overline{f(\lambda)}$. One can check that an inverse spectral density $f^{-1}(\lambda)$ is also self-adjoint: $f^{-1}(\lambda)=(f^{-1}(\lambda))^*$ and $(f^{-1}(\lambda))^{\top}=\overline{f^{-1}(\lambda)}$.

Let $\mathcal{O}$ denote a $q\times T$ or $T\times q$ matrix with zero entries. The following lemmas provide factorizations of the operators $\mathbf{P}_{\overline{\mu}}$ and $\mathbf{T}_{\overline{\mu}}$, which contain coefficients of factorizations (3.29)–(3.31).

Lemma 3.1.2. *Let factorization* (3.29) *take place and let the $q\times T$ matrix function $\Psi_{\overline{\mu}}(e^{-i\lambda})$ satisfy equation $\Psi_{\overline{\mu}}(e^{-i\lambda})\Phi_{\overline{\mu}}(e^{-i\lambda})=E_q$. Define the linear operators $\Psi_{\overline{\mu}}$ and $\Theta_{\overline{\mu}}$ in the space ℓ_2 by the matrices with the matrix entries $(\Psi_{\overline{\mu}})_{k,j}=\psi_{\overline{\mu}}(k-j)$, $(\Theta_{\overline{\mu}})_{k,j}=\theta_{\overline{\mu}}(k-j)$ for $0\le j\le k$ and $(\Psi_{\overline{\mu}})_{k,j}=\mathcal{O}$, $(\Theta_{\overline{\mu}})_{k,j}=\mathcal{O}$ for $0\le k<j$. Then* (a) *the linear operator $\mathbf{P}_{\overline{\mu}}$ and* (b) *the inverse operator $(\mathbf{P}_{\overline{\mu}})^{-1}$ admit the factorizations*

$$\mathbf{P}_{\overline{\mu}}=(\Psi_{\overline{\mu}})^{\top}\overline{\Psi}_{\overline{\mu}},\quad (\mathbf{P}_{\overline{\mu}})^{-1}=\overline{\Theta}_{\overline{\mu}}(\Theta_{\overline{\mu}})^{\top}.$$

Proof. Making use of factorization (3.31), we obtain the relation

$$\begin{aligned}
&\frac{|\beta^{(d)}(i\lambda)|^2}{|\chi_{\overline{\mu}}^{(d)}(e^{-i\lambda})|^2}((f(\lambda)+|\beta^{(d)}(i\lambda)|^2 g(\lambda))^{-1})^{\top}\\
&=\sum_{m=-\infty}^{\infty} P_{\overline{\mu}}(m)e^{i\lambda m}=(\Psi_{\overline{\mu}}(e^{-i\lambda}))^{\top}\overline{\Psi_{\overline{\mu}}(e^{-i\lambda})}\\
&=\sum_{m=-\infty}^{-1}\sum_{k=-m}^{\infty}\psi_{\overline{\mu}}^{\top}(k)\overline{\psi}_{\overline{\mu}}(k+m)e^{i\lambda m}\sum_{m=0}^{\infty}\sum_{k=0}^{\infty}\psi_{\overline{\mu}}^{\top}(k)\overline{\psi}_{\overline{\mu}}(k+m)e^{i\lambda m}.
\end{aligned}$$

Thus,

$$\begin{aligned}
P_{\overline{\mu}}(m)&=\sum_{k=0}^{\infty}\psi_{\overline{\mu}}^{\top}(k)\overline{\psi}_{\overline{\mu}}(k+m), \quad m\ge 0,\\
P_{\overline{\mu}}(-m)&=(P_{\overline{\mu}}(m))^{*}, \quad m\ge 0.
\end{aligned}$$

In the case $i\ge j$, we have

$$P_{i,j}^{\mu}=P_{\overline{\mu}}(i-j)=\sum_{l=i}^{\infty}\psi_{\overline{\mu}}^{\top}(l-i)\overline{\psi}_{\overline{\mu}}(l-j)=((\Psi_{\overline{\mu}})^{\top}\overline{\Psi}_{\overline{\mu}})_{i,j},$$

and in the case $i<j$, we have

$$P_{i,j}^{\mu}=P_{\overline{\mu}}(i-j)=\overline{P_{\overline{\mu}}(j-i)}=\sum_{l=j}^{\infty}\psi_{\overline{\mu}}^{\top}(l-i)\overline{\psi}_{\overline{\mu}}(l-j)=((\Psi_{\overline{\mu}})^{\top}\overline{\Psi}_{\overline{\mu}})_{i,j},$$

which proves statement (a).

From factorizations (3.29) and (3.31), we obtain

$$E_q=\Psi_{\overline{\mu}}(e^{-i\lambda})\Theta_{\overline{\mu}}(e^{-i\lambda})=\sum_{j=0}^{\infty}\left(\sum_{k=0}^{j}\psi_{\overline{\mu}}(k)\theta_{\overline{\mu}}(j-k)\right)e^{-i\lambda j}. \tag{3.32}$$

Thus, we derive the relations

$$\mathrm{diag}_T(\delta_{i,j})=\sum_{k=0}^{i-j}\psi_{\overline{\mu}}(k)\theta_{\overline{\mu}}(i-j-k)=\sum_{p=i}^{j}\psi_{\overline{\mu}}(i-p)\theta_{\overline{\mu}}(p-j)=(\Psi_{\overline{\mu}}\Theta_{\overline{\mu}})_{i,j},$$

which imply $\Psi_{\overline{\mu}}\Theta_{\overline{\mu}}=E_q$. Using $\mathbf{P}_{\overline{\mu}}=(\Psi_{\overline{\mu}})^{\top}\overline{\Psi}_{\overline{\mu}}$ we get $(\Theta_{\overline{\mu}})^{\top}\mathbf{P}_{\overline{\mu}}=\overline{\Psi}_{\overline{\mu}}$ and $\mathbf{P}_{\overline{\mu}}\overline{\Theta}_{\overline{\mu}}=(\Psi_{\overline{\mu}})^{\top}$ or $(\Theta_{\overline{\mu}})^{\top}=\overline{\Psi}_{\overline{\mu}}(\mathbf{P}_{\overline{\mu}})^{-1}$ and $\overline{\Theta}_{\overline{\mu}}=(\mathbf{P}_{\overline{\mu}})^{-1}(\Psi_{\overline{\mu}})^{\top}$. The last two relations imply

$$\overline{\Theta}_{\overline{\mu}}(\Theta_{\overline{\mu}})^{\top}=(\mathbf{P}_{\overline{\mu}})^{-1}(\Psi_{\overline{\mu}})^{\top}\overline{\Psi}_{\overline{\mu}}(\mathbf{P}_{\overline{\mu}})^{-1}=(\mathbf{P}_{\overline{\mu}})^{-1}\mathbf{P}_{\overline{\mu}}(\mathbf{P}_{\overline{\mu}})^{-1}=(\mathbf{P}_{\overline{\mu}})^{-1},$$

which proves statement (b). □

Lemma 3.1.3. *Let factorizations* (3.29) *and* (3.30) *take place. Then the operator* $\mathbf{T}_{\overline{\mu}}$ *admits the representation*

$$\mathbf{T}_{\overline{\mu}} = (\Psi_{\overline{\mu}})^{\top}\mathbf{Z}_{\overline{\mu}},$$

where $\mathbf{Z}_{\overline{\mu}}$ *is a linear operator in the space* ℓ_2 *defined by a matrix with the entries*

$$(\mathbf{Z}_{\overline{\mu}})_{k,j} = \sum_{l=j}^{\infty} \overline{\psi}_{\overline{\mu}}(l-j)\overline{g}(l-k), \quad g(k) = \sum_{m=\max\{0,-k\}}^{\infty} \phi(m)\phi^*(k+m), \quad k,j \geq 0.$$

Proof. Factorizations (3.30) and (3.31) and Remark 3.1.2 imply

$$\begin{aligned}
&\frac{|\beta^{(d)}(i\lambda)|^2}{|\chi_{\overline{\mu}}^{(d)}(e^{-i\lambda})|^2}(g(\lambda)(f(\lambda)+|\beta^{(d)}(i\lambda)|^2 g(\lambda))^{-1})^{\top} \\
&= (\Psi_{\overline{\mu}}(e^{-i\lambda}))^{\top}\overline{\Psi}_{\overline{\mu}}(e^{-i\lambda})\overline{g}(\lambda) \\
&= \sum_{l=0}^{\infty} \Psi_{\overline{\mu}}^{\top}(l)e^{-i\lambda l}\sum_{j=0}^{\infty} Z_{\overline{\mu}}(j)e^{i\lambda j} = \sum_{k\in\mathbb{Z}}\sum_{l=0}^{\infty} \Psi_{\overline{\mu}}^{\top}(l)Z_{\overline{\mu}}(l+k)e^{i\lambda k}.
\end{aligned}$$

Then

$$(\mathbf{T}_{\overline{\mu}})_{k,j} = \mathbf{T}_{\overline{\mu}}(k-j) = \sum_{m=k}^{\infty} \Psi_{\overline{\mu}}^{\top}(m-k)Z_{\overline{\mu}}(m-j) = (\Psi_{\overline{\mu}}^{\top}\mathbf{Z}_{\overline{\mu}})_{k,j}.$$

The representation for the entries $(\mathbf{Z}_{\overline{\mu}})_{k,j} = Z_{\overline{\mu}}(k-j)$ follows from

$$\sum_{j=0}^{\infty} Z_{\overline{\mu}}(j)e^{i\lambda j} = \overline{\Psi}_{\overline{\mu}}(e^{-i\lambda})\overline{g}(\lambda) = \sum_{k\in\mathbb{Z}}\sum_{l=0}^{\infty} \overline{\Psi}_{\overline{\mu}}(l)\overline{g}_{\overline{\mu}}(l-k)e^{i\lambda k}.$$ □

Remark 3.1.3. Lemma 3.1.2 and Lemma 3.1.3 imply a factorization

$$(\mathbf{P}_{\overline{\mu}})^{-1}\mathbf{T}_{\overline{\mu}}\mathbf{a}_{\overline{\mu}} = \overline{\Theta}_{\overline{\mu}}(\Theta_{\overline{\mu}})^{\top}(\Psi_{\overline{\mu}})^{\top}\mathbf{Z}_{\overline{\mu}}\mathbf{a}_{\overline{\mu}} = \overline{\Theta}_{\overline{\mu}}\mathbf{Z}_{\overline{\mu}}\mathbf{a}_{\overline{\mu}} = \overline{\Theta}_{\overline{\mu}}\mathbf{e}_{\overline{\mu}},$$

where $\mathbf{e}_{\overline{\mu}} := \mathbf{Z}_{\overline{\mu}}\mathbf{a}_{\overline{\mu}}$.

Assuming that factorizations (3.29)–(3.31) take place and making use of Remark 3.1.3, spectral characteristic (3.12) and MSE (3.13) can be presented in terms of the coefficients of these factorizations. We make the following transformations:

$$\begin{aligned}
&\frac{|\beta^{(d)}(i\lambda)|^2}{|\chi_{\overline{\mu}}^{(d)}(e^{-i\lambda})|^2}((f(\lambda)+|\beta^{(d)}(i\lambda)|^2 g(\lambda))^{-1})^{\top}\left(\sum_{k=0}^{\infty}(\mathbf{P}_{\overline{\mu}}^{-1}\mathbf{T}_{\overline{\mu}}\mathbf{a}_{\overline{\mu}})_k e^{ik\lambda}\right) \\
&= \left(\sum_{k=0}^{\infty} \psi_{\overline{\mu}}^{\top}(k)e^{-i\lambda k}\right)\sum_{j=0}^{\infty}\sum_{k=0}^{\infty} \overline{\psi}_{\overline{\mu}}(j)(\overline{\Theta}_{\overline{\mu}}\mathbf{e}_{\overline{\mu}})_k e^{i\lambda(k+j)}
\end{aligned}$$

$$
\begin{aligned}
&= \left(\sum_{k=0}^{\infty} \psi_{\overline{\mu}}^{\top}(k) e^{-i\lambda k} \right) \sum_{m=0}^{\infty} \sum_{p=0}^{m} \sum_{k=p}^{m} \overline{\psi}_{\overline{\mu}}(m-k) \overline{\theta}_{\overline{\mu}}(k-p) \vec{e}_{\overline{\mu}}(p) e^{i\lambda m} \\
&= \left(\sum_{k=0}^{\infty} \psi_{\overline{\mu}}^{\top}(k) e^{-i\lambda k} \right) \sum_{m=0}^{\infty} \sum_{p=0}^{\infty} \operatorname{diag}(\delta_{m,p}) \vec{e}_{\overline{\mu}}(p) e^{i\lambda m} \\
&= \left(\sum_{k=0}^{\infty} \psi_{\overline{\mu}}^{\top}(k) e^{-i\lambda k} \right) \sum_{m=0}^{\infty} \vec{e}_{\overline{\mu}}(m) e^{i\lambda m},
\end{aligned}
\tag{3.33}
$$

where $\vec{e}_{\overline{\mu}}(m) = (\mathbf{Z}_{\overline{\mu}} \mathbf{a}_{\overline{\mu}})_m$, $m \geq 0$, is the mth vector entry of the vector $\mathbf{e}_{\overline{\mu}} = \mathbf{Z}_{\overline{\mu}} \mathbf{a}_{\overline{\mu}}$.

Using factorizations (3.30) and (3.31) we can check the following transformations:

$$
\begin{aligned}
&\frac{|\beta^{(d)}(i\lambda)|^2}{|\chi_{\overline{\mu}}^{(d)}(e^{-i\lambda})|^2} ((f(\lambda) + |\beta^{(d)}(i\lambda)|^2 g(\lambda))^{-1})^{\top} (g(\lambda))^{\top} \vec{A}_{\overline{\mu}}(e^{i\lambda}) \\
&\quad = \left(\sum_{k=0}^{\infty} \psi_{\overline{\mu}}^{\top}(k) e^{-i\lambda k} \right) \left(\sum_{k=-\infty}^{\infty} Z_{\overline{\mu}}(k) e^{i\lambda k} \right) \sum_{j=0}^{\infty} \vec{a}_{\overline{\mu}}(j) e^{i\lambda j} \\
&\quad = \left(\sum_{k=0}^{\infty} \psi_{\overline{\mu}}^{\top}(k) e^{-i\lambda k} \right) \sum_{m=-\infty}^{\infty} \sum_{j=0}^{\infty} Z_{\overline{\mu}}(m-j) \vec{a}_{\overline{\mu}}(j) e^{i\lambda m} \\
&\quad = \left(\sum_{k=0}^{\infty} \psi_{\overline{\mu}}^{\top}(k) e^{-i\lambda k} \right) \sum_{m=-\infty}^{\infty} \vec{e}_{\overline{\mu}}(m) e^{i\lambda m}.
\end{aligned}
\tag{3.34}
$$

Making use of (3.33) and (3.34), formula (3.18) for the spectral characteristic $\vec{h}_{\overline{\mu}}^{2}(\lambda)$ of the optimal estimate $\widehat{A}\vec{\eta}$ of the functionals $A\vec{\eta}$ can be presented as

$$
\begin{aligned}
\vec{h}_{\overline{\mu}}^{2}(\lambda) &= \frac{\chi_{\overline{\mu}}^{(d)}(e^{-i\lambda})}{\beta^{(d)}(i\lambda)} \left(\sum_{k=0}^{\infty} \psi_{\overline{\mu}}^{\top}(k) e^{-i\lambda k} \right) \sum_{m=1}^{\infty} \vec{e}_{\overline{\mu}}(-m) e^{-i\lambda m} \\
&= \frac{\chi_{\overline{\mu}}^{(d)}(e^{-i\lambda})}{\beta^{(d)}(i\lambda)} \left(\sum_{k=0}^{\infty} \psi_{\overline{\mu}}^{\top}(k) e^{-i\lambda k} \right) \\
&\quad \times \sum_{m=1}^{\infty} \sum_{j=0}^{\infty} \sum_{l=0}^{\infty} \overline{\psi}_{\overline{\mu}}(l) \overline{g}(m+j+l) \vec{a}_{\overline{\mu}}(j) e^{-i\lambda m} \\
&= \frac{\chi_{\overline{\mu}}^{(d)}(e^{-i\lambda})}{\beta^{(d)}(i\lambda)} \left(\sum_{k=0}^{\infty} \psi_{\overline{\mu}}^{\top}(k) e^{-i\lambda k} \right) \sum_{m=1}^{\infty} (\overline{\psi}_{\overline{\mu}} \mathbf{C}_{\overline{\mu},g})_m e^{-i\lambda m} \\
&= \frac{\chi_{\overline{\mu}}^{(d)}(e^{-i\lambda})}{\beta^{(d)}(i\lambda)} \Psi_{\overline{\mu}}^{\top}(e^{-i\lambda}) \vec{C}_{\overline{\mu},g}(e^{-i\lambda}),
\end{aligned}
$$

where $\overline{\psi}_{\overline{\mu}} = (\overline{\psi}_{\overline{\mu}}(0), \overline{\psi}_{\overline{\mu}}(1), \overline{\psi}_{\overline{\mu}}(2), \dots)$,

$$(\overline{\psi}_{\overline{\mu}}\mathbf{C}_{\overline{\mu},g})_m = \sum_{k=0}^{\infty} \overline{\psi}_{\overline{\mu}}(k)\mathbf{c}_{\overline{\mu},g}(k+m),$$

$$\mathbf{c}_{\overline{\mu},g}(m) = \sum_{k=0}^{\infty} \overline{g}(m+k)\vec{a}_{\overline{\mu}}(k) = \sum_{l=0}^{\infty} \overline{\phi}(l) \sum_{k=0}^{\infty} \phi^{\top}(l+m+k)\vec{a}_{\overline{\mu}}(k)$$

$$= \sum_{l=0}^{\infty} \overline{\phi}(l)(\widetilde{\Phi}\mathbf{a}_{\overline{\mu}})_{l+m},$$

and

$$(\widetilde{\Phi}\mathbf{a}_{\overline{\mu}})_m = \sum_{k=0}^{\infty} \phi^{\top}(m+k)\vec{a}_{\overline{\mu}}(k).$$

Making use of factorization of the inverse operator $(\mathbf{P}_{\overline{\mu}})^{-1}$ and following the transformation steps of (3.33) we obtain the formula for the spectral characteristic $\vec{h}^{1}_{\overline{\mu}}(\lambda)$ of the optimal estimate $\widehat{B}\vec{\zeta}$ of the functionals $B\vec{\zeta}$:

$$\vec{h}^{1}_{\overline{\mu}}(\lambda) = \frac{\chi_{\overline{\mu}}^{(d)}(e^{-i\lambda})}{\beta^{(d)}(i\lambda)}(\vec{B}_{\overline{\mu}}(e^{i\lambda}) - \frac{|\beta^{(d)}(i\lambda)|^2}{|\chi_{\overline{\mu}}^{(d)}(e^{-i\lambda})|^2}((f(\lambda) + |\beta^{(d)}(i\lambda)|^2 g(\lambda))^{-1})^{\top}$$

$$\times \left(\sum_{k=0}^{\infty}((\mathbf{P}_{\overline{\mu}}^{-1}D^{\overline{\mu}}\mathbf{a})_k e^{ik\lambda})\right)$$

$$= \frac{\chi_{\overline{\mu}}^{(d)}(e^{-i\lambda})}{\beta^{(d)}(i\lambda)}\left(\vec{B}_{\overline{\mu}}(e^{i\lambda}) - \Psi_{\overline{\mu}}^{\top}(e^{-i\lambda}) \sum_{m=0}^{\infty}\sum_{p=0}^{\infty} \theta_{\overline{\mu}}^{\top}(p)\vec{b}_{\overline{\mu}}(b+m)e^{i\lambda m}\right)$$

$$= \frac{\chi_{\overline{\mu}}^{(d)}(e^{-i\lambda})}{\beta^{(d)}(i\lambda)}(\vec{B}_{\overline{\mu}}(e^{i\lambda}) - \Psi_{\overline{\mu}}^{\top}(e^{-i\lambda})\vec{r}_{\overline{\mu}}(e^{i\lambda}))$$

$$= \frac{\chi_{\overline{\mu}}^{(d)}(e^{-i\lambda})}{\beta^{(d)}(i\lambda)}\Psi_{\overline{\mu}}^{\top}(e^{-i\lambda})\vec{C}_{\overline{\mu},1}(e^{-i\lambda}),$$

where

$$\vec{r}_{\overline{\mu}}(e^{i\lambda}) = \sum_{m=0}^{\infty}(\theta_{\overline{\mu}}^{\top}D^{\mu}\mathbf{A})_m e^{i\lambda m} = \sum_{m=0}^{\infty}\sum_{p=0}^{\infty} \theta_{\overline{\mu}}^{\top}(p)\vec{b}_{\overline{\mu}}(b+m)e^{i\lambda m},$$

$$\vec{C}_{\overline{\mu},1}(e^{-i\lambda}) = \sum_{m=1}^{\infty}(\theta_{\overline{\mu}}^{\top}\widetilde{\mathbf{B}}_{\overline{\mu}})_m e^{-i\lambda m} = \sum_{m=1}^{\infty}\sum_{p=m}^{\infty} \theta_{\overline{\mu}}^{\top}(p)\vec{b}_{\overline{\mu}}(p-m)e^{-i\lambda m}$$

$$= \sum_{m=1}^{\infty}\sum_{p=0}^{\infty} \theta_{\overline{\mu}}^{\top}(m+p)\vec{b}_{\overline{\mu}}(p)e^{-i\lambda m},$$

where vector $\theta_{\overline{\mu}}^{\top} = ((\theta_{\overline{\mu}}(0))^{\top}, (\theta_{\overline{\mu}}(1))^{\top}, (\theta_{\overline{\mu}}(2))^{\top}, \ldots)$, $\mathbf{A}$ is a linear symmetric operator determined by the matrix with the vector entries $(\mathbf{A})_{k,j} = \vec{a}(k+j)$, $k,j \geq 0$, and $\widetilde{\mathbf{B}}_{\overline{\mu}}$ is a linear operator which is determined by a matrix with the vector entries $(\widetilde{\mathbf{B}}_{\overline{\mu}})_{k,j} = \vec{b}_{\overline{\mu}}(k-j)$ for $0 \le j \le k$ and $(\widetilde{\mathbf{B}}_{\overline{\mu}})_{k,j} = \vec{0}$ for $0 \le k < j$.

In this case the spectral characteristic $\vec{h}_{\overline{\mu}}(\lambda)$ of the estimate $\widehat{A}\xi$ can be calculated by the formula

$$\begin{aligned}\vec{h}_{\overline{\mu}}(\lambda) &= \frac{\chi_{\overline{\mu}}^{(d)}(e^{-i\lambda})}{\beta^{(d)}(i\lambda)}\left(\sum_{k=0}^{\infty}\psi_{\overline{\mu}}^{\top}(k)e^{-i\lambda k}\right)\sum_{m=1}^{\infty}(\theta_{\overline{\mu}}^{\top}\widetilde{\mathbf{B}}_{\overline{\mu}} - \overline{\psi}_{\overline{\mu}}\mathbf{C}_{\overline{\mu},g})_m e^{-i\lambda m}\\ &= \frac{\chi_{\overline{\mu}}^{(d)}(e^{-i\lambda})}{\beta^{(d)}(i\lambda)}\Psi_{\overline{\mu}}^{\top}(e^{-i\lambda})(\vec{C}_{\overline{\mu},1}(e^{-i\lambda}) - \vec{C}_{\overline{\mu},g}(e^{-i\lambda}))\\ &= \vec{B}_{\overline{\mu}}(e^{i\lambda})\frac{\chi_{\overline{\mu}}^{(d)}(e^{-i\lambda})}{\beta^{(d)}(i\lambda)} - \widetilde{\vec{h}}_{\overline{\mu}}(\lambda),\end{aligned} \tag{3.35}$$

where

$$\begin{aligned}\widetilde{\vec{h}}_{\overline{\mu}}(\lambda) &= \frac{\chi_{\overline{\mu}}^{(d)}(e^{-i\lambda})}{\beta^{(d)}(i\lambda)}\left(\sum_{k=0}^{\infty}\psi_{\overline{\mu}}^{\top}(k)e^{-i\lambda k}\right)\left(\sum_{m=0}^{\infty}(\theta_{\overline{\mu}}^{\top}D^{\mu}\mathbf{A})_m e^{i\lambda m} + \sum_{m=1}^{\infty}(\overline{\psi}_{\overline{\mu}}\mathbf{C}_{\overline{\mu},g})_m e^{-i\lambda m}\right)\\ &= \frac{\chi_{\overline{\mu}}^{(d)}(e^{-i\lambda})}{\beta^{(d)}(i\lambda)}\Psi_{\overline{\mu}}^{\top}(e^{-i\lambda})(\vec{r}_{\overline{\mu}}(e^{i\lambda}) + \vec{C}_{\overline{\mu},g}(e^{-i\lambda})).\end{aligned}$$

The MSE of the estimate $\widehat{A}\xi$ is calculated by the formula

$$\begin{aligned}\Delta(f,g;\widehat{A}\xi) &= \Delta(f,g;\widehat{H}\xi) = \mathsf{E}|H\xi - \widehat{H}\xi|^2\\ &= \frac{1}{2\pi}\int_{-\pi}^{\pi}(\vec{A}(e^{i\lambda}))^{\top}g(\lambda)\overline{\vec{A}(e^{i\lambda})}d\lambda\\ &+ \frac{1}{2\pi}\int_{-\pi}^{\pi}(\widetilde{\vec{h}}_{\overline{\mu}}(e^{i\lambda}))^{\top}(f(\lambda) + |\beta^{(d)}(i\lambda)|^2 g(\lambda))\overline{\widetilde{\vec{h}}_{\overline{\mu}}(e^{i\lambda})}d\lambda\\ &- \frac{1}{2\pi}\int_{-\pi}^{\pi}\frac{\beta^{(d)}(i\lambda)}{\chi_{\overline{\mu}}^{(d)}(e^{-i\lambda})}(\widetilde{\vec{h}}_{\overline{\mu}}(e^{i\lambda}))^{\top}g(\lambda)\overline{A_{\overline{\mu}}(e^{i\lambda})}d\lambda\\ &- \frac{1}{2\pi}\int_{-\pi}^{\pi}\overline{\frac{\beta^{(d)}(i\lambda)}{\chi_{\overline{\mu}}^{(d)}(e^{-i\lambda})}}(A_{\overline{\mu}}(e^{i\lambda}))^{\top}g(\lambda)\overline{\widetilde{\vec{h}}_{\overline{\mu}}(e^{i\lambda})}d\lambda\\ &= \|\Phi^{\top}\mathbf{a}_{\overline{\mu}}\|^2 + \|\widetilde{\Phi}\mathbf{a}_{\overline{\mu}}\|_1^2 + \langle\theta_{\overline{\mu}}^{\top}D^{\mu}\mathbf{A} - \overline{\psi}_{\overline{\mu}}\mathbf{C}_{\overline{\mu},g}, \theta_{\overline{\mu}}^{\top}D^{\mu}\mathbf{A}\rangle\\ &- \langle\theta_{\overline{\mu}}^{\top}D^{\mu}\mathbf{A}, \mathbf{Z}_{\overline{\mu}}\mathbf{a}_{\overline{\mu}}\rangle - \langle\mathbf{Z}_{\overline{\mu}}\mathbf{a}_{\overline{\mu}}, \overline{\psi}_{\overline{\mu}}\mathbf{C}_{\overline{\mu},g}\rangle_1,\end{aligned} \tag{3.36}$$

where $\|\mathbf{x}\|_1^2 = \langle\mathbf{x},\mathbf{x}\rangle_1$, $\langle\mathbf{x},\mathbf{y}\rangle_1 = \sum_{k=1}^{\infty}(\vec{x}(k))^{\top}\overline{\vec{y}(k)}$ for the vectors $\mathbf{x} = ((\vec{x}(0))^{\top}, (\vec{x}(1))^{\top}, (\vec{x}(2))^{\top}, \ldots)^{\top}$, $\mathbf{y} = ((\vec{y}(0))^{\top}, (\vec{y}(1))^{\top}, (\vec{y}(2))^{\top}, \ldots)^{\top}$.

The obtained results are summarized in the form of the following theorem.

Theorem 3.1.3. *Let $\vec{\xi}(m)$, $m \in \mathbb{Z}$, be a vector stochastic sequence which determines the stationary GM increment sequence $\chi_{\overline{\mu},\overline{s}}^{(d)}(\vec{\xi}(m))$ and let $\vec{\eta}(m)$, $m \in \mathbb{Z}$, be a vector stationary stochastic sequence, uncorrelated with $\vec{\xi}(m)$. Suppose that the coefficients $\vec{a}(k)$, $k \geq 0$, satisfy conditions* (3.2)–(3.3) *and the spectral densities $f(\lambda)$ and $g(\lambda)$ of the stochastic sequences $\vec{\xi}(m)$ and $\vec{\eta}(m)$ admit canonical factorizations* (3.29)–(3.31)*. Then the spectral characteristic $\vec{h}_{\overline{\mu}}(\lambda)$ and the MSE $\Delta(f,g;\widehat{A}\vec{\xi})$ of the optimal estimate $\widehat{A}\vec{\xi}$ of the functional $A\vec{\xi}$ based on observations of the sequence $\vec{\xi}(m)+\vec{\eta}(m)$ at points $m=-1,-2,\dots$ can be calculated by formulas* (3.35) *and* (3.36)*, respectively.*

3.1.4 Extrapolation of stochastic sequences with periodically stationary increment

Consider the problem of mean square optimal linear estimation of the functionals

$$A\vartheta = \sum_{k=0}^{\infty} a^{(\vartheta)}(k)\vartheta(k), \quad A_M\vartheta = \sum_{k=0}^{N} a^{(\vartheta)}(k)\vartheta(k), \tag{3.37}$$

which depend on unobserved values of the stochastic sequence $\vartheta(m)$ with periodically stationary increments. Estimates are based on observations of the sequence $\zeta(m) = \vartheta(m)+\eta(m)$ at points $m=-1,-2,\dots$.

The functional $A\vartheta$ can be represented in the form

$$\begin{aligned} A\vartheta &= \sum_{k=0}^{\infty} a^{(\vartheta)}(k)\vartheta(k) = \sum_{m=0}^{\infty}\sum_{p=1}^{T} a^{(\vartheta)}(mT+p-1)\vartheta(mT+p-1) \\ &= \sum_{m=0}^{\infty}\sum_{p=1}^{T} a_p(m)\xi_p(m) = \sum_{m=0}^{\infty} (\vec{a}(m))^{\top}\vec{\xi}(m) \\ &= A\vec{\xi}, \end{aligned}$$

where

$$\vec{\xi}(m) = (\xi_1(m),\xi_2(m),\dots,\xi_T(m))^{\top}, \quad \xi_p(m) = \vartheta(mT+p-1),\ p=1,2,\dots,T, \tag{3.38}$$

$$\vec{a}(m) = (a_1(m),a_2(m),\dots,a_T(m))^{\top}, \quad a_p(m) = a^{(\vartheta)}(mT+p-1),\ p=1,2,\dots,T. \tag{3.39}$$

Theorem 3.1.4. *Let a stochastic sequence $\vartheta(k)$ with periodically stationary GM increments generate by formula* (3.38) *a multidimensional stochastic sequence $\vec{\xi}(m)$ which determines a stationary stochastic GM increment sequence $\chi_{\overline{\mu},\overline{s}}^{(d)}(\vec{\xi}(m))$ with the spectral density matrix $f(\lambda)$. Let $\vec{\eta}(m)$, $m \in \mathbb{Z}$, $\vec{\eta}(m) = (\eta_1(m),\eta_2(m),\dots,\eta_T(m))^{\top}$, $\eta_p(m) = \eta(mT+p-1)$, $p=1,2,\dots,T$, be an uncorrelated with the sequence $\vec{\xi}(m)$ multidimensional stationary stochastic sequence with an absolutely continuous spectral function $G(\lambda)$ which has spectral density matrix $g(\lambda)$. Let minimality condition* (3.4) *be satisfied. Let coefficients $\vec{a}(k)$, $k \geqslant 0$, determined by formula* (3.39) *satisfy conditions* (3.2)–(3.3)*. Then the* *optimal linear estimate $\widehat{A}\vartheta$ of the functional $A\vartheta$ based on observations of the sequence $\zeta(m) = \vartheta(m) + \eta(m)$ at points $m = -1, -2, \dots$ is calculated by formula* (3.10). *The spectral characteristic $\vec{h}_{\overline{\mu}}(\lambda)$ and the MSE $\Delta(f, g; \widehat{A}\zeta)$ of the optimal estimate $\widehat{A}\zeta$ are calculated by formulas* (3.12) *and* (3.13), *respectively.*

The functional $A_M\vartheta$ can be represented in the form

$$\begin{aligned} A_M\vartheta &= \sum_{k=0}^{M} a^{(\vartheta)}(k)\vartheta(k) = \sum_{m=0}^{N}\sum_{p=1}^{T} a^{(\vartheta)}(mT+p-1)\vartheta(mT+p-1) \\ &= \sum_{m=0}^{N}\sum_{p=1}^{T} a_p(m)\xi_p(m) = \sum_{m=0}^{N} (\vec{a}(m))^{\top}\vec{\xi}(m) \\ &= A_N\vec{\xi}, \end{aligned}$$

where $N = [\frac{M}{T}]$, the sequence $\vec{\xi}(m)$ is determined by formula (3.38),

$$\begin{aligned} &\vec{a}(m) = (a_1(m), a_2(m), \dots, a_T(m))^{\top}, \\ &a_p(m) = a^{\vartheta}(mT+p-1), \quad 0 \le m \le N,\ 1 \le p \le T,\ mT+p-1 \le M, \\ &a_p(N) = 0, \quad M+1 \le NT+p-1 \le (N+1)T-1,\ 1 \le p \le T. \end{aligned} \tag{3.40}$$

Making use of the introduced notations and statements of Theorem 3.1.2 we can claim that the following corollary holds.

Corollary 3.1.5. *Suppose that the conditions of Theorem* 3.1.4 *are satisfied. Let coefficients $\vec{a}(k)$, $k = \overline{0, N}$, be determined by formula* (3.40). *The optimal linear estimate $\widehat{A}_M\zeta$ of the functional $A_M\zeta = A_N\vec{\xi}$ based on observations of the sequence $\zeta(m) = \vartheta(m) + \eta(m)$ at points $m = -1, -2, \dots$ is calculated by formula* (3.23). *The spectral characteristic $\vec{h}_{\overline{\mu},N}(\lambda)$ and the MSE $\Delta(f; \widehat{A}_M\zeta)$ are calculated by formulas* (3.24) *and* (3.25), *respectively.*

As a corollary from the proposed theorem, one can obtain the mean square optimal estimate of an unobserved value $\vartheta(M)$, $M \ge 0$, of a stochastic sequence $\vartheta(m)$ with periodically stationary GM increments based on observations of the sequence $\zeta(m) = \vartheta(m) + \eta(m)$ at points $m = -1, -2, \dots$. Making use of the notations $\vartheta(M) = \vartheta_p(N) = (\vec{\xi}(N))^{\top}\vec{\delta}_p$, $N = [\frac{M}{T}]$, $p = M + 1 - NT$, and the obtained results we can conclude that the following corollary holds.

Corollary 3.1.6. *Suppose that the conditions of Theorem* 3.1.4 *are satisfied. The optimal linear estimate $\widehat{\vartheta}(M)$ of the unobserved value $\vartheta(M)$, $M \ge 0$, of a stochastic sequence $\vartheta(m)$ with periodically stationary GM increments based on observations of the sequence $\zeta(m) = \vartheta(m) + \eta(m)$ at points $m = -1, -2, \dots$ is calculated by formula* (3.26)*. The spectral characteristic $\vec{h}_{\overline{\mu},N,p}(\lambda)$ of the estimate is calculated by formula* (3.27). *The MSE of the optimal estimate is calculated by formula* (3.1.3).

3.1.5 Extrapolation of cointegrated vector stochastic sequences

Consider two seasonal vector stochastic sequences $\{\vec{\xi}(m), m \in \mathbb{Z}\}$ and $\{\vec{\zeta}(m), m \in \mathbb{Z}\}$ with absolutely continuous spectral functions $F(\lambda)$ and $P(\lambda)$ and spectral densities $f(\lambda)$ and $p(\lambda)$, respectively. Assume that both of them have the same order d and seasonal vector $\overline{s}$.

Definition 3.1.1. Within this subsection, a pair of seasonal vector stochastic sequences $\{\vec{\xi}(m), m \in \mathbb{Z}\}$ and $\{\vec{\zeta}(m), m \in \mathbb{Z}\}$ are called seasonally cointegrated if there exists a constant $\alpha \neq 0$ such that the linear combination $\vec{\zeta}(m) - \alpha\vec{\xi}(m)$ is a stationary vector stochastic sequence.

Under the extrapolation of seasonally cointegrated stochastic sequences we understand the mean square optimal linear estimation of the functionals

$$A\vec{\xi} = \sum_{k=0}^{\infty} (\vec{a}(k))^{\top} \vec{\xi}(k), \quad A_N\vec{\xi} = \sum_{k=0}^{N} (\vec{a}(k))^{\top} \vec{\xi}(k),$$

which depend on the unobserved values of the stochastic sequence $\vec{\xi}(m)$ based on observations of the stochastic sequence $\vec{\zeta}(m)$ at points $m = -1, -2, \ldots$. Applying the results of previous sections, the estimates can be found under the assumption that the vector sequences $\vec{\xi}(m)$ and $\vec{\zeta}(m) - \alpha\vec{\xi}(m)$ are uncorrelated.

Let the minimality condition hold:

$$\int_{-\pi}^{\pi} \operatorname{Tr}\left[\frac{|\beta^{(d)}(i\lambda)|^2}{|\chi_{\overline{\mu}}^{(d)}(e^{-i\lambda})|^2} p^{-1}(\lambda) \right] d\lambda < \infty. \tag{3.41}$$

Determine operators $\mathbf{P}_{\overline{\mu}}^{\alpha}$, $\mathbf{T}_{\overline{\mu}}^{\alpha}$, $\mathbf{Q}^{\alpha}$ with the help of the Fourier coefficients

$$P_{k,j}^{\overline{\mu},\alpha} = \frac{1}{2\pi} \int_{-\pi}^{\pi} e^{-i\lambda(k-j)} \frac{|\beta^{(d)}(i\lambda)|^2}{|\chi_{\overline{\mu}}^{(d)}(e^{-i\lambda})|^2} ((p(\lambda))^{-1})^{\top} d\lambda,$$

$$T_{k,j}^{\overline{\mu},\beta} = \frac{1}{2\pi} \int_{-\pi}^{\pi} e^{-i\lambda(k-j)} \frac{1}{|\chi_{\overline{\mu}}^{(d)}(e^{-i\lambda})|^2} ((p(\lambda) - \alpha^2 f(\lambda))(p(\lambda))^{-1})^{\top} d\lambda,$$

$$Q_{k,j}^{\alpha} = \frac{1}{2\pi} \int_{-\pi}^{\pi} e^{-i\lambda(k-j)} \frac{1}{|\beta^{(d)}(i\lambda)|^2} (f(\lambda)(p(\lambda))^{-1}(p(\lambda) - \alpha^2 f(\lambda)))^{\top} d\lambda$$

of the functions

$$\frac{1}{|\chi_{\overline{\mu}}^{(d)}(e^{-i\lambda})|^2} ((p(\lambda) - \alpha^2 f(\lambda))(p(\lambda))^{-1})^{\top},$$

$$\frac{1}{|\chi_{\overline{\mu}}^{(d)}(e^{-i\lambda})|^2} (f(\lambda)(p(\lambda))^{-1}(p(\lambda) - \alpha^2 f(\lambda)))^{\top},$$

$$\frac{|\beta^{(d)}(i\lambda)|^2}{|\chi_{\overline{\mu}}^{(d)}(e^{-i\lambda})|^2}((p(\lambda))^{-1})^{\top} \tag{3.42}$$

in the same way as we defined operators $\mathbf{P}_{\overline{\mu}}, \mathbf{T}_{\overline{\mu}}, \mathbf{Q}$. Theorem 3.1.1 implies that the spectral characteristic $\vec{h}_{\overline{\mu}}^{\alpha}(\lambda)$ of the optimal estimate

$$\widehat{A}\vec{\xi} = \int_{-\pi}^{\pi} (\vec{h}_{\overline{\mu}}^{\alpha}(\lambda))^{\top} d\vec{Z}_{\zeta^{(d)}}(\lambda) - \sum_{k=-d(\gamma)}^{-1} (\vec{v}_{\overline{\mu}}(k))^{\top} \vec{\zeta}(k) \tag{3.43}$$

of the functional $A\vec{\xi}$ is calculated by the formula

$$\begin{aligned}(\vec{h}_{\overline{\mu}}^{\alpha}(\lambda))^{\top} &= (\vec{B}_{\overline{\mu}}(e^{i\lambda}))^{\top} \frac{\chi_{\overline{\mu}}^{(d)}(e^{-i\lambda})}{\beta^{(d)}(i\lambda)} \\ &\quad - \frac{1}{\chi_{\overline{\mu}}^{(d)}(e^{-i\lambda})\beta^{(d)}(i\lambda)} (\vec{A}_{\overline{\mu}}(e^{i\lambda}))^{\top} (p(\lambda) - \alpha^2 f(\lambda))(p(\lambda))^{-1} \\ &\quad - \frac{\overline{\beta^{(d)}(i\lambda)}}{\chi_{\overline{\mu}}^{(d)}(e^{-i\lambda})} (\vec{C}_{\alpha}^{\overline{\mu}}(e^{i\lambda}))^{\top} (p(\lambda))^{-1},\end{aligned} \tag{3.44}$$

where

$$\vec{C}_{\alpha}^{\overline{\mu}}(e^{i\lambda}) = \sum_{k=0}^{\infty} ((\mathbf{P}_{\overline{\mu}}^{\alpha})^{-1} D^{\overline{\mu}}\mathbf{a} - (\mathbf{P}_{\overline{\mu}}^{\alpha})^{-1} \mathbf{T}_{\overline{\mu}}^{\alpha} \mathbf{a}_{\overline{\mu}})_k e^{i\lambda k}.$$

The MSE of the estimate $\widehat{A}\vec{\xi}$ is calculated by the formula

$$\begin{aligned}\Delta(f,g;\widehat{A}\vec{\xi}) &= \frac{1}{2\pi} \int_{-\pi}^{\pi} \left(\frac{1}{|\beta^{(d)}(i\lambda)|^2} (\vec{A}_{\overline{\mu}}(e^{i\lambda}))^{\top} (p(\lambda) - \alpha^2 f(\lambda)) + (\vec{C}_{\alpha}^{\overline{\mu}}(e^{i\lambda}))^{\top} \right) \\ &\quad \times \frac{|\beta^{(d)}(i\lambda)|^2}{|\chi_{\overline{\mu}}^{(d)}(e^{-i\lambda})|^2} (p(\lambda))^{-1} f(\lambda) (p(\lambda))^{-1} \\ &\quad \times \left(\frac{1}{|\beta^{(d)}(i\lambda)|^2} (p(\lambda) - \alpha^2 f(\lambda)) \overline{\vec{A}(e^{i\lambda})} + \overline{\vec{C}_{\alpha}^{\overline{\mu}}(e^{i\lambda})} \right) d\lambda \\ &\quad + \frac{1}{2\pi} \int_{-\pi}^{\pi} ((\vec{A}_{\overline{\mu}}(e^{i\lambda}))^{\top} f(\lambda) - |\beta^{(d)}(i\lambda)|^2 (\vec{C}_{\alpha}^{\overline{\mu}}(e^{i\lambda}))^{\top}) \\ &\quad \times \frac{1}{|\chi_{\overline{\mu}}^{(d)}(e^{-i\lambda})|^2} \frac{1}{|\beta^{(d)}(i\lambda)|^2} (p(\lambda))^{-1} (p(\lambda) - \alpha^2 f(\lambda)) (p(\lambda))^{-1} \\ &\quad \times (f(\lambda) \overline{\vec{A}_{\overline{\mu}}(e^{i\lambda})} - |\beta^{(d)}(i\lambda)|^2 \overline{\vec{C}_{\alpha}^{\overline{\mu}}(e^{i\lambda})}) d\lambda \\ &= \langle D^{\overline{\mu}}\mathbf{a} - \mathbf{T}_{\overline{\mu}}^{\alpha} \mathbf{a}_{\overline{\mu}}, (\mathbf{P}_{\overline{\mu}}^{\alpha})^{-1} D^{\overline{\mu}}\mathbf{a} - (\mathbf{P}_{\overline{\mu}}^{\alpha})^{-1} \mathbf{T}_{\overline{\mu}}^{\alpha} \mathbf{a}_{\overline{\mu}} \rangle + \langle \mathbf{Q}^{\alpha}\mathbf{a}, \mathbf{a} \rangle.\end{aligned} \tag{3.45}$$

Theorem 3.1.5. *Let $\vec{\xi}(m)$, $m \in \mathbb{Z}$, and $\vec{\zeta}(m)$, $m \in \mathbb{Z}$, be seasonally cointegrated stochastic sequences with the spectral densities $f(\lambda)$ and $p(\lambda)$, respectively. Suppose that the spectral density $p(\lambda)$ satisfies the minimality condition* (3.41) *and the coefficients $\vec{a}(k)$, $k \geq 0$, satisfy conditions* (3.2)–(3.3). *If the stochastic sequences $\vec{\xi}(m)$ and $\vec{\zeta}(m) - \alpha\vec{\xi}(m)$ are uncorrelated, then the spectral characteristic $\vec{h}^{\alpha}_{\overline{\mu}}(\lambda)$ and the MSE $\Delta(f,g;\widehat{A}\xi)$ of the optimal estimate $\widehat{A}\vec{\xi}$* (3.43) *of the functional $A\vec{\xi}$ based on observations of the sequence $\vec{\zeta}(m)$ at points $m = -1, -2, \ldots$ are calculated by formulas* (3.44) *and* (3.45), *respectively.*

Let operators $\mathbf{P}^{\alpha}_{\overline{\mu}}$, $\mathbf{T}^{\alpha}_{\overline{\mu},N}$, $\mathbf{Q}^{\alpha}_{N}$ be defined by the Fourier coefficients of the functions (3.42) in the same way as we defined operators $\mathbf{P}_{\overline{\mu}}$, $\mathbf{T}_{\overline{\mu},N}$, $\mathbf{Q}_{N}$. Theorem 3.1.2 implies the following theorem.

Theorem 3.1.6. *Let $\vec{\xi}(m)$, $m \in \mathbb{Z}$, and $\vec{\zeta}(m)$, $m \in \mathbb{Z}$, be seasonally cointegrated stochastic sequences with the spectral densities $f(\lambda)$ and $p(\lambda)$, respectively. Suppose that the spectral density $p(\lambda)$ satisfies the minimality condition* (3.41). *If the stochastic sequences $\vec{\xi}(m)$ and $\vec{\zeta}(m) - \vec{\alpha}\xi(m)$ are uncorrelated, then the spectral characteristic $\vec{h}^{\alpha}_{\overline{\mu},N}(\lambda)$ of the optimal estimate*

$$\widehat{A}_N\vec{\xi} = \int_{-\pi}^{\pi} \vec{h}^{\alpha}_{\overline{\mu},N}(\lambda) d\vec{Z}_{\zeta^{(d)}}(\lambda) - \sum_{k=-d(\gamma)}^{-1} (\vec{v}_{\overline{\mu},N}(k))^{\top}\vec{\zeta}(k)$$

of the functional $A_N\vec{\xi}$ based on observations of the sequence $\vec{\zeta}(m)$ at points $m = -1, -2, \ldots$ is calculated by the formula

$$\begin{aligned}(\vec{h}^{\alpha}_{\overline{\mu},N}(\lambda))^{\top} = {} & (\vec{B}_{\overline{\mu},N}(e^{i\lambda}))^{\top}\frac{\chi^{(d)}_{\overline{\mu}}(e^{-i\lambda})}{\beta^{(d)}(i\lambda)} \\ & - \frac{1}{\chi^{(d)}_{\overline{\mu}}(e^{-i\lambda})\beta^{(d)}(i\lambda)}(\vec{A}_{\overline{\mu},N}(e^{i\lambda}))^{\top}(p(\lambda) - \alpha^2 f(\lambda))(p(\lambda))^{-1} \\ & - \frac{\overline{\beta^{(d)}(i\lambda)}}{\chi^{(d)}_{\overline{\mu}}(e^{-i\lambda})}(\vec{C}^{\overline{\mu},N}_{\alpha}(e^{i\lambda}))^{\top}(p(\lambda))^{-1},\end{aligned}$$

where

$$\vec{C}^{\overline{\mu},N}_{\alpha}(e^{i\lambda}) = \sum_{k=0}^{\infty}((\mathbf{P}^{\alpha}_{\overline{\mu}})^{-1}D^{\overline{\mu}}_{N}\mathbf{a}_N - (\mathbf{P}^{\alpha}_{\overline{\mu}})^{-1}\mathbf{T}^{\alpha}_{\overline{\mu},N}\mathbf{a}_{\overline{\mu},N})_k e^{i\lambda k}.$$

The MSE of the estimate $\widehat{A}_N\xi$ is calculated by the formula

$$\Delta(f,g;\widehat{A}_N\vec{\xi}) = \frac{1}{2\pi}\int_{-\pi}^{\pi}\left(\frac{1}{|\beta^{(d)}(i\lambda)|^2}(\vec{A}_{\overline{\mu},N}(e^{i\lambda}))^{\top}(p(\lambda) - \alpha^2 f(\lambda)) + (\vec{C}^{\overline{\mu},N}_{\alpha}(e^{i\lambda}))^{\top}\right)$$

$$\times \frac{|\beta^{(d)}(i\lambda)|^2}{|\chi_{\overline{\mu}}^{(d)}(e^{-i\lambda})|^2}(p(\lambda))^{-1}f(\lambda)(p(\lambda))^{-1}$$
$$\times\left(\frac{1}{|\beta^{(d)}(i\lambda)|^2}(p(\lambda)-a^2f(\lambda))\overline{\vec{A}_{\overline{\mu},N}(e^{i\lambda})}+\overline{\vec{C}_a^{\overline{\mu},N}(e^{i\lambda})}\right)d\lambda$$
$$+\frac{1}{2\pi}\int_{-\pi}^{\pi}\left(\frac{1}{|\beta^{(d)}(i\lambda)|^2}(\vec{A}_{\overline{\mu},N}(e^{i\lambda}))^{\top}f(\lambda)-(\vec{C}_a^{\overline{\mu},N}(e^{i\lambda}))^{\top}\right)$$
$$\times \frac{|\beta^{(d)}(i\lambda)|^2}{|\chi_{\overline{\mu}}^{(d)}(e^{-i\lambda})|^2}(p(\lambda))^{-1}(p(\lambda)-a^2f(\lambda))(p(\lambda))^{-1}$$
$$\times\left(\frac{1}{|\beta^{(d)}(i\lambda)|^2}f(\lambda)\overline{\vec{A}_{\overline{\mu},N}(e^{i\lambda})}-\overline{\vec{C}_a^{\overline{\mu},N}(e^{i\lambda})}\right)d\lambda$$
$$=\langle D_N^{\overline{\mu}}\mathbf{a}_N-\mathbf{T}_{\overline{\mu},N}^{a}\mathbf{a}_{\overline{\mu},N},(\mathbf{P}_{\overline{\mu}}^{a})^{-1}D_N^{\overline{\mu}}\mathbf{a}_N-(\mathbf{P}_{\overline{\mu}}^{a})^{-1}\mathbf{T}_{\overline{\mu},N}^{a}\mathbf{a}_{\overline{\mu},N}\rangle+\langle\mathbf{Q}_N^{a}\mathbf{a}_N,\mathbf{a}_N\rangle.$$

Let the spectral densities $f(\lambda)$ and $p(\lambda)$ admit the factorizations

$$\frac{|\chi_{\overline{\mu}}^{(d)}(e^{-i\lambda})|^2}{|\beta^{(d)}(i\lambda)|^2}p(\lambda)=\Theta_{\overline{\mu}}^{a}(e^{-i\lambda})(\Theta_{\overline{\mu}}^{a}(e^{-i\lambda}))^{*},\quad \Theta_{\overline{\mu}}^{a}(e^{-i\lambda})=\sum_{k=0}^{\infty}\theta_{\overline{\mu}}^{a}(k)e^{-i\lambda k},\tag{3.46}$$
$$\frac{|\beta^{(d)}(i\lambda)|^2}{|\chi_{\overline{\mu}}^{(d)}(e^{-i\lambda})|^2}(p(\lambda))^{-1}=(\Psi_{\overline{\mu}}^{a}(e^{-i\lambda}))^{*}\Psi_{\overline{\mu}}^{a}(e^{-i\lambda}),\quad \Psi_{\overline{\mu}}^{a}(e^{-i\lambda})=\sum_{k=0}^{\infty}\psi_{\overline{\mu}}^{a}(k)e^{-i\lambda k},\tag{3.47}$$
$$|\beta^{(d)}(i\lambda)|^{-2}(p(\lambda)-a^2f(\lambda))=\Phi^{a}(e^{-i\lambda})(\Phi^{a}(e^{-i\lambda}))^{*},\quad \Phi^{a}(e^{-i\lambda})=\sum_{k=0}^{\infty}\phi^{a}(k)e^{-i\lambda k}.\tag{3.48}$$

Let operators and vectors $\mathbf{C}_{\overline{\mu},g}^{a}$, $\widetilde{\Phi}^{a}$, Φ^{a}, $\mathbf{Z}_{\overline{\mu}}^{a}$, $\theta_{\overline{\mu}}^{a}$, $\overline{\psi}_{\overline{\mu}}^{a}$ be defined by coefficients of the canonical factorizations (3.46)–(3.48) in the same way as operators and vectors $\mathbf{C}_{\overline{\mu},g}$, $\widetilde{\Phi}$, Φ, $\mathbf{Z}_{\overline{\mu}}$, $\theta_{\overline{\mu}}$, $\overline{\psi}_{\overline{\mu}}$ were defined. From Theorem 3.1.3 we obtain that the spectral characteristic $\vec{h}_{\overline{\mu}}^{a}(\lambda)$ of the optimal estimate $\widehat{A}\vec{\xi}$ of the functional $A\vec{\xi}$ can be calculated by the formula

$$\vec{h}_{\overline{\mu}}^{a}(\lambda)=\frac{\chi_{\overline{\mu}}^{(d)}(e^{-i\lambda})}{\beta^{(d)}(i\lambda)}\left(\sum_{k=0}^{\infty}(\psi_{\overline{\mu}}^{a}(k))^{\top}e^{-i\lambda k}\right)\sum_{m=1}^{\infty}((\theta_{\overline{\mu}}^{a})^{\top}\widetilde{\mathbf{B}}_{\overline{\mu}}-\overline{\psi}_{\overline{\mu}}^{a}\mathbf{C}_{\overline{\mu},g}^{a})_m e^{-i\lambda m}.\tag{3.49}$$

The MSE of the estimate $\widehat{A}\vec{\xi}$ is calculated by the formula

$$\begin{aligned}\Delta(f,g;\widehat{A}\vec{\xi})&=\|(\Phi^{a})^{\top}\mathbf{a}_{\overline{\mu}}\|^2+\|\widetilde{\Phi}^{a}\mathbf{a}_{\overline{\mu}}\|_1^2\\&\quad+\langle(\theta_{\overline{\mu}}^{a})^{\top}D^{\mu}\mathbf{A}-\overline{\psi}_{\overline{\mu}}^{a}\mathbf{C}_{\overline{\mu},g}^{a},(\theta_{\overline{\mu}}^{a})^{\top}D^{\mu}\mathbf{A}\rangle\\&\quad-\langle(\theta_{\overline{\mu}}^{a})^{\top}D^{\mu}\mathbf{A},\mathbf{Z}_{\overline{\mu}}^{a}\mathbf{a}_{\overline{\mu}}\rangle-\langle\mathbf{Z}_{\overline{\mu}}^{a}\mathbf{a}_{\overline{\mu}},\overline{\psi}_{\overline{\mu}}^{a}\mathbf{C}_{\overline{\mu},g}^{a}\rangle_1.\end{aligned}\tag{3.50}$$

Theorem 3.1.7. *Let $\vec{\xi}(m)$, $m\in\mathbb{Z}$, and $\vec{\zeta}(m)$, $m\in\mathbb{Z}$, be seasonally cointegrated stochastic vector sequences with the spectral densities $f(\lambda)$ and $p(\lambda)$ which admit the canoni-*

cal factorizations (3.46)–(3.48). *Suppose that the coefficients* $\vec{a}(k)$, $k \geq 0$, *satisfy conditions* (3.2)–(3.3). *Then the spectral characteristic* $\vec{h}^{\beta}_{\overline{\mu}}(\lambda)$ *and the MSE* $\Delta(f,g;\widehat{A}\vec{\xi})$ *of the optimal estimate* $\widehat{A}\vec{\xi}$ *of the functional* $A\vec{\xi}$ *based on observations of the sequence* $\vec{\zeta}(m)$ *at points* $m = -1, -2, \ldots$ *are calculated by formulas* (3.49) *and* (3.50), *respectively.*

3.2 Minimax (robust) method of extrapolation

3.2.1 Least favorable spectral densities and minimax-robust spectral characteristic

Values of the MSEs and the spectral characteristics of the optimal estimates of the functionals $A\vec{\xi}$ and $A_N\vec{\xi}$ depending on the unobserved values of a stochastic sequence $\vec{\xi}(m)$ which determine a stationary stochastic GM increment sequence $\chi^{(d)}_{\overline{\mu},\overline{s}}(\vec{\xi}(m))$ with the spectral density matrix $f(\lambda)$ based on observations of the sequence $\vec{\xi}(m)+\vec{\eta}(m)$ at points $m = -1, -2, \ldots$ can be calculated by formulas (3.12)–(3.13) and (3.24)–(3.25), respectively, under the condition that spectral densities $f(\lambda)$ and $g(\lambda)$ of stochastic sequences $\vec{\xi}(m)$ and $\vec{\eta}(m)$ are exactly known.

In practical cases, however, spectral densities of sequences usually are not exactly known. If in such cases a set $\mathcal{D} = \mathcal{D}_f \times \mathcal{D}_g$ of admissible spectral densities is defined, the minimax (robust) approach to estimation of linear functionals depending on unobserved values of stochastic sequences with stationary GM increments may be applied. This method consists in finding an estimate that minimizes the maximal values of the MSEs for all spectral densities from a given class $\mathcal{D} = \mathcal{D}_f \times \mathcal{D}_g$ of admissible spectral densities simultaneously.

To formalize this approach we present the following definitions.

Definition 3.2.1. For a given class of spectral densities $\mathcal{D} = \mathcal{D}_f \times \mathcal{D}_g$ the spectral densities $f^0(\lambda) \in \mathcal{D}_f$ and $g^0(\lambda) \in \mathcal{D}_g$ are called least favorable in the class $\mathcal{D}$ for the optimal linear forecasting of the functional $A\vec{\xi}$ if the following relation holds:

$$\Delta(f^0,g^0) = \Delta(h(f^0,g^0);f^0,g^0) = \max_{(f,g)\in\mathcal{D}_f\times\mathcal{D}_g} \Delta(h(f,g);f,g).$$

Definition 3.2.2. For a given class of spectral densities $\mathcal{D} = \mathcal{D}_f \times \mathcal{D}_g$ the spectral characteristic $h^0(\lambda)$ of the optimal linear estimate of the functional $A\vec{\xi}$ is called minimax-robust if the following conditions are satisfied:

$$h^0(\lambda) \in H_{\mathcal{D}} = \bigcap_{(f,g)\in\mathcal{D}_f\times\mathcal{D}_g} L_2^{-1}(f(\lambda) + |\beta^{(d)}(i\lambda)|^2 g(\lambda)),$$

$$\min_{h\in H_{\mathcal{D}}} \max_{(f,g)\in\mathcal{D}_f\times\mathcal{D}_g} \Delta(h;f,g) = \max_{(f,g)\in\mathcal{D}_f\times\mathcal{D}_g} \Delta(h^0;f,g).$$

Taking into account the introduced definitions and the derived relations we can verify that the following lemmas hold.

Lemma 3.2.1. *Spectral densities $f^0 \in \mathcal{D}_f$ and $g^0 \in \mathcal{D}_g$ which satisfy condition (3.4) are least favorable in the class $\mathcal{D} = \mathcal{D}_f \times \mathcal{D}_g$ for the optimal linear forecasting of the functional $A\vec{\xi}$ if operators $\mathbf{P}^0_{\overline{\mu}}$, $\mathbf{T}^0_{\overline{\mu}}$, $\mathbf{Q}^0$ determined by the Fourier coefficients of the functions*

$$\frac{|\beta^{(d)}(i\lambda)|^2}{|\chi^{(d)}_{\overline{\mu}}(e^{-i\lambda})|^2}(g^0(\lambda)(f^0(\lambda)+|\beta^{(d)}(i\lambda)|^2 g^0(\lambda))^{-1})^\top,$$

$$\frac{|\beta^{(d)}(i\lambda)|^2}{|\chi^{(d)}_{\overline{\mu}}(e^{-i\lambda})|^2}((f^0(\lambda)+|\beta^{(d)}(i\lambda)|^2 g^0(\lambda))^{-1})^\top,$$

$$(f^0(\lambda)(f^0(\lambda)+|\beta^{(d)}(i\lambda)|^2 g^0(\lambda))^{-1} g^0(\lambda))^\top$$

determine a solution of the constrained optimization problem

$$\begin{aligned}&\max_{(f,g)\in\mathcal{D}_f\times\mathcal{D}_g}(\langle D^{\overline{\mu}}\mathbf{a}-\mathbf{T}_{\overline{\mu}}\mathbf{a}_{\overline{\mu}}, \mathbf{P}^{-1}_{\overline{\mu}}D^{\overline{\mu}}\mathbf{a}-\mathbf{P}^{-1}_{\overline{\mu}}\mathbf{T}_{\overline{\mu}}\mathbf{a}_{\overline{\mu}}\rangle+\langle\mathbf{Q}\mathbf{a},\mathbf{a}\rangle)\\ &=\langle D^{\overline{\mu}}\mathbf{a}-\mathbf{T}^0_{\overline{\mu}}\mathbf{a}_{\overline{\mu}}, (\mathbf{P}^0_{\overline{\mu}})^{-1}D^{\overline{\mu}}\mathbf{a}-(\mathbf{P}^0_{\overline{\mu}})^{-1}\mathbf{T}^0_{\overline{\mu}}\mathbf{a}_{\overline{\mu}}\rangle+\langle\mathbf{Q}^0\mathbf{a},\mathbf{a}\rangle.\end{aligned} \tag{3.51}$$

The minimax spectral characteristic $h^0 = \vec{h}_{\overline{\mu}}(f^0, g^0)$ is calculated by formula (3.12) *if $\vec{h}_{\overline{\mu}}(f^0, g^0) \in H_{\mathcal{D}}$.*

Lemma 3.2.2. *The spectral densities $f^0 \in \mathcal{D}_f$ and $g^0 \in \mathcal{D}_g$ which admit canonical factorizations* (1.14), (3.29), *and* (3.30) *are least favorable densities in the class $\mathcal{D}$ for the optimal linear forecasting of the functional $A\vec{\xi}$ based on observations of the sequence $\vec{\xi}(m) + \vec{\eta}(m)$ at points $m = -1, -2, \dots$ if the matrix coefficients of canonical factorizations* (3.29) *and* (3.30) *determine a solution to the constrained optimization problem*

$$\begin{aligned}&\|\Phi^\top\mathbf{a}_{\overline{\mu}}\|^2+\|\widetilde{\Phi}\mathbf{a}_{\overline{\mu}}\|_1^2+\langle\theta^\top_{\overline{\mu}}D^{\mu}\mathbf{A}-\overline{\psi}_{\overline{\mu}}\mathbf{C}_{\mu,g},\theta^\top_{\overline{\mu}}D^{\mu}\mathbf{A}\rangle\\ &-\langle\theta^\top_{\overline{\mu}}D^{\mu}\mathbf{A},\mathbf{Z}_{\overline{\mu}}\mathbf{a}_{\overline{\mu}}\rangle-\langle\mathbf{Z}_{\overline{\mu}}\mathbf{a}_{\overline{\mu}},\overline{\psi}_{\overline{\mu}}\mathbf{C}_{\mu,g}\rangle_1\to\sup,\end{aligned} \tag{3.52}$$

$$f(\lambda)=\frac{|\beta^{(d)}(i\lambda)|^2}{|\chi^{(d)}_{\overline{\mu}}(e^{-i\lambda})|^2}\Theta_{\overline{\mu}}(e^{-i\lambda})\Theta^*_{\overline{\mu}}(e^{-i\lambda})-|\beta^{(d)}(i\lambda)|^2\Phi(e^{-i\lambda})\Phi^*(e^{-i\lambda})\in\mathcal{D}_f,$$

$$g(\lambda)=\Phi(e^{-i\lambda})\Phi^*(e^{-i\lambda})\in\mathcal{D}_g.$$

The minimax spectral characteristic $h^0 = \vec{h}_{\overline{\mu}}(f^0, g^0)$ is calculated by formula (3.35) *if $\vec{h}_{\overline{\mu}}(f^0, g^0) \in H_{\mathcal{D}}$.*

Lemma 3.2.3. *The spectral density $g^0 \in \mathcal{D}_g$ which admits canonical factorizations* (3.29) *and* (3.30) *with the known spectral density $f(\lambda)$ is the least favorable in the class $\mathcal{D}_g$ for the optimal linear forecasting of the functional $A\xi$ based on observations of the sequence $\vec{\xi}(m) + \vec{\eta}(m)$ at points $m = -1, -2, \dots$ if the matrix coefficients of the canonical factorizations*

$$f(\lambda) + |\beta^{(d)}(i\lambda)|^2 g^0(\lambda) = \frac{|\beta^{(d)}(i\lambda)|^2}{|\chi_{\overline{\mu}}^{(d)}(e^{-i\lambda})|^2} \left(\sum_{k=0}^{\infty} \theta_{\overline{\mu}}^0(k) e^{-i\lambda k} \right) \left(\sum_{k=0}^{\infty} \theta_{\overline{\mu}}^0(k) e^{-i\lambda k} \right)^* \tag{3.53}$$

and

$$g^0(\lambda) = \left(\sum_{k=0}^{\infty} \phi^0(k) e^{-i\lambda k} \right) \left(\sum_{k=0}^{\infty} \phi^0(k) e^{-i\lambda k} \right)^* \tag{3.54}$$

determine a solution to the constrained optimization problem

$$\begin{aligned} &\|\Phi^\top \mathbf{a}_{\overline{\mu}}\|^2 + \|\widetilde{\Phi}\mathbf{a}_{\overline{\mu}}\|_1^2 + \langle \theta_{\overline{\mu}}^\top D^\mu \mathbf{A} - \overline{\psi}_{\overline{\mu}} \mathbf{C}_{\mu,g}, \theta_{\overline{\mu}}^\top D^\mu \mathbf{A} \rangle \\ &- \langle \theta_{\overline{\mu}}^\top D^\mu \mathbf{A}, \mathbf{Z}_{\overline{\mu}} \mathbf{a}_{\overline{\mu}} \rangle - \langle \mathbf{Z}_{\overline{\mu}} \mathbf{a}_{\overline{\mu}}, \overline{\psi}_{\overline{\mu}} \mathbf{C}_{\mu,g} \rangle_1 \to \sup, \\ &g(\lambda) = \Phi(e^{-i\lambda}) \Phi^*(e^{-i\lambda}) \in \mathcal{D}_g. \end{aligned} \tag{3.55}$$

The minimax spectral characteristic $h^0 = \vec{h}_{\overline{\mu}}(f, g^0)$ *is calculated by formula* (3.35) *if* $\vec{h}_{\overline{\mu}}(f, g^0) \in H_{\mathcal{D}}$.

Lemma 3.2.4. *The spectral density* $f^0 \in \mathcal{D}_f$ *which admits canonical factorizations* (1.14) *and* (3.29) *with the known spectral density* $g(\lambda)$ *is the least favorable spectral density in the class* $\mathcal{D}_f$ *for the optimal linear forecasting of the functional* $A\vec{\xi}$ *based on observations of the sequence* $\vec{\xi}(m) + \vec{\eta}(m)$ *at points* $m = -1, -2, \dots$ *if matrix coefficients of the canonical factorization*

$$f^0(\lambda) + |\beta^{(d)}(i\lambda)|^2 g(\lambda) = \frac{|\beta^{(d)}(i\lambda)|^2}{|\chi_{\overline{\mu}}^{(d)}(e^{-i\lambda})|^2} \left(\sum_{k=0}^{\infty} \theta_{\overline{\mu}}^0(k) e^{-i\lambda k} \right) \left(\sum_{k=0}^{\infty} \theta_{\overline{\mu}}^0(k) e^{-i\lambda k} \right)^* \tag{3.56}$$

and the equation $\Psi_{\overline{\mu}}^0(e^{-i\lambda}) \Theta_{\overline{\mu}}^0(e^{-i\lambda}) = E_q$ *determine a solution to the constrained optimization problem*

$$\begin{aligned} &\langle \theta_{\overline{\mu}}^\top D^\mu \mathbf{A} - \overline{\psi}_{\overline{\mu}} \mathbf{C}_{\mu,g}, \theta_{\overline{\mu}}^\top D^\mu \mathbf{A} \rangle - \langle \theta_{\overline{\mu}}^\top D^\mu \mathbf{A}, \mathbf{Z}_{\overline{\mu}} \mathbf{a}_{\overline{\mu}} \rangle - \langle \mathbf{Z}_{\overline{\mu}} \mathbf{a}_{\overline{\mu}}, \overline{\psi}_{\overline{\mu}} \mathbf{C}_{\mu,g} \rangle_1 \to \sup, \\ &f(\lambda) = \frac{|\beta^{(d)}(i\lambda)|^2}{|\chi_{\overline{\mu}}^{(d)}(e^{-i\lambda})|^2} \Theta_{\overline{\mu}}(e^{-i\lambda}) \Theta_{\overline{\mu}}^*(e^{-i\lambda}) - |\beta^{(d)}(i\lambda)|^2 \Phi(e^{-i\lambda}) \Phi^*(e^{-i\lambda}) \in \mathcal{D}_f \end{aligned} \tag{3.57}$$

for the fixed matrix coefficients $\{\phi(k) : k \geq 0\}$. *The minimax spectral characteristic* $h^0 = \vec{h}_{\overline{\mu}}(f^0, g)$ *is calculated by formula* (3.35) *if* $\vec{h}_{\overline{\mu}}(f^0, g) \in H_{\mathcal{D}}$.

For more detailed analysis of properties of the least favorable spectral densities and minimax-robust spectral characteristics we observe that the minimax spectral characteristic h^0 and the least favorable spectral densities (f^0, g^0) form a saddle point of the function $\Delta(h; f, g)$ on the set $H_{\mathcal{D}} \times \mathcal{D}$.

The saddle point inequalities

$$\Delta(h;f^0,g^0) \geq \Delta(h^0;f^0,g^0) \geq \Delta(h^0;f,g) \quad \forall f \in \mathcal{D}_f,\ \forall g \in \mathcal{D}_g,\ \forall h \in H_{\mathcal{D}}$$

hold if $h^0 = \vec{h}_{\overline{\mu}}(f^0,g^0)$ and $\vec{h}_{\overline{\mu}}(f^0,g^0) \in H_{\mathcal{D}}$, where (f^0,g^0) is a solution of the constrained optimization problem

$$\widetilde{\Delta}(f,g) = -\Delta(\vec{h}_{\overline{\mu}}(f^0,g^0);f,g) \to \inf, \quad (f,g) \in \mathcal{D}, \tag{3.58}$$

where the functional $\Delta(\vec{h}_{\overline{\mu}}(f^0,g^0);f,g)$ is calculated by the formula

$$\begin{aligned}
&\Delta(\vec{h}_{\overline{\mu}}(f^0,g^0);f,g)\\
&= \frac{1}{2\pi}\int_{-\pi}^{\pi} \frac{|\beta^{(d)}(i\lambda)|^2}{|\chi_{\overline{\mu}}^{(d)}(e^{-i\lambda})|^2}(\mathbf{C}_{\overline{\mu}}^{f0}(e^{i\lambda}))^{\top}(f^0(\lambda)+|\beta^{(d)}(i\lambda)|^2 g^0(\lambda))^{-1}\\
&\quad\times f(\lambda)\,(f^0(\lambda)+|\beta^{(d)}(i\lambda)|^2 g^0(\lambda))^{-1}\overline{\mathbf{C}_{\overline{\mu}}^{f0}(e^{i\lambda})}d\lambda\\
&\quad+\frac{1}{2\pi}\int_{-\pi}^{\pi}\left(\frac{|\beta^{(d)}(i\lambda)|^2}{|\chi_{\overline{\mu}}^{(d)}(e^{-i\lambda})|^2}\right)^2(\mathbf{C}_{\overline{\mu}}^{g0}(e^{i\lambda}))^{\top}(f^0(\lambda)+|\beta^{(d)}(i\lambda)|^2 g^0(\lambda))^{-1}\\
&\quad\times g(\lambda)\,(f^0(\lambda)+|\beta^{(d)}(i\lambda)|^2 g^0(\lambda))^{-1}\overline{\mathbf{C}_{\overline{\mu}}^{g0}(e^{i\lambda})}d\lambda,
\end{aligned}$$

where

$$\begin{aligned}
\mathbf{C}_{\overline{\mu}}^{f0}(e^{i\lambda}) &= \overline{g^0(\lambda)\vec{A}_{\overline{\mu}}(e^{i\lambda})} + \sum_{k=0}^{\infty}((\mathbf{P}_{\overline{\mu}}^0)^{-1}D^{\overline{\mu}}\mathbf{a} - (\mathbf{P}_{\overline{\mu}}^0)^{-1}\mathbf{T}_{\overline{\mu}}^0\mathbf{a}_{\overline{\mu}})_k e^{i\lambda k},\\
\mathbf{C}_{\overline{\mu}}^{g0}(e^{i\lambda}) &= \frac{|\chi_{\overline{\mu}}^{(d)}(e^{-i\lambda})|^2}{|\beta^{(d)}(i\lambda)|^2}\overline{f^0(\lambda)\vec{A}(e^{i\lambda})}\\
&\quad-\chi_{\overline{\mu}}^{(d)}(e^{-i\lambda})\sum_{k=0}^{\infty}((\mathbf{P}_{\overline{\mu}}^0)^{-1}D^{\overline{\mu}}\mathbf{a} - (\mathbf{P}_{\overline{\mu}}^0)^{-1}\mathbf{T}_{\overline{\mu}}^0\mathbf{a}_{\overline{\mu}})_k e^{i\lambda k},
\end{aligned}$$

or it is calculated by the formula

$$\begin{aligned}
&\Delta(\vec{h}_{\overline{\mu}}(f^0,g^0);f,g)\\
&= \frac{1}{2\pi}\int_{-\pi}^{\pi}\frac{|\chi_{\overline{\mu}}^{(d)}(e^{-i\lambda})|^2}{|\beta^{(d)}(i\lambda)|^2}(\mathbf{r}_{\overline{\mu},f}^0(e^{-i\lambda}))^{\top}\Psi_{\overline{\mu}}^0(e^{-i\lambda}))f(\lambda)(\Psi_{\overline{\mu}}^0(e^{-i\lambda}))^{*}\overline{\mathbf{r}_{\overline{\mu},f}^0(e^{-i\lambda})}d\lambda\\
&\quad+\frac{1}{2\pi}\int_{-\pi}^{\pi}(\mathbf{r}_{\overline{\mu},g}^0(e^{-i\lambda}))^{\top}\Psi_{\overline{\mu}}^0(e^{-i\lambda})g(\lambda)(\Psi_{\overline{\mu}}^0(e^{-i\lambda}))^{*}\overline{\mathbf{r}_{\overline{\mu},g}^0(e^{-i\lambda})}d\lambda,
\end{aligned}$$

where

$$\mathbf{r}^0_{\overline{\mu},f}(e^{-i\lambda}) = \sum_{m=0}^{\infty} ((\theta^0)^{\top}_{\overline{\mu}} D^{\mu}\mathbf{A})_m e^{i\lambda m} + \sum_{m=1}^{\infty} (\overline{\psi}^0_{\overline{\mu}} \mathbf{C}^0_{\mu,g})_m e^{-i\lambda m},$$

$$\mathbf{r}^0_{\overline{\mu},g}(e^{-i\lambda}) = \chi^{(d)}_{\overline{\mu}}(e^{-i\lambda})\left(\sum_{m=0}^{\infty} ((\theta^0)^{\top}_{\overline{\mu}} D^{\mu}\mathbf{A})_m e^{i\lambda m} + \sum_{m=1}^{\infty} (\overline{\psi}^0_{\overline{\mu}} \mathbf{C}^0_{\mu,g})_m e^{-i\lambda m}\right)$$
$$- (\Theta^0_{\overline{\mu}}(e^{-i\lambda}))^{\top} A(e^{i\lambda}).$$

The constrained optimization problem (3.58) is equivalent to the unconstrained optimization problem

$$\Delta_{\mathcal{D}}(f,g) = \widetilde{\Delta}(f,g) + \delta(f,g|\mathcal{D}_f \times \mathcal{D}_g) \to \inf, \tag{3.59}$$

where $\delta(f,g|\mathcal{D}_f \times \mathcal{D}_g)$ is the indicator function of the set $\mathcal{D} = \mathcal{D}_f \times \mathcal{D}_g$. Solution (f^0,g^0) to this unconstrained optimization problem is characterized by the condition $0 \in \partial\Delta_{\mathcal{D}}(f^0,g^0)$, where $\partial\Delta_{\mathcal{D}}(f^0,g^0)$ is the subdifferential of the functional $\Delta_{\mathcal{D}}(f,g)$ at point $(f^0,g^0) \in \mathcal{D} = \mathcal{D}_f \times \mathcal{D}_g$, that is, the set of all continuous linear functionals Λ on $L_1 \times L_1$ which satisfy the inequality $\Delta_{\mathcal{D}}(f,g) - \Delta_{\mathcal{D}}(f^0,g^0) \geq \Lambda((f,g)-(f^0,g^0))$, $(f,g) \in \mathcal{D}$ (see [146, 180] for more details). This condition makes it possible to find the least favorable spectral densities in some special classes of spectral densities $\mathcal{D} = \mathcal{D}_f \times \mathcal{D}_g$.

In the case of cointegrated vector sequences (in terms of Section 3.1.5) we have the following optimization problem for determining the least favorable spectral densities:

$$\Delta_{\mathcal{D}}(f,p) = \widetilde{\Delta}(f,p) + \delta(f,p|\mathcal{D}_f \times \mathcal{D}_p) \to \inf, \tag{3.60}$$
$$\widetilde{\Delta}(f,p) = -\Delta(\vec{h}^{\alpha}_{\overline{\mu}}(f^0,p^0);f,p),$$

$$\Delta(\vec{h}^{\alpha}_{\overline{\mu}}(f^0,p^0);f,p)$$
$$= \frac{1}{2\pi}\int_{-\pi}^{\pi} \frac{|\beta^{(d)}(i\lambda)|^2}{|\chi^{(d)}_{\overline{\mu}}(e^{-i\lambda})|^2} (\mathbf{c}^{f0}_{\overline{\mu},\alpha}(e^{i\lambda}))^{\top} (p^0(\lambda))^{-1} f(\lambda)\,(p^0(\lambda))^{-1}\overline{\mathbf{c}^{f0}_{\overline{\mu},\alpha}(e^{i\lambda})}d\lambda$$
$$+ \frac{1}{2\pi}\int_{-\pi}^{\pi} \frac{|\beta^{(d)}(i\lambda)|^2}{|\chi^{(d)}_{\overline{\mu}}(e^{-i\lambda})|^2} (\mathbf{c}^{p0}_{\overline{\mu},\alpha}(e^{i\lambda}))^{\top} (p^0(\lambda))^{-1}$$
$$\times (p(\lambda) - \alpha^2 f(\lambda))\,(p^0(\lambda))^{-1}\overline{\mathbf{c}^{p0}_{\overline{\mu},\alpha}(e^{i\lambda})}d\lambda,$$

where

$$\mathbf{c}^{f0}_{\overline{\mu},\alpha}(e^{i\lambda}) = \frac{1}{|\beta^{(d)}(i\lambda)|^2}\overline{(p^0(\lambda) - \alpha^2 f^0(\lambda))}\vec{A}_{\overline{\mu}}(e^{i\lambda}) + \vec{C}^{\alpha,0}_{\overline{\mu}}(e^{i\lambda}),$$
$$\mathbf{c}^{p0}_{\overline{\mu},\alpha}(e^{i\lambda}) = \frac{1}{|\beta^{(d)}(i\lambda)|^2}\overline{f^0(\lambda)}\vec{A}_{\overline{\mu}}(e^{i\lambda}) - \vec{C}^{\alpha,0}_{\overline{\mu}}(e^{i\lambda}),$$

and

$$\vec{C}^{\alpha,0}_{\overline{\mu}}(e^{i\lambda}) = \sum_{k=0}^{\infty} (((\mathbf{P}^{\alpha}_{\overline{\mu}})^0)^{-1}(D^{\overline{\mu}}\mathbf{a} - (\mathbf{T}^{\alpha}_{\overline{\mu}})^0\mathbf{a}_{\overline{\mu}}))_k e^{i\lambda k}.$$

A solution (f^0, p^0) of this optimization problem is characterized by the condition $0 \in \partial\Delta_{\mathcal{D}}(f^0, p^0)$.

The form of the functionals $\Delta(\vec{h}_{\overline{\mu}}(f^0, g^0); f, g)$ and $\Delta(\vec{h}_{\overline{\mu}}(f^0, p^0); f, p)$ is convenient for application of the Lagrange method of indefinite multipliers for finding a solution to the problem (3.59). Making use of the method of Lagrange multipliers and the form of subdifferentials of the indicator functions $\delta(f, g|\mathcal{D}_f \times \mathcal{D}_g)$ and $\delta(f, g|\mathcal{D}_f \times \mathcal{D}_p)$ of the sets $\mathcal{D}_f \times \mathcal{D}_g$ and $\mathcal{D}_f \times \mathcal{D}_p$ of spectral densities, we describe relations that determine least favorable spectral densities in some special classes of spectral densities (see [115, 146] for additional details).

3.2.2 Least favorable spectral densities in classes $\mathcal{D}_0 \times \mathcal{D}_V^U$

Consider the extrapolation problem for the functional $A\vec{\xi}$ which depends on unobserved values of a sequence $\vec{\xi}(m)$ with stationary increments based on observations of the sequence $\vec{\xi}(m) + \vec{\eta}(m)$ at points $m = -1, -2, \dots$ under the condition that the sets of admissible spectral densities $\mathcal{D}_0^k$, $\mathcal{D}_V^{Uk}$, $k = 1, 2, 3, 4$, are defined as follows:

$$\mathcal{D}_0^1 = \left\{ f(\lambda) \left| \frac{1}{2\pi} \int_{-\pi}^{\pi} \frac{|\chi_{\overline{\mu}}^{(d)}(e^{-i\lambda})|^2}{|\beta^{(d)}(i\lambda)|^2} f(\lambda) d\lambda = P \right. \right\},$$

$$\mathcal{D}_0^2 = \left\{ f(\lambda) \left| \frac{1}{2\pi} \int_{-\pi}^{\pi} \frac{|\chi_{\overline{\mu}}^{(d)}(e^{-i\lambda})|^2}{|\beta^{(d)}(i\lambda)|^2} \operatorname{Tr}[f(\lambda)] d\lambda = p \right. \right\},$$

$$\mathcal{D}_0^3 = \left\{ f(\lambda) \left| \frac{1}{2\pi} \int_{-\pi}^{\pi} \frac{|\chi_{\overline{\mu}}^{(d)}(e^{-i\lambda})|^2}{|\beta^{(d)}(i\lambda)|^2} f_{kk}(\lambda) d\lambda = p_k, k = \overline{1, T} \right. \right\},$$

$$\mathcal{D}_0^4 = \left\{ f(\lambda) \left| \frac{1}{2\pi} \int_{-\pi}^{\pi} \frac{|\chi_{\overline{\mu}}^{(d)}(e^{-i\lambda})|^2}{|\beta^{(d)}(i\lambda)|^2} \langle B_1, f(\lambda)\rangle d\lambda = p \right. \right\}$$

and

$$\mathcal{D}_V^{U1} = \left\{ g(\lambda) \left| V(\lambda) \le g(\lambda) \le U(\lambda), \frac{1}{2\pi} \int_{-\pi}^{\pi} g(\lambda) d\lambda = Q \right. \right\},$$

$$\mathcal{D}_V^{U2} = \left\{ g(\lambda) \left| \operatorname{Tr}[V(\lambda)] \le \operatorname{Tr}[g(\lambda)] \le \operatorname{Tr}[U(\lambda)], \frac{1}{2\pi} \int_{-\pi}^{\pi} \operatorname{Tr}[g(\lambda)] d\lambda = q \right. \right\},$$

$$\mathcal{D}_V^{U3} = \left\{ g(\lambda) \left| v_{kk}(\lambda) \le g_{kk}(\lambda) \le u_{kk}(\lambda), \frac{1}{2\pi} \int_{-\pi}^{\pi} g_{kk}(\lambda) d\lambda = q_k, k = \overline{1, T} \right. \right\},$$

$$\mathcal{D}_V^{U4} = \left\{ g(\lambda) \left| \langle B_2, V(\lambda)\rangle \le \langle B_2, g(\lambda)\rangle \le \langle B_2, U(\lambda)\rangle, \frac{1}{2\pi} \int_{-\pi}^{\pi} \langle B_2, g(\lambda)\rangle d\lambda = q \right. \right\}.$$

Here spectral densities $V(\lambda)$ and $U(\lambda)$ are known and fixed, $p, p_k, q, q_k, k = \overline{1,T}$, are given numbers, and P, B_1, Q, B_2 are given positive-definite Hermitian matrices.

From the condition $0 \in \partial\Delta_{\mathcal{D}}(f^0, g^0)$ we find the following equations which determine the least favorable spectral densities for these given sets of admissible spectral densities.

For the first set of admissible spectral densities $\mathcal{D}_0^1 \times \mathcal{D}_V^{U1}$, we have

$$(\mathbf{C}_{\overline{\mu}}^{f0}(e^{i\lambda}))(\mathbf{C}_{\overline{\mu}}^{f0}(e^{i\lambda}))^* = \left(\frac{|\chi_{\overline{\mu}}^{(d)}(e^{-i\lambda})|^2}{|\beta^{(d)}(i\lambda)|^2}(f^0(\lambda) + |\beta^{(d)}(i\lambda)|^2 g^0(\lambda))\right) \times \vec{\alpha}_f \cdot \vec{\alpha}_f^* \left(\frac{|\chi_{\overline{\mu}}^{(d)}(e^{-i\lambda})|^2}{|\beta^{(d)}(i\lambda)|^2}(f^0(\lambda) + |\beta^{(d)}(i\lambda)|^2 g^0(\lambda))\right), \tag{3.61}$$

$$(\mathbf{C}_{\overline{\mu}}^{g0}(e^{i\lambda}))(\mathbf{C}_{\overline{\mu}}^{g0}(e^{i\lambda}))^* = \left(\frac{|\chi_{\overline{\mu}}^{(d)}(e^{-i\lambda})|^2}{|\beta^{(d)}(i\lambda)|^2}(f^0(\lambda) + |\beta^{(d)}(i\lambda)|^2 g^0(\lambda))\right) \times (\vec{\alpha}_g \cdot \vec{\alpha}_g^* + \Gamma_1(\lambda) + \Gamma_2(\lambda)) \times \left(\frac{|\chi_{\overline{\mu}}^{(d)}(e^{-i\lambda})|^2}{|\beta^{(d)}(i\lambda)|^2}(f^0(\lambda) + |\beta^{(d)}(i\lambda)|^2 g^0(\lambda))\right), \tag{3.62}$$

where $\vec{\alpha}_f$ and $\vec{\alpha}_g$ are vectors of Lagrange multipliers, the matrix $\Gamma_1(\lambda) \leq 0$ and $\Gamma_1(\lambda) = 0$ if $g^0(\lambda) > V(\lambda)$, and the matrix $\Gamma_2(\lambda) \geq 0$ and $\Gamma_2(\lambda) = 0$ if $g^0(\lambda) < U(\lambda)$.

For the second set of admissible spectral densities $\mathcal{D}_0^2 \times \mathcal{D}_V^{U2}$, we have

$$(\mathbf{C}_{\overline{\mu}}^{f0}(e^{i\lambda}))(\mathbf{C}_{\overline{\mu}}^{f0}(e^{i\lambda}))^* = \alpha_f^2 \left(\frac{|\chi_{\overline{\mu}}^{(d)}(e^{-i\lambda})|^2}{|\beta^{(d)}(i\lambda)|^2}(f^0(\lambda) + |\beta^{(d)}(i\lambda)|^2 g^0(\lambda))\right)^2, \tag{3.63}$$

$$(\mathbf{C}_{\overline{\mu}}^{g0}(e^{i\lambda}))(\mathbf{C}_{\overline{\mu}}^{g0}(e^{i\lambda}))^* = (\alpha_g^2 + \gamma_1(\lambda) + \gamma_2(\lambda)) \times \left(\frac{|\chi_{\overline{\mu}}^{(d)}(e^{-i\lambda})|^2}{|\beta^{(d)}(i\lambda)|^2}(f^0(\lambda) + |\beta^{(d)}(i\lambda)|^2 g^0(\lambda))\right)^2, \tag{3.64}$$

where α^2, α_g^2 are Lagrange multipliers, the function $\gamma_1(\lambda) \leq 0$ and $\gamma_1(\lambda) = 0$ if $\mathrm{Tr}[g^0(\lambda)] > \mathrm{Tr}[V(\lambda)]$, and the function $\gamma_2(\lambda) \geq 0$ and $\gamma_2(\lambda) = 0$ if $\mathrm{Tr}[g^0(\lambda)] < \mathrm{Tr}[U(\lambda)]$.

For the third set of admissible spectral densities $\mathcal{D}_0^3 \times \mathcal{D}_V^{U3}$, we have

$$(\mathbf{C}_{\overline{\mu}}^{f0}(e^{i\lambda}))(\mathbf{C}_{\overline{\mu}}^{f0}(e^{i\lambda}))^* = \left(\frac{|\chi_{\overline{\mu}}^{(d)}(e^{-i\lambda})|^2}{|\beta^{(d)}(i\lambda)|^2}(f^0(\lambda) + |\beta^{(d)}(i\lambda)|^2 g^0(\lambda))\right) \times \{\alpha_{fk}^2 \delta_{kl}\}_{k,l=1}^T \left(\frac{|\chi_{\overline{\mu}}^{(d)}(e^{-i\lambda})|^2}{|\beta^{(d)}(i\lambda)|^2}(f^0(\lambda) + |\beta^{(d)}(i\lambda)|^2 g^0(\lambda))\right), \tag{3.65}$$

$$(\mathbf{C}_{\overline{\mu}}^{g0}(e^{i\lambda}))(\mathbf{C}_{\overline{\mu}}^{g0}(e^{i\lambda}))^* = \left(\frac{|\chi_{\overline{\mu}}^{(d)}(e^{-i\lambda})|^2}{|\beta^{(d)}(i\lambda)|^2}(f^0(\lambda) + |\beta^{(d)}(i\lambda)|^2 g^0(\lambda))\right) \times \{(\alpha_{gk}^2 + \gamma_{1k}(\lambda) + \gamma_{2k}(\lambda))\delta_{kl}\}_{k,l=1}^T$$

$$\times\Bigg(\frac{|\chi_{\overline{\mu}}^{(d)}(e^{-i\lambda})|^2}{|\beta^{(d)}(i\lambda)|^2}(f^0(\lambda)+\big|\beta^{(d)}(i\lambda)\big|^2 g^0(\lambda))\Bigg), \tag{3.66}$$

where $\alpha_{fk}^2, \alpha_{gk}^2$ are Lagrange multipliers, δ_{kl} are Kronecker symbols, functions $\gamma_{1k}(\lambda)\le 0$ and $\gamma_{1k}(\lambda) = 0$ if $g_{kk}^0(\lambda) > v_{kk}(\lambda)$, and functions $\gamma_{2k}(\lambda) \ge 0$ and $\gamma_{2k}(\lambda) = 0$ if $g_{kk}^0(\lambda) < u_{kk}(\lambda)$.

For the fourth set of admissible spectral densities $\mathcal{D}_0^4 \times \mathcal{D}_V^{U4}$, we have

$$\begin{aligned}
(\mathbf{C}_{\overline{\mu}}^{f0}(e^{i\lambda}))(\mathbf{C}_{\overline{\mu}}^{f0}(e^{i\lambda}))^* &= \alpha_f^2\Bigg(\frac{|\chi_{\overline{\mu}}^{(d)}(e^{-i\lambda})|^2}{|\beta^{(d)}(i\lambda)|^2}(f^0(\lambda)+\big|\beta^{(d)}(i\lambda)\big|^2 g^0(\lambda))\Bigg)\\
&\quad\times B_1^\top\Bigg(\frac{|\chi_{\overline{\mu}}^{(d)}(e^{-i\lambda})|^2}{|\beta^{(d)}(i\lambda)|^2}(f^0(\lambda)+\big|\beta^{(d)}(i\lambda)\big|^2 g^0(\lambda))\Bigg),
\end{aligned} \tag{3.67}$$

$$\begin{aligned}
(\mathbf{C}_{\overline{\mu}}^{g0}(e^{i\lambda}))(\mathbf{C}_{\overline{\mu}}^{g0}(e^{i\lambda}))^* &= (\alpha_g^2+\gamma_1'(\lambda)+\gamma_2'(\lambda))\\
&\quad\times\Bigg(\frac{|\chi_{\overline{\mu}}^{(d)}(e^{-i\lambda})|^2}{|\beta^{(d)}(i\lambda)|^2}(f^0(\lambda)+\big|\beta^{(d)}(i\lambda)\big|^2 g^0(\lambda))\Bigg)\\
&\quad\times B_2^\top\Bigg(\frac{|\chi_{\overline{\mu}}^{(d)}(e^{-i\lambda})|^2}{|\beta^{(d)}(i\lambda)|^2}(f^0(\lambda)+\big|\beta^{(d)}(i\lambda)\big|^2 g^0(\lambda))\Bigg),
\end{aligned} \tag{3.68}$$

where α_f^2, α_g^2 are Lagrange multipliers, functions $\gamma_1'(\lambda)\le 0$ and $\gamma_1'(\lambda)=0$ if $\langle B_2, g^0(\lambda)\rangle > \langle B_2, V(\lambda)\rangle$, and functions $\gamma_2'(\lambda)\ge 0$ and $\gamma_2'(\lambda)=0$ if $\langle B_2, g^0(\lambda)\rangle < \langle B_2, U(\lambda)\rangle$.

The following theorem holds.

Theorem 3.2.1. *The least favorable spectral densities $f^0(\lambda)$ and $g^0(\lambda)$ in the classes $\mathcal{D}_0^k \times \mathcal{D}_V^{Uk}$, $k = 1,2,3,4$, for the optimal linear extrapolation of the functional $A\vec{\xi}$ from observations of the vector sequence $\vec{\xi}(m)+\vec{\eta}(m)$ at points $m = -1,-2,\ldots$ are determined by equations* (3.61)–(3.62), (3.63)–(3.64), (3.65)–(3.66), *and* (3.67)–(3.68), *respectively, the minimality condition* (3.4), *the constrained optimization problem* (3.51), *and restrictions on densities from the corresponding classes $\mathcal{D}_0^k$, $\mathcal{D}_V^{Uk}$, $k = 1,2,3,4$. The minimax-robust spectral characteristic of the optimal estimate of the functional $A\vec{\xi}$ is determined by formula* (3.12).

If the spectral densities $f(\lambda)$ and $g(\lambda)$ admit canonical factorizations (1.14), (3.29), and (3.30), we can derive the following equation for the least favorable spectral densities.

For the first set of admissible spectral densities $\mathcal{D}_0^1 \times \mathcal{D}_V^{U1}$, we have

$$(\mathbf{r}_{\overline{\mu},f}^0(e^{i\lambda}))(\mathbf{r}_{\overline{\mu},f}^0(e^{i\lambda}))^* = (\Theta_{\overline{\mu}}^0(e^{-i\lambda}))^\top \vec{\alpha}_f\cdot\vec{\alpha}_f^*\,\overline{\Theta_{\overline{\mu}}^0(e^{-i\lambda})}, \tag{3.69}$$

$$(\mathbf{r}_{\overline{\mu},g}^0(e^{i\lambda}))(\mathbf{r}_{\overline{\mu},g}^0(e^{i\lambda}))^* = (\Theta_{\overline{\mu}}^0(e^{-i\lambda}))^\top (\vec{\alpha}_g\cdot\vec{\alpha}_g^* + \Gamma_1(\lambda)+\Gamma_2(\lambda))\overline{\Theta_{\overline{\mu}}^0(e^{-i\lambda})}, \tag{3.70}$$

where $\vec{\alpha}_f$ and $\vec{\alpha}_g$ are vectors of Lagrange multipliers, the matrix $\Gamma_1(\lambda)\le 0$ and $\Gamma_1(\lambda)=0$ if $g^0(\lambda) > V(\lambda)$, and the matrix $\Gamma_2(\lambda)\ge 0$ and $\Gamma_2(\lambda) = 0$ if $g^0(\lambda) < U(\lambda)$.

For the second set of admissible spectral densities $\mathcal{D}_0^2 \times \mathcal{D}_V^{U2}$, we have

$$(\mathbf{r}_{\overline{\mu},f}^0(e^{i\lambda}))(\mathbf{r}_{\overline{\mu},f}^0(e^{i\lambda}))^* = \alpha_f^2(\Theta_{\overline{\mu}}^0(e^{-i\lambda}))^\top \overline{\Theta_{\overline{\mu}}^0(e^{-i\lambda})}, \tag{3.71}$$

$$(\mathbf{r}_{\overline{\mu},g}^0(e^{i\lambda}))(\mathbf{r}_{\overline{\mu},g}^0(e^{i\lambda}))^* = (\alpha_g^2 + \gamma_1(\lambda) + \gamma_2(\lambda))(\Theta_{\overline{\mu}}^0(e^{-i\lambda}))^\top \overline{\Theta_{\overline{\mu}}^0(e^{-i\lambda})}, \tag{3.72}$$

where α_f^2, α_g^2 are Lagrange multipliers, the function $\gamma_1(\lambda) \leq 0$ and $\gamma_1(\lambda) = 0$ if $\mathrm{Tr}[g^0(\lambda)] > \mathrm{Tr}[V(\lambda)]$, and the function $\gamma_2(\lambda) \geq 0$ and $\gamma_2(\lambda) = 0$ if $\mathrm{Tr}[g^0(\lambda)] < \mathrm{Tr}[U(\lambda)]$.

For the third set of admissible spectral densities $\mathcal{D}_0^3 \times \mathcal{D}_V^{U3}$, we have

$$(\mathbf{r}_{\overline{\mu},f}^0(e^{i\lambda}))(\mathbf{r}_{\overline{\mu},f}^0(e^{i\lambda}))^* = (\Theta_{\overline{\mu}}^0(e^{-i\lambda}))^\top \{\alpha_{fk}^2 \delta_{kl}\}_{k,l=1}^T \overline{\Theta_{\overline{\mu}}^0(e^{-i\lambda})}, \tag{3.73}$$

$$(\mathbf{r}_{\overline{\mu},g}^0(e^{i\lambda}))(\mathbf{r}_{\overline{\mu},g}^0(e^{i\lambda}))^* = (\Theta_{\overline{\mu}}^0(e^{-i\lambda}))^\top \{(\alpha_{gk}^2 + \gamma_{1k}(\lambda) + \gamma_{2k}(\lambda))\delta_{kl}\}_{k,l=1}^T \overline{\Theta_{\overline{\mu}}^0(e^{-i\lambda})}, \tag{3.74}$$

where $\alpha_{fk}^2, \alpha_{gk}^2$ are Lagrange multipliers, δ_{kl} are Kronecker symbols, functions $\gamma_{1k}(\lambda) \leq 0$ and $\gamma_{1k}(\lambda) = 0$ if $g_{kk}^0(\lambda) > v_{kk}(\lambda)$, and functions $\gamma_{2k}(\lambda) \geq 0$ and $\gamma_{2k}(\lambda) = 0$ if $g_{kk}^0(\lambda) < u_{kk}(\lambda)$.

For the fourth set of admissible spectral densities $\mathcal{D}_0^4 \times \mathcal{D}_V^{U4}$, we have

$$(\mathbf{r}_{\overline{\mu},f}^0(e^{i\lambda}))(\mathbf{r}_{\overline{\mu},f}^0(e^{i\lambda}))^* = \alpha_f^2(\Theta_{\overline{\mu}}^0(e^{-i\lambda}))^\top B_1 \overline{\Theta_{\overline{\mu}}^0(e^{-i\lambda})}, \tag{3.75}$$

$$(\mathbf{r}_{\overline{\mu},g}^0(e^{i\lambda}))(\mathbf{r}_{\overline{\mu},g}^0(e^{i\lambda}))^* = (\alpha_g^2 + \gamma_1'(\lambda) + \gamma_2'(\lambda))(\Theta_{\overline{\mu}}^0(e^{-i\lambda}))^\top B_2 \overline{\Theta_{\overline{\mu}}^0(e^{-i\lambda})}, \tag{3.76}$$

where α_f^2, α_g^2 are Lagrange multipliers, functions $\gamma_1'(\lambda) \leq 0$ and $\gamma_1'(\lambda) = 0$ if $\langle B_2, g^0(\lambda)\rangle > \langle B_2, V(\lambda)\rangle$, and functions $\gamma_2'(\lambda) \geq 0$ and $\gamma_2'(\lambda) = 0$ if $\langle B_2, g^0(\lambda)\rangle < \langle B_2, U(\lambda)\rangle$.

The following theorem holds.

Theorem 3.2.2. *The least favorable spectral densities $f^0(\lambda)$ and $g^0(\lambda)$ in the classes $\mathcal{D}_0^k \times \mathcal{D}_V^{Uk}$, $k = 1, 2, 3, 4$, for the optimal linear extrapolation of the functional $A\vec{\xi}$ from observations of the vector sequence $\vec{\xi}(m) + \vec{\eta}(m)$ at points $m = -1, -2, \ldots$ are determined by canonical factorizations* (1.14), (3.29), *and* (3.30), *equations* (3.69)–(3.70), (3.71)–(3.72), (3.73)–(3.74), *and* (3.75)–(3.76), *respectively, the constrained optimization problem* (3.52), *and restrictions on densities from the corresponding classes $\mathcal{D}_0^k$, $\mathcal{D}_V^{Uk}$, $k = 1, 2, 3, 4$. The minimax-robust spectral characteristic of the optimal estimate of the functional $A\vec{\xi}$ is determined by formula* (3.12).

Theorem 3.2.3. *If the spectral density $g(\lambda)$ is known, the least favorable spectral density $f^0(\lambda)$ in the classes $\mathcal{D}_0^k$, $k = 1, 2, 3, 4$, for the optimal linear extrapolation of the functional $A\vec{\xi}$ from observations of the sequence $\vec{\xi}(m) + \vec{\eta}(m)$ at points $m = -1, -2, \ldots$ is determined by canonical factorizations* (1.14) *and* (3.29), *equations* (3.69), (3.71), (3.73), *and* (3.75), *respectively, the constrained optimization problem* (3.57), *and restrictions on density from the corresponding classes $\mathcal{D}_0^k$, $k = 1, 2, 3, 4$. The minimax-robust spectral characteristic of the optimal estimate of the functional $A\vec{\xi}$ is determined by formula* (3.12).

3.2.3 Least favorable spectral densities in classes $\mathcal{D}_0 \times \mathcal{D}_V^U$ for cointegrated vector sequences

Consider the extrapolation problem for the functional $A\vec{\xi}$ which depends on unobserved values of a sequence $\vec{\xi}(m)$ with stationary increments based on observations of the vector sequence $\vec{\zeta}(m)$, cointegrated with $\vec{\xi}(m)$ in terms of Definition 3.1.1, at points $m = -1, -2, \ldots$ for the sets of admissible spectral densities $\mathcal{D}_0^k$, $\mathcal{D}_V^{Uk}$, $k = 1, 2, 3, 4$, where the sets $\mathcal{D}_0^k$ are defined in Section 3.2.2 and the sets $\mathcal{D}_V^{Uk}$ are the following:

$$\mathcal{D}_V^{U1} = \left\{ p(\lambda) \,\middle|\, V(\lambda) \le p(\lambda) \le U(\lambda), \frac{1}{2\pi}\int_{-\pi}^{\pi} \frac{|\chi_{\overline{\mu}}^{(d)}(e^{-i\lambda})|^2}{|\beta^{(d)}(i\lambda)|^2} p(\lambda)d\lambda = Q \right\},$$

$$\mathcal{D}_V^{U2} = \left\{ p(\lambda) \,\middle|\, \mathrm{Tr}[V(\lambda)] \le \mathrm{Tr}[p(\lambda)] \le \mathrm{Tr}[U(\lambda)], \frac{1}{2\pi}\int_{-\pi}^{\pi} \frac{|\chi_{\overline{\mu}}^{(d)}(e^{-i\lambda})|^2}{|\beta^{(d)}(i\lambda)|^2} \mathrm{Tr}[p(\lambda)]d\lambda = q \right\},$$

$$\mathcal{D}_V^{U3} = \left\{ p(\lambda) \,\middle|\, v_{kk}(\lambda) \le p_{kk}(\lambda) \le u_{kk}(\lambda), \frac{1}{2\pi}\int_{-\pi}^{\pi} \frac{|\chi_{\overline{\mu}}^{(d)}(e^{-i\lambda})|^2}{|\beta^{(d)}(i\lambda)|^2} p_{kk}(\lambda)d\lambda = q_k, k = \overline{1,T} \right\},$$

$$\mathcal{D}_V^{U4} = \left\{ p(\lambda) \,\middle|\, \langle B_2, V(\lambda)\rangle \le \langle B_2, p(\lambda)\rangle \le \langle B_2, U(\lambda)\rangle, \frac{1}{2\pi}\int_{-\pi}^{\pi} \frac{|\chi_{\overline{\mu}}^{(d)}(e^{-i\lambda})|^2}{|\beta^{(d)}(i\lambda)|^2} \langle B_2, p(\lambda)\rangle d\lambda = q \right\}.$$

Here spectral densities $V(\lambda)$ and $U(\lambda)$ are known and fixed, q, q_k, $k = \overline{1,T}$, are given numbers, and Q, B_2 are given positive-definite Hermitian matrices.

The condition $0 \in \partial\Delta_{\mathcal{D}}(f^0, p^0)$ implies the following equations which determine the least favorable spectral densities for these given sets of admissible spectral densities.

For the first set of admissible spectral densities $\mathcal{D}_0^1 \times \mathcal{D}_V^{U1}$, we have

$$\begin{aligned}(\mathbf{C}^{f0}_{\overline{\mu},a}(e^{i\lambda}))(\mathbf{C}^{f0}_{\overline{\mu},a}(e^{i\lambda}))^* - \alpha^2(\mathbf{C}^{p0}_{\overline{\mu},a}(e^{i\lambda}))(\mathbf{C}^{p0}_{\overline{\mu},a}(e^{i\lambda}))^* \\ = \left(\frac{|\chi_{\overline{\mu}}^{(d)}(e^{-i\lambda})|^2}{|\beta^{(d)}(i\lambda)|^2} p^0(\lambda)\right)\vec{\alpha}_f \cdot \vec{\alpha}_f^* \left(\frac{|\chi_{\overline{\mu}}^{(d)}(e^{-i\lambda})|^2}{|\beta^{(d)}(i\lambda)|^2} p^0(\lambda)\right),\end{aligned} \tag{3.77}$$

$$\begin{aligned}(\mathbf{C}^{p0}_{\overline{\mu},a}(e^{i\lambda}))(\mathbf{C}^{p0}_{\overline{\mu},a}(e^{i\lambda}))^* = \left(\frac{|\chi_{\overline{\mu}}^{(d)}(e^{-i\lambda})|^2}{|\beta^{(d)}(i\lambda)|^2} p^0(\lambda)\right) \\ \times (\vec{\alpha}_g \cdot \vec{\alpha}_g^* + \Gamma_1(\lambda) + \Gamma_2(\lambda))\left(\frac{|\chi_{\overline{\mu}}^{(d)}(e^{-i\lambda})|^2}{|\beta^{(d)}(i\lambda)|^2} p^0(\lambda)\right),\end{aligned} \tag{3.78}$$

where $\vec{\alpha}_f$ and $\vec{\alpha}_g$ are vectors of Lagrange multipliers, the matrix $\Gamma_1(\lambda) \le 0$ and $\Gamma_1(\lambda) = 0$ if $g^0(\lambda) > V(\lambda)$, and the matrix $\Gamma_2(\lambda) \ge 0$ and $\Gamma_2(\lambda) = 0$ if $p_0(\lambda) < U(\lambda)$.

For the second set of admissible spectral densities $\mathcal{D}_0^2 \times \mathcal{D}_V^{U2}$, we have

$$(\mathbf{C}^{f0}_{\overline{\mu},a}(e^{i\lambda}))(\mathbf{C}^{f0}_{\overline{\mu},a}(e^{i\lambda}))^* - \alpha^2(\mathbf{C}^{p0}_{\overline{\mu},a}(e^{i\lambda}))(\mathbf{C}^{p0}_{\overline{\mu},a}(e^{i\lambda}))^* = \alpha_f^2\left(\frac{|\chi_{\overline{\mu}}^{(d)}(e^{-i\lambda})|^2}{|\beta^{(d)}(i\lambda)|^2} p^0(\lambda)\right)^2, \tag{3.79}$$

$$(\mathbf{C}^{p0}_{\overline{\mu},a}(e^{i\lambda}))(\mathbf{C}^{p0}_{\overline{\mu},a}(e^{i\lambda}))^* = (\alpha_g^2 + \gamma_1(\lambda) + \gamma_2(\lambda))\left(\frac{|\chi^{(d)}_{\overline{\mu}}(e^{-i\lambda})|^2}{|\beta^{(d)}(i\lambda)|^2}p^0(\lambda)\right)^2, \tag{3.80}$$

where α_f^2, α_g^2 are Lagrange multipliers, the function $\gamma_1(\lambda) \leq 0$ and $\gamma_1(\lambda) = 0$ if $\mathrm{Tr}[p^0(\lambda)] > \mathrm{Tr}[V(\lambda)]$, and the function $\gamma_2(\lambda) \geq 0$ and $\gamma_2(\lambda) = 0$ if $\mathrm{Tr}[p^0(\lambda)] < \mathrm{Tr}[U(\lambda)]$.

For the third set of admissible spectral densities $\mathcal{D}_0^3 \times \mathcal{D}_V^{U3}$, we have

$$\begin{aligned}(\mathbf{C}^{f0}_{\overline{\mu},a}(e^{i\lambda}))(\mathbf{C}^{f0}_{\overline{\mu},a}(e^{i\lambda}))^* &- \alpha^2(\mathbf{C}^{p0}_{\overline{\mu},a}(e^{i\lambda}))(\mathbf{C}^{p0}_{\overline{\mu},a}(e^{i\lambda}))^*\\ &= \left(\frac{|\chi^{(d)}_{\overline{\mu}}(e^{-i\lambda})|^2}{|\beta^{(d)}(i\lambda)|^2}p^0(\lambda)\right)\{\alpha_{fk}^2\delta_{kl}\}_{k,l=1}^T\left(\frac{|\chi^{(d)}_{\overline{\mu}}(e^{-i\lambda})|^2}{|\beta^{(d)}(i\lambda)|^2}p^0(\lambda)\right),\end{aligned} \tag{3.81}$$

$$\begin{aligned}(\mathbf{C}^{p0}_{\overline{\mu},a}(e^{i\lambda}))(\mathbf{C}^{p0}_{\overline{\mu},a}(e^{i\lambda}))^* &= \left(\frac{|\chi^{(d)}_{\overline{\mu}}(e^{-i\lambda})|^2}{|\beta^{(d)}(i\lambda)|^2}p^0(\lambda)\right)\\ &\times\{(\alpha_{gk}^2 + \gamma_{1k}(\lambda) + \gamma_{2k}(\lambda))\delta_{kl}\}_{k,l=1}^T\left(\frac{|\chi^{(d)}_{\overline{\mu}}(e^{-i\lambda})|^2}{|\beta^{(d)}(i\lambda)|^2}p^0(\lambda)\right),\end{aligned} \tag{3.82}$$

where $\alpha_{fk}^2, \alpha_{gk}^2$ are Lagrange multipliers, δ_{kl} are Kronecker symbols, functions $\gamma_{1k}(\lambda) \leq 0$ and $\gamma_{1k}(\lambda) = 0$ if $p^0_{kk}(\lambda) > v_{kk}(\lambda)$, and functions $\gamma_{2k}(\lambda) \geq 0$ and $\gamma_{2k}(\lambda) = 0$ if $p^0_{kk}(\lambda) < u_{kk}(\lambda)$.

For the fourth set of admissible spectral densities $\mathcal{D}_0^4 \times \mathcal{D}_V^{U4}$, we have

$$\begin{aligned}(\mathbf{C}^{f0}_{\overline{\mu},a}(e^{i\lambda}))(\mathbf{C}^{f0}_{\overline{\mu},a}(e^{i\lambda}))^* &- \alpha^2(\mathbf{C}^{p0}_{\overline{\mu},a}(e^{i\lambda}))(\mathbf{C}^{p0}_{\overline{\mu},a}(e^{i\lambda}))^*\\ &= \alpha_f^2\left(\frac{|\chi^{(d)}_{\overline{\mu}}(e^{-i\lambda})|^2}{|\beta^{(d)}(i\lambda)|^2}p^0(\lambda)\right)B_1^\top\left(\frac{|\chi^{(d)}_{\overline{\mu}}(e^{-i\lambda})|^2}{|\beta^{(d)}(i\lambda)|^2}p^0(\lambda)\right),\end{aligned} \tag{3.83}$$

$$\begin{aligned}(\mathbf{C}^{p0}_{\overline{\mu},a}(e^{i\lambda}))(\mathbf{C}^{p0}_{\overline{\mu},a}(e^{i\lambda}))^* &= (\alpha_g^2 + \gamma_1'(\lambda) + \gamma_2'(\lambda))\\ &\times\left(\frac{|\chi^{(d)}_{\overline{\mu}}(e^{-i\lambda})|^2}{|\beta^{(d)}(i\lambda)|^2}p^0(\lambda)\right)B_2^\top\left(\frac{|\chi^{(d)}_{\overline{\mu}}(e^{-i\lambda})|^2}{|\beta^{(d)}(i\lambda)|^2}p^0(\lambda)\right),\end{aligned} \tag{3.84}$$

where α_f^2, α_g^2, are Lagrange multipliers, functions $\gamma_1'(\lambda) \leq 0$ and $\gamma_1'(\lambda) = 0$ if $\langle B_2, p^0(\lambda)\rangle > \langle B_2, V(\lambda)\rangle$, and functions $\gamma_2'(\lambda) \geq 0$ and $\gamma_2'(\lambda) = 0$ if $\langle B_2, p^0(\lambda)\rangle < \langle B_2, U(\lambda)\rangle$.

The following theorem holds.

Theorem 3.2.4. *The least favorable spectral densities $f^0(\lambda)$ and $p^0(\lambda)$, in the classes $\mathcal{D}_0^k \times \mathcal{D}_V^{Uk}$, $k = 1, 2, 3, 4$, for the optimal linear extrapolation of the functional $A\vec{\xi}$ from observations of the vector sequence $\vec{\zeta}(m)$, cointegrated with $\vec{\xi}(m)$ in terms of Definition* 3.1.1, *at points $m = -1, -2, \ldots$ are determined by equations* (3.77)–(3.78), (3.79)–(3.80), (3.81)–(3.82), *and* (3.83)–(3.84), *respectively, the minimality condition* (3.41), *the constrained optimization problem* (3.51) *with $g(\lambda) := |\beta^{(d)}(i\lambda)|^{-2}(p(\lambda) - \alpha^2 f(\lambda))$, and restrictions on densities from the corresponding classes $\mathcal{D}_0^k$, $\mathcal{D}_V^{Uk}$, $k = 1, 2, 3, 4$. The minimax-robust spectral characteristic of the optimal estimate of the functional $A\vec{\xi}$ is determined by formula* (3.44).

3.2.4 Least favorable spectral densities in classes $\mathcal{D}_\varepsilon \times \mathcal{D}_{1\delta}$

Consider the extrapolation problem for the functional $A\vec{\xi}$ which depends on unobserved values of a sequence $\vec{\xi}(m)$ with stationary increments based on observations of the sequence $\vec{\xi}(m) + \vec{\eta}(m)$ at points $m = -1, -2, \ldots$ under the condition that the sets of admissible spectral densities $\mathcal{D}_\varepsilon^k$, $\mathcal{D}_{1\delta}^k$, $k = 1, 2, 3, 4$, are defined as follows:

$$\mathcal{D}_\varepsilon^1 = \left\{ f(\lambda) \left| \operatorname{Tr}[f(\lambda)] = (1-\varepsilon)\operatorname{Tr}[f_1(\lambda)] + \varepsilon \operatorname{Tr}[W(\lambda)], \right. \right.$$

$$\left. \frac{1}{2\pi}\int_{-\pi}^{\pi} \frac{|\chi_{\overline{\mu}}^{(d)}(e^{-i\lambda})|^2}{|\beta^{(d)}(i\lambda)|^2} \operatorname{Tr}[f(\lambda)]d\lambda = p \right\},$$

$$\mathcal{D}_\varepsilon^2 = \left\{ f(\lambda) \left| f_{kk}(\lambda) = (1-\varepsilon) f_{kk}^1(\lambda) + \varepsilon w_{kk}(\lambda), \right. \right.$$

$$\left. \frac{1}{2\pi}\int_{-\pi}^{\pi} \frac{|\chi_{\overline{\mu}}^{(d)}(e^{-i\lambda})|^2}{|\beta^{(d)}(i\lambda)|^2} f_{kk}(\lambda)d\lambda = p_k, k = \overline{1,T} \right\},$$

$$\mathcal{D}_\varepsilon^3 = \left\{ f(\lambda) \left| \langle B_1, f(\lambda)\rangle = (1-\varepsilon)\langle B_1, f_1(\lambda)\rangle + \varepsilon \langle B_1, W(\lambda)\rangle, \right. \right.$$

$$\left. \frac{1}{2\pi}\int_{-\pi}^{\pi} \frac{|\chi_{\overline{\mu}}^{(d)}(e^{-i\lambda})|^2}{|\beta^{(d)}(i\lambda)|^2} \langle B_1, f(\lambda)\rangle d\lambda = p \right\},$$

$$\mathcal{D}_\varepsilon^4 = \left\{ f(\lambda) \left| f(\lambda) = (1-\varepsilon) f_1(\lambda) + \varepsilon W(\lambda), \frac{1}{2\pi}\int_{-\pi}^{\pi} \frac{|\chi_{\overline{\mu}}^{(d)}(e^{-i\lambda})|^2}{|\beta^{(d)}(i\lambda)|^2} f(\lambda)d\lambda = P \right. \right\}$$

and

$$\mathcal{D}_{1\delta}^1 = \left\{ g(\lambda) \left| \frac{1}{2\pi}\int_{-\pi}^{\pi} \left|\operatorname{Tr}(g(\lambda) - g_1(\lambda))\right| d\lambda \le \delta \right. \right\},$$

$$\mathcal{D}_{1\delta}^2 = \left\{ g(\lambda) \left| \frac{1}{2\pi}\int_{-\pi}^{\pi} \left|g_{kk}(\lambda) - g_{kk}^1(\lambda)\right| d\lambda \le \delta_k, k = \overline{1,T} \right. \right\},$$

$$\mathcal{D}_{1\delta}^3 = \left\{ g(\lambda) \left| \frac{1}{2\pi}\int_{-\pi}^{\pi} \left|\langle B_2, g(\lambda) - g_1(\lambda)\rangle\right| d\lambda \le \delta \right. \right\},$$

$$\mathcal{D}_{1\delta}^4 = \left\{ g(\lambda) \left| \frac{1}{2\pi}\int_{-\pi}^{\pi} \left|g_{ij}(\lambda) - g_{ij}^1(\lambda)\right| d\lambda \le \delta_i^j, i, j = \overline{1,T} \right. \right\}.$$

Here $f_1(\lambda)$ and $g_1(\lambda)$ are fixed spectral densities, $W(\lambda)$ is an unknown spectral density, p, p_k, $k = \overline{1,T}$, are given numbers, P is a given positive-definite Hermitian matrix, and δ, δ_k, $k = \overline{1,T}$, and δ_i^j, $i,j = \overline{1,T}$, are given numbers.

From the condition $0 \in \partial\Delta_{\mathcal{D}}(f^0, g^0)$ we find the following equations which determine the least favorable spectral densities for these given sets of admissible spectral densities.

For the first set of admissible spectral densities $\mathcal{D}_\varepsilon^1 \times \mathcal{D}_{1\delta}^1$, we have

$$(\mathbf{C}_{\overline{\mu}}^{f0}(e^{i\lambda}))(\mathbf{C}_{\overline{\mu}}^{f0}(e^{i\lambda}))^* = (\alpha_f^2 + \gamma_1(\lambda))\left(\frac{|\chi_{\overline{\mu}}^{(d)}(e^{-i\lambda})|^2}{|\beta^{(d)}(i\lambda)|^2}(f^0(\lambda) + |\beta^{(d)}(i\lambda)|^2 g^0(\lambda))\right)^2, \tag{3.85}$$

$$(\mathbf{C}_{\overline{\mu}}^{g0}(e^{i\lambda}))(\mathbf{C}_{\overline{\mu}}^{g0}(e^{i\lambda}))^* = \alpha_g^2\gamma_2(\lambda)\left(\frac{|\chi_{\overline{\mu}}^{(d)}(e^{-i\lambda})|^2}{|\beta^{(d)}(i\lambda)|^2}(f^0(\lambda) + |\beta^{(d)}(i\lambda)|^2 g^0(\lambda))\right)^2, \tag{3.86}$$

$$\frac{1}{2\pi}\int_{-\pi}^{\pi} |\mathrm{Tr}(g^0(\lambda) - g_1(\lambda))| d\lambda = \delta, \tag{3.87}$$

where α_f^2, α_g^2 are Lagrange multipliers, the function $\gamma_1(\lambda) \le 0$ and $\gamma_1(\lambda) = 0$ if $\mathrm{Tr}[f^0(\lambda)] > (1-\varepsilon)\,\mathrm{Tr}[f_1(\lambda)]$, the function $|\gamma_2(\lambda)| \le 1$, and

$$\gamma_2(\lambda) = \mathrm{sign}(\mathrm{Tr}(g^0(\lambda) - g_1(\lambda))) : \ \mathrm{Tr}(g^0(\lambda) - g_1(\lambda)) \ne 0.$$

For the second set of admissible spectral densities $\mathcal{D}_\varepsilon^2 \times \mathcal{D}_{1\delta}^2$, we have

$$\begin{aligned}(\mathbf{C}_{\overline{\mu}}^{f0}(e^{i\lambda}))(\mathbf{C}_{\overline{\mu}}^{f0}(e^{i\lambda}))^* &= \left(\frac{|\chi_{\overline{\mu}}^{(d)}(e^{-i\lambda})|^2}{|\beta^{(d)}(i\lambda)|^2}(f^0(\lambda) + |\beta^{(d)}(i\lambda)|^2 g^0(\lambda))\right)\{(\alpha_{fk}^2 + \gamma_k^1(\lambda))\delta_{kl}\}_{k,l=1}^T \\ &\times \left(\frac{|\chi_{\overline{\mu}}^{(d)}(e^{-i\lambda})|^2}{|\beta^{(d)}(i\lambda)|^2}(f^0(\lambda) + |\beta^{(d)}(i\lambda)|^2 g^0(\lambda))\right),\end{aligned} \tag{3.88}$$

$$\begin{aligned}(\mathbf{C}_{\overline{\mu}}^{g0}(e^{i\lambda}))(\mathbf{C}_{\overline{\mu}}^{g0}(e^{i\lambda}))^* &= \left(\frac{|\chi_{\overline{\mu}}^{(d)}(e^{-i\lambda})|^2}{|\beta^{(d)}(i\lambda)|^2}(f^0(\lambda) + |\beta^{(d)}(i\lambda)|^2 g^0(\lambda))\right)\{\alpha_{gk}^2\gamma_k^2(\lambda)\delta_{kl}\}_{k,l=1}^T \\ &\times \left(\frac{|\chi_{\overline{\mu}}^{(d)}(e^{-i\lambda})|^2}{|\beta^{(d)}(i\lambda)|^2}(f^0(\lambda) + |\beta^{(d)}(i\lambda)|^2 g^0(\lambda))\right),\end{aligned} \tag{3.89}$$

$$\frac{1}{2\pi}\int_{-\pi}^{\pi} |g_{kk}^0(\lambda) - g_{kk}^1(\lambda)| d\lambda = \delta_k, \tag{3.90}$$

where α_{fk}^2, α_{gk}^2 are Lagrange multipliers, functions $\gamma_k^1(\lambda) \le 0$ and $\gamma_k^1(\lambda) = 0$ if $f_{kk}^0(\lambda) > (1-\varepsilon)f_{kk}^1(\lambda)$, functions $|\gamma_k^2(\lambda)| \le 1$, and

$$\gamma_k^2(\lambda) = \mathrm{sign}(g_{kk}^0(\lambda) - g_{kk}^1(\lambda)) : \ g_{kk}^0(\lambda) - g_{kk}^1(\lambda) \ne 0, \quad k = \overline{1,T}.$$

For the third set of admissible spectral densities $\mathcal{D}_\varepsilon^3 \times \mathcal{D}_{1\delta}^3$, we have

$$(\mathbf{C}_{\overline{\mu}}^{f0}(e^{i\lambda}))(\mathbf{C}_{\overline{\mu}}^{f0}(e^{i\lambda}))^* = (\alpha_f^2 + \gamma_1'(\lambda))\left(\frac{|\chi_{\overline{\mu}}^{(d)}(e^{-i\lambda})|^2}{|\beta^{(d)}(i\lambda)|^2}(f^0(\lambda) + |\beta^{(d)}(i\lambda)|^2 g^0(\lambda))\right)$$
$$\times B_1^\top\left(\frac{|\chi_{\overline{\mu}}^{(d)}(e^{-i\lambda})|^2}{|\beta^{(d)}(i\lambda)|^2}(f^0(\lambda) + |\beta^{(d)}(i\lambda)|^2 g^0(\lambda))\right), \tag{3.91}$$

$$(\mathbf{C}_{\overline{\mu}}^{g0}(e^{i\lambda}))(\mathbf{C}_{\overline{\mu}}^{g0}(e^{i\lambda}))^* = \alpha_g^2\gamma_2'(\lambda)\left(\frac{|\chi_{\overline{\mu}}^{(d)}(e^{-i\lambda})|^2}{|\beta^{(d)}(i\lambda)|^2}(f^0(\lambda) + |\beta^{(d)}(i\lambda)|^2 g^0(\lambda))\right)$$
$$\times B_2^\top\left(\frac{|\chi_{\overline{\mu}}^{(d)}(e^{-i\lambda})|^2}{|\beta^{(d)}(i\lambda)|^2}(f^0(\lambda) + |\beta^{(d)}(i\lambda)|^2 g^0(\lambda))\right), \tag{3.92}$$

$$\frac{1}{2\pi}\int_{-\pi}^{\pi} |\langle B_2, g^0(\lambda) - g_1(\lambda)\rangle| d\lambda = \delta, \tag{3.93}$$

where α_f^2, α_g^2 are Lagrange multipliers, function $\gamma_1'(\lambda) \le 0$ and $\gamma_1'(\lambda) = 0$ if $\langle B_1, f^0(\lambda)\rangle > (1-\varepsilon)\langle B_1, f_1(\lambda)\rangle$, function $|\gamma_2'(\lambda)| \le 1$, and

$$\gamma_2'(\lambda) = \operatorname{sign}\langle B_2, g^0(\lambda) - g_1(\lambda)\rangle : \ \langle B_2, g^0(\lambda) - g_1(\lambda)\rangle \neq 0.$$

For the fourth set of admissible spectral densities $\mathcal{D}_\varepsilon^4 \times \mathcal{D}_{1\delta}^4$, we have

$$(\mathbf{C}_{\overline{\mu}}^{f0}(e^{i\lambda}))(\mathbf{C}_{\overline{\mu}}^{f0}(e^{i\lambda}))^* = \left(\frac{|\chi_{\overline{\mu}}^{(d)}(e^{-i\lambda})|^2}{|\beta^{(d)}(i\lambda)|^2}(f^0(\lambda) + |\beta^{(d)}(i\lambda)|^2 g^0(\lambda))\right)(\vec{\alpha}_f \cdot \vec{\alpha}_f^* - \Gamma(\lambda))$$
$$\times\left(\frac{|\chi_{\overline{\mu}}^{(d)}(e^{-i\lambda})|^2}{|\beta^{(d)}(i\lambda)|^2}(f^0(\lambda) + |\beta^{(d)}(i\lambda)|^2 g^0(\lambda))\right), \tag{3.94}$$

$$(\mathbf{C}_{\overline{\mu}}^{g0}(e^{i\lambda}))(\mathbf{C}_{\overline{\mu}}^{g0}(e^{i\lambda}))^* = \left(\frac{|\chi_{\overline{\mu}}^{(d)}(e^{-i\lambda})|^2}{|\beta^{(d)}(i\lambda)|^2}(f^0(\lambda) + |\beta^{(d)}(i\lambda)|^2 g^0(\lambda))\right)\{\alpha_{g,ij}\gamma_{ij}(\lambda)\}_{i,j=1}^T$$
$$\times\left(\frac{|\chi_{\overline{\mu}}^{(d)}(e^{-i\lambda})|^2}{|\beta^{(d)}(i\lambda)|^2}(f^0(\lambda) + |\beta^{(d)}(i\lambda)|^2 g^0(\lambda))\right), \tag{3.95}$$

$$\frac{1}{2\pi}\int_{-\pi}^{\pi} |g_{ij}^0(\lambda) - g_{ij}^1(\lambda)| d\lambda = \delta_i^j, \tag{3.96}$$

where $\vec{\alpha}_f$, $\alpha_{g,ij}$ are Lagrange multipliers, function $\Gamma(\lambda) \le 0$ and $\Gamma(\lambda) = 0$ if $f^0(\lambda) > (1-\varepsilon)f_1(\lambda)$, functions $|\gamma_{ij}(\lambda)| \le 1$, and

$$\gamma_{ij}(\lambda) = \frac{g_{ij}^0(\lambda) - g_{ij}^1(\lambda)}{|g_{ij}^0(\lambda) - g_{ij}^1(\lambda)|} : \ g_{ij}^0(\lambda) - g_{ij}^1(\lambda) \neq 0, \quad i,j = \overline{1,T}.$$

The following theorem holds.

Theorem 3.2.5. *The least favorable spectral densities $f^0(\lambda)$ and $g^0(\lambda)$ in classes $\mathcal{D}_\varepsilon^k \times \mathcal{D}_{1\delta}^k$, $k = 1,2,3,4$, for the optimal linear extrapolation of the functional $A\vec{\xi}$ from observations of the vector sequence $\vec{\xi}(m) + \vec{\eta}(m)$ at points $m = -1,-2,\dots$ are determined by equations* (3.85)–(3.87), (3.88)–(3.90), (3.91)–(3.93), *and* (3.94)–(3.96), *respectively, the minimality condition* (3.4), *the constrained optimization problem* (3.51), *and restrictions on densities from the corresponding classes $\mathcal{D}_\varepsilon^k$, $\mathcal{D}_{1\delta}^k$, $k = 1,2,3,4$. The minimax-robust spectral characteristic of the optimal estimate of the functional $A\vec{\xi}$ is determined by formula* (3.12).

Let the spectral densities $f(\lambda)$ and $g(\lambda)$ admit canonical factorizations (1.14), (3.29), and (3.30). Then we derive the following equation for the least favorable spectral densities.

For the first set of admissible spectral densities $\mathcal{D}_\varepsilon^1 \times \mathcal{D}_{1\delta}^1$, we have

$$(\mathbf{r}^0_{\overline{\mu},f}(e^{i\lambda}))(\mathbf{r}^0_{\overline{\mu},f}(e^{i\lambda}))^* = (\alpha_f^2 + \gamma_1(\lambda))(\Theta^0_{\overline{\mu}}(e^{-i\lambda}))^\top \overline{\Theta^0_{\overline{\mu}}(e^{-i\lambda})}, \tag{3.97}$$

$$(\mathbf{r}^0_{\overline{\mu},g}(e^{i\lambda}))(\mathbf{r}^0_{\overline{\mu},g}(e^{i\lambda}))^* = \alpha_g^2\gamma_2(\lambda)(\Theta^0_{\overline{\mu}}(e^{-i\lambda}))^\top \overline{\Theta^0_{\overline{\mu}}(e^{-i\lambda})}, \tag{3.98}$$

$$\frac{1}{2\pi}\int_{-\pi}^{\pi} |\mathrm{Tr}(g^0(\lambda) - g_1(\lambda))| d\lambda = \delta, \tag{3.99}$$

where α_f^2, α_g^2 are Lagrange multipliers, the function $\gamma_1(\lambda) \le 0$ and $\gamma_1(\lambda) = 0$ if $\mathrm{Tr}[f^0(\lambda)] > (1-\varepsilon)\,\mathrm{Tr}[f_1(\lambda)]$, the function $|\gamma_2(\lambda)| \le 1$, and

$$\gamma_2(\lambda) = \mathrm{sign}(\mathrm{Tr}(g^0(\lambda) - g_1(\lambda))) : \ \mathrm{Tr}(g^0(\lambda) - g_1(\lambda)) \ne 0.$$

For the second set of admissible spectral densities $\mathcal{D}_\varepsilon^2 \times \mathcal{D}_{1\delta}^2$, we have

$$(\mathbf{r}^0_{\overline{\mu},f}(e^{i\lambda}))(\mathbf{r}^0_{\overline{\mu},f}(e^{i\lambda}))^* = (\Theta^0_{\overline{\mu}}(e^{-i\lambda}))^\top \{(\alpha_{fk}^2 + \gamma_k^1(\lambda))\delta_{kl}\}_{k,l=1}^T \overline{\Theta^0_{\overline{\mu}}(e^{-i\lambda})}, \tag{3.100}$$

$$(\mathbf{r}^0_{\overline{\mu},g}(e^{i\lambda}))(\mathbf{r}^0_{\overline{\mu},g}(e^{i\lambda}))^* = (\Theta^0_{\overline{\mu}}(e^{-i\lambda}))^\top \{\alpha_{gk}^2\gamma_k^2(\lambda)\delta_{kl}\}_{k,l=1}^T \overline{\Theta^0_{\overline{\mu}}(e^{-i\lambda})}, \tag{3.101}$$

$$\frac{1}{2\pi}\int_{-\pi}^{\pi} |g^0_{kk}(\lambda) - g^1_{kk}(\lambda)| d\lambda = \delta_k, \tag{3.102}$$

where α_{fk}^2, α_{gk}^2 are Lagrange multipliers, functions $\gamma_k^1(\lambda) \le 0$ and $\gamma_k^1(\lambda) = 0$ if $f^0_{kk}(\lambda) > (1-\varepsilon)f^1_{kk}(\lambda)$, functions $|\gamma_k^2(\lambda)| \le 1$, and

$$\gamma_k^2(\lambda) = \mathrm{sign}(g^0_{kk}(\lambda) - g^1_{kk}(\lambda)) : \ g^0_{kk}(\lambda) - g^1_{kk}(\lambda) \ne 0, \ k = \overline{1,T}.$$

For the third set of admissible spectral densities $\mathcal{D}_\varepsilon^3 \times \mathcal{D}_{1\delta}^3$, we have

$$(\mathbf{r}^0_{\overline{\mu},f}(e^{i\lambda}))(\mathbf{r}^0_{\overline{\mu},f}(e^{i\lambda}))^* = (\alpha_f^2 + \gamma_1'(\lambda))(\Theta^0_{\overline{\mu}}(e^{-i\lambda}))^\top B_1 \overline{\Theta^0_{\overline{\mu}}(e^{-i\lambda})}, \tag{3.103}$$

$$(\mathbf{r}^0_{\overline{\mu},g}(e^{i\lambda}))(\mathbf{r}^0_{\overline{\mu},g}(e^{i\lambda}))^* = \alpha_g^2\gamma_2'(\lambda)(\Theta^0_{\overline{\mu}}(e^{-i\lambda}))^\top B_2 \overline{\Theta^0_{\overline{\mu}}(e^{-i\lambda})}, \tag{3.104}$$

$$\frac{1}{2\pi}\int_{-\pi}^{\pi}\left|\left\langle B_2, g^0(\lambda)-g_1(\lambda)\right\rangle\right|d\lambda=\delta, \tag{3.105}$$

where α_f^2, α_g^2 are Lagrange multipliers, function $\gamma_1'(\lambda) \leq 0$ and $\gamma_1'(\lambda) = 0$ if $\langle B_1, f^0(\lambda)\rangle > (1-\varepsilon)\langle B_1, f_1(\lambda)\rangle$, function $|\gamma_2'(\lambda)| \leq 1$, and

$$\gamma_2'(\lambda) = \operatorname{sign}\left\langle B_2, g^0(\lambda)-g_1(\lambda)\right\rangle : \ \left\langle B_2, g^0(\lambda)-g_1(\lambda)\right\rangle \neq 0.$$

For the fourth set of admissible spectral densities $\mathcal{D}_\varepsilon^4 \times \mathcal{D}_{1\delta}^4$, we have

$$(\mathbf{r}_{\overline{\mu},f}^0(e^{i\lambda}))(\mathbf{r}_{\overline{\mu},f}^0(e^{i\lambda}))^* = (\Theta_{\overline{\mu}}^0(e^{-i\lambda}))^\top(\vec{\alpha}_f\cdot\vec{\alpha}_f^* + \Gamma(\lambda))\overline{\Theta_{\overline{\mu}}^0(e^{-i\lambda})}, \tag{3.106}$$

$$(\mathbf{r}_{\overline{\mu},g}^0(e^{i\lambda}))(\mathbf{r}_{\overline{\mu},g}^0(e^{i\lambda}))^* = (\Theta_{\overline{\mu}}^0(e^{-i\lambda}))^\top\{\alpha_{g,ij}\gamma_{ij}(\lambda)\}_{i,j=1}^T\overline{\Theta_{\overline{\mu}}^0(e^{-i\lambda})}, \tag{3.107}$$

$$\frac{1}{2\pi}\int_{-\pi}^{\pi}|g_{ij}^0(\lambda)-g_{ij}^1(\lambda)|d\lambda=\delta_i^j, \tag{3.108}$$

where $\vec{\alpha}_f$, $\alpha_{g,ij}$ are Lagrange multipliers, function $\Gamma(\lambda) \leq 0$ and $\Gamma(\lambda) = 0$ if $f^0(\lambda) > (1-\varepsilon)f_1(\lambda)$, functions $|\gamma_{ij}(\lambda)| \leq 1$, and

$$\gamma_{ij}(\lambda) = \frac{g_{ij}^0(\lambda)-g_{ij}^1(\lambda)}{|g_{ij}^0(\lambda)-g_{ij}^1(\lambda)|} : \ g_{ij}^0(\lambda)-g_{ij}^1(\lambda) \neq 0, \ i,j=\overline{1,T}.$$

The following theorem holds.

Theorem 3.2.6. *The least favorable spectral densities $f^0(\lambda)$ and $g^0(\lambda)$ in the classes $\mathcal{D}_\varepsilon^k \times \mathcal{D}_{1\delta}^k$, $k = 1,2,3,4$, for the optimal linear extrapolation of the functional $A\vec{\xi}$ from observations of the vector sequence $\vec{\xi}(m) + \vec{\eta}(m)$ at points $m = -1,-2,\dots$ are determined by canonical factorizations* (1.14), (3.29), *and* (3.30), *equations* (3.97)–(3.99), (3.100)–(3.102), (3.103)–(3.105), *and* (3.106)–(3.108), *respectively, the constrained optimization problem* (3.52), *and restrictions on densities from the corresponding classes $\mathcal{D}_\varepsilon^k$, $\mathcal{D}_{1\delta}^k$, $k = 1,2,3,4$. The minimax-robust spectral characteristic of the optimal estimate of the functional $A\vec{\xi}$ is determined by formula* (3.12).

Theorem 3.2.7. *If the spectral density $g(\lambda)$ is known, the least favorable spectral density $f^0(\lambda)$ in the classes $\mathcal{D}_\varepsilon^k$, $k = 1,2,3,4$, for the optimal linear extrapolation of the functional $A\vec{\xi}$ from observations of the sequence $\vec{\xi}(m) + \vec{\eta}(m)$ at points $m = -1,-2,\dots$ is determined by canonical factorizations* (1.14) *and* (3.29), *equations* (3.97), (3.100), (3.103), *and* (3.106), *respectively, the constrained optimization problem* (3.57), *and restrictions on density from the corresponding classes $\mathcal{D}_\varepsilon^k$, $k = 1,2,3,4$. The minimax-robust spectral characteristic of the optimal estimate of the functional $A\vec{\xi}$ is determined by formula* (3.12).

3.2.5 Least favorable densities for cointegrated sequences in classes $\mathcal{D}_\varepsilon \times \mathcal{D}_{1\delta}$

Consider the minimax extrapolation problem for the functional $A\vec{\xi}$ which depends on unobserved values of a vector sequence $\vec{\xi}(m)$ with GM increments based on observations of the vector sequence $\vec{\zeta}(m)$, cointegrated with $\vec{\xi}(m)$ in terms of Definition 3.1.1, at points $m = -1, -2, \dots$ for the sets of admissible spectral densities $\mathcal{D}_\varepsilon^k$, $k = 1, 2, 3, 4$, defined in Section 3.2.4 and $\mathcal{D}_{1\delta}^k$, $k = 1, 2, 3, 4$, defined as follows:

$$\mathcal{D}_{1\delta}^1 = \left\{ p(\lambda) \left| \frac{1}{2\pi} \int_{-\pi}^{\pi} \frac{|\chi_{\overline{\mu}}^{(d)}(e^{-i\lambda})|^2}{|\beta^{(d)}(i\lambda)|^2} |\operatorname{Tr}(p(\lambda) - p_1(\lambda))| d\lambda \le \delta \right. \right\},$$

$$\mathcal{D}_{1\delta}^2 = \left\{ p(\lambda) \left| \frac{1}{2\pi} \int_{-\pi}^{\pi} \frac{|\chi_{\overline{\mu}}^{(d)}(e^{-i\lambda})|^2}{|\beta^{(d)}(i\lambda)|^2} |p_{kk}(\lambda) - p_{kk}^1(\lambda)| d\lambda \le \delta_k, k = \overline{1, T} \right. \right\},$$

$$\mathcal{D}_{1\delta}^3 = \left\{ p(\lambda) \left| \frac{1}{2\pi} \int_{-\pi}^{\pi} \frac{|\chi_{\overline{\mu}}^{(d)}(e^{-i\lambda})|^2}{|\beta^{(d)}(i\lambda)|^2} |\langle B_2, p(\lambda) - p_1(\lambda) \rangle| d\lambda \le \delta \right. \right\},$$

$$\mathcal{D}_{1\delta}^4 = \left\{ p(\lambda) \left| \frac{1}{2\pi} \int_{-\pi}^{\pi} \frac{|\chi_{\overline{\mu}}^{(d)}(e^{-i\lambda})|^2}{|\beta^{(d)}(i\lambda)|^2} |p_{ij}(\lambda) - p_{ij}^1(\lambda)| d\lambda \le \delta_i^j, i, j = \overline{1, T} \right. \right\}.$$

Here $p_1(\lambda)$ is a fixed spectral density and δ, δ_k, $k = \overline{1, T}$, and δ_i^j, $i, j = \overline{1, T}$, are given numbers.

The condition $0 \in \partial \Delta_{\mathcal{D}}(f^0, g^0)$ implies the following equations which determine the least favorable spectral densities for these given sets of admissible spectral densities.

For the first set of admissible spectral densities $\mathcal{D}_\varepsilon^1 \times \mathcal{D}_{1\delta}^1$, we have

$$\begin{aligned}(\mathbf{C}_{\overline{\mu},a}^{f0}(e^{i\lambda}))(\mathbf{C}_{\overline{\mu},a}^{f0}(e^{i\lambda}))^* - \alpha^2(\mathbf{C}_{\overline{\mu},a}^{p0}(e^{i\lambda}))(\mathbf{C}_{\overline{\mu},a}^{p0}(e^{i\lambda}))^* \\ = (\alpha_f^2 + \gamma_1(\lambda))\left(\frac{|\chi_{\overline{\mu}}^{(d)}(e^{-i\lambda})|^2}{|\beta^{(d)}(i\lambda)|^2} p^0(\lambda) \right)^2,\end{aligned} \tag{3.109}$$

$$(\mathbf{C}_{\overline{\mu},a}^{p0}(e^{i\lambda}))(\mathbf{C}_{\overline{\mu},a}^{p0}(e^{i\lambda}))^* = \alpha_g^2 \gamma_2(\lambda) \left(\frac{|\chi_{\overline{\mu}}^{(d)}(e^{-i\lambda})|^2}{|\beta^{(d)}(i\lambda)|^2} p^0(\lambda) \right)^2, \tag{3.110}$$

$$\frac{1}{2\pi} \int_{-\pi}^{\pi} \frac{|\chi_{\overline{\mu}}^{(d)}(e^{-i\lambda})|^2}{|\beta^{(d)}(i\lambda)|^2} |\operatorname{Tr}(p^0(\lambda) - p_1(\lambda))| d\lambda = \delta, \tag{3.111}$$

where α_f^2, α_g^2 are Lagrange multipliers, the function $\gamma_1(\lambda) \le 0$ and $\gamma_1(\lambda) = 0$ if $\operatorname{Tr}[f^0(\lambda)] > (1 - \varepsilon) \operatorname{Tr}[f_1(\lambda)]$, the function $|\gamma_2(\lambda)| \le 1$, and

$$\gamma_2(\lambda) = \operatorname{sign}(\operatorname{Tr}[p^0(\lambda) - p_1(\lambda)]) : \ \operatorname{Tr}[p_0(\lambda) - p_1(\lambda)] \ne 0.$$

For the second set of admissible spectral densities $\mathcal{D}_{\varepsilon}^2 \times \mathcal{D}_{1\delta}^2$, we have

$$(\mathbf{C}_{\overline{\mu},a}^{f0}(e^{i\lambda}))(\mathbf{C}_{\overline{\mu},a}^{f0}(e^{i\lambda}))^* - \alpha^2(\mathbf{C}_{\overline{\mu},a}^{p0}(e^{i\lambda}))(\mathbf{C}_{\overline{\mu},a}^{p0}(e^{i\lambda}))^*$$

$$= \left(\frac{|\chi_{\overline{\mu}}^{(d)}(e^{-i\lambda})|^2}{|\beta^{(d)}(i\lambda)|^2} p^0(\lambda)\right)\{(\alpha_{fk}^2 + \gamma_k^1(\lambda))\delta_{kl}\}_{k,l=1}^T \left(\frac{|\chi_{\overline{\mu}}^{(d)}(e^{-i\lambda})|^2}{|\beta^{(d)}(i\lambda)|^2} p^0(\lambda)\right), \tag{3.112}$$

$$(\mathbf{C}_{\overline{\mu},a}^{p0}(e^{i\lambda}))(\mathbf{C}_{\overline{\mu},a}^{p0}(e^{i\lambda}))^* = \left(\frac{|\chi_{\overline{\mu}}^{(d)}(e^{-i\lambda})|^2}{|\beta^{(d)}(i\lambda)|^2} p^0(\lambda)\right)$$

$$\times \{\alpha_{gk}^2\gamma_k^2(\lambda)\delta_{kl}\}_{k,l=1}^T \left(\frac{|\chi_{\overline{\mu}}^{(d)}(e^{-i\lambda})|^2}{|\beta^{(d)}(i\lambda)|^2} p^0(\lambda)\right), \tag{3.113}$$

$$\frac{1}{2\pi}\int_{-\pi}^{\pi} \frac{|\chi_{\overline{\mu}}^{(d)}(e^{-i\lambda})|^2}{|\beta^{(d)}(i\lambda)|^2} |p_{kk}^0(\lambda) - p_{kk}^1(\lambda)| d\lambda = \delta_k, \tag{3.114}$$

where α_{fk}^2, α_{gk}^2 are Lagrange multipliers, functions $\gamma_k^1(\lambda) \leq 0$ and $\gamma_k^1(\lambda) = 0$ if $f_{kk}^0(\lambda) > (1-\varepsilon)f_{kk}^1(\lambda)$, functions $|\gamma_k^2(\lambda)| \leq 1$, and

$$\gamma_k^2(\lambda) = \operatorname{sign}(p_{kk}^0(\lambda) - p_{kk}^1(\lambda)) : \; p_{kk}^0(\lambda) - p_{kk}^1(\lambda) \neq 0, \quad k = \overline{1,T}.$$

For the third set of admissible spectral densities $\mathcal{D}_{\varepsilon}^3 \times \mathcal{D}_{1\delta}^3$, we have

$$(\mathbf{C}_{\overline{\mu},a}^{f0}(e^{i\lambda}))(\mathbf{C}_{\overline{\mu},a}^{f0}(e^{i\lambda}))^* - \alpha_f^2(\mathbf{C}_{\overline{\mu},a}^{p0}(e^{i\lambda}))(\mathbf{C}_{\overline{\mu},a}^{p0}(e^{i\lambda}))^*$$

$$= (\alpha^2 + \gamma_1'(\lambda))\left(\frac{|\chi_{\overline{\mu}}^{(d)}(e^{-i\lambda})|^2}{|\beta^{(d)}(i\lambda)|^2} p^0(\lambda)\right) B_1^{\top} \left(\frac{|\chi_{\overline{\mu}}^{(d)}(e^{-i\lambda})|^2}{|\beta^{(d)}(i\lambda)|^2} p^0(\lambda)\right), \tag{3.115}$$

$$(\mathbf{C}_{\overline{\mu},a}^{p0}(e^{i\lambda}))(\mathbf{C}_{\overline{\mu},a}^{p0}(e^{i\lambda}))^* = \alpha_g^2\gamma_2'(\lambda)\left(\frac{|\chi_{\overline{\mu}}^{(d)}(e^{-i\lambda})|^2}{|\beta^{(d)}(i\lambda)|^2} p^0(\lambda)\right)$$

$$\times B_2^{\top} \left(\frac{|\chi_{\overline{\mu}}^{(d)}(e^{-i\lambda})|^2}{|\beta^{(d)}(i\lambda)|^2} p^0(\lambda)\right), \tag{3.116}$$

$$\frac{1}{2\pi}\int_{-\pi}^{\pi} \frac{|\chi_{\overline{\mu}}^{(d)}(e^{-i\lambda})|^2}{|\beta^{(d)}(i\lambda)|^2} |\langle B_2, p^0(\lambda) - p_1(\lambda)\rangle| d\lambda = \delta, \tag{3.117}$$

where α_f^2, α_g^2 are Lagrange multipliers, function $\gamma_1'(\lambda) \leq 0$ and $\gamma_1'(\lambda) = 0$ if $\langle B_1, f^0(\lambda)\rangle > (1-\varepsilon)\langle B_1, f_1(\lambda)\rangle$, function $|\gamma_2'(\lambda)| \leq 1$, and

$$\gamma_2'(\lambda) = \operatorname{sign}\langle B_2, p^0(\lambda) - p_1(\lambda)\rangle : \; \langle B_2, p^0(\lambda) - p_1(\lambda)\rangle \neq 0.$$

For the fourth set of admissible spectral densities $\mathcal{D}_\varepsilon^4 \times \mathcal{D}_{1\delta}^4$, we have

$$(\mathbf{C}^{f0}_{\overline{\mu},a}(e^{i\lambda}))(\mathbf{C}^{f0}_{\overline{\mu},a}(e^{i\lambda}))^* - a^2(\mathbf{C}^{p0}_{\overline{\mu},a}(e^{i\lambda}))(\mathbf{C}^{p0}_{\overline{\mu},a}(e^{i\lambda}))^*$$
$$= \left(\frac{|\chi_{\overline{\mu}}^{(d)}(e^{-i\lambda})|^2}{|\beta^{(d)}(i\lambda)|^2} p^0(\lambda)\right)(\vec{\alpha}_f \cdot \vec{\alpha}_f^* + \Gamma(\lambda))\left(\frac{|\chi_{\overline{\mu}}^{(d)}(e^{-i\lambda})|^2}{|\beta^{(d)}(i\lambda)|^2} p^0(\lambda)\right), \tag{3.118}$$

$$(\mathbf{C}^{p0}_{\overline{\mu},a}(e^{i\lambda}))(\mathbf{C}^{p0}_{\overline{\mu},a}(e^{i\lambda}))^* = \left(\frac{|\chi_{\overline{\mu}}^{(d)}(e^{-i\lambda})|^2}{|\beta^{(d)}(i\lambda)|^2} p^0(\lambda)\right)$$
$$\times \{\alpha_{g,ij}\gamma_{ij}(\lambda)\}_{i,j=1}^T \left(\frac{|\chi_{\overline{\mu}}^{(d)}(e^{-i\lambda})|^2}{|\beta^{(d)}(i\lambda)|^2} p^0(\lambda)\right), \tag{3.119}$$

$$\frac{1}{2\pi}\int_{-\pi}^{\pi} \frac{|\chi_{\overline{\mu}}^{(d)}(e^{-i\lambda})|^2}{|\beta^{(d)}(i\lambda)|^2} |p^0_{ij}(\lambda) - p^1_{ij}(\lambda)| d\lambda = \delta_i^j, \tag{3.120}$$

where $\vec{\alpha}_f$, $\alpha_{g,ij}$ are Lagrange multipliers, function $\Gamma(\lambda) \le 0$ and $\Gamma(\lambda) = 0$ if $f^0(\lambda) > (1-\varepsilon)f_1(\lambda)$, functions $|\gamma_{ij}(\lambda)| \le 1$, and

$$\gamma_{ij}(\lambda) = \frac{p^0_{ij}(\lambda) - p^1_{ij}(\lambda)}{|p^0_{ij}(\lambda) - p^1_{ij}(\lambda)|} : \; p^0_{ij}(\lambda) - p^1_{ij}(\lambda) \neq 0, \quad i,j = \overline{1,T}.$$

The following theorem holds.

Theorem 3.2.8. *The least favorable spectral densities $f^0(\lambda)$ and $p^0(\lambda)$ in the classes $\mathcal{D}_\varepsilon^k \times \mathcal{D}_{1\delta}^k$, $k = 1,2,3,4$, for the optimal linear extrapolation of the functional $A\vec{\xi}$ from observations of the vector sequence $\vec{\zeta}(m)$ cointegrated with $\vec{\xi}(m)$ in terms of Definition* 3.1.1 *at points $m = -1,-2,\dots$ are determined by equations* (3.109)–(3.111), (3.112)–(3.114), (3.115)–(3.117), *and* (3.118)–(3.120), *respectively, the minimality condition* (3.41), *the constrained optimization problem* (3.51) *with $g(\lambda) := |\beta^{(d)}(i\lambda)|^{-2}(p(\lambda) - a^2 f(\lambda))$, and restrictions on densities from the corresponding classes $\mathcal{D}_\varepsilon^k$, $\mathcal{D}_{1\delta}^k$, $k = 1,2,3,4$. The minimax-robust spectral characteristic of the optimal estimate of the functional $A\vec{\xi}$ is determined by formula* (3.44).

Conclusions

In this chapter, we describe methods of solution of the extrapolation problem for linear functionals which are constructed from unobserved values of a sequence with periodically stationary long memory multi-seasonal increments. Estimates are based on observations of the sequence with a periodically stationary noise sequence. Estimates are obtained by representing the sequence under investigation as a multidimensional sequence with stationary GM increments.

The problem is investigated in the case of spectral certainty, where the spectral densities of the sequences are exactly known. In this case, we propose an approach based on the Hilbert space projection method. We derive formulas for calculating the spectral characteristics and the MSEs of the optimal estimates of the functionals.

In the case of spectral uncertainty, where the spectral density is not exactly known while, instead, a set of admissible spectral densities is specified, the minimax-robust method is applied. We propose a representation of the MSE in the form of a linear functional in L_1 space with respect to spectral densities, which allows us to solve the corresponding constrained optimization problem and describe the minimax (robust) estimates of the functionals. Formulas that determine the least favorable spectral densities and minimax (robust) spectral characteristic of the optimal linear estimates of the functionals are derived for a collection of specific classes of admissible spectral densities: $\mathcal{D}_0 \times \mathcal{D}_V^U$ and $\mathcal{D}_\varepsilon \times \mathcal{D}_{1\delta}$. Here D_0 are classes of spectral densities with the moment restrictions, D_V^U are classes describing the "strip" models of spectral densities, D_ε are classes describing the "ε-contaminated" models of a fixed bounded spectral density, and $D_{1\delta}$ are classes describing the "δ-neighborhood" models in the space L_1 of a fixed bounded spectral density.

The main results of this chapter have been published in [120]. The corresponding results for minimax extrapolation of sequences with periodically stationary increments (a particular case of GM increments) observed with noise are described in the paper by Luz and Moklyachuk [124]. The results presented in this chapter rely on and develop prior studies of minimax extrapolation by Moklyachuk, Masyutka, and Sidei [130] for multidimensional stationary sequences, Moklyachuk and Golichenko [60] for periodically correlated sequences, and Luz and Moklyachuk [110] for sequences with stationary increments.

4 Interpolation of sequences with periodically stationary increments observed with or without noise

In this chapter we deal with the interpolation problem for stochastic sequences with periodically stationary long memory multi-seasonal increments. Estimates are based on observations of the sequence with a periodically stationary noise sequence.

To describe the problem precisely consider a stochastic sequence $\vartheta(m)$ with periodically stationary GM increments and a periodically stationary noise sequence $\eta(m)$ uncorrelated with $\vartheta(m)$. Consider a multidimensional stochastic sequence with stationary GM increments $\vec{\xi}(m)$ generated by transformation (1.2) and a multidimensional stationary stochastic sequence $\vec{\eta}(m)$ uncorrelated with the sequence $\vec{\xi}(m)$ and corresponding to the sequence $\eta(m)$. Let the stationary GM increment sequence $\chi_{\overline{\mu},\overline{s}}^{(d)}(\vec{\xi}(m)) = \{\chi_{\overline{\mu},\overline{s}}^{(d)}(\xi_p(m))\}_{p=1}^{T}$ and the stationary stochastic sequence $\vec{\eta}(m)$ have absolutely continuous spectral functions $F(\lambda)$ and $G(\lambda)$ with the spectral densities $f(\lambda) = \{f_{ij}(\lambda)\}_{i,j=1}^{T}$ and $g(\lambda) = \{g_{ij}(\lambda)\}_{i,j=1}^{T}$, respectively.

Without loss of generality we will assume that the mean values of the increment sequences $\mathsf{E}\chi_{\overline{\mu},\overline{s}}^{(d)}(\vec{\xi}(m)) = 0$, $\mathsf{E}\vec{\eta}(m) = 0$, and $\overline{\mu} > \overline{0}$.

As the *interpolation problem* we understand the mean square optimal linear estimation of the functional

$$A_M\vartheta = \sum_{k=0}^{N} a^{(\vartheta)}(k)\vartheta(k), \tag{4.1}$$

which depends on unobserved values of the stochastic sequence $\vartheta(m)$ with periodically stationary GM increments. The estimate is based on observations of the sequence $\zeta(m) = \vartheta(m) + \eta(m)$ at points of the set $Z \setminus \{0, 1, 2, \ldots, N\}$, where the periodically stationary noise sequence $\eta(m)$ is uncorrelated with $\vartheta(m)$. In terms of vector sequences $\vec{\xi}(m)$ and $\vec{\eta}(m)$, we consider the problem of mean square optimal linear estimation of the functional

$$A_N\vec{\xi} = \sum_{k=0}^{N} (\vec{a}(k))^{\top}\vec{\xi}(k),$$

which depends on the unobserved values of the stochastic sequence $\vec{\xi}(k) = \{\xi_p(k)\}_{p=1}^{T}$ with stationary GM increments. Estimates are based on observations of the sequence $\vec{\zeta}(k) = \vec{\xi}(k) + \vec{\eta}(k)$ at points of the set $\mathbb{Z} \setminus \{0, 1, 2 \ldots, N\}$.

https://doi.org/10.1515/9783111325620-004

4.1 Hilbert space projection method of interpolation

4.1.1 Interpolation of multidimensional stochastic sequences with stationary GM increment based on observations with stationary vector-valued noise

Assume that spectral densities $f(\lambda)$ and $g(\lambda)$ of the sequences satisfy the minimality condition

$$\int_{-\pi}^{\pi} \frac{|\beta^{(d)}(i\lambda)|^2}{|\chi_{\overline{\mu}}^{(d)}(e^{-i\lambda})|^2} \operatorname{Tr}[(f(\lambda) + |\beta^{(d)}(i\lambda)|^2 g(\lambda))^{-1}]d\lambda < \infty. \tag{4.2}$$

The latter condition is the necessary and sufficient one under which the MSEs of estimates of functional $A_N\vec{\xi}$ are not equal to 0.

We apply the Hilbert space estimation technique, which can be described as a three-stage procedure:

(i) define a target element (to be estimated) of the space $H = L_2(\Omega, \mathcal{F}, \mathsf{P})$ of random variables γ which have zero mean values and finite variances, $\mathsf{E}\gamma = 0$, $\mathsf{E}|\gamma|^2 < \infty$, endowed with the inner product $\langle \gamma_1, \gamma_2 \rangle = \mathsf{E}\gamma_1\overline{\gamma_2}$,

(ii) define a subspace of H generated by observations, and

(iii) find an estimate of the target element as an orthogonal projection of the target element on the defined subspace.

Stage (i)

The functional $A_N\vec{\xi}$ does not belong to the space H. The following lemma, which is directly obtained from Corollary 3.1.2, describes a representation of the functional as the sum of a functional with finite second moment belonging to H and a functional depending on observed values of the sequence $\vec{\zeta}(k)$ and considered an initial value.

Lemma 4.1.1. *The functional $A_N\vec{\xi}$ admits a representation*

$$A_N\vec{\xi} = A_N\vec{\zeta} - A_N\vec{\eta} = H_N\vec{\xi} - V_N\vec{\zeta}, \tag{4.3}$$

where

$$H_N\vec{\xi} := B_N\chi\vec{\zeta} - A_N\vec{\eta},$$

$$A_N\vec{\zeta} = \sum_{k=0}^{N} (\vec{a}(k))^{\top}\vec{\zeta}(k), \quad A_N\vec{\eta} = \sum_{k=0}^{N} (\vec{a}(k))^{\top}\vec{\eta}(k),$$

$$B_N\chi\vec{\zeta} = \sum_{k=0}^{N} (\vec{b}_{\overline{\mu},N}(k))^{\top}\chi_{\overline{\mu},\overline{s}}^{(d)}(\vec{\zeta}(k)), \quad V_N\vec{\zeta} = \sum_{k=-d(\gamma)}^{-1} (\vec{v}_{\overline{\mu},N}(k))^{\top}\vec{\zeta}(k),$$

where the coefficients

$$\vec{b}_{\overline{\mu},N}(k) = \{b_{\overline{\mu},N,p}(k)\}_{p=1}^{T}, \quad k = 0,1,\dots,N,$$

and

$$\vec{v}_{\overline{\mu},N}(k) = \{v_{\overline{\mu},N,p}(k)\}_{p=1}^{T}, \quad k = -1,-2,\dots,-d(\gamma),$$

are calculated by the formulas

$$\vec{v}_{\overline{\mu},N}(k) = \sum_{l=0}^{\min\{N,k+d(\gamma)\}} \operatorname{diag}_T(e_\nu(l-k))\vec{b}_{\overline{\mu},N}(l), \quad k = -1,-2,\dots,-d(\gamma), \tag{4.4}$$

$$\vec{b}_{\overline{\mu},N}(k) = \sum_{m=k}^{N} \operatorname{diag}_T(d_{\overline{\mu}}(m-k))\vec{a}(m) = (D_N^{\overline{\mu}}\mathbf{a}_N)_k, \quad k = 0,1,\dots,N, \tag{4.5}$$

where the vector

$$\mathbf{a}_N = ((\vec{a}(0))^\top, (\vec{a}(1))^\top, \dots, (\vec{a}(N))^\top)^\top,$$

$D_N^{\overline{\mu}}$ *is the linear transformation determined by a matrix with the entries* $(D_N^{\overline{\mu}})(k,j) = \operatorname{diag}_T(d_{\overline{\mu}}(j-k))$ *if* $0 \le k \le j \le N$ *and* $(D_N^{\overline{\mu}})(k,j) = \mathcal{O}$ *if* $0 \le j < k \le N$, $\operatorname{diag}_T(x)$ *denotes a* $T \times T$ *diagonal matrix with* x *on its diagonal, and coefficients* $\{d_{\overline{\mu}}(k) : k \ge 0\}$ *are determined by the relationship*

$$\sum_{k=0}^{\infty} d_{\overline{\mu}}(k)x^k = \prod_{i=1}^{r}\left(\sum_{j_i=0}^{\infty} x^{\mu_i s_i j_i}\right)^{d_i}.$$

The functional $H_N\vec{\xi}$ from representation (4.3) has finite variance and the functional $V_N\vec{\zeta}$ depends on the known observations of the stochastic sequence $\vec{\zeta}(k)$ at points $k = -d(\gamma), -d(\gamma)+1, \dots, -1$. Therefore, estimates $\widehat{A}_N\vec{\xi}$ and $\widehat{H}_N\vec{\xi}$ of the functionals $A_N\vec{\xi}$ and $H_N\vec{\xi}$ and the MSEs $\Delta(f,g;\widehat{A}_N\vec{\xi}) = \mathsf{E}|A_N\vec{\xi} - \widehat{A}_N\vec{\xi}|^2$ and $\Delta(f,g;\widehat{H}_N\vec{\xi}) = \mathsf{E}|H_N\vec{\xi} - \widehat{H}_N\vec{\xi}|^2$ of the estimates $\widehat{A}_N\vec{\xi}$ and $\widehat{H}_N\vec{\xi}$ satisfy the following relations:

$$\widehat{A}_N\vec{\xi} = \widehat{H}_N\vec{\xi} - V_N\vec{\zeta}, \tag{4.6}$$

$$\Delta(f,g;\widehat{A}_N\vec{\xi}) = \mathsf{E}|A_N\vec{\xi} - \widehat{A}_N\vec{\xi}|^2 = \mathsf{E}|H_N\vec{\xi} - \widehat{H}_N\vec{\xi}|^2 = \Delta(f,g;\widehat{H}_N\vec{\xi}). \tag{4.7}$$

For these reasons the estimation problem for the functional $A_N\vec{\xi}$ is equivalent to the one for the functional $H_N\vec{\xi}$. This problem can be solved by applying the Hilbert space projection method.

The functional $H_N\vec{\xi}$ admits the spectral representation

$$H_N\vec{\xi} = \int_{-\pi}^{\pi} (\vec{B}_{\overline{\mu},N}(e^{i\lambda}))^\top \frac{\chi_{\overline{\mu}}^{(d)}(e^{-i\lambda})}{\beta^{(d)}(i\lambda)} d\vec{Z}_{\xi^{(d)}+\eta^{(d)}}(\lambda) - \int_{-\pi}^{\pi} (\vec{A}_N(e^{i\lambda}))^\top d\vec{Z}_\eta(\lambda),$$

where

$$\vec{B}_{\overline{\mu},N}(e^{i\lambda}) = \sum_{k=0}^{N} \vec{b}_{\overline{\mu},N}(k)e^{i\lambda k} = \sum_{k=0}^{N} (D_N^{\overline{\mu}}\mathbf{a}_N)_k e^{i\lambda k}, \quad \vec{A}_N(e^{i\lambda}) = \sum_{k=0}^{N} \vec{a}(k)e^{i\lambda k}.$$

Stage (ii)

We introduce the following notations. We define the closed linear subspace

$$H^{0-}(\vec{\xi}^{(d)}_{\overline{\mu},\overline{s}} + \vec{\eta}^{(d)}_{\overline{\mu},\overline{s}}) = \overline{\operatorname{span}}\{\chi^{(d)}_{\overline{\mu},\overline{s}}(\xi_p(k)) + \chi^{(d)}_{\overline{\mu},\overline{s}}(\eta_p(k)) : k = -1, -2, -3, \ldots, \; p = 1, \ldots, T\}$$

in the Hilbert space $H = L_2(\Omega, \mathcal{F}, \mathsf{P})$ of random variables γ with zero mean value, $\mathsf{E}\gamma = 0$, finite variance, $\mathsf{E}|\gamma|^2 < \infty$, and the inner product $(\gamma_1; \gamma_2) = \mathsf{E}\gamma_1\overline{\gamma_2}$.

We define also the closed linear subspace

$$H^{N+}(\vec{\xi}^{(d)}_{-\overline{\mu},\overline{s}} + \vec{\eta}^{(d)}_{-\overline{\mu},\overline{s}}) = \overline{\operatorname{span}}\{\chi^{(d)}_{-\overline{\mu},\overline{s}}(\xi_p(k)) + \chi^{(d)}_{-\overline{\mu},\overline{s}}(\eta_p(k)) : k = N+1, \ldots, \; p = 1, \ldots, T\}$$

in the Hilbert space $H = L_2(\Omega, \mathcal{F}, \mathsf{P})$. The defined subspaces are generated by the observed GM increments.

Denote by $L_2^{0-}(f(\lambda) + |\beta^{(d)}(i\lambda)|^2 g(\lambda))$ and $L_2^{N+}(f(\lambda) + |\beta^{(d)}(i\lambda)|^2 g(\lambda))$ the closed linear subspaces of the Hilbert space $L_2(f(\lambda) + |\beta^{(d)}(i\lambda)|^2 g(\lambda))$ of vector-valued functions with the inner product

$$\langle g_1; g_2\rangle = \int_{-\pi}^{\pi} (g_1(\lambda))^\top (f(\lambda) + |\beta^{(d)}(i\lambda)|^2 g(\lambda))\overline{g_2(\lambda)}d\lambda,$$

which is generated by the functions

$$e^{i\lambda k}\chi^{(d)}_{\overline{\mu}}(e^{-i\lambda})\frac{1}{\beta^{(d)}(i\lambda)}\vec{\delta}_l, \quad l = 1, \ldots, T, \quad k \leq -1 \text{ and } k \geq N+1,$$

respectively, where $\vec{\delta}_l = \{\delta_{lp}\}_{p=1}^{T}$, where δ_{lp} are Kronecker symbols.

Remark 4.1.1. A relation

$$\chi^{(d)}_{\overline{\mu},\overline{s}}(\vec{\xi}(k)) + \chi^{(d)}_{\overline{\mu},\overline{s}}(\vec{\eta}(k)) = \int_{-\pi}^{\pi} e^{i\lambda k}\chi^{(d)}_{\overline{\mu}}(e^{-i\lambda})\frac{1}{\beta^{(d)}(i\lambda)}d\vec{Z}_{\xi^{(d)}+\eta^{(d)}}(\lambda)$$

yields a map between elements

$$e^{i\lambda k}\chi^{(d)}_{\overline{\mu}}(e^{-i\lambda})\frac{1}{\beta^{(d)}(i\lambda)}\vec{\delta}_l$$

from the space

$$L_2^{0-}(f(\lambda)+|\beta^{(d)}(i\lambda)|^2 g(\lambda)) \oplus L_2^{(N+d(\gamma))+}(f(\lambda)+|\beta^{(d)}(i\lambda)|^2 g(\lambda))$$

and elements $\chi_{\overline{\mu},\overline{s}}^{(d)}(\xi_l(k)) + \chi_{\overline{\mu},\overline{s}}^{(d)}(\eta_l(k))$ from the space

$$\begin{aligned} H^{0-}(\vec{\xi}^{(d)}_{\overline{\mu},\overline{s}} + \vec{\eta}^{(d)}_{\overline{\mu},\overline{s}}) \oplus H^{N+}(\vec{\xi}^{(d)}_{-\overline{\mu},\overline{s}} + \vec{\eta}^{(d)}_{-\overline{\mu},\overline{s}}) \\ = H^{0-}(\vec{\xi}^{(d)}_{\overline{\mu},\overline{s}} + \vec{\eta}^{(d)}_{\overline{\mu},\overline{s}}) \oplus H^{(N+d(\gamma))+}(\vec{\xi}^{(d)}_{\overline{\mu},\overline{s}} + \vec{\eta}^{(d)}_{\overline{\mu},\overline{s}}). \end{aligned}$$

Stage (iii)

At this stage we find the mean square optimal estimate $\widehat{H}_N\vec{\xi}$ as an orthogonal projection of the element $H_N\vec{\xi}$ on the subspace $H^{0-}(\vec{\xi}^{(d)}_{\overline{\mu},\overline{s}} + \vec{\eta}^{(d)}_{\overline{\mu},\overline{s}}) \oplus H^{(N+d(\gamma))+}(\vec{\xi}^{(d)}_{\overline{\mu},\overline{s}} + \vec{\eta}^{(d)}_{\overline{\mu},\overline{s}})$:

$$\widehat{H}_N\vec{\xi} = \mathrm{Proj}_{H^{0-}(\vec{\xi}^{(d)}_{\overline{\mu},\overline{s}} + \vec{\eta}^{(d)}_{\overline{\mu},\overline{s}}) \oplus H^{(N+d(\gamma))+}(\vec{\xi}^{(d)}_{\overline{\mu},\overline{s}} + \vec{\eta}^{(d)}_{\overline{\mu},\overline{s}})} H_N\vec{\xi}.$$

Relation (4.6) implies that an optimal linear estimate $\widehat{A}\vec{\xi}$ of the functional $A\vec{\xi}$ can be represented in the form

$$\widehat{A}\vec{\xi} = \mathrm{Proj}_{H^{0-}(\vec{\xi}^{(d)}_{\overline{\mu},\overline{s}} + \vec{\eta}^{(d)}_{\overline{\mu},\overline{s}}) \oplus H^{(N+d(\gamma))+}(\vec{\xi}^{(d)}_{\overline{\mu},\overline{s}} + \vec{\eta}^{(d)}_{\overline{\mu},\overline{s}})} H_N\vec{\xi} - V_N\vec{\zeta}$$

or in the form

$$\widehat{A}_N\vec{\xi} = \int_{-\pi}^{\pi} (\vec{h}_{\overline{\mu},N}(\lambda))^{\top} d\vec{Z}_{\xi^{(d)}+\eta^{(d)}}(\lambda) - \sum_{k=-d(\gamma)}^{-1} (\vec{v}_{\overline{\mu},N}(k))^{\top}(\vec{\xi}(k)+\vec{\eta}(k)), \tag{4.8}$$

where $\vec{h}_{\overline{\mu},N}(\lambda) = \{h_{\overline{\mu},N,p}(\lambda)\}_{p=1}^{T}$ is the spectral characteristic of the optimal estimate $\widehat{H}_N\vec{\xi}$.

For all $0 \le k,j \le N+d(\gamma)$, define the Fourier coefficients of the corresponding functions

$$T^{\overline{\mu}}_{k,j} = \frac{1}{2\pi}\int_{-\pi}^{\pi} e^{-i\lambda(k-j)} \frac{|\beta^{(d)}(i\lambda)|^2}{|\chi^{(d)}_{\overline{\mu}}(e^{-i\lambda})|^2} (g(\lambda)(f(\lambda)+|\beta^{(d)}(i\lambda)|^2 g(\lambda))^{-1})^{\top} d\lambda,$$

$$P^{\overline{\mu}}_{k,j} = \frac{1}{2\pi}\int_{-\pi}^{\pi} e^{-i\lambda(k-j)} \frac{|\beta^{(d)}(i\lambda)|^2}{|\chi^{(d)}_{\overline{\mu}}(e^{-i\lambda})|^2} ((f(\lambda)+|\beta^{(d)}(i\lambda)|^2 g(\lambda))^{-1})^{\top} d\lambda,$$

$$Q_{k,j} = \frac{1}{2\pi}\int_{-\pi}^{\pi} e^{-i\lambda(k-j)} (f(\lambda)(f(\lambda)+|\beta^{(d)}(i\lambda)|^2 g(\lambda))^{-1} g(\lambda))^{\top} d\lambda,$$

and matrices $\mathbf{P}^{\overline{\mu}}_N$ and $\mathbf{T}^{\overline{\mu}}_N$ of dimension $(N+d(\gamma)+1)T \times (N+d(\gamma)+1)T$ with $T\times T$ matrix entries $(\mathbf{P}^{\overline{\mu}}_N)_{j,k} = P^{\overline{\mu}}_{j,k}$ and $(\mathbf{T}^{\overline{\mu}}_N)_{j,k} = T^{\overline{\mu}}_{j,k}$, $0 \le j,k \le N+d(\gamma)$, and a matrix $\mathbf{Q}_N$ of dimension $(N+1)T \times (N+1)T$ with the $T\times T$ matrix entries $(\mathbf{Q}_N)_{j,k} = Q_{j,k}$, $0 \le j,k \le N$.

Define coefficients $\{\vec{a}_{\overline{\mu},N}(m) : 0 \leq m \leq N + d(\gamma)\}$ by the formulas

$$\vec{a}_{\overline{\mu},N}(m) = \vec{a}_{-\overline{\mu},N}(m - d(\gamma)), \quad 0 \leq m \leq N + d(\gamma),$$

and

$$\vec{a}_{-\overline{\mu},N}(m) = \sum_{l=\max\{m,0\}}^{\min\{m+d(\gamma),N\}} e_\gamma(l-m)\vec{a}(l), \quad -d(\gamma) \leq m \leq N, \tag{4.9}$$

a vector of dimension $(N + d(\gamma) + 1)T$,

$$\mathbf{a}_N^{\overline{\mu}} = ((\vec{a}_{\overline{\mu},N}(0))^\top, (\vec{a}_{\overline{\mu},N}(1))^\top, (\vec{a}_{\overline{\mu},N}(2))^\top, \dots, (\vec{a}_{\overline{\mu},N}(N + d(\gamma)))^\top)^\top,$$

and a functional

$$\vec{A}_{\overline{\mu},N}(e^{i\lambda}) = \vec{A}_N(e^{i\lambda})\overline{\chi_{\overline{\mu}}^{(d)}(e^{-i\lambda})} = \sum_{k=0}^{N+d(\gamma)} \vec{a}_{\overline{\mu},N}(k)e^{ik\lambda}.$$

Denote by $[D_N^{\overline{\mu}}\mathbf{a}_N]_{+d(\gamma)}$ a vector of dimension $(N + d(\gamma) + 1)T$ which is constructed by adding $d(\gamma)T$ zeros to the vector $D_N^{\overline{\mu}}\mathbf{a}_N$ of dimension $(N + 1)T$.

A solution of the stated interpolation problem is formulated by the following theorem.

Theorem 4.1.1. *Let $\{\vec{\xi}(m), m \in \mathbb{Z}\}$ be a stochastic sequence which defines the stationary GM increment sequence $\chi_{\overline{\mu},\overline{s}}^{(d)}(\vec{\xi}(m)) = \{\chi_{\overline{\mu},\overline{s}}^{(d)}(\xi_p(m))\}_{p=1}^T$ with the absolutely continuous spectral function $F(\lambda)$ which has spectral density $f(\lambda)$. Let $\{\vec{\eta}(m), m \in \mathbb{Z}\}$ be an uncorrelated with the sequence $\vec{\xi}(m)$ stationary stochastic sequence with an absolutely continuous spectral function $G(\lambda)$ which has spectral density $g(\lambda)$. Let minimality condition (4.2) be satisfied. The optimal linear estimate $\widehat{A}_N\vec{\xi}$ of the functional $A_N\vec{\xi}$ which depends on the unknown values of elements $\vec{\xi}(k)$, $k = 0, 1, 2, \dots, N$, from observations of the sequence $\vec{\xi}(m) + \vec{\eta}(m)$ at points of the set $Z \setminus \{0, 1, 2, \dots, N\}$ is calculated by formula (4.8), where the spectral characteristic $\vec{h}_{\overline{\mu},N}(\lambda)$ is calculated by the formula*

$$\begin{aligned}(\vec{h}_{\overline{\mu},N}(\lambda))^\top &= (\vec{B}_{\overline{\mu},N}(e^{i\lambda}))^\top \frac{\chi_{\overline{\mu}}^{(d)}(e^{-i\lambda})}{\beta^{(d)}(i\lambda)} \\ &\quad - (\vec{A}_N(e^{i\lambda}))^\top g(\lambda)\overline{\beta^{(d)}(i\lambda)}(f(\lambda) + |\beta^{(d)}(i\lambda)|^2 g(\lambda))^{-1} \\ &\quad - (\vec{C}_{\overline{\mu},N}(e^{i\lambda}))^\top \frac{\overline{\beta^{(d)}(i\lambda)}}{\chi_{\overline{\mu}}^{(d)}(e^{-i\lambda})}(f(\lambda) + |\beta^{(d)}(i\lambda)|^2 g(\lambda))^{-1},\end{aligned} \tag{4.10}$$

where

$$\vec{C}_{\overline{\mu},N}(e^{i\lambda}) = \sum_{k=0}^{N+d(\gamma)} ((\mathbf{P}_N^{\overline{\mu}})^{-1}[D_N^{\overline{\mu}}\mathbf{a}_N]_{+d(\gamma)} - (\mathbf{P}_N^{\overline{\mu}})^{-1}\mathbf{T}_N^{\overline{\mu}}\mathbf{a}_N^{\overline{\mu}})_k e^{i\lambda k}. \tag{4.11}$$

The MSE $\Delta(f,g;\widehat{A}_N\vec{\xi})$ *is calculated by the formula*

$$
\begin{aligned}
\Delta(f,g;\widehat{A}_N\vec{\xi}) &= \Delta(f,g;\widehat{H}_N\vec{\xi}) = \mathsf{E}|H_N\vec{\xi}-\widehat{H}_N\vec{\xi}|^2 \\
&= \frac{1}{2\pi}\int_{-\pi}^{\pi}\frac{|\beta^{(d)}(i\lambda)|^2}{|\chi_{\overline{\mu}}^{(d)}(e^{-i\lambda})|^2}((\vec{A}_{\overline{\mu},N}(e^{i\lambda}))^{\top}g(\lambda)+(\vec{C}_{\overline{\mu},N}(e^{i\lambda}))^{\top}) \\
&\quad\times(f(\lambda)+|\beta^{(d)}(i\lambda)|^2g(\lambda))^{-1}f(\lambda)\,(f(\lambda)+|\beta^{(d)}(i\lambda)|^2g(\lambda))^{-1} \\
&\quad\times(g(\lambda)\overline{\vec{A}_{\overline{\mu},N}(e^{i\lambda})}+\overline{\vec{C}_{\overline{\mu},N}(e^{i\lambda})})d\lambda \\
&\quad+\frac{1}{2\pi}\int_{-\pi}^{\pi}\frac{1}{|\chi_{\overline{\mu}}^{(d)}(e^{-i\lambda})|^2}((\vec{A}_{\overline{\mu},N}(e^{i\lambda}))^{\top}f(\lambda)-|\beta^{(d)}(i\lambda)|^2(\vec{C}_{\overline{\mu},N}(e^{i\lambda}))^{\top}) \\
&\quad\times(f(\lambda)+|\beta^{(d)}(i\lambda)|^2g(\lambda))^{-1}g(\lambda)\,(f(\lambda)+|\beta^{(d)}(i\lambda)|^2g(\lambda))^{-1} \\
&\quad\times(\overline{\vec{A}_{\overline{\mu},N}(e^{i\lambda})}-|\beta^{(d)}(i\lambda)|^2\overline{\vec{C}_{\overline{\mu},N}(e^{i\lambda})})d\lambda \\
&= \langle[D_N^{\overline{\mu}}\mathbf{a}_N]_{+d(\gamma)}-\mathbf{T}_N^{\overline{\mu}}\mathbf{a}_N^{\overline{\mu}},(\mathbf{P}_N^{\overline{\mu}})^{-1}[D_N^{\overline{\mu}}\mathbf{a}_N]_{+d(\gamma)}-(\mathbf{P}_N^{\overline{\mu}})^{-1}\mathbf{T}_N^{\overline{\mu}}\mathbf{a}_N^{\overline{\mu}}\rangle \\
&\quad+\langle\mathbf{Q}_N\mathbf{a}_N,\mathbf{a}_N\rangle.
\end{aligned}
\tag{4.12}
$$

Proof. The projection $\widehat{H}_N\vec{\xi}$ is determined by two conditions:
(1) $\widehat{H}_N\vec{\xi}\in H^{0-}(\vec{\xi}_{\overline{\mu},\overline{s}}^{(d)}+\vec{\eta}_{\overline{\mu},\overline{s}}^{(d)})\oplus H^{(N+d(\gamma))+}(\vec{\xi}_{\overline{\mu},\overline{s}}^{(d)}+\vec{\eta}_{\overline{\mu},\overline{s}}^{(d)})$;
(2) $(H_N\vec{\xi}-\widehat{H}_N\vec{\xi})\perp H^{0-}(\vec{\xi}_{\overline{\mu},\overline{s}}^{(d)}+\vec{\eta}_{\overline{\mu},\overline{s}}^{(d)})\oplus H^{(N+d(\gamma))+}(\vec{\xi}_{\overline{\mu},\overline{s}}^{(d)}+\vec{\eta}_{\overline{\mu},\overline{s}}^{(d)})$.

The second condition implies the following relation, which holds for all $k\leq -1$, $k\geq N+d(\gamma)+1$, and $l=1,\dots,T$:

$$
\begin{aligned}
&\int_{-\pi}^{\pi}\left(\left(\vec{B}_{\overline{\mu},N}(e^{i\lambda})\frac{\chi_{\overline{\mu}}^{(d)}(e^{-i\lambda})}{\beta^{(d)}(i\lambda)}-\vec{h}_{\overline{\mu},N}(\lambda)\right)^{\top}(f(\lambda)+|\beta^{(d)}(i\lambda)|^2g(\lambda))\right. \\
&\quad\left.-(\vec{A}_N(e^{i\lambda}))^{\top}g(\lambda)\overline{\beta^{(d)}(i\lambda)}\right)\frac{\overline{\chi_{\overline{\mu}}^{(d)}(e^{-i\lambda})}}{\overline{\beta^{(d)}(i\lambda)}}\vec{\delta}_le^{-i\lambda k}d\lambda=0.
\end{aligned}
$$

This relation implies a representation (4.10) for the spectral characteristic $\vec{h}_{\overline{\mu},N}(\lambda)$, where

$$
\vec{C}_{\overline{\mu},N}(e^{i\lambda})=\sum_{k=0}^{N+d(\gamma)}\vec{c}_{\overline{\mu},N}(k)e^{ik\lambda},
$$

where $\vec{c}_{\overline{\mu},N}(k)=\{c_{\overline{\mu},N,p}(k)\}_{p=1}^{T}$, $k=0,1,\dots,N+d(\gamma)$, are unknown coefficients to be found.

It follows from condition (1) that the following equations should be satisfied for $0\leq j\leq N+d(\gamma)$:

$$\int_{-\pi}^{\pi}\Big((\vec{B}_{\overline{\mu},N}(e^{i\lambda}))^{\top}-(\vec{A}_{\overline{\mu},N}(e^{i\lambda}))^{\top}g(\lambda)\frac{|\beta^{(d)}(i\lambda)|^2}{|\chi_{\overline{\mu}}^{(d)}(e^{-i\lambda})|^2}(f(\lambda)+|\beta^{(d)}(i\lambda)|^2g(\lambda))^{-1}$$

$$-(\vec{C}_{\overline{\mu},N}(e^{i\lambda}))^{\top}\frac{|\beta^{(d)}(i\lambda)|^2}{|\chi_{\overline{\mu}}^{(d)}(e^{-i\lambda})|^2}(f(\lambda)+|\beta^{(d)}(i\lambda)|^2g(\lambda))^{-1}\Big)e^{-i\lambda j}d\lambda=0. \tag{4.13}$$

Making use of the defined Fourier coefficients $T_{k,j}^{\overline{\mu}}$ and $P_{k,j}$, relation (4.13) can be presented as a system of $N+d(\gamma)+1$ linear equations determining the unknown coefficients $\vec{c}_{\overline{\mu},N}(k)$, $0\leq k\leq N+d(\gamma)$:

$$\begin{cases}\vec{b}_{\overline{\mu},N}(j)-\sum_{m=0}^{N+d(\gamma)}T_{j,m}^{\overline{\mu}}\vec{a}_{\overline{\mu},N}(m)=\sum_{k=0}^{N+d(\gamma)}P_{j,k}^{\overline{\mu}}\vec{c}_{\overline{\mu},N}(k), & 0\leq j\leq N,\\ -\sum_{m=0}^{N+d(\gamma)}T_{j,m}^{\overline{\mu}}\vec{a}_{\overline{\mu},N}(m)=\sum_{k=0}^{N+d(\gamma)}P_{j,k}^{\overline{\mu}}\vec{c}_{\overline{\mu},N}(k), & N+1\leq j\leq N+d(\gamma),\end{cases} \tag{4.14}$$

where coefficients $\{\vec{a}_{\overline{\mu},N}(m):0\leq m\leq N+d(\gamma)\}$ are determined with the help of coefficients (4.9).

Making use of the definition of the vector $[D_N^{\overline{\mu}}\mathbf{a}_N]_{+d(\gamma)}$, the system (4.14) can be represented in matrix form as

$$[D_N^{\overline{\mu}}\mathbf{a}_N]_{+d(\gamma)}-\mathbf{T}_N^{\overline{\mu}}\mathbf{a}_N^{\overline{\mu}}=\mathbf{P}_N^{\overline{\mu}}\mathbf{c}_N^{\overline{\mu}},$$

where

$$\mathbf{c}_N^{\overline{\mu}}=((\vec{c}_{\overline{\mu},N}(0))^{\top},(\vec{c}_{\overline{\mu},N}(1))^{\top},(\vec{c}_{\overline{\mu},N}(2))^{\top},\dots,(\vec{c}_{\overline{\mu},N}(N+d(\gamma)))^{\top})^{\top}$$

is a vector of dimension $(N+d(\gamma)+1)T$.

Thus, the coefficients $\vec{c}_{\overline{\mu},N}(k)$, $0\leq k\leq N+d(\gamma)$, are determined by the formula

$$\vec{c}_{\overline{\mu},N}(k)=((\mathbf{P}_N^{\overline{\mu}})^{-1}[D_N^{\overline{\mu}}\mathbf{a}_N]_{+d(\gamma)}-(\mathbf{P}_N^{\overline{\mu}})^{-1}\mathbf{T}_N^{\overline{\mu}}\mathbf{a}_N^{\overline{\mu}})_k,$$

where $((\mathbf{P}_N^{\overline{\mu}})^{-1}[D_N^{\overline{\mu}}\mathbf{a}_N]_{+d(\gamma)}-(\mathbf{P}_N^{\overline{\mu}})^{-1}\mathbf{T}_N^{\overline{\mu}}\mathbf{a}_N^{\overline{\mu}})_k$, $0\leq k\leq N+d(\gamma)$, is the kth element of the vector $(\mathbf{P}_N^{\overline{\mu}})^{-1}[D_N^{\overline{\mu}}\mathbf{a}_N]_{+d(\gamma)}-(\mathbf{P}_N^{\overline{\mu}})^{-1}\mathbf{T}_N^{\overline{\mu}}\mathbf{a}_N^{\overline{\mu}}$.

The derived expressions justify formulas (4.10) and (4.12) for the spectral characteristic $\vec{h}_{\overline{\mu},N}(\lambda)$ and the value of the MSE $\Delta(f,g;\widehat{A}_N\xi)$ of the estimate $\widehat{A}_N\xi$ of the functional $A_N\xi$. □

Corollary 4.1.1. *The spectral characteristic* $\vec{h}_{\overline{\mu},N}(\lambda)$ (4.10) *admits the representation* $\vec{h}_{\overline{\mu},N}(\lambda)=\vec{h}_{\overline{\mu},N}^1(\lambda)-\vec{h}_{\overline{\mu},N}^2(\lambda)$, *where*

$$(\vec{h}_{\overline{\mu},N}^1(\lambda))^{\top}=(\vec{B}_{\overline{\mu},N}(e^{i\lambda}))^{\top}\frac{\chi_{\overline{\mu}}^{(d)}(e^{-i\lambda})}{\beta^{(d)}(i\lambda)}$$

$$-\frac{\overline{\beta^{(d)}(i\lambda)}}{\chi_{\overline{\mu}}^{(d)}(e^{-i\lambda})}\left(\sum_{k=0}^{N+d(\gamma)}((\mathbf{P}_N^{\overline{\mu}})^{-1}[D_N^{\overline{\mu}}\mathbf{a}_N]_{+d(\gamma)})_k e^{i\lambda k}\right)^{\top}(f(\lambda)+|\beta^{(d)}(i\lambda)|^2 g(\lambda))^{-1}, \tag{4.15}$$

$$(\vec{h}_{\overline{\mu},N}^2(\lambda))^{\top} = (\vec{A}_N(e^{i\lambda}))^{\top}\overline{\beta^{(d)}(i\lambda)}g(\lambda)(f(\lambda)+|\beta^{(d)}(i\lambda)|^2 g(\lambda))^{-1}$$
$$-\frac{\overline{\beta^{(d)}(i\lambda)}}{\chi_{\overline{\mu}}^{(d)}(e^{-i\lambda})}\left(\sum_{k=0}^{N+d(\gamma)}((\mathbf{P}_N^{\overline{\mu}})^{-1}\mathbf{T}_N^{\overline{\mu}}\mathbf{a}_N^{\overline{\mu}})_k e^{i\lambda k}\right)^{\top}(f(\lambda)+|\beta^{(d)}(i\lambda)|^2 g(\lambda))^{-1}. \tag{4.16}$$

Here $\vec{h}_{\overline{\mu},N}^1(\lambda)$ and $\vec{h}_{\overline{\mu},N}^2(\lambda)$ are the spectral characteristics of the optimal estimates $\widehat{B}_N\chi\vec{\zeta}$ and $\widehat{A}_N\vec{\eta}$ of the functionals $B_N\chi\vec{\zeta}$ and $A_N\vec{\eta}$, respectively, based on observations $\vec{\xi}(k)+\vec{\eta}(k)$ at points of the set $Z\setminus\{0,1,2,\dots,N\}$.

4.1.2 Interpolation of multidimensional stationary GM increments from observations without noise

In this subsection, we consider a corollary for solution of the interpolation problem for the case of observations without noise. Let the stationary GM increment sequence $\chi_{\overline{\mu},\overline{s}}^{(d)}(\vec{\xi}(m)) = \{\chi_{\overline{\mu},\overline{s}}^{(d)}(\xi_p(m))\}_{p=1}^T$ have the absolutely continuous spectral functions $F(\lambda)$ with the spectral densities $f(\lambda) = \{f_{ij}(\lambda)\}_{i,j=1}^T$ satisfying the minimality condition

$$\int_{-\pi}^{\pi}\frac{|\beta^{(d)}(i\lambda)|^2}{|\chi_{\overline{\mu}}^{(d)}(e^{-i\lambda})|^2}\operatorname{Tr}[f^{-1}(\lambda)]d\lambda<\infty. \tag{4.17}$$

By *interpolation problem without noise* we understand the problem of mean square optimal linear estimation of the functional

$$A_N\vec{\xi} = \sum_{k=0}^{N}(\vec{a}(k))^{\top}\vec{\xi}(k), \tag{4.18}$$

which depends on the unobserved values of the stochastic sequence $\vec{\xi}(k) = \{\xi_p(k)\}_{p=1}^T$ with stationary GM increments based on observations of the sequence $\vec{\xi}(k)$ at points $k\in\mathbb{Z}\setminus\{0,1,2\dots,N\}$.

Lemma 4.1.1 provides a representation of the functional $A_N\vec{\xi}$ as the sum of a functional $B_N\chi\vec{\xi}$ from the space $H = L_2(\Omega,\mathcal{F},\mathsf{P})$ and a functional $V_N\vec{\xi}$ of observed values $\vec{\xi}(k)$, $k=-1,-2,\dots,-d(\gamma)$. The following equality holds:

$$\widehat{A}_N\vec{\xi} = \widehat{B}_N\chi\vec{\xi} - V_N\vec{\xi}. \tag{4.19}$$

We denote by $\Delta(f,\widehat{A}_N\vec{\xi}) := \mathsf{E}|A_N\vec{\xi}-\widehat{A}_N\vec{\xi}|^2$ and $\Delta(f,\widehat{B}_N\chi\vec{\xi}) := \mathsf{E}|B_N\chi\vec{\xi}-\widehat{B}_N\chi\vec{\xi}|^2$ the MSE of the optimal estimates $\widehat{A}_N\vec{\xi}$ of the functional $A_N\vec{\xi}$ and the optimal estimate $\widehat{B}_N\chi\vec{\xi}$ of the functional $B_N\chi\vec{\xi}$, respectively. Then

$$\begin{aligned}\Delta(f;\widehat{A}_N\vec{\xi}) &= \mathsf{E}|A_N\vec{\xi}-\widehat{A}_N\vec{\xi}|^2 = \mathsf{E}|B_N\chi\vec{\xi}-V\vec{\xi}-\widehat{B}_N\chi\vec{\xi}+V_N\vec{\xi}|^2\\ &= \mathsf{E}|B_N\chi\vec{\xi}-\widehat{B}_N\chi\vec{\xi}|^2 = \Delta(f;\widehat{B}_N\chi\vec{\xi}).\end{aligned}$$

The spectral representation of the functional $B_N\chi\vec{\xi}$ is

$$B_N\chi\vec{\xi} = \int_{-\pi}^{\pi}(\vec{B}_{\overline{\mu},N}(e^{i\lambda}))^{\top}\frac{\chi_{\overline{\mu}}^{(d)}(e^{-i\lambda})}{\beta^{(d)}(i\lambda)}d\vec{Z}_{\xi^{(d)}}(\lambda),$$

where

$$\vec{B}_{\overline{\mu},N}(e^{i\lambda}) = \sum_{k=0}^{N}\vec{b}_{\overline{\mu},N}(k)e^{i\lambda k} = \sum_{k=0}^{\infty}(D_N^{\overline{\mu}}\mathbf{a}_N)_k e^{i\lambda k}.$$

The linear estimate $\widehat{A}_N\vec{\xi}$ of the functional $A_N\vec{\xi}$ can be represented in the form

$$\widehat{A}_N\vec{\xi} = \int_{-\pi}^{\pi}(\vec{h}_{\overline{\mu},N}^{\xi}(\lambda))^{\top}d\vec{Z}_{\xi^{(d)}}(\lambda) - \sum_{k=-d(\gamma)}^{-1}(\vec{v}_{\overline{\mu},N}(k))^{\top}(\vec{\xi}(k)), \tag{4.20}$$

where $\vec{h}_{\overline{\mu},N}^{\xi}(\lambda) = \{h_{\overline{\mu},N,p}^{\xi}(\lambda)\}_{p=1}^{T}$ is the spectral characteristic of the optimal estimate $\widehat{B}_N\vec{\xi}$.

The estimate $\widehat{B}_N\vec{\xi}$ can be found as

$$\widehat{B}_N\vec{\xi} = \mathrm{Proj}_{H^{0-}(\vec{\xi}_{\overline{\mu},\overline{s}}^{(d)})\oplus H^{(N+d(\gamma))+}(\vec{\xi}_{\overline{\mu},\overline{s}}^{(d)})}B_N\vec{\xi},$$

where

$$\begin{aligned}&H^{0-}(\vec{\xi}_{\overline{\mu},\overline{s}}^{(d)})\oplus H^{(N+d(\gamma))+}(\vec{\xi}_{\overline{\mu},\overline{s}}^{(d)})\\ &= \overline{\mathrm{span}}\{\chi_{\overline{\mu},\overline{s}}^{(d)}(\xi_p(k)) : k\in\mathbb{Z}\setminus\{0,1,\dots,N+d(\gamma)\};\ p=1,\dots,T\}.\end{aligned}$$

Or, equivalently, the spectral characteristic $\vec{h}_{\overline{\mu},N}^{\xi}(\lambda)$ can be found as

$$\vec{h}_{\overline{\mu},N}^{\xi}(\lambda) = \mathrm{Proj}_{L_2^{0-}(f)\oplus L_2^{(N+d(\gamma))+}(f)}\vec{B}_{\overline{\mu},N}(e^{i\lambda}),$$

where

$$\begin{aligned}&L_2^{0-}(f)\oplus L_2^{(N+d(\gamma))+}(f)\\ &= \overline{\mathrm{span}}\left\{e^{i\lambda k}(1-e^{-i\lambda\tau})^d\frac{1}{(i\lambda)^d}\vec{\delta}_p : k\in\mathbb{Z}\setminus\{0,1,\dots,N+d(\gamma)\};\ p=1,\dots,T\right\}\end{aligned}$$

is a closed linear subspace of the Hilbert space $L_2(f)$ of vector-valued functions endowed with the inner product $\langle g_1; g_2\rangle = \int_{-\pi}^{\pi}(g_1(\lambda))^{\top} f(\lambda)\overline{g_2(\lambda)}d\lambda$. Then equality (4.19) can be written in the form

$$\widehat{A}_N\vec{\xi} = \mathrm{Proj}_{H^{0-}(\vec{\xi}^{(d)}_{\overline{\mu},\overline{s}})\oplus H^{(N+d(\gamma))+}(\vec{\xi}^{(d)}_{\overline{\mu},\overline{s}})} B_N\vec{\xi} - V_N\vec{\xi}.$$

Define for $0 \le j, k \le N + d(\gamma)$ the Fourier coefficients

$$F^{\overline{\mu}}_{j,k} = \frac{1}{2\pi}\int_{-\pi}^{\pi} e^{-i\lambda(j-k)}\frac{|\beta^{(d)}(i\lambda)|^2}{|\chi^{(d)}_{\overline{\mu}}(e^{-i\lambda})|^2}(f^{-1}(\lambda))^{\top}d\lambda$$

and a corresponding matrix $\mathbf{F}^{\overline{\mu}}_N$ of dimension $(N + d(\gamma) + 1)T \times (N + d(\gamma) + 1)T$ with $T \times T$ matrix entries $(\mathbf{F}^{\overline{\mu}}_N)_{j,k} = F^{\overline{\mu}}_{j,k}$, $0 \le j, k \le N + d(\gamma)$.

Then we can formulate the following corollary from Theorem 4.1.1.

Corollary 4.1.2. *Let $\{\vec{\xi}(m), m \in \mathbb{Z}\}$ be a stochastic sequence which defines the stationary GM increment sequence $\chi^{(d)}_{\overline{\mu},\overline{s}}(\vec{\xi}(m)) = \{\chi^{(d)}_{\overline{\mu},\overline{s}}(\xi_p(m))\}_{p=1}^{T}$ with the absolutely continuous spectral function $F(\lambda)$ which has spectral density $f(\lambda)$. Let the minimality condition* (4.17) *be satisfied. The optimal linear estimate $\widehat{A}_N\vec{\xi}$ of the functional $A_N\vec{\xi}$ which depends on the unknown values of elements $\vec{\xi}(k)$, $k = 0, 1, 2, \dots, N$, from observations of the sequence $\vec{\xi}(m)$ at points of the set $\mathbb{Z} \setminus \{0, 1, 2, \dots, N\}$ is calculated by formula* (4.20). *The spectral characteristic $\vec{h}^{\xi}_{\overline{\mu},N}(\lambda)$ of the optimal estimate $\widehat{A}_N\vec{\xi}$ is calculated by the formula*

$$(\vec{h}^{\xi}_{\overline{\mu},N}(\lambda))^{\top} = (\vec{B}_{\overline{\mu},N}(e^{i\lambda}))^{\top}\frac{\chi^{(d)}_{\overline{\mu}}(e^{-i\lambda})}{\beta^{(d)}(i\lambda)} - (\vec{C}^{\xi}_{\overline{\mu},N}(e^{i\lambda}))^{\top}\frac{\overline{\beta^{(d)}(i\lambda)}}{\chi^{(d)}_{\overline{\mu}}(e^{-i\lambda})}f^{-1}(\lambda), \tag{4.21}$$

where

$$\vec{C}^{\xi}_{\overline{\mu},N}(e^{i\lambda}) = \sum_{k=0}^{N+d(\gamma)}((\mathbf{F}^{\overline{\mu}}_N)^{-1}[D^{\overline{\mu}}_N\mathbf{a}_N]_{+d(\gamma)})_k e^{i\lambda k}.$$

The value of the MSE $\Delta(f; \widehat{A}_N\vec{\xi})$ is calculated by the formula

$$\begin{aligned}\Delta(f;\widehat{A}_N\vec{\xi}) &= \Delta(f;\widehat{B}_N\vec{\xi}) = \mathsf{E}|B_N\vec{\xi} - \widehat{B}_N\vec{\xi}|^2\\ &= \frac{1}{2\pi}\int_{-\pi}^{\pi}\frac{|\beta^{(d)}(i\lambda)|^2}{|\chi^{(d)}_{\overline{\mu}}(e^{-i\lambda})|^2}\left(\sum_{k=0}^{N+d(\gamma)}((\mathbf{F}^{\overline{\mu}}_N)^{-1}[D^{\overline{\mu}}_N\mathbf{a}_N]_{+d(\gamma)})_k e^{i\lambda k}\right)^{\top}\\ &\quad\times f(\lambda)\overline{\sum_{k=0}^{N+d(\gamma)}((\mathbf{F}^{\overline{\mu}}_N)^{-1}[D^{\overline{\mu}}_N\mathbf{a}_N]_{+d(\gamma)})_k e^{i\lambda k}}d\lambda\\ &= \langle[D^{\overline{\mu}}_N\mathbf{a}_N]_{+d(\gamma)}, (\mathbf{F}^{\overline{\mu}}_N)^{-1}[D^{\overline{\mu}}_N\mathbf{a}_N]_{+d(\gamma)}\rangle.\end{aligned} \tag{4.22}$$

4.1.3 Interpolation of sequences with FM increments

Let $d_j = R_j + D_j$, $j = 0,1,\dots,r$, denote the fractional differentiation order and let R_j, $j = 0,1,\dots,r$, be non-negative integer numbers. Set the increment step $\overline{\mu} = (1,1,\dots,1)$. Let $1 < s_1 < \cdots < s_r$. Consider a stochastic vector sequence $\vec{\xi}(m)$ with stationary FM increments $\chi^{(R+D)}_{\overline{1},\overline{s}}(\vec{\xi}(m))$ having a bounded from zero and from infinity spectral density $\widetilde{f}_{\overline{1}}(\lambda)$. Assume that the numbers R_j, D_j, $j = 0,1,\dots,r$, can be chosen in such way that $-1/2 < D_\nu < 1/2$ for all $\nu \in \mathcal{M}$, where the numbers D_ν and the set are defined in Lemma 1.2.2. Then Theorem 1.2.1 implies that the GM increment sequence $\chi^{(R)}_{\overline{1},\overline{s}}(\vec{\xi}(m))$ is stationary.

Corollary 4.1.3. *A solution of the interpolation problem for the multidimensional sequence $\vec{\xi}(m)$ with stationary FM increments based on its observations with a stationary noise can be obtained from Theorem 4.1.1 for the GM increment sequence $\chi^{(R)}_{\overline{1},\overline{s}}(\vec{\xi}(m))$ making use of the Fourier coefficients*

$$T^{\overline{1}}_{k,j} = \frac{1}{2\pi}\int_{-\pi}^{\pi} e^{-i\lambda(k-j)}\frac{|\beta^{(R)}(i\lambda)|^2}{|\chi^{(R)}_{\overline{1}}(e^{-i\lambda})|^2}(g(\lambda)p^{-1}(\lambda))^{\top}d\lambda,\quad 0\le k,j\le N+d(\gamma),$$

$$P^{\overline{1}}_{k,j} = \frac{1}{2\pi}\int_{-\pi}^{\pi} e^{-i\lambda(k-j)}\frac{|\beta^{(R)}(i\lambda)|^2}{|\chi^{(R)}_{\overline{1}}(e^{-i\lambda})|^2}(p^{-1}(\lambda))^{\top}d\lambda,\quad 0\le k,j\le N+d(\gamma),$$

$$Q_{k,j} = \frac{1}{2\pi}\int_{-\pi}^{\pi} e^{-i\lambda(k-j)}(|\chi^{(D)}_{\overline{1}}(e^{-i\lambda})|^{-2}\widetilde{f}(\lambda)p^{-1}(\lambda)g(\lambda))^{\top}d\lambda,\quad 0\le k,j\le N+d(\gamma),$$

where

$$p(\lambda) = |\chi^{(D)}_{\overline{1}}(e^{-i\lambda})|^{-2}\widetilde{f}(\lambda) + |\beta^{(R)}(i\lambda)|^2 g(\lambda),$$
$$\widetilde{f}(\lambda) = |\beta^{(R)}(i\lambda)|^2|\chi^{(R)}_{\overline{1}}(e^{-i\lambda})|^{-2}\widetilde{f}_{\overline{1}}(\lambda),$$

under the conditions of Theorem 1.2.1.

Corollary 4.1.4. *A solution of the interpolation problem for the multidimensional sequence $\vec{\xi}(m)$ with stationary FM increments based on its observations without noise can be obtained from Theorem 4.1.2 for the GM increment sequence $\chi^{(R)}_{\overline{1},\overline{s}}(\vec{\xi}(m))$ making use of the Fourier coefficients*

$$F^{\overline{1}}(k,j) = \frac{1}{2\pi}\int_{-\pi}^{\pi} e^{i\lambda(j-k)}|\chi^{(D)}_{\overline{1}}(e^{-i\lambda})|^2(\widetilde{f}^{-1}_{\overline{1}}(\lambda))^{\top}d\lambda,\quad 0\le k,j\le N+d(\gamma)$$

4.1.4 Interpolation of stochastic sequences with periodically stationary GM increment with and without periodically stationary noise

Consider the problem of mean square optimal linear estimation of the functional

$$A_M\vartheta = \sum_{k=0}^{N} a^{(\vartheta)}(k)\vartheta(k), \tag{4.23}$$

which depends on unobserved values of the stochastic sequence $\vartheta(m)$ with periodically stationary GM increments. The estimate is based on observations of the sequence $\zeta(m) = \vartheta(m)+\eta(m)$ at points of the set $Z\setminus\{0,1,2,\dots,N\}$, where the periodically stationary noise sequence $\eta(m)$ is uncorrelated with $\vartheta(m)$.

The functional $A_M\vartheta$ can be represented in the form

$$\begin{aligned} A_M\vartheta &= \sum_{k=0}^{M} a^{(\vartheta)}(k)\vartheta(k) = \sum_{m=0}^{N}\sum_{p=1}^{T} a^{(\vartheta)}(mT+p-1)\vartheta(mT+p-1) \\ &= \sum_{m=0}^{N}\sum_{p=1}^{T} a_p(m)\xi_p(m) = \sum_{m=0}^{N} (\vec{a}(m))^{\top}\vec{\xi}(m) = A_N\vec{\xi}, \end{aligned}$$

where $N = [\frac{M}{T}]$, the sequence $\vec{\xi}(m)$ is determined by the formula

$$\vec{\xi}(m) = (\xi_1(m),\xi_2(m),\dots,\xi_T(m))^{\top}, \quad \xi_p(m) = \vartheta(mT+p-1), \quad p = 1,2,\dots,T, \tag{4.24}$$

$$\begin{aligned} &\vec{a}(m) = (a_1(m),a_2(m),\dots,a_T(m))^{\top}, \\ &a_p(m) = a^{\vartheta}(mT+p-1), \quad 0\le m\le N,\ 1\le p\le T,\ mT+p-1\le M, \\ &a_p(N) = 0, \quad M+1\le NT+p-1\le (N+1)T-1,\ 1\le p\le T. \end{aligned} \tag{4.25}$$

Making use of the introduced notations and statements of Theorem 4.1.1 we can claim that the following theorem holds.

Theorem 4.1.2. *Let a stochastic sequence $\vartheta(k)$ with periodically stationary GM increments generate by formula* (4.24) *a multidimensional stochastic sequence $\vec{\xi}(m)$ which determines a stationary GM increment sequence $\chi_{\overline{\mu},\overline{s}}^{(d)}(\vec{\xi}(m)) = \{\chi_{\overline{\mu},\overline{s}}^{(d)}(\xi_p(m))\}_{p=1}^{T}$ with the spectral density matrix $f(\lambda) = \{f_{ij}(\lambda)\}_{i,j=1}^{T}$. Let $\{\vec{\eta}(m), m\in\mathbb{Z}\}$, $\vec{\eta}(m) = (\eta_1(m),\eta_2(m),\dots,\eta_T(m))^{\top}$, $\eta_p(m) = \eta(mT+p-1)$, $p = 1,2,\dots,T$, be an uncorrelated with the sequence $\vec{\xi}(m)$ multidimensional stationary stochastic sequence with an absolutely continuous spectral function $G(\lambda)$ which has spectral density $g(\lambda)$. Let the minimality condition* (4.2) *be satisfied. Let coefficients $\vec{a}(k)$, $k = \overline{0,N}$, be determined by formula* (4.25). *The optimal linear estimate $\widehat{A}_M\zeta$ of the functional $A_M\zeta = A_N\vec{\xi}$ based on observations of the sequence $\zeta(m) = \vartheta(m)+\eta(m)$ at points of the set $Z\setminus\{0,1,2,\dots,N\}$ is calculated by formula* (4.8). *The spectral characteristic $\vec{h}_{\overline{\mu},N}(\lambda)$ and the value of the MSE $\Delta(f,g;\widehat{A}_M\zeta)$ are calculated by formulas* (4.10) *and* (4.12), *respectively.*

Consider the case when the estimate of the functional $A_M\vartheta$ is based on observations of the sequence $\vartheta(m)$ at points of the set $Z \setminus \{0,1,2,\ldots,N\}$ without noise.

Theorem 4.1.3. *Let a stochastic sequence $\vartheta(k)$ with periodically stationary GM increments generate by formula (4.24) a multidimensional stochastic sequence $\vec{\xi}(m)$ which determines a stationary GM increment sequence $\chi_{\overline{\mu},\overline{s}}^{(d)}(\vec{\xi}(m)) = \{\chi_{\overline{\mu},\overline{s}}^{(d)}(\xi_p(m))\}_{p=1}^T$ with the spectral density matrix $f(\lambda) = \{f_{ij}(\lambda)\}_{i,j=1}^T$. Let the minimality condition (4.17) be satisfied. Let coefficients $\vec{a}(k)$, $k = \overline{0,N}$, be determined by formula (4.25). The optimal linear estimate $\widehat{A}_M\zeta$ of the functional $A_M\zeta = A_N\vec{\xi}$ based on observations of the sequence $\vartheta(m)$ at points of the set $Z \setminus \{0,1,2,\ldots,N\}$ is calculated by formula (4.20). The spectral characteristic $\vec{h}_{\overline{\mu},N}^{\vartheta}(\lambda) = \{h_{\overline{\mu},N,p}^{\vartheta}(\lambda)\}_{p=1}^T$ and the value of the MSE $\Delta(f;\widehat{A}_M\zeta)$ are calculated by formulas (4.21) and (4.22), respectively.*

4.2 Minimax (robust) method of interpolation

4.2.1 Least favorable spectral densities and minimax-robust spectral characteristic

The values of the MSEs and the spectral characteristics of the optimal estimate of the functional $A_N\vec{\xi}$ depending on the unobserved values of a stochastic sequence $\vec{\xi}(m)$ which determine a stationary GM increment sequence $\chi_{\overline{\mu},\overline{s}}^{(d)}(\vec{\xi}(m))$ with the spectral density matrix $f(\lambda)$ based on observations of the sequence $\vec{\xi}(m) + \vec{\eta}(m)$ at points $Z \setminus \{0,1,2,\ldots,N\}$ can be calculated by formulas (4.10), (4.11), and (4.12), respectively, in the case where the spectral density matrices $f(\lambda)$ and $g(\lambda)$ of the target sequence and the noise are exactly known.

In most practical cases, however, complete information about the spectral density matrices is not available. If in such cases a set $\mathcal{D} = \mathcal{D}_f \times \mathcal{D}_g$ of admissible spectral densities is defined, the minimax-robust approach to estimation of linear functionals depending on unobserved values of stochastic sequences with stationary increments may be applied.

This method consists in finding an estimate that minimizes the maximal values of the MSEs for all spectral densities from a given class $\mathcal{D} = \mathcal{D}_f \times \mathcal{D}_g$ of admissible spectral densities simultaneously. This method will be applied in the case of concrete classes of spectral densities.

To formalize this approach we recall the following definitions [134].

Definition 4.2.1. For a given class of spectral densities $\mathcal{D} = \mathcal{D}_f \times \mathcal{D}_g$ the spectral densities $f^0(\lambda) \in \mathcal{D}_f$ and $g^0(\lambda) \in \mathcal{D}_g$ are called least favorable in the class $\mathcal{D}$ for the optimal linear estimation of the functional $A_N\vec{\xi}$ if the following relation holds:

$$\Delta(f^0,g^0) = \Delta(h(f^0,g^0);f^0,g^0) = \max_{(f,g)\in\mathcal{D}_f\times\mathcal{D}_g} \Delta(h(f,g);f,g).$$

Definition 4.2.2. For a given class of spectral densities $\mathcal{D} = \mathcal{D}_f \times \mathcal{D}_g$ the spectral characteristic $h^0(\lambda)$ of the optimal linear estimate of the functional $A_N\vec{\xi}$ is called minimax-robust if the following conditions are satisfied:

$$h^0(\lambda) \in H_{\mathcal{D}} = \bigcap_{(f,g)\in\mathcal{D}_f\times\mathcal{D}_g} L_2^{0-}(f(\lambda) + |\beta^{(d)}(i\lambda)|^2 g(\lambda)) \oplus L_2^{(N+d(\gamma))+}(f(\lambda) + |\beta^{(d)}(i\lambda)|^2 g(\lambda)),$$

$$\min_{h\in H_{\mathcal{D}}} \max_{(f,g)\in\mathcal{D}_f\times\mathcal{D}_g} \Delta(h;f,g) = \max_{(f,g)\in\mathcal{D}_f\times\mathcal{D}_g} \Delta(h^0;f,g).$$

Taking into account the introduced definitions and the relations derived in the previous sections we can verify that the following lemma holds.

Lemma 4.2.1. *The spectral densities $f^0 \in \mathcal{D}_f$, $g^0 \in \mathcal{D}_g$ which satisfy the minimality condition* (4.2) *are least favorable in the class $\mathcal{D} = \mathcal{D}_f \times \mathcal{D}_g$ for the optimal linear estimation of the functional $A_N\vec{\xi}$ based on observations of the sequence $\xi(m) + \eta(m)$ at points $m \in \mathbb{Z} \setminus \{0, 1, 2, \ldots, N\}$ if the matrices $(\mathbf{T}_N^{\overline{\mu}})^0$, $(\mathbf{P}_N^{\overline{\mu}})^0$, $(\mathbf{Q}_N)^0$ whose elements are defined by the Fourier coefficients of the functions*

$$\frac{|\beta^{(d)}(i\lambda)|^2}{\chi_{\overline{\mu}}^{(d)}(e^{-i\lambda})}(g^0(\lambda)(f^0(\lambda) + |\beta^{(d)}(i\lambda)|^2 g^0(\lambda))^{-1})^\top,$$

$$\frac{|\beta^{(d)}(i\lambda)|^2}{|\chi_{\overline{\mu}}^{(d)}(e^{-i\lambda})|^2}((f^0(\lambda) + |\beta^{(d)}(i\lambda)|^2 g^0(\lambda))^{-1})^\top,$$

$$(f^0(\lambda)(f^0(\lambda) + |\beta^{(d)}(i\lambda)|^2 g^0(\lambda))^{-1} g^0(\lambda))^\top$$

determine a solution of the constrained optimization problem

$$\max_{(f,g)\in\mathcal{D}_f\times\mathcal{D}_g} (\langle \mathbf{P}_N^{\overline{\mu}} \mathbf{c}_N^{\overline{\mu}}, \mathbf{c}_N^{\overline{\mu}}\rangle + \langle \mathbf{Q}_N \mathbf{a}_N, \mathbf{a}_N\rangle) = \langle (\mathbf{P}_N^{\overline{\mu}})^0 (\mathbf{c}_N^{\overline{\mu}})^0, (\mathbf{c}_N^{\overline{\mu}})^0\rangle + \langle \mathbf{Q}_N^0 \mathbf{a}_N, \mathbf{a}_N\rangle, \tag{4.26}$$

where

$$\mathbf{c}_N^{\overline{\mu}} := (\mathbf{P}_N^{\overline{\mu}})^{-1}([D_N^{\overline{\mu}} \mathbf{a}_N]_{+d(\gamma)} - \mathbf{T}_N^{\overline{\mu}} \mathbf{a}_N^{\overline{\mu}}),$$
$$(\mathbf{c}_N^{\overline{\mu}})^0 := ((\mathbf{P}_N^{\overline{\mu}})^0)^{-1}([D_N^{\overline{\mu}} \mathbf{a}_N]_{+d(\gamma)} - (\mathbf{T}_N^{\overline{\mu}})^0 \mathbf{a}_N^{\overline{\mu}}).$$

The minimax spectral characteristic $h^0 = \vec{h}_{\overline{\mu},N}(f^0, g^0)$ is calculated by formula (4.10) *if $\vec{h}_{\overline{\mu},N}(f^0, g^0) \in H_{\mathcal{D}}$.*

For the interpolation problem without noise, the corresponding definitions formalizing the minimax approach and lemmas concluded from them are the following [134].

Definition 4.2.3. For a given class of spectral densities $\mathcal{D}$ the spectral density $f^0(\lambda) \in \mathcal{D}$ is called least favorable in the class $\mathcal{D}$ for the optimal linear interpolation of the functional $A_N\vec{\xi}$ if the following relation holds:

$$\Delta(f^0) = \Delta(h(f^0); f^0) = \max_{f \in \mathcal{D}} \Delta(h(f); f).$$

Definition 4.2.4. For a given class of spectral densities $\mathcal{D}$ the spectral characteristic $h^0(\lambda)$ of the optimal linear estimate of the functional $A_N\vec{\xi}$ is called minimax-robust if the following conditions are satisfied:

$$h^0(\lambda) \in H_{\mathcal{D}} = \bigcap_{f \in \mathcal{D}_f} L_2^{0-}(f) \oplus L_2^{(N+d(\gamma))+}(f),$$

$$\min_{h \in H_{\mathcal{D}}} \max_{f \in \mathcal{D}} \Delta(h; f) = \max_{f \in \mathcal{D}} \Delta(h^0; f).$$

Lemma 4.2.2. *The spectral density $f^0 \in \mathcal{D}$ which satisfies the minimality condition* (4.17) *is the least favorable one in the class $\mathcal{D}$ for the optimal linear interpolation of the functional $A_N\vec{\xi}$ based on observations of the sequence $\xi(m)$ at points $m \in \mathbb{Z} \setminus \{0, 1, 2, \ldots, N\}$ if the matrix $(\mathbf{F}_N^{\overline{\mu}})^0$ whose elements are defined by the Fourier coefficients of the function*

$$\frac{|\beta^{(d)}(i\lambda)|^2}{|\chi_{\overline{\mu}}^{(d)}(e^{-i\lambda})|^2}\left((f^0(\lambda))^{-1}\right)^{\top}$$

determine a solution of the constrained optimization problem

$$\begin{aligned} &\max_{f \in \mathcal{D}} \left(\left\langle [D_N^{\overline{\mu}} \mathbf{a}_N]_{+d(\gamma)}, (\mathbf{F}_N^{\overline{\mu}})^{-1} [D_N^{\overline{\mu}} \mathbf{a}_N]_{+d(\gamma)} \right\rangle\right) \\ &= \left\langle [D_N^{\overline{\mu}} \mathbf{a}_N]_{+d(\gamma)}, ((\mathbf{F}_N^{\overline{\mu}})^0)^{-1} [D_N^{\overline{\mu}} \mathbf{a}_N]_{+d(\gamma)} \right\rangle. \end{aligned} \tag{4.27}$$

The minimax spectral characteristic $h^0 = \vec{h}_{\overline{\mu},N}^{\xi}(f^0)$ is calculated by formula (4.21) *if $\vec{h}_{\overline{\mu},N}^{\xi}(f^0) \in H_{\mathcal{D}}$.*

More detailed analysis of properties of the least favorable spectral densities and the minimax-robust spectral characteristics shows that the minimax spectral characteristic h^0 and the least favorable spectral densities f^0 and g^0 (or the least favorable spectral density f^0 in the case of observations without noise) form a saddle point of the function $\Delta(h; f, g)$ (or the function $\Delta(h; f)$) on the set $H_{\mathcal{D}} \times \mathcal{D}$. The saddle point inequalities

$$\Delta(h; f^0, g^0) \geq \Delta(h^0; f^0, g^0) \geq \Delta(h^0; f, g) \quad \forall (f, g) \in \mathcal{D}, \forall h \in H_{\mathcal{D}}$$

or

$$\Delta(h; f^0) \geq \Delta(h^0; f^0) \geq \Delta(h^0; f) \quad \forall f \in \mathcal{D}, \forall h \in H_{\mathcal{D}}$$

hold if $h^0 = \vec{h}_{\overline{\mu},N}(f^0,g^0)$, $\vec{h}_{\overline{\mu},N}(f^0,g^0) \in H_{\mathcal{D}}$ (or $h^0 = \vec{h}^{\xi}_{\overline{\mu},N}(f^0)$, $\vec{h}^{\xi}_{\overline{\mu},N}(f^0) \in H_{\mathcal{D}}$) and (f^0,g^0) (or f^0) is a solution of the constrained optimization problem

$$\widetilde{\Delta}(f,g) = -\Delta(\vec{h}_{\overline{\mu},N}(f^0,g^0);f,g) \to \inf, \quad (f,g) \in \mathcal{D}, \tag{4.28}$$

or

$$\widetilde{\Delta}(f) = -\Delta(\vec{h}^{\xi}_{\overline{\mu},N}(f^0);f) \to \inf, \quad f \in \mathcal{D}, \tag{4.29}$$

where the functional $\Delta(\vec{h}_{\overline{\mu},N}(f^0,g^0);f,g)$ is calculated by the formula

$$\begin{aligned}\Delta(\vec{h}_{\overline{\mu},N}(f^0,g^0);f,g) = {} & \frac{1}{2\pi}\int_{-\pi}^{\pi} \frac{|\beta^{(d)}(i\lambda)|^2}{|\chi^{(d)}_{\overline{\mu}}(e^{i\lambda})|^2} (\mathbf{C}^{f0}_{\overline{\mu},N}(e^{i\lambda}))^{\top} (f^0(\lambda) + |\beta^{(d)}(i\lambda)|^2 g^0(\lambda))^{-1} \\ & \times f(\lambda)\,(f^0(\lambda) + |\beta^{(d)}(i\lambda)|^2 g^0(\lambda))^{-1} \overline{\mathbf{C}^{f0}_{\overline{\mu},N}(e^{i\lambda})} d\lambda \\ & + \frac{1}{2\pi}\int_{-\pi}^{\pi} \left(\frac{|\beta^{(d)}(i\lambda)|^2}{|\chi^{(d)}_{\overline{\mu}}(e^{i\lambda})|^2}\right)^2 (\mathbf{C}^{g0}_{\overline{\mu},N}(e^{i\lambda}))^{\top} (f^0(\lambda) + |\beta^{(d)}(i\lambda)|^2 g^0(\lambda))^{-1} \\ & \times g(\lambda)\,(f^0(\lambda) + |\beta^{(d)}(i\lambda)|^2 g^0(\lambda))^{-1} \overline{\mathbf{C}^{g0}_{\overline{\mu},N}(e^{i\lambda})} d\lambda,\end{aligned}$$

with

$$\mathbf{C}^{f0}_{\overline{\mu},N}(e^{i\lambda}) := \chi^{(d)}_{\overline{\mu}}(e^{i\lambda})\overline{g^0(\lambda)}\vec{A}_N(e^{-i\lambda}) - \vec{C}^{0}_{\overline{\mu},N}(e^{i\lambda}),$$

$$\mathbf{C}^{g0}_{\overline{\mu},N}(e^{i\lambda}) := \frac{|\chi^{(d)}_{\overline{\mu}}(e^{-i\lambda})|^2}{|\beta^{(d)}(i\lambda)|^2}\overline{f^0(\lambda)}\vec{A}_N(e^{-i\lambda}) + \chi^{(d)}_{\overline{\mu}}(e^{-i\lambda})\vec{C}^{0}_{\overline{\mu},N}(e^{i\lambda}),$$

where

$$\vec{C}^{0}_{\overline{\mu},N}(e^{i\lambda}) = \sum_{k=0}^{N+d(\gamma)} (((\mathbf{P}^{\overline{\mu}}_N)^0)^{-1}[D^{\overline{\mu}}_N \mathbf{a}_N]_{+d(\gamma)} - ((\mathbf{P}^{\overline{\mu}}_N)^0)^{-1}(\mathbf{T}^{\overline{\mu}}_N)^0 \mathbf{a}^{\overline{\mu}}_N)_k e^{i\lambda k},$$

or the functional $\Delta(\vec{h}^{\xi}_{\overline{\mu},N}(f^0);f)$, in the case of observations without noise, is calculated by the formula

$$\begin{aligned}\Delta(\vec{h}^{\xi}_{\overline{\mu},N}(f^0);f) = {} & \frac{1}{2\pi}\int_{-\pi}^{\pi} \frac{|\beta^{(d)}(i\lambda)|^2}{|\chi^{(d)}_{\overline{\mu}}(e^{-i\lambda})|^2} \left(\sum_{k=0}^{N+d(\gamma)} (((\mathbf{F}^{\overline{\mu}}_N)^0)^{-1}[D^{\overline{\mu}}_N \mathbf{a}_N]_{+d(\gamma)})_k e^{i\lambda k}\right)^{\top} \\ & \times (f^0(\lambda))^{-1} f(\lambda)\,(f^0(\lambda))^{-1} \overline{\sum_{k=0}^{N+d(\gamma)} (((\mathbf{F}^{\overline{\mu}}_N)^0)^{-1}[D^{\overline{\mu}}_N \mathbf{a}_N]_{+d(\gamma)})_k e^{i\lambda k}} d\lambda.\end{aligned}$$

The constrained optimization problems (4.28) and (4.29) are equivalent to the unconstrained optimization problem

$$\Delta_{\mathcal{D}}(f,g) = \widetilde{\Delta}(f,g) + \delta(f,g|\mathcal{D}) \to \inf \tag{4.30}$$

and

$$\Delta_{\mathcal{D}}(f) = \widetilde{\Delta}(f) + \delta(f|\mathcal{D}) \to \inf, \tag{4.31}$$

where $\delta(f,g|\mathcal{D})$ and $\delta(f|\mathcal{D})$ are the indicator functions of the sets $\mathcal{D} = \mathcal{D}_f \times \mathcal{D}_g$ and $\mathcal{D} = \mathcal{D}_f$, namely, $\delta(f,g|\mathcal{D}) = 0$ if $(f,g) \in \mathcal{D}$ and $\delta(f,g|\mathcal{D}) = +\infty$ if $(f,g) \notin \mathcal{D}$, and, respectively, $\delta(f|\mathcal{D}) = 0$ if $f \in \mathcal{D}$ and $\delta(f|\mathcal{D}) = +\infty$ if $f \notin \mathcal{D}$. The conditions $0 \in \partial\Delta_{\mathcal{D}}(f^0, g^0)$ and $0 \in \partial\Delta_{\mathcal{D}}(f^0)$ characterize the solutions (f^0, g^0) and f^0 of the stated unconstrained optimization problems. This condition is the necessary and sufficient condition under which the points (f^0, g^0) and f^0 belong to the sets of minimums of the convex functionals $\Delta_{\mathcal{D}}(f,g)$ and $\Delta_{\mathcal{D}}(f)$, respectively [146, 180]. Thus, it allows us to find equations which determine the least favorable spectral densities in some special classes of spectral densities $\mathcal{D}$.

The form of the functionals $\Delta(\vec{h}_{\overline{\mu},N}(f^0, g^0); f, g)$ and $\Delta(\vec{h}^{\xi}_{\overline{\mu},N}(f^0); f, g)$ is suitable for application of the Lagrange method of indefinite multipliers to the constrained optimization problems (4.28) and (4.29). Thus, the complexity of the problems is reduced to finding the subdifferentials of the indicator functions of the sets of admissible spectral densities. In the following subsections, the solutions of the problems (4.30) and (4.31) are illustrated for concrete sets of admissible spectral densities.

4.2.2 Least favorable spectral densities in classes $\mathcal{D}_0 \times \mathcal{D}_\varepsilon$

Consider the minimax interpolation problem for the functional $A_N\vec{\xi}$ depending on the unobserved values of the stochastic sequence $\vec{\xi}(m)$ which determine a stationary GM increment sequence $\chi^{(d)}_{\overline{\mu},\overline{s}}(\vec{\xi}(m))$ for the following sets of admissible spectral densities $\mathcal{D}_0^k$, $k = 1, 2, 3, 4$:

$$\mathcal{D}_0^1 = \left\{ f(\lambda) \left| \frac{1}{2\pi} \int_{-\pi}^{\pi} \frac{|\chi^{(d)}_{\overline{\mu}}(e^{-i\lambda})|^2}{|\beta^{(d)}(i\lambda)|^2} f(\lambda) d\lambda = P \right. \right\},$$

$$\mathcal{D}_0^2 = \left\{ f(\lambda) \left| \frac{1}{2\pi} \int_{-\pi}^{\pi} \frac{|\chi^{(d)}_{\overline{\mu}}(e^{-i\lambda})|^2}{|\beta^{(d)}(i\lambda)|^2} \operatorname{Tr}[f(\lambda)] d\lambda = p \right. \right\},$$

$$\mathcal{D}_0^3 = \left\{ f(\lambda) \left| \frac{1}{2\pi} \int_{-\pi}^{\pi} \frac{|\chi^{(d)}_{\overline{\mu}}(e^{-i\lambda})|^2}{|\beta^{(d)}(i\lambda)|^2} f_{kk}(\lambda) d\lambda = p_k, k = \overline{1,T} \right. \right\},$$

$$\mathcal{D}_0^4 = \left\{ f(\lambda) \left| \frac{1}{2\pi} \int_{-\pi}^{\pi} \frac{|\chi^{(d)}_{\overline{\mu}}(e^{-i\lambda})|^2}{|\beta^{(d)}(i\lambda)|^2} \langle B_1, f(\lambda) \rangle d\lambda = p \right. \right\},$$

where p, p_k, $k = \overline{1,T}$, are given numbers, P, B_1 are given positive-definite Hermitian matrices, and we have the following sets of admissible spectral densities $\mathcal{D}_\varepsilon^k$, $k = 1,2,3,4$, for the stationary noise sequence $\vec{\eta}(m)$:

$$\mathcal{D}_\varepsilon^1 = \left\{ g(\lambda) \,\middle|\, \mathrm{Tr}[g(\lambda)] = (1-\varepsilon)\,\mathrm{Tr}[g_1(\lambda)] + \varepsilon\,\mathrm{Tr}[W(\lambda)], \frac{1}{2\pi}\int_{-\pi}^{\pi} \mathrm{Tr}[g(\lambda)]d\lambda = q \right\},$$

$$\mathcal{D}_\varepsilon^2 = \left\{ g(\lambda) \,\middle|\, g_{kk}(\lambda) = (1-\varepsilon)g_{kk}^1(\lambda) + \varepsilon w_{kk}(\lambda), \frac{1}{2\pi}\int_{-\pi}^{\pi} g_{kk}(\lambda)d\lambda = q_k, k = \overline{1,T} \right\},$$

$$\mathcal{D}_\varepsilon^3 = \left\{ g(\lambda) \,\middle|\, \langle B_2, g(\lambda)\rangle = (1-\varepsilon)\langle B_2, g_1(\lambda)\rangle + \varepsilon\langle B_2, W(\lambda)\rangle, \frac{1}{2\pi}\int_{-\pi}^{\pi} \langle B_2, g(\lambda)\rangle d\lambda = q \right\},$$

$$\mathcal{D}_\varepsilon^4 = \left\{ g(\lambda) \,\middle|\, g(\lambda) = (1-\varepsilon)g_1(\lambda) + \varepsilon W(\lambda), \frac{1}{2\pi}\int_{-\pi}^{\pi} g(\lambda)d\lambda = Q \right\},$$

where $g_1(\lambda)$ is a fixed spectral density, $W(\lambda)$ is an unknown spectral density, q, q_k, $k = \overline{1,T}$, are given numbers, and Q is a given positive-definite Hermitian matrix.

From the condition $0 \in \partial\Delta_{\mathcal{D}}(f^0, g^0)$, we find the following equations which determine the least favorable spectral densities for these given sets of admissible spectral densities.

For the first pair of sets of admissible spectral densities $\mathcal{D}_0^1 \times \mathcal{D}_\varepsilon^1$, we have

$$\begin{aligned}(\mathbf{C}^{f0}_{\overline{\mu},N}(e^{i\lambda}))(\mathbf{C}^{f0}_{\overline{\mu},N}(e^{i\lambda}))^* &= \frac{|\chi^{(d)}_{\overline{\mu}}(e^{-i\lambda})|^2}{|\beta^{(d)}(i\lambda)|^2}(f^0(\lambda) + |\beta^{(d)}(i\lambda)|^2 g^0(\lambda)) \\ &\quad \times \vec{\alpha}\cdot\vec{\alpha}^* \frac{|\chi^{(d)}_{\overline{\mu}}(e^{-i\lambda})|^2}{|\beta^{(d)}(i\lambda)|^2}(f^0(\lambda) + |\beta^{(d)}(i\lambda)|^2 g^0(\lambda)), \end{aligned} \tag{4.32}$$

$$\begin{aligned}(\mathbf{C}^{g0}_{\overline{\mu},N}(e^{i\lambda}))(\mathbf{C}^{g0}_{\overline{\mu},N}(e^{i\lambda}))^* &= (\alpha^2 + \gamma_1(\lambda)) \\ &\quad \times \left(\frac{|\chi^{(d)}_{\overline{\mu}}(e^{-i\lambda})|^2}{|\beta^{(d)}(i\lambda)|^2}(f^0(\lambda) + |\beta^{(d)}(i\lambda)|^2 g^0(\lambda)) \right)^2, \end{aligned} \tag{4.33}$$

where α^2, $\vec{\alpha}$ are Lagrange multipliers and the function $\gamma_1(\lambda) \le 0$ and $\gamma_1(\lambda) = 0$ if $\mathrm{Tr}[g^0(\lambda)] > (1-\varepsilon)\,\mathrm{Tr}[g_1(\lambda)]$.

For the second pair of sets of admissible spectral densities $\mathcal{D}_{f0}^2 \times \mathcal{D}_\varepsilon^2$, we have

$$(\mathbf{C}^{f0}_{\overline{\mu},N}(e^{i\lambda}))(\mathbf{C}^{f0}_{\overline{\mu},N}(e^{i\lambda}))^* = \alpha^2 \left(\frac{|\chi^{(d)}_{\overline{\mu}}(e^{-i\lambda})|^2}{|\beta^{(d)}(i\lambda)|^2}(f^0(\lambda) + |\beta^{(d)}(i\lambda)|^2 g^0(\lambda)) \right)^2, \tag{4.34}$$

$$(\mathbf{C}^{g0}_{\overline{\mu},N}(e^{i\lambda}))(\mathbf{C}^{g0}_{\overline{\mu},N}(e^{i\lambda}))^* = \frac{|\chi^{(d)}_{\overline{\mu}}(e^{-i\lambda})|^2}{|\beta^{(d)}(i\lambda)|^2}(f^0(\lambda)+|\beta^{(d)}(i\lambda)|^2 g^0(\lambda))$$
$$\times\{(\alpha_k^2+\gamma_k^1(\lambda))\delta_{kl}\}_{k,l=1}^T \frac{|\chi^{(d)}_{\overline{\mu}}(e^{-i\lambda})|^2}{|\beta^{(d)}(i\lambda)|^2}(f^0(\lambda)+|\beta^{(d)}(i\lambda)|^2 g^0(\lambda)), \tag{4.35}$$

where α^2, α_k^2 are Lagrange multipliers and functions $\gamma_k^1(\lambda)\le 0$ and $\gamma_k^1(\lambda)=0$ if $g^0_{kk}(\lambda) > (1-\varepsilon)g^1_{kk}(\lambda)$.

For the third pair of sets of admissible spectral densities $\mathcal{D}^3_{f0}\times\mathcal{D}^3_{\varepsilon}$, we have

$$(\mathbf{C}^{f0}_{\overline{\mu},N}(e^{i\lambda}))(\mathbf{C}^{f0}_{\overline{\mu},N}(e^{i\lambda}))^* = \frac{|\chi^{(d)}_{\overline{\mu}}(e^{-i\lambda})|^2}{|\beta^{(d)}(i\lambda)|^2}(f^0(\lambda)+|\beta^{(d)}(i\lambda)|^2 g^0(\lambda))$$
$$\times\{\alpha_k^2\delta_{kl}\}_{k,l=1}^T \frac{|\chi^{(d)}_{\overline{\mu}}(e^{-i\lambda})|^2}{|\beta^{(d)}(i\lambda)|^2}(f^0(\lambda)+|\beta^{(d)}(i\lambda)|^2 g^0(\lambda)) \tag{4.36}$$

$$(\mathbf{C}^{g0}_{\overline{\mu},N}(e^{i\lambda}))(\mathbf{C}^{g0}_{\overline{\mu},N}(e^{i\lambda}))^* = (\alpha^2+\gamma_1'(\lambda))\frac{|\chi^{(d)}_{\overline{\mu}}(e^{-i\lambda})|^2}{|\beta^{(d)}(i\lambda)|^2}(f^0(\lambda)+|\beta^{(d)}(i\lambda)|^2 g^0(\lambda))$$
$$\times B_2^{\top}\frac{|\chi^{(d)}_{\overline{\mu}}(e^{-i\lambda})|^2}{|\beta^{(d)}(i\lambda)|^2}(f^0(\lambda)+|\beta^{(d)}(i\lambda)|^2 g^0(\lambda)), \tag{4.37}$$

where α_k^2, α^2 are Lagrange multipliers, the function $\gamma_1'(\lambda)\le 0$ and $\gamma_1'(\lambda)=0$ if $\langle B_2, g^0(\lambda)\rangle > (1-\varepsilon)\langle B_2, g_1(\lambda)\rangle$, and δ_{kl} are Kronecker symbols.

For the fourth pair of sets of admissible spectral densities $\mathcal{D}^4_{f0}\times\mathcal{D}^4_{\varepsilon}$, we have

$$(\mathbf{C}^{f0}_{\overline{\mu},N}(e^{i\lambda}))(\mathbf{C}^{f0}_{\overline{\mu},N}(e^{i\lambda}))^* = \alpha^2\frac{|\chi^{(d)}_{\overline{\mu}}(e^{-i\lambda})|^2}{|\beta^{(d)}(i\lambda)|^2}(f^0(\lambda)+|\beta^{(d)}(i\lambda)|^2 g^0(\lambda))$$
$$\times B_1^{\top}\frac{|\chi^{(d)}_{\overline{\mu}}(e^{-i\lambda})|^2}{|\beta^{(d)}(i\lambda)|^2}(f^0(\lambda)+|\beta^{(d)}(i\lambda)|^2 g^0(\lambda)), \tag{4.38}$$

$$(\mathbf{C}^{g0}_{\overline{\mu},N}(e^{i\lambda}))(\mathbf{C}^{g0}_{\overline{\mu},N}(e^{i\lambda}))^* = \frac{|\chi^{(d)}_{\overline{\mu}}(e^{-i\lambda})|^2}{|\beta^{(d)}(i\lambda)|^2}(f^0(\lambda)+|\beta^{(d)}(i\lambda)|^2 g^0(\lambda))$$
$$\times(\vec{\alpha}\cdot\vec{\alpha}^*+\Gamma(\lambda))\frac{|\chi^{(d)}_{\overline{\mu}}(e^{-i\lambda})|^2}{|\beta^{(d)}(i\lambda)|^2}(f^0(\lambda)+|\beta^{(d)}(i\lambda)|^2 g^0(\lambda)) \tag{4.39}$$

where α^2, $\vec{\alpha}$ are Lagrange multipliers and the function $\Gamma(\lambda)\le 0$ and $\Gamma(\lambda)=0$ if $g^0(\lambda) > (1-\varepsilon)g_1(\lambda)$.

The following theorem holds.

Theorem 4.2.1. *The least favorable spectral densities $f^0(\lambda)$ and $g^0(\lambda)$ in the classes $\mathcal{D}_0^k\times\mathcal{D}_{\varepsilon}^k$, $k=1,2,3,4$, for the optimal linear interpolation of the functional $A_N\vec{\xi}$ are determined* *by pairs of equations* (4.32)–(4.33), (4.34)–(4.35), (4.36)–(4.37), *and* (4.38)–(4.39), *the minimality condition* (4.2), *the constrained optimization problem* (4.26), *and restrictions on densities from the corresponding classes* $\mathcal{D}_0^k \times \mathcal{D}_\varepsilon^k$, $k = 1, 2, 3, 4$. *The minimax-robust spectral characteristic* $\vec{h}_{\overline{\mu},N}(f^0, g^0)$ *of the optimal estimate of the functional* $A_N\vec{\xi}$ *is determined by formula* (4.10).

4.2.3 Least favorable spectral densities in classes $\mathcal{D}_0$

Consider the minimax interpolation problem for the functional $A_N\vec{\xi}$ depending on the unobserved values of the stochastic sequence $\vec{\xi}(m)$ which determine a stationary GM increment sequence $\chi^{(d)}_{\overline{\mu},\overline{s}}(\vec{\xi}(m))$ from observations without noise for the sets of admissible spectral densities $\mathcal{D}_0^k$, $k = 1, 2, 3, 4$, defined in Section 4.2.2.

The condition $0 \in \partial\Delta_{\mathcal{D}}(f^0)$ implies the following equations which determine the least favorable spectral densities for these given sets of admissible spectral densities $\mathcal{D}_0^1$ to $\mathcal{D}_0^4$, respectively:

$$\left(\sum_{k=0}^{N+d(\gamma)} (((\mathbf{F}_N^{\overline{\mu}})^0)^{-1}[D_N^{\overline{\mu}}\mathbf{a}_N]_{+d(\gamma)})_k e^{i\lambda k}\right)\left(\sum_{k=0}^{N+d(\gamma)} (((\mathbf{F}_N^{\overline{\mu}})^0)^{-1}[D_N^{\overline{\mu}}\mathbf{a}_N]_{+d(\gamma)})_k e^{i\lambda k}\right)^*$$
$$= \frac{|\chi_{\overline{\mu}}^{(d)}(e^{-i\lambda})|^2}{|\beta^{(d)}(i\lambda)|^2} f^0(\lambda)\vec{\alpha}\cdot\vec{\alpha}^* \frac{|\chi_{\overline{\mu}}^{(d)}(e^{-i\lambda})|^2}{|\beta^{(d)}(i\lambda)|^2} f^0(\lambda), \tag{4.40}$$

$$\left(\sum_{k=0}^{N+d(\gamma)} (((\mathbf{F}_N^{\overline{\mu}})^0)^{-1}[D_N^{\overline{\mu}}\mathbf{a}_N]_{+d(\gamma)})_k e^{i\lambda k}\right)\left(\sum_{k=0}^{N+d(\gamma)} (((\mathbf{F}_N^{\overline{\mu}})^0)^{-1}[D_N^{\overline{\mu}}\mathbf{a}_N]_{+d(\gamma)})_k e^{i\lambda k}\right)^*$$
$$= \alpha^2\left(\frac{|\chi_{\overline{\mu}}^{(d)}(e^{-i\lambda})|^2}{|\beta^{(d)}(i\lambda)|^2} f^0(\lambda)\right)^2, \tag{4.41}$$

$$\left(\sum_{k=0}^{N+d(\gamma)} (((\mathbf{F}_N^{\overline{\mu}})^0)^{-1}[D_N^{\overline{\mu}}\mathbf{a}_N]_{+d(\gamma)})_k e^{i\lambda k}\right)\left(\sum_{k=0}^{N+d(\gamma)} (((\mathbf{F}_N^{\overline{\mu}})^0)^{-1}[D_N^{\overline{\mu}}\mathbf{a}_N]_{+d(\gamma)})_k e^{i\lambda k}\right)^*$$
$$= \frac{|\chi_{\overline{\mu}}^{(d)}(e^{-i\lambda})|^2}{|\beta^{(d)}(i\lambda)|^2} f^0(\lambda)\{\alpha_k^2\delta_{kl}\}_{k,l=1}^T \frac{|\chi_{\overline{\mu}}^{(d)}(e^{-i\lambda})|^2}{|\beta^{(d)}(i\lambda)|^2} f^0(\lambda), \tag{4.42}$$

$$\left(\sum_{k=0}^{N+d(\gamma)} (((\mathbf{F}_N^{\overline{\mu}})^0)^{-1}[D_N^{\overline{\mu}}\mathbf{a}_N]_{+d(\gamma)})_k e^{i\lambda k}\right)\left(\sum_{k=0}^{N+d(\gamma)} (((\mathbf{F}_N^{\overline{\mu}})^0)^{-1}[D_N^{\overline{\mu}}\mathbf{a}_N]_{+d(\gamma)})_k e^{i\lambda k}\right)^*$$
$$= \beta^2 \frac{|\chi_{\overline{\mu}}^{(d)}(e^{-i\lambda})|^2}{|\beta^{(d)}(i\lambda)|^2} f^0(\lambda) B_1^\top \frac{|\chi_{\overline{\mu}}^{(d)}(e^{-i\lambda})|^2}{|\beta^{(d)}(i\lambda)|^2} f^0(\lambda), \tag{4.43}$$

where α^2, $\vec{\alpha}$, α_k^2, and β^2 are Lagrange multipliers.

The following theorem holds.

Theorem 4.2.2. *The least favorable spectral density $f^0(\lambda)$ in the classes $\mathcal{D}_0^k$, $k = 1, 2, 3, 4$, for the optimal linear interpolation of the functional $A_N\vec{\xi}$ is determined by pairs of equations (4.40)–(4.43), the minimality condition (4.17), the constrained optimization problem (4.27), and restrictions on densities from the corresponding classes $\mathcal{D}_0^k$, $k = 1, 2, 3, 4$. The minimax-robust spectral characteristic $\vec{h}_{\overline{\mu},N}^{\xi}(f^0)$ of the optimal estimate of the functional $A_N\vec{\xi}$ is determined by formula (4.21).*

4.2.4 Least favorable spectral densities in classes $\mathcal{D}_{1\delta} \times \mathcal{D}_V^U$

Consider the minimax interpolation problem for the functional $A_N\vec{\xi}$ depending on the unobserved values of the stochastic sequence $\vec{\xi}(m)$ which determine a stationary GM increment sequence $\chi_{\overline{\mu},\overline{s}}^{(d)}(\vec{\xi}(m))$ for the following sets of admissible spectral densities $\mathcal{D}_{1\delta}^k$, $k = 1, 2, 3, 4$:

$$\mathcal{D}_{1\delta}^1 = \left\{ f(\lambda) \left| \frac{1}{2\pi} \int_{-\pi}^{\pi} \frac{|\chi_{\overline{\mu}}^{(d)}(e^{-i\lambda})|^2}{|\beta^{(d)}(i\lambda)|^2} |\operatorname{Tr}(f(\lambda) - f_1(\lambda))| d\lambda \le \delta \right.\right\},$$

$$\mathcal{D}_{1\delta}^2 = \left\{ f(\lambda) \left| \frac{1}{2\pi} \int_{-\pi}^{\pi} \frac{|\chi_{\overline{\mu}}^{(d)}(e^{-i\lambda})|^2}{|\beta^{(d)}(i\lambda)|^2} |f_{kk}(\lambda) - f_{kk}^1(\lambda)| d\lambda \le \delta_k, k = \overline{1,T} \right.\right\},$$

$$\mathcal{D}_{1\delta}^3 = \left\{ f(\lambda) \left| \frac{1}{2\pi} \int_{-\pi}^{\pi} \frac{|\chi_{\overline{\mu}}^{(d)}(e^{-i\lambda})|^2}{|\beta^{(d)}(i\lambda)|^2} |\langle B_1, f(\lambda) - f_1(\lambda)\rangle| d\lambda \le \delta \right.\right\},$$

$$\mathcal{D}_{1\delta}^4 = \left\{ f(\lambda) \left| \frac{1}{2\pi} \int_{-\pi}^{\pi} \frac{|\chi_{\overline{\mu}}^{(d)}(e^{-i\lambda})|^2}{|\beta^{(d)}(i\lambda)|^2} |f_{ij}(\lambda) - f_{ij}^1(\lambda)| d\lambda \le \delta_i^j, i, j = \overline{1,T} \right.\right\},$$

where $f_1(\lambda)$ is a fixed spectral density, B_1 is a given positive-definite Hermitian matrix, δ, δ_k, $k = \overline{1,T}$, and δ_i^j, $i, j = \overline{1,T}$, are given numbers, and we have the following sets of admissible spectral densities $\mathcal{D}_V^{U^k}$, $k = 1, 2, 3, 4$, for the stationary noise sequence $\vec{\eta}(m)$:

$$\mathcal{D}_V^{U^1} = \left\{ g(\lambda) \left| V(\lambda) \le g(\lambda) \le U(\lambda), \frac{1}{2\pi} \int_{-\pi}^{\pi} g(\lambda) d\lambda = Q \right.\right\},$$

$$\mathcal{D}_V^{U^2} = \left\{ g(\lambda) \left| \operatorname{Tr}[V(\lambda)] \le \operatorname{Tr}[g(\lambda)] \le \operatorname{Tr}[U(\lambda)], \frac{1}{2\pi} \int_{-\pi}^{\pi} \operatorname{Tr}[g(\lambda)] d\lambda = q \right.\right\},$$

$$\mathcal{D}_V^{U^3} = \left\{ g(\lambda) \left| v_{kk}(\lambda) \le g_{kk}(\lambda) \le u_{kk}(\lambda), \frac{1}{2\pi} \int_{-\pi}^{\pi} g_{kk}(\lambda) d\lambda = q_k, k = \overline{1,T} \right.\right\},$$

$$\mathcal{D}_V^{U^4} = \left\{ g(\lambda) \left| \langle B_2, V(\lambda)\rangle \le \langle B_2, g(\lambda)\rangle \le \langle B_2, U(\lambda)\rangle, \frac{1}{2\pi} \int_{-\pi}^{\pi} \langle B_2, g(\lambda)\rangle d\lambda = q \right.\right\},$$

where the spectral densities $V(\lambda)$ and $U(\lambda)$ are known and fixed, $q, q_k, k = \overline{1,T}$, are given numbers, and Q, B_2 are given positive-definite Hermitian matrices.

From the condition $0 \in \partial\Delta_{\mathcal{D}}(f^0, g^0)$ we find the following equations which determine the least favorable spectral densities for these given sets of admissible spectral densities.

For the first pair of sets of admissible spectral densities $\mathcal{D}_{1\delta}^1 \times \mathcal{D}_V^{U1}$, we have

$$(\mathbf{C}^{f0}_{\overline{\mu},N}(e^{i\lambda}))(\mathbf{C}^{f0}_{\overline{\mu},N}(e^{i\lambda}))^* = \beta^2\gamma_2(\lambda)\left(\frac{|\chi_{\overline{\mu}}^{(d)}(e^{-i\lambda})|^2}{|\beta^{(d)}(i\lambda)|^2}(f^0(\lambda) + |\beta^{(d)}(i\lambda)|^2 g^0(\lambda))\right)^2, \tag{4.44}$$

$$\begin{aligned}(\mathbf{C}^{g0}_{\overline{\mu},N}(e^{i\lambda}))(\mathbf{C}^{g0}_{\overline{\mu},N}(e^{i\lambda}))^* &= \frac{|\chi_{\overline{\mu}}^{(d)}(e^{-i\lambda})|^2}{|\beta^{(d)}(i\lambda)|^2}(f^0(\lambda) + |\beta^{(d)}(i\lambda)|^2 g^0(\lambda)) \\ &\times (\vec{\beta}\cdot\vec{\beta}^* + \Gamma_1(\lambda) + \Gamma_2(\lambda))\frac{|\chi_{\overline{\mu}}^{(d)}(e^{-i\lambda})|^2}{|\beta^{(d)}(i\lambda)|^2}(f^0(\lambda) + |\beta^{(d)}(i\lambda)|^2 g^0(\lambda)),\end{aligned} \tag{4.45}$$

$$\frac{1}{2\pi}\int_{-\pi}^{\pi}\frac{|\chi_{\overline{\mu}}^{(d)}(e^{-i\lambda})|^2}{|\beta^{(d)}(i\lambda)|^2}|\mathrm{Tr}(f^0(\lambda) - f_1(\lambda))|d\lambda = \delta, \tag{4.46}$$

where $\beta^2, \vec{\beta}$ are Lagrange multipliers, the function $\Gamma_1(\lambda) \leq 0$ and $\Gamma_1(\lambda) = 0$ if $g^0(\lambda) > V(\lambda)$, the function $\Gamma_2(\lambda) \geq 0$ and $\Gamma_2(\lambda) = 0$ if $g^0(\lambda) < U(\lambda)$, the function $|\gamma_2(\lambda)| \leq 1$, and

$$\gamma_2(\lambda) = \mathrm{sign}(\mathrm{Tr}(f^0(\lambda) - f_1(\lambda))) : \ \mathrm{Tr}(f^0(\lambda) - f_1(\lambda)) \neq 0.$$

For the second pair of sets of admissible spectral densities $\mathcal{D}_{1\delta}^2 \times \mathcal{D}_V^{U2}$, we have

$$\begin{aligned}(\mathbf{C}^{f0}_{\overline{\mu},N}(e^{i\lambda}))(\mathbf{C}^{f0}_{\overline{\mu},N}(e^{i\lambda}))^* &= \frac{|\chi_{\overline{\mu}}^{(d)}(e^{-i\lambda})|^2}{|\beta^{(d)}(i\lambda)|^2}(f^0(\lambda) + |\beta^{(d)}(i\lambda)|^2 g^0(\lambda)) \\ &\times \{\beta_k^2\gamma_k^2(\lambda)\delta_{kl}\}_{k,l=1}^T\frac{|\chi_{\overline{\mu}}^{(d)}(e^{-i\lambda})|^2}{|\beta^{(d)}(i\lambda)|^2}(f^0(\lambda) + |\beta^{(d)}(i\lambda)|^2 g^0(\lambda)),\end{aligned} \tag{4.47}$$

$$\begin{aligned}(\mathbf{C}^{g0}_{\overline{\mu},N}(e^{i\lambda}))(\mathbf{C}^{g0}_{\overline{\mu},N}(e^{i\lambda}))^* &= (\beta^2 + \gamma_1(\lambda) + \gamma_2(\lambda)) \\ &\times \left(\frac{|\chi_{\overline{\mu}}^{(d)}(e^{-i\lambda})|^2}{|\beta^{(d)}(i\lambda)|^2}(f^0(\lambda) + |\beta^{(d)}(i\lambda)|^2 g^0(\lambda))\right)^2,\end{aligned} \tag{4.48}$$

$$\frac{1}{2\pi}\int_{-\pi}^{\pi}\frac{|\chi_{\overline{\mu}}^{(d)}(e^{-i\lambda})|^2}{|\beta^{(d)}(i\lambda)|^2}|f_{kk}^0(\lambda) - f_{kk}^1(\lambda)|d\lambda = \delta_k, \quad k = \overline{1,T}, \tag{4.49}$$

where β^2, β_k^2 are Lagrange multipliers, δ_{kl} are Kronecker symbols, the function $\gamma_1(\lambda) \leq 0$ and $\gamma_1(\lambda) = 0$ if $\mathrm{Tr}[g^0(\lambda)] > \mathrm{Tr}[V(\lambda)]$, the function $\gamma_2(\lambda) \geq 0$ and $\gamma_2(\lambda) = 0$ if $\mathrm{Tr}[g^0(\lambda)] < \mathrm{Tr}[U(\lambda)]$, the functions $|\gamma_k^2(\lambda)| \leq 1$, and

$$\gamma_k^2(\lambda) = \mathrm{sign}(f_{kk}^0(\lambda) - f_{kk}^1(\lambda)) : \; f_{kk}^0(\lambda) - f_{kk}^1(\lambda) \neq 0, \quad k = \overline{1,T}.$$

For the third pair of sets of admissible spectral densities $\mathcal{D}_{1\delta}^3 \times \mathcal{D}_V^{U3}$, we have

$$\begin{aligned}(\mathbf{C}_{\overline{\mu},N}^{f0}(e^{i\lambda}))(\mathbf{C}_{\overline{\mu},N}^{f0}(e^{i\lambda}))^* &= \frac{|\chi_{\overline{\mu}}^{(d)}(e^{-i\lambda})|^2}{|\beta^{(d)}(i\lambda)|^2}(f^0(\lambda) + |\beta^{(d)}(i\lambda)|^2 g^0(\lambda)) \\ &\times \beta^2\gamma_2'(\lambda)B_1^\top \frac{|\chi_{\overline{\mu}}^{(d)}(e^{-i\lambda})|^2}{|\beta^{(d)}(i\lambda)|^2}(f^0(\lambda) + |\beta^{(d)}(i\lambda)|^2 g^0(\lambda)),\end{aligned} \tag{4.50}$$

$$\begin{aligned}(\mathbf{C}_{\overline{\mu},N}^{g0}(e^{i\lambda}))(\mathbf{C}_{\overline{\mu},N}^{g0}(e^{i\lambda}))^* &= \frac{|\chi_{\overline{\mu}}^{(d)}(e^{-i\lambda})|^2}{|\beta^{(d)}(i\lambda)|^2}(f^0(\lambda) + |\beta^{(d)}(i\lambda)|^2 g^0(\lambda)) \\ &\times \{(\beta_k^2 + \gamma_{1k}(\lambda) + \gamma_{2k}(\lambda))\delta_{kl}\}_{k,l=1}^T \\ &\times \frac{|\chi_{\overline{\mu}}^{(d)}(e^{-i\lambda})|^2}{|\beta^{(d)}(i\lambda)|^2}(f^0(\lambda) + |\beta^{(d)}(i\lambda)|^2 g^0(\lambda)),\end{aligned} \tag{4.51}$$

$$\frac{1}{2\pi}\int_{-\pi}^{\pi} \frac{|\chi_{\overline{\mu}}^{(d)}(e^{-i\lambda})|^2}{|\beta^{(d)}(i\lambda)|^2} |\langle B_1, f^0(\lambda) - f_1(\lambda)\rangle| d\lambda = \delta, \tag{4.52}$$

where β^2, β_k^2 are Lagrange multipliers, δ_{kl} are Kronecker symbols, the function $\gamma_{1k}(\lambda) \leq 0$ and $\gamma_{1k}(\lambda) = 0$ if $g_{kk}^0(\lambda) > v_{kk}(\lambda)$, the function $\gamma_{2k}(\lambda) \geq 0$ and $\gamma_{2k}(\lambda) = 0$ if $g_{kk}^0(\lambda) < u_{kk}(\lambda)$, the function $|\gamma_2'(\lambda)| \leq 1$, and

$$\gamma_2'(\lambda) = \mathrm{sign}\langle B_1, f^0(\lambda) - f_1(\lambda)\rangle : \; \langle B_1, f^0(\lambda) - f_1(\lambda)\rangle \neq 0.$$

For the fourth pair of sets of admissible spectral densities $\mathcal{D}_{1\delta}^4 \times \mathcal{D}_V^{U4}$, we have

$$\begin{aligned}(\mathbf{C}_{\overline{\mu},N}^{f0}(e^{i\lambda}))(\mathbf{C}_{\overline{\mu},N}^{f0}(e^{i\lambda}))^* &= \frac{|\chi_{\overline{\mu}}^{(d)}(e^{-i\lambda})|^2}{|\beta^{(d)}(i\lambda)|^2}(f^0(\lambda) + |\beta^{(d)}(i\lambda)|^2 g^0(\lambda)) \\ &\times \{\beta_{ij}\gamma_{ij}(\lambda)\}_{i,j=1}^T \frac{|\chi_{\overline{\mu}}^{(d)}(e^{-i\lambda})|^2}{|\beta^{(d)}(i\lambda)|^2}(f^0(\lambda) + |\beta^{(d)}(i\lambda)|^2 g^0(\lambda)),\end{aligned} \tag{4.53}$$

$$\begin{aligned}(\mathbf{C}_{\overline{\mu},N}^{g0}(e^{i\lambda}))(\mathbf{C}_{\overline{\mu},N}^{g0}(e^{i\lambda}))^* &= (\beta^2 + \gamma_1'(\lambda) + \gamma_2'(\lambda)) \\ &\times \frac{|\chi_{\overline{\mu}}^{(d)}(e^{-i\lambda})|^2}{|\beta^{(d)}(i\lambda)|^2}(f^0(\lambda) + |\beta^{(d)}(i\lambda)|^2 g^0(\lambda)) \\ &\times B_2^\top \frac{|\chi_{\overline{\mu}}^{(d)}(e^{-i\lambda})|^2}{|\beta^{(d)}(i\lambda)|^2}(f^0(\lambda) + |\beta^{(d)}(i\lambda)|^2 g^0(\lambda)),\end{aligned} \tag{4.54}$$

$$\frac{1}{2\pi}\int_{-\pi}^{\pi}\frac{|\chi_{\overline{\mu}}^{(d)}(e^{-i\lambda})|^2}{|\beta^{(d)}(i\lambda)|^2}|f_{ij}^0(\lambda)-f_{ij}^1(\lambda)|d\lambda=\delta_i^j,\quad i,j=\overline{1,T},\tag{4.55}$$

where β_{ij},β^2 are Lagrange multipliers, the function $\gamma_1'(\lambda)\le 0$ and $\gamma_1'(\lambda)=0$ if $\langle B_2,g^0(\lambda)\rangle>\langle B_2,V(\lambda)\rangle$, the function $\gamma_2'(\lambda)\ge 0$ and $\gamma_2'(\lambda)=0$ if $\langle B_2,g^0(\lambda)\rangle<\langle B_2,U(\lambda)\rangle$, functions $|\gamma_{ij}(\lambda)|\le 1$, and

$$\gamma_{ij}(\lambda)=\frac{f_{ij}^0(\lambda)-f_{ij}^1(\lambda)}{|f_{ij}^0(\lambda)-f_{ij}^1(\lambda)|}\ :\ f_{ij}^0(\lambda)-f_{ij}^1(\lambda)\ne 0,\quad i,j=\overline{1,T}.$$

The following theorem holds.

Theorem 4.2.3. *The least favorable spectral densities $f^0(\lambda)$ and $g^0(\lambda)$ in the classes $\mathcal{D}_{1\delta}^k\times\mathcal{D}_V^{U^k}$, $k=1,2,3,4$, for the optimal linear interpolation of the functional $A_N\vec{\xi}$ are determined by pairs of equations* (4.44) *and* (4.46), (4.47) *and* (4.49), (4.50) *and* (4.52), *and* (4.53) *and* (4.55), *the minimality condition* (4.2), *the constrained optimization problem* (4.26), *and restrictions on densities from the corresponding classes $\mathcal{D}_{1\delta}^k\times\mathcal{D}_V^{U^k}$, $k=1,2,3,4$. The minimax-robust spectral characteristic $\vec{h}_{\overline{\mu},N}(f^0,g^0)$ of the optimal estimate of the functional $A_N\vec{\xi}$ is determined by formula* (4.10).

4.2.5 Least favorable spectral densities in classes $\mathcal{D}_{1\delta}$

Consider the minimax interpolation problem for the functional $A_N\vec{\xi}$ depending on the unobserved values of the stochastic sequence $\vec{\xi}(m)$ which determine a stationary GM increment sequence $\chi_{\overline{\mu},\overline{s}}^{(d)}(\vec{\xi}(m))$ from observations without noise for the sets of admissible spectral densities $\mathcal{D}_{1\delta}^k$, $k=1,2,3,4$, defined in Section 4.2.4.

The condition $0\in\partial\Delta_{\mathcal{D}}(f^0)$ implies the following equations which determine the least favorable spectral densities for these given sets of admissible spectral densities $\mathcal{D}_{1\delta}^1$ to $\mathcal{D}_{1\delta}^4$, respectively:

$$\left(\sum_{k=0}^{N+d(\gamma)}(((\mathbf{F}_N^{\overline{\mu}})^0)^{-1}[D_N^{\overline{\mu}}\mathbf{a}_N]_{+d(\gamma)})_k e^{i\lambda k}\right)\left(\sum_{k=0}^{N+d(\gamma)}(((\mathbf{F}_N^{\overline{\mu}})^0)^{-1}[D_N^{\overline{\mu}}\mathbf{a}_N]_{+d(\gamma)})_k e^{i\lambda k}\right)^*$$
$$=\alpha^2\gamma(\lambda)\left(\frac{|\chi_{\overline{\mu}}^{(d)}(e^{-i\lambda})|^2}{|\beta^{(d)}(i\lambda)|^2}f^0(\lambda)\right)^2,\tag{4.56}$$

$$\left(\sum_{k=0}^{N+d(\gamma)}(((\mathbf{F}_N^{\overline{\mu}})^0)^{-1}[D_N^{\overline{\mu}}\mathbf{a}_N]_{+d(\gamma)})_k e^{i\lambda k}\right)\left(\sum_{k=0}^{N+d(\gamma)}(((\mathbf{F}_N^{\overline{\mu}})^0)^{-1}[D_N^{\overline{\mu}}\mathbf{a}_N]_{+d(\gamma)})_k e^{i\lambda k}\right)^*$$
$$=\frac{|\chi_{\overline{\mu}}^{(d)}(e^{-i\lambda})|^2}{|\beta^{(d)}(i\lambda)|^2}f^0(\lambda)\{\alpha_k^2\gamma_k^2(\lambda)\delta_{kl}\}_{k,l=1}^T\frac{|\chi_{\overline{\mu}}^{(d)}(e^{-i\lambda})|^2}{|\beta^{(d)}(i\lambda)|^2}f^0(\lambda),\tag{4.57}$$

$$\left(\sum_{k=0}^{N+d(\gamma)}(((\mathbf{F}_N^{\overline{\mu}})^0)^{-1}[D_N^{\overline{\mu}}\mathbf{a}_N]_{+d(\gamma)})_k e^{i\lambda k}\right)\left(\sum_{k=0}^{N+d(\gamma)}(((\mathbf{F}_N^{\overline{\mu}})^0)^{-1}[D_N^{\overline{\mu}}\mathbf{a}_N]_{+d(\gamma)})_k e^{i\lambda k}\right)^* \\ = \frac{|\chi_{\overline{\mu}}^{(d)}(e^{-i\lambda})|^2}{|\beta^{(d)}(i\lambda)|^2} f^0(\lambda)\alpha'^2\gamma'(\lambda)B_1^{\top}\frac{|\chi_{\overline{\mu}}^{(d)}(e^{-i\lambda})|^2}{|\beta^{(d)}(i\lambda)|^2} f^0(\lambda), \tag{4.58}$$

$$\left(\sum_{k=0}^{N+d(\gamma)}(((\mathbf{F}_N^{\overline{\mu}})^0)^{-1}[D_N^{\overline{\mu}}\mathbf{a}_N]_{+d(\gamma)})_k e^{i\lambda k}\right)\left(\sum_{k=0}^{N+d(\gamma)}(((\mathbf{F}_N^{\overline{\mu}})^0)^{-1}[D_N^{\overline{\mu}}\mathbf{a}_N]_{+d(\gamma)})_k e^{i\lambda k}\right)^* \\ = \frac{|\chi_{\overline{\mu}}^{(d)}(e^{-i\lambda})|^2}{|\beta^{(d)}(i\lambda)|^2} f^0(\lambda)\{\alpha_{ij}\gamma_{ij}(\lambda)\}_{i,j=1}^{T}\frac{|\chi_{\overline{\mu}}^{(d)}(e^{-i\lambda})|^2}{|\beta^{(d)}(i\lambda)|^2} f^0(\lambda), \tag{4.59}$$

with the following restriction on the spectral density $f^0(\lambda)$:

$$\frac{1}{2\pi}\int_{-\pi}^{\pi}\frac{|\chi_{\overline{\mu}}^{(d)}(e^{-i\lambda})|^2}{|\beta^{(d)}(i\lambda)|^2}|f_{kk}^0(\lambda)-f_{kk}^1(\lambda)|d\lambda=\delta_k,\quad k=\overline{1,T}, \tag{4.60}$$

$$\frac{1}{2\pi}\int_{-\pi}^{\pi}\frac{|\chi_{\overline{\mu}}^{(d)}(e^{-i\lambda})|^2}{|\beta^{(d)}(i\lambda)|^2}|\mathrm{Tr}(f^0(\lambda)-f_1(\lambda))|d\lambda=\delta, \tag{4.61}$$

$$\frac{1}{2\pi}\int_{-\pi}^{\pi}\frac{|\chi_{\overline{\mu}}^{(d)}(e^{-i\lambda})|^2}{|\beta^{(d)}(i\lambda)|^2}|\langle B_1,f^0(\lambda)-f_1(\lambda)\rangle|d\lambda=\delta, \tag{4.62}$$

$$\frac{1}{2\pi}\int_{-\pi}^{\pi}\frac{|\chi_{\overline{\mu}}^{(d)}(e^{-i\lambda})|^2}{|\beta^{(d)}(i\lambda)|^2}|f_{ij}^0(\lambda)-f_{ij}^1(\lambda)|d\lambda=\delta_i^j,\quad i,j=\overline{1,T}, \tag{4.63}$$

where α^2, α_k^2, α'^2, and α_{ij} are Lagrange multipliers, the function $|\gamma(\lambda)|\leq 1$,

$$\gamma(\lambda)=\mathrm{sign}(\mathrm{Tr}(f^0(\lambda)-f_1(\lambda))):\ \mathrm{Tr}(f^0(\lambda)-f_1(\lambda))\neq 0,$$

the function $|\gamma_k^2(\lambda)|\leq 1$,

$$\gamma_k^2(\lambda)=\mathrm{sign}(f_{kk}^0(\lambda)-f_{kk}^1(\lambda)):\ f_{kk}^0(\lambda)-f_{kk}^1(\lambda)\neq 0,\quad k=\overline{1,T},$$

the function $|\gamma'(\lambda)|\leq 1$,

$$\gamma'(\lambda)=\mathrm{sign}\langle B_1,f^0(\lambda)-f_1(\lambda)\rangle:\ \langle B_1,f^0(\lambda)-f_1(\lambda)\rangle\neq 0,$$

the function $|\gamma_{ij}(\lambda)|\leq 1$,

$$\gamma_{ij}(\lambda)=\frac{f_{ij}^0(\lambda)-f_{ij}^1(\lambda)}{|f_{ij}^0(\lambda)-f_{ij}^1(\lambda)|}:\ f_{ij}^0(\lambda)-f_{ij}^1(\lambda)\neq 0,\quad i,j=\overline{1,T},$$

and δ_{kl} are Kronecker symbols.

The following theorem holds.

Theorem 4.2.4. *The least favorable spectral density* $f^0(\lambda)$ *in the classes* $\mathcal{D}_{1\delta}^k$, $k = 1, 2, 3, 4$, *for the optimal linear interpolation without noise of the functional* $A_N\vec{\xi}$ *are determined by pairs of equations* (4.56) *and* (4.61), (4.57) *and* (4.60), (4.58) *and* (4.62), *and* (4.59) *and* (4.63), *minimality condition* (4.17), *the constrained optimization problem* (4.27), *and restrictions on the density from the corresponding classes* $\mathcal{D}_{1\delta}^k$, $k = 1, 2, 3, 4$. *The minimax-robust spectral characteristic* $\vec{h}_{\overline{\mu},N}^{\xi}(f^0)$ *of the optimal estimate of the functional* $A_N\vec{\xi}$ *is determined by formula* (4.21).

Conclusions

In this chapter, we describe methods of solution of the interpolation problem for linear functionals which are constructed from unobserved values of a sequence with periodically stationary long memory multi-seasonal increments. Estimates are based on observations of the sequence with a periodically stationary noise sequence. Estimates are obtained by representing the sequence under investigation as a multidimensional sequence with stationary GM increments.

The Hilbert space projection method is applied to derive formulas for calculating the spectral characteristics and the MSEs of the optimal estimates of the functionals in the case of spectral certainty, where the spectral densities of the sequences are exactly known.

In the case of spectral uncertainty, where the spectral density is not exactly known while, instead, a set of admissible spectral densities is specified, the minimax-robust method is applied. We propose a representation of the MSE in the form of a linear functional in L_1 space with respect to spectral densities, which allows us to solve the corresponding constrained optimization problem and describe the minimax (robust) estimates of the functionals. Formulas that determine the least favorable spectral densities and minimax (robust) spectral characteristic of the optimal linear estimates of the functionals are derived for a collection of specific classes of admissible spectral densities. This method is illustrated with the help of concrete sets of admissible spectral densities: $\mathcal{D}_0 \times \mathcal{D}_\varepsilon$ and $\mathcal{D}_{1\delta} \times \mathcal{D}_V^U$. Here D_0 are classes of spectral densities with the moment restrictions, D_ε are classes describing the "ε-contaminated" models of a fixed bounded spectral density, $D_{1\delta}$ are classes describing the "δ-neighborhood" models in the space L_1 of a fixed bounded spectral density, and D_V^U are classes describing the "strip" models of spectral densities.

The main results of this chapter have been published in [121]. The corresponding results for minimax interpolation of sequences with periodically stationary increments (a particular case of GM increments) observed without and with noise are described in the papers by Kozak and Moklyachuk [95] and Luz and Moklyachuk [116], respectively. The results presented in this chapter rely on and develop prior studies of the minimax interpolation by Moklyachuk and Masyutka [152] for multidimensional stationary sequences, Golichenko (Dubovets'ka), Masyutka, and Moklyachuk [39] for periodically correlated sequences, and Luz and Moklyachuk [104, 111] for sequences with stationary increments.

5 Filtering of sequences with periodically stationary increments

In this chapter, we deal with the filtering problem for stochastic sequences with periodically stationary long memory multi-seasonal increments based on observations of the sequence with a periodically stationary noise sequence. The problem is investigated with the help of representation of the considered sequences in the form of a multidimensional stochastic sequence with stationary GM increments and a multidimensional stationary noise sequence.

5.1 Hilbert space projection method of filtering

5.1.1 Filtering of multidimensional stochastic sequences with stationary GM increment

Consider a multidimensional stochastic sequence $\vec{\xi}(m)$ with stationary GM increments constructed from transformation (1.2) and a multidimensional stationary stochastic sequence $\vec{\eta}(m)$ uncorrelated with the sequence $\vec{\xi}(m)$. Let the stationary GM increment sequence $\chi_{\overline{\mu},\overline{s}}^{(d)}(\vec{\xi}(m)) = \{\chi_{\overline{\mu},\overline{s}}^{(d)}(\xi_p(m))\}_{p=1}^{T}$ and the stationary sequence $\vec{\eta}(m)$ have absolutely continuous spectral functions $F(\lambda)$ and $G(\lambda)$ with the spectral densities $f(\lambda) = \{f_{ij}(\lambda)\}_{i,j=1}^{T}$ and $g(\lambda) = \{g_{ij}(\lambda)\}_{i,j=1}^{T}$, respectively. Without loss of generality assume that $\mathsf{E}\chi_{\overline{\mu},\overline{s}}^{(d)}(\vec{\xi}(m)) = \vec{0}$, $\mathsf{E}\vec{\eta}(m) = \vec{0}$, and $\overline{\mu} > \overline{0}$.

Filtering problem

Consider the problem of mean square optimal linear estimation of the functional

$$A\vec{\xi} = \sum_{k=0}^{\infty} (\vec{a}(k))^{\top} \vec{\xi}(-k), \tag{5.1}$$

which depends on unobserved values of a stochastic sequence $\vec{\xi}(k) = \{\xi_p(k)\}_{p=1}^{T}$ with stationary GM increments. Estimates are based on observations of the sequence $\vec{\zeta}(k) = \vec{\xi}(k) + \vec{\eta}(k)$ at points $k = 0, -1, -2, \ldots$.

We will suppose that the following conditions are satisfied:

– conditions on coefficients $\vec{a}(k) = \{a_p(k)\}_{p=1}^{T}$, $k \geq 0$,

$$\sum_{k=0}^{\infty} \|\vec{a}(k)\| < \infty, \quad \sum_{k=0}^{\infty} (k+1)\|\vec{a}(k)\|^2 < \infty, \tag{5.2}$$

https://doi.org/10.1515/9783111325620-005

– *the minimality condition* on the spectral densities $f(\lambda)$ and $g(\lambda)$,

$$\int_{-\pi}^{\pi} \mathrm{Tr}\left[\frac{|\beta^{(d)}(i\lambda)|^2}{|\chi_{\overline{\mu}}^{(d)}(e^{-i\lambda})|^2}\left(f(\lambda)+|\beta^{(d)}(i\lambda)|^2 g(\lambda)\right)^{-1}\right] d\lambda < \infty. \tag{5.3}$$

The second condition (5.3) is the necessary and sufficient one under which the MSE of the optimal estimate of functional $A\vec{\xi}$ is not equal to 0.

Represent the functional $A\vec{\xi}$ as $A\vec{\xi} = A\vec{\zeta} - A\vec{\eta}$, where

$$A\vec{\zeta} = \sum_{k=0}^{\infty} (\vec{a}(k))^{\top} \vec{\zeta}(-k), \quad A\vec{\eta} = \sum_{k=0}^{\infty} (\vec{a}(k))^{\top} \vec{\eta}(-k).$$

Note that under conditions (5.2), the functional $A\vec{\eta}$ belongs to the space H. To find the optimal estimate of the functional $A\vec{\xi}$, it is sufficient to find the estimate of the functional $A\vec{\eta}$

$$\widehat{A}\vec{\xi} = A\vec{\zeta} - \widehat{A}\vec{\eta}, \tag{5.4}$$

since the functional $A\vec{\zeta}$ depends on values of the stochastic sequence $\zeta(k)$ which are observed. The MSEs of the optimal estimates $\widehat{A}\vec{\xi}$ and $\widehat{A}\vec{\eta}$ are connected as

$$\begin{aligned}\Delta(f,g;\widehat{A}\vec{\xi}) &= \mathsf{E}|A\vec{\xi}-\widehat{A}\vec{\xi}|^2 = \mathsf{E}|A\vec{\zeta}-A\vec{\eta}-A\vec{\zeta}+\widehat{A}\vec{\eta}|^2\\ &= \mathsf{E}|A\vec{\eta}-\widehat{A}\vec{\eta}|^2 = \Delta(f,g;\widehat{A}\vec{\eta}).\end{aligned}$$

The set of the observed GM increments $\{\chi_{\overline{\mu},\overline{s}}^{(d)}(\vec{\xi}(k)) + \chi_{\overline{\mu},\overline{s}}^{(d)}(\vec{\eta}(k)) : k \le 0\}$, $\overline{\mu} > \vec{0}$ generates a closed linear subspace

$$H^0(\vec{\xi}^{(d)} + \vec{\eta}^{(d)}) = \overline{\mathrm{span}}\{\chi_{\overline{\mu},\overline{s}}^{(d)}(\xi_p(k)) + \chi_{\overline{\mu},\overline{s}}^{(d)}(\eta_p(k)) : k \le 0,\ p = 1,\dots,T\}$$

of the Hilbert space $H = L_2(\Omega, \mathfrak{F}, \mathsf{P})$. The functions

$$e^{i\lambda k}\chi_{\overline{\mu}}^{(d)}(e^{-i\lambda})\frac{1}{\beta^{(d)}(i\lambda)}\vec{\delta}_l, \quad l = 1,\dots,T,\ k \le 0,$$

generate a closed linear subspace

$$L_2^0(f(\lambda) + |\beta^{(d)}(i\lambda)|^2 g(\lambda)) = \overline{\mathrm{span}}\left\{e^{i\lambda k}\chi_{\overline{\mu}}^{(d)}(e^{-i\lambda})\frac{1}{\beta^{(d)}(i\lambda)}\vec{\delta}_l, l = 1,\dots,T,\ k \le 0\right\}$$

of the Hilbert space $L_2(f(\lambda) + |\beta^{(d)}(i\lambda)|^2 g(\lambda))$. Here $\vec{\delta}_l = \{\delta_{lp}\}_{p=1}^{T}$.

Remark 5.1.1. The relation

$$\chi_{\overline{\mu},\overline{s}}^{(d)}(\vec{\xi}(k)) + \chi_{\overline{\mu},\overline{s}}^{(d)}(\vec{\eta}(k)) = \int_{-\pi}^{\pi} e^{i\lambda k}\chi_{\overline{\mu}}^{(d)}(e^{-i\lambda})\frac{1}{\beta^{(d)}(i\lambda)}\, d\vec{Z}_{\xi^{(d)}+\eta^{(d)}}(\lambda)$$

implies a map between the elements $e^{i\lambda k}\chi_{\overline{\mu}}^{(d)}(e^{-i\lambda})(\beta^{(d)}(i\lambda))^{-1}\vec{\delta}_l$ from the space $L_2(f(\lambda)+|\beta^{(d)}(i\lambda)|^2 g(\lambda))$ and the elements $\chi_{\overline{\mu},\overline{s}}^{(d)}(\xi_p(k))+\chi_{\overline{\mu},\overline{s}}^{(d)}(\eta_p(k))$ from the space $H(\vec{\xi}^{(d)}+\vec{\eta}^{(d)})$, respectively.

The estimate $\widehat{A}\vec{\eta}$ is found as an orthogonal projection of the functional $A\vec{\eta}$ from the space $H=L_2(\Omega,\mathfrak{F},\mathsf{P})$ on the subspace $H(\vec{\xi}^{(d)}+\vec{\eta}^{(d)})$:

$$\widehat{A}\vec{\eta}=\mathrm{Proj}_{H(\vec{\xi}^{(d)}+\vec{\eta}^{(d)})}A\vec{\eta}.$$

Then relation (5.4) implies that the optimal estimate $\widehat{A}\vec{\xi}$ of the functional $A\vec{\xi}$ allows a representation

$$\widehat{A}\vec{\xi}=A\vec{\zeta}-\mathrm{Proj}_{H(\vec{\xi}^{(d)}+\vec{\eta}^{(d)})}A\vec{\eta}$$

or representation

$$\widehat{A}\vec{\xi}=\sum_{k=0}^{\infty}(\vec{a}(k))^{\top}(\vec{\xi}(-k)+\vec{\eta}(-k))-\int_{-\pi}^{\pi}(\vec{h}_{\overline{\mu}}(\lambda))^{\top}d\vec{Z}_{\xi^{(d)}+\eta^{(d)}}(\lambda), \tag{5.5}$$

where $\vec{h}_{\overline{\mu}}(\lambda)=\{h_p(\lambda)\}_{p=1}^{T}$ is the spectral characteristic of the optimal estimate $\widehat{A}\vec{\eta}$ of the functional $A\vec{\eta}$.

Define the following matrix-valued Fourier coefficients:

$$S_{k,j}^{\overline{\mu}}=\frac{1}{2\pi}\int_{-\pi}^{\pi}e^{-i\lambda(k+j)}\frac{|\beta^{(d)}(i\lambda)|^2}{|\chi_{\overline{\mu}}^{(d)}(e^{-i\lambda})|^2}\left(g(\lambda)(f(\lambda)+|\beta^{(d)}(i\lambda)|^2g(\lambda))^{-1}\right)^{\top}d\lambda,\quad k\geq 0,\ j\geq -d(\gamma),$$

$$P_{k,j}^{\overline{\mu}}=\frac{1}{2\pi}\int_{-\pi}^{\pi}e^{-i\lambda(k-j)}\frac{|\beta^{(d)}(i\lambda)|^2}{|\chi_{\overline{\mu}}^{(d)}(e^{-i\lambda})|^2}\left((f(\lambda)+|\beta^{(d)}(i\lambda)|^2g(\lambda))^{-1}\right)^{\top}d\lambda,\quad k,j\geq 0,$$

$$Q_{k,j}=\frac{1}{2\pi}\int_{-\pi}^{\pi}e^{-i\lambda(k-j)}\left(f(\lambda)(f(\lambda)+|\beta^{(d)}(i\lambda)|^2g(\lambda))^{-1}g(\lambda)\right)^{\top}d\lambda,\quad k,j\geq 0.$$

Define the vectors

$$\mathbf{a}=((\vec{a}(0))^{\top},(\vec{a}(1))^{\top},(\vec{a}(2))^{\top},\dots)^{\top}$$

and

$$\mathbf{a}_{\overline{\mu}}=((\vec{a}_{\overline{\mu}}(0))^{\top},(\vec{a}_{\overline{\mu}}(1))^{\top},(\vec{a}_{\overline{\mu}}(2))^{\top},\dots)^{\top},$$

where the coefficients $\vec{a}_{\overline{\mu}}(k)=\vec{a}_{-\overline{\mu}}(k-d(\gamma))$, $k\geq 0$,

$$\vec{a}_{-\overline{\mu}}(m)=\sum_{l=\max\{m,0\}}^{m+d(\gamma)}e_{\gamma}(l-m)\vec{a}(l),\quad m\geq -d(\gamma). \tag{5.6}$$

Denote by $\mathbf{P}_{\overline{\mu}}$, $\mathbf{S}_{\overline{\mu}}$, and $\mathbf{Q}$ matrices with the matrix-valued entries $(\mathbf{P}_{\overline{\mu}})_{l,k} = P^{\overline{\mu}}_{l,k}$, $(\mathbf{S}_{\overline{\mu}})_{l,k} = S^{\overline{\mu}}_{l+1,k-d(\gamma)}$, and $(\mathbf{Q})_{l,k} = Q_{l,k}$, $l, k \geq 0$.

Theorem 5.1.1. *A solution $\widehat{A}\vec{\xi}$ of the filtering problem for the linear functional $A\vec{\xi}$ of the values of a vector-valued stochastic sequence $\vec{\xi}(m)$ with stationary GM increments under conditions* (5.2) *and* (5.3) *is calculated by formula* (5.5). *The spectral characteristic $\vec{h}_{\overline{\mu}}(\lambda)$ and the value of the MSE $\Delta(f, g; \widehat{A}\vec{\xi})$ are calculated by the formulas*

$$(\vec{h}_{\overline{\mu}}(\lambda))^{\top} = (\chi^{(d)}_{\overline{\mu}}(e^{i\lambda})(\vec{A}(e^{-i\lambda}))^{\top} g(\lambda) - (\vec{C}_{\overline{\mu}}(e^{i\lambda}))^{\top}) \times (f(\lambda) + |\beta^{(d)}(i\lambda)|^2 g(\lambda))^{-1} \frac{\overline{\beta^{(d)}(i\lambda)}}{\chi^{(d)}_{\overline{\mu}}(e^{i\lambda})}, \tag{5.7}$$

where

$$\vec{A}(e^{-i\lambda}) = \sum_{k=0}^{\infty} \vec{a}(k) e^{-i\lambda k}, \quad \vec{C}_{\overline{\mu}}(e^{i\lambda}) = \sum_{k=0}^{\infty} (\mathbf{P}^{-1}_{\overline{\mu}} \mathbf{S}_{\overline{\mu}} \mathbf{a}_{\overline{\mu}})_k e^{i\lambda(k+1)},$$

and

$$\begin{aligned}
\Delta(f, g; \widehat{A}\vec{\xi}) &= \Delta(f, g; \widehat{A}\vec{\eta}) = \mathsf{E}|A\vec{\eta} - \widehat{A}\vec{\eta}|^2 \\
&= \frac{1}{2\pi} \int_{-\pi}^{\pi} \frac{1}{|\chi^{(d)}_{\overline{\mu}}(e^{-i\lambda})|^2} (\chi^{(d)}_{\overline{\mu}}(e^{i\lambda})(\vec{A}(e^{-i\lambda}))^{\top} f(\lambda) + |\beta^{(d)}(i\lambda)|^2 (\vec{C}_{\overline{\mu}}(e^{i\lambda}))^{\top}) \\
&\quad \times (f(\lambda) + |\beta^{(d)}(i\lambda)|^2 g(\lambda))^{-1} g(\lambda) (f(\lambda) + |\beta^{(d)}(i\lambda)|^2 g(\lambda))^{-1} \\
&\quad \times (\chi^{(d)}_{\overline{\mu}}(e^{-i\lambda}) f(\lambda) \overline{\vec{A}(e^{-i\lambda})} + |\beta^{(d)}(i\lambda)|^2 \overline{\vec{C}_{\overline{\mu}}(e^{i\lambda})}) d\lambda \\
&\quad + \frac{1}{2\pi} \int_{-\pi}^{\pi} \frac{|\beta^{(d)}(i\lambda)|^2}{|\chi^{(d)}_{\overline{\mu}}(e^{-i\lambda})|^2} (\chi^{(d)}_{\overline{\mu}}(e^{i\lambda})(\vec{A}(e^{-i\lambda}))^{\top} g(\lambda) - (\vec{C}_{\overline{\mu}}(e^{i\lambda}))^{\top}) \\
&\quad \times (f(\lambda) + |\beta^{(d)}(i\lambda)|^2 g(\lambda))^{-1} f(\lambda) (f(\lambda) + |\beta^{(d)}(i\lambda)|^2 g(\lambda))^{-1} \\
&\quad \times (\chi^{(d)}_{\overline{\mu}}(e^{-i\lambda}) g(\lambda) \overline{\vec{A}(e^{-i\lambda})} - \overline{\vec{C}_{\overline{\mu}}(e^{i\lambda})}) d\lambda \\
&= \langle \mathbf{S}_{\overline{\mu}} \mathbf{a}_{\overline{\mu}}, \mathbf{P}^{-1}_{\overline{\mu}} \mathbf{S}_{\overline{\mu}} \mathbf{a}_{\overline{\mu}} \rangle + \langle \mathbf{Q}\mathbf{a}, \mathbf{a} \rangle.
\end{aligned} \tag{5.8}$$

Proof. The projection $\widehat{A}\vec{\eta}$ is characterized by the following conditions:

(1) $\widehat{A}\vec{\eta} \in H^0(\vec{\xi}^{(d)} + \vec{\eta}^{(d)})$;

(2) $(A\vec{\eta} - \widehat{A}\vec{\eta}) \perp H^0(\vec{\xi}^{(d)} + \vec{\eta}^{(d)})$.

The second condition $(A\vec{\eta} - \widehat{A}\vec{\eta}) \perp H^0(\xi^{(d)} + \eta^{(d)})$ implies the following relation, which holds for all $k \leq 0$ and $l = 1, \ldots, T$:

$$\mathsf{E}(A\vec{\eta}-\widehat{A}\vec{\eta})^{\top}\overline{(\chi^{(d)}_{\overline{\mu},\overline{s}}(\vec{\xi}(m))+\chi^{(d)}_{\overline{\mu},\overline{s}}(\vec{\eta}(m)))}$$
$$=\frac{1}{2\pi}\int_{-\pi}^{\pi}(\vec{A}(e^{-i\lambda})-\beta^{(d)}(i\lambda)\vec{h}_{\overline{\mu}}(\lambda))^{\top}g(\lambda)\vec{\delta}_l e^{-i\lambda k}\chi^{(d)}_{\overline{\mu}}(e^{i\lambda})d\lambda$$
$$-\frac{1}{2\pi}\int_{-\pi}^{\pi}(\vec{h}_{\overline{\mu}}(\lambda))^{\top}f(\lambda)\vec{\delta}_l e^{-i\lambda k}\chi^{(d)}_{\overline{\mu}}(e^{i\lambda})\frac{1}{\beta^{(d)}(i\lambda)}d\lambda=0.$$

Thus, for all $k\leq 0$, the function $\vec{h}_{\overline{\mu}}(\lambda)$ satisfies the relation

$$\int_{-\pi}^{\pi}((\vec{A}(e^{-i\lambda}))^{\top}g(\lambda)\overline{\beta^{(d)}(i\lambda)}-(\vec{h}_{\overline{\mu}}(\lambda))^{\top}(f(\lambda)+|\beta^{(d)}(i\lambda)|^2g(\lambda)))\frac{\chi^{(d)}_{\overline{\mu}}(e^{i\lambda})}{\beta^{(d)}(i\lambda)}\vec{\delta}_l e^{-i\lambda k}d\lambda=0.$$

The latter relation implies that the spectral characteristic $\vec{h}_{\overline{\mu}}(\lambda)$ allows a representation of the form (5.7) with

$$\vec{C}_{\overline{\mu}}(e^{i\lambda})=\sum_{k=0}^{\infty}\vec{c}_{\overline{\mu}}(k)e^{i\lambda(k+1)},$$

where $\vec{c}_{\overline{\mu}}(k)=\{c_{\overline{\mu}p}(k)\}_{p=1}^{T}$, $k\geq 0$, are unknown coefficients to be found.

The condition $\widehat{A}\vec{\eta}\in H^0(\xi^{(d)}+\eta^{(d)})$ implies the following representation of the spectral characteristic $\vec{h}_{\overline{\mu}}(\lambda)$:

$$\vec{h}_{\overline{\mu}}(\lambda)=\vec{h}(\lambda)\chi^{(d)}_{\overline{\mu}}(e^{-i\lambda})\frac{1}{\beta^{(d)}(i\lambda)},\qquad \vec{h}(\lambda)=\sum_{k=0}^{\infty}\vec{s}(k)e^{-i\lambda k},$$

which leads to the following relation, which holds for all $l\geq 1$:

$$\int_{-\pi}^{\pi}(\chi^{(d)}_{\overline{\mu}}(e^{i\lambda})(\vec{A}(e^{-i\lambda}))^{\top}g(\lambda)-(\vec{C}_{\overline{\mu}}(e^{i\lambda}))^{\top})$$
$$\times\frac{|\beta^{(d)}(i\lambda)|^2}{|\chi^{(d)}_{\overline{\mu}}(e^{-i\lambda})|^2}(f(\lambda)+|\beta^{(d)}(i\lambda)|^2g(\lambda))^{-1}e^{-i\lambda l}d\lambda=\vec{0}. \tag{5.9}$$

In terms of the matrix-valued Fourier coefficients $S^{\overline{\mu}}_{k,j}$ for $k\geq 0, j\geq -d(\gamma)$ and $P^{\overline{\mu}}_{k,j}$, $Q_{k,j}$ for $k,j\geq 0$, relation (5.9) is presented as a system of linear equations

$$\sum_{m=-d(\gamma)}^{\infty}S^{\overline{\mu}}_{l+1,m}\vec{a}_{-\overline{\mu}}(m)=\sum_{k=0}^{\infty}P^{\overline{\mu}}_{l+1,k+1}\vec{c}_{\overline{\mu}}(k),\quad l\geq 0,$$

or in matrix form as

$$\mathbf{S}_{\overline{\mu}}\mathbf{a}_{\overline{\mu}}=\mathbf{P}_{\overline{\mu}}\mathbf{c}_{\overline{\mu}},$$

where $\mathbf{c}_{\overline{\mu}} = ((\vec{c}_{\overline{\mu}}(0))^\top, (\vec{c}_{\overline{\mu}}(1))^\top, (\vec{c}_{\mu}(2))^\top, \dots)^\top$. Making use of the described relations we can conclude that the unknown coefficients $\vec{c}_{\overline{\mu}}(k)$, $k \geq 0$, are calculated by the formula

$$\vec{c}_{\overline{\mu}}(k) = (\mathbf{P}_{\overline{\mu}}^{-1}\mathbf{S}_{\overline{\mu}}\mathbf{a}_{\overline{\mu}})_k, \quad k \geq 0,$$

where $(\mathbf{P}_{\overline{\mu}}^{-1}\mathbf{S}_{\overline{\mu}}\mathbf{a}_{\overline{\mu}})_k$, $k \geq 0$, is the kth T-dimensional vector element of the vector $\mathbf{P}_{\overline{\mu}}^{-1}\mathbf{S}_{\overline{\mu}}\mathbf{a}_{\overline{\mu}}$.

The derived expressions justify the use of formulas (5.7) and (5.8) for calculating the spectral characteristic $\vec{h}_{\overline{\mu}}(\lambda)$ and the value of the MSE $\Delta(f, g; \widehat{A}\vec{\xi})$ of the estimate $\widehat{A}\vec{\xi}$ of the functional $A\vec{\xi}$. □

The filtering problem for the functional $A_N\vec{\xi}$ is solved directly using Theorem 5.1.1 by putting $\vec{a}(k) = \vec{0}$ for $k > N$. To solve the filtering problem for the pth coordinate of the single vector $\vec{\xi}(-N)$, we put $\vec{a}(N) = \vec{\delta}_p$, $\vec{a}(k) = \vec{0}$ for $k \neq N$.

Define a matrix $\mathbf{S}_{\overline{\mu},N}$ with the $T \times T$ matrix entries $(\mathbf{S}_{\overline{\mu},N})_{l,m} = S^{\overline{\mu}}_{l+1,m-d(\gamma)}$ for $l \geq 0$, $0 \leq m \leq N + d(\gamma)$ and $(\mathbf{S}_{\overline{\mu},N})_{l,m} = \mathcal{O}$ for $l \geq 0$, $m > N + d(\gamma)$. Define another matrix $\mathbf{Q}_N$ with the $T \times T$ matrix entries $(\mathbf{Q}_N)_{l,k} = Q_{l,k}$, $0 \leq l, k \leq N$, and $(\mathbf{Q}_N)_{l,k} = \mathcal{O}$ otherwise. The following corollaries hold.

Corollary 5.1.1. *A solution $\widehat{A}_N\vec{\xi}$ of the filtering problem for the linear functional $A_N\vec{\xi}$ of the values of a vector-valued stochastic sequence $\vec{\xi}(m)$ with stationary GM increments under condition* (5.2) *is calculated by the formula*

$$\widehat{A}_N\vec{\xi} = \sum_{k=0}^{N} (\vec{a}(k))^\top (\vec{\xi}(-k) + \vec{\eta}(-k)) - \int_{-\pi}^{\pi} (h_{\overline{\mu},N}(\lambda))^\top d\vec{Z}_{\xi^{(d)}+\eta^{(d)}}(\lambda). \tag{5.10}$$

The spectral characteristic $\vec{h}_{\overline{\mu},N}(\lambda)$ of the optimal estimate $\widehat{A}_N\vec{\xi}$ is calculated by the formula

$$\begin{aligned}(\vec{h}_{\overline{\mu},N}(\lambda))^\top &= (\chi^{(d)}_{\overline{\mu}}(e^{i\lambda})(\vec{A}_N(e^{-i\lambda}))^\top g(\lambda) - (\vec{C}^{\overline{\mu}}_N(e^{i\lambda}))^\top)\\ &\quad \times (f(\lambda) + |\beta^{(d)}(i\lambda)|^2 g(\lambda))^{-1} \frac{\overline{\beta^{(d)}(i\lambda)}}{\chi^{(d)}_{\overline{\mu}}(e^{i\lambda})},\end{aligned} \tag{5.11}$$

where

$$\vec{A}_N(e^{i\lambda}) = \sum_{k=0}^{N} \vec{a}(k) e^{-i\lambda k}, \quad \vec{C}^{\overline{\mu}}_N(e^{i\lambda}) = \sum_{k=0}^{\infty} (\mathbf{P}_{\overline{\mu}}^{-1}\mathbf{S}_{\overline{\mu},N}\mathbf{a}_{\overline{\mu},N})_k e^{i\lambda(k+1)},$$

and the vector

$$\mathbf{a}_{\overline{\mu},N} = (\vec{a}_{\overline{\mu},N}(0), \vec{a}_{\overline{\mu},N}(1), \dots, \vec{a}_{\overline{\mu},N}(N + d(\gamma)), \vec{0}^\top, \dots)^\top,$$

with the entries

$$\vec{a}_{\overline{\mu},N}(k) = \vec{a}_{-\overline{\mu},N}(k - d(\gamma)), \quad 0 \le k \le N + d(\gamma),$$
$$\vec{a}_{-\overline{\mu},N}(m) = \sum_{l=\max\{m,0\}}^{\min\{m+d(\gamma),N\}} e_\gamma(l-m)\vec{a}(l), \quad -d(\gamma) \le m \le N. \tag{5.12}$$

The MSE $\Delta(f,g;\widehat{A}_N\vec{\xi})$ is calculated by the formula

$$\Delta(f,g;\widehat{A}_N\vec{\xi}) = \Delta(f,g;\widehat{A}\vec{\eta}_N) = \mathsf{E}|A\vec{\eta}_N - \widehat{A}\vec{\eta}_N|^2 = \langle \mathbf{S}_{\overline{\mu},N}\mathbf{a}_{\overline{\mu},N}, \mathbf{P}_{\overline{\mu}}^{-1}\mathbf{S}_{\overline{\mu},N}\mathbf{a}_{\overline{\mu},N}\rangle + \langle \mathbf{Q}_N\mathbf{a}_N, \mathbf{a}_N\rangle, \tag{5.13}$$

where $\mathbf{a}_N = ((\vec{a}(0))^\top, (\vec{a}(1))^\top, (\vec{a}(2))^\top, \dots, (\vec{a}(N))^\top, \vec{0}^\top, \dots)^\top$.

Corollary 5.1.2. *The optimal linear estimate $\widehat{\xi}_p(-N)$ of an unobserved value $\xi_p(-N)$, $N \ge 0$, of the stochastic vector sequence $\vec{\xi}(m)$ with stationary GM increments based on observations of the sequence $\vec{\xi}(m) + \vec{\eta}(m)$ at points $m = 0, -1, -2, \dots$, where the noise sequence $\vec{\eta}(m)$ is uncorrelated with $\vec{\xi}(m)$, is calculated by the formula*

$$\widehat{\xi}_p(-N) = (\xi_p(-N) + \eta_p(-N)) - \int_{-\pi}^{\pi} (\vec{h}_{\overline{\mu},N,p}(\lambda))^\top d\vec{Z}_{\xi^{(d)}+\eta^{(d)}}(\lambda). \tag{5.14}$$

The spectral characteristic of the estimate $\widehat{\xi}_p(-N)$ is calculated by the formula

$$(\vec{h}_{\overline{\mu},N,p}(\lambda))^\top = (e^{-i\lambda N}\chi_{\overline{\mu}}^{(d)}(e^{i\lambda})(\vec{\delta}_p)^\top g(\lambda) - (\vec{C}_{N,p}^{\overline{\mu}}(e^{i\lambda}))^\top)$$
$$\times (f(\lambda) + |\beta^{(d)}(i\lambda)|^2 g(\lambda))^{-1} \frac{\overline{\beta^{(d)}(i\lambda)}}{\chi_{\overline{\mu}}^{(d)}(e^{i\lambda})}, \tag{5.15}$$

where

$$\vec{C}_{N,p}^{\overline{\mu}}(e^{i\lambda}) = \sum_{k=0}^{\infty} (\mathbf{P}_{\overline{\mu}}^{-1}\mathbf{S}_{\mu,N}\vec{\mathbf{a}}_{\overline{\mu},N,p})_k e^{i\lambda(k+1)},$$

and the vector

$$\vec{\mathbf{a}}_{\overline{\mu},N,p} = (\vec{0}^\top, \dots, \vec{0}^\top, (\vec{a}_{\overline{\mu},N,p}(N))^\top, (\vec{a}_{\overline{\mu},N,p}(N+1))^\top, \dots, (\vec{a}_{\overline{\mu},N,p}(N+d(\gamma)))^\top, \vec{0}^\top, \dots)^\top,$$

with the entries

$$\vec{a}_{\overline{\mu},N,p}(k) = \vec{a}_{-\overline{\mu},N,p}(k - d(\gamma)), \quad N \le k \le N + d(\gamma),$$
$$\vec{a}_{-\overline{\mu},N,p}(m) = e_\gamma(N-m)\vec{\delta}_p, \quad N - d(\gamma) \le m \le N.$$

The MSE of the estimate $\widehat{\xi}_p(-N)$ is calculated by the formula

$$\Delta(f,g;\widehat{\xi}_p(-N)) = \Delta(f,g;\widehat{\eta}_p(-N)) = \mathsf{E}|\eta_p(-N) - \widehat{\eta}_p(-N)|^2$$
$$= \langle \mathbf{S}_{\overline{\mu},N}\mathbf{a}_{\overline{\mu},N,p}, \mathbf{P}_{\overline{\mu}}^{-1}\mathbf{S}_{\overline{\mu},N}\mathbf{a}_{\overline{\mu},N,p}\rangle + \langle Q_{0,0}\vec{\delta}_p, \vec{\delta}_p\rangle. \tag{5.16}$$

5.1.2 Filtering of sequences with FM increments

Let $d_j = R_j + D_j, j = 0, 1, \ldots, r$, denote the fractional differentiation order and let R_j, $j = 0, 1, \ldots, r$, be non-negative integer numbers. Set the increment step $\overline{\mu} = (1, 1, \ldots, 1)$. Let $1 < s_1 < \cdots < s_r$. Consider a stochastic vector sequence $\vec{\xi}(m)$ with stationary FM increments $\chi^{(R+D)}_{\overline{1},\overline{s}}(\vec{\xi}(m))$ having a bounded from zero and from infinity spectral density $\widetilde{f}_{\overline{1}}(\lambda)$. Assume that the numbers $R_j, D_j, j = 0, 1, \ldots, r$, can be chosen in such way that $-1/2 < D_\nu < 1/2$ for all $\nu \in \mathcal{M}$, where the numbers D_ν and the set are defined in Lemma 1.2.2. Then Theorem 1.2.1 implies that the GM increment sequence $\chi^{(R)}_{\overline{1},\overline{s}}(\vec{\xi}(m))$ is stationary.

Corollary 5.1.3. *A solution of the filtering problem for the multidimensional sequence $\vec{\xi}(m)$ with the stationary FM increments based on its observations with stationary noise can be obtained from Theorem* 5.1.1, *Corollary* 5.1.1, *and Corollary* 5.1.2 *for the GM increment sequence $\chi^{(R)}_{\overline{1},\overline{s}}(\vec{\xi}(m))$ making use of the Fourier coefficients*

$$S^{\overline{1}}_{k,j} = \frac{1}{2\pi}\int_{-\pi}^{\pi} e^{-i\lambda(k+j)}\frac{|\beta^{(R)}(i\lambda)|^2}{|\chi^{(R)}_{\overline{1}}(e^{-i\lambda})|^2}(g(\lambda)p^{-1}(\lambda))^\top d\lambda, \quad k \geq 0,\ j \geq -d(\gamma),$$

$$P^{\overline{1}}_{k,j} = \frac{1}{2\pi}\int_{-\pi}^{\pi} e^{-i\lambda(k-j)}\frac{|\beta^{(R)}(i\lambda)|^2}{|\chi^{(R)}_{\overline{1}}(e^{-i\lambda})|^2}(p^{-1}(\lambda))^\top d\lambda, \quad k, j \geq 0,$$

$$Q_{k,j} = \frac{1}{2\pi}\int_{-\pi}^{\pi} e^{-i\lambda(k-j)}\left(|\chi^{(D)}_{\overline{1}}(e^{-i\lambda})|^{-2}\widetilde{f}(\lambda)p^{-1}(\lambda)g(\lambda)\right)^\top d\lambda, \quad k, j \geq 0,$$

where

$$p(\lambda) = |\chi^{(D)}_{\overline{1}}(e^{-i\lambda})|^{-2}\widetilde{f}(\lambda) + |\beta^{(R)}(i\lambda)|^2 g(\lambda),$$
$$\widetilde{f}(\lambda) = |\beta^{(R)}(i\lambda)|^2 |\chi^{(R)}_{\overline{1}}(e^{-i\lambda})|^{-2}\widetilde{f}_{\overline{1}}(\lambda),$$

under the conditions of Theorem 1.2.1.

5.1.3 Filtering based on factorizations of the spectral densities

The main goal of this subsection is to derive the estimates of the functional $A\vec{\xi}$ in terms of the coefficients of the canonical factorizations of the spectral densities $f(\lambda)$, $g(\lambda)$, and $f(\lambda) + |\beta^{(d)}(i\lambda)|^2 g(\lambda)$.

Let the following canonical factorizations take place:

$$\frac{|\chi^{(d)}_{\overline{\mu}}(e^{-i\lambda})|^2}{|\beta^{(d)}(i\lambda)|^2}(f(\lambda) + |\beta^{(d)}(i\lambda)|^2 g(\lambda)) = \Theta_{\overline{\mu}}(e^{-i\lambda})\Theta^*_{\overline{\mu}}(e^{-i\lambda}), \quad \Theta_{\overline{\mu}}(e^{-i\lambda}) = \sum_{k=0}^{\infty}\theta_{\overline{\mu}}(k)e^{-i\lambda k}, \tag{5.17}$$

$$g(\lambda) = \sum_{k=-\infty}^{\infty} g(k)e^{i\lambda k} = \Phi(e^{-i\lambda})\Phi^*(e^{-i\lambda}), \quad \Phi(e^{-i\lambda}) = \sum_{k=0}^{\infty} \phi(k)e^{-i\lambda k}. \tag{5.18}$$

Define the matrix-valued function $\Psi_{\overline{\mu}}(e^{-i\lambda}) = \{\Psi_{\overline{\mu},ij}(e^{-i\lambda})\}_{i=\overline{1,q}}^{j=\overline{1,T}}$ by the equation

$$\Psi_{\overline{\mu}}(e^{-i\lambda})\Theta_{\overline{\mu}}(e^{-i\lambda}) = E_q,$$

where E_q is an identity $q \times q$ matrix. One can check that the following factorization takes place:

$$\frac{|\beta^{(d)}(i\lambda)|^2}{|\chi_{\overline{\mu}}^{(d)}(e^{-i\lambda})|^2}(f(\lambda) + |\beta^{(d)}(i\lambda)|^2 g(\lambda))^{-1} = \Psi_{\overline{\mu}}^*(e^{-i\lambda})\Psi_{\overline{\mu}}(e^{-i\lambda}), \quad \Psi_{\overline{\mu}}(e^{-i\lambda}) = \sum_{k=0}^{\infty} \psi_{\overline{\mu}}(k)e^{-i\lambda k}. \tag{5.19}$$

Recall the lemmas providing representations of $\mathbf{P}_{\overline{\mu}}$ and $\mathbf{P}_{\overline{\mu}}^{-1}\mathbf{S}_{\overline{\mu}}\mathbf{a}_{\overline{\mu}}$, which contain coefficients of factorizations (5.17)–(5.19).

Lemma 5.1.1. *Let factorization* (5.17) *take place and let $q \times T$ matrix function $\Psi_{\overline{\mu}}(e^{-i\lambda})$ satisfy the equation $\Psi_{\overline{\mu}}(e^{-i\lambda})\Phi_{\overline{\mu}}(e^{-i\lambda}) = E_q$. Define the linear operators $\Psi_{\overline{\mu}}$ and $\Theta_{\overline{\mu}}$ in the space ℓ_2 by the matrices with the matrix entries $(\Psi_{\overline{\mu}})_{k,j} = \psi_{\overline{\mu}}(k-j)$, $(\Theta_{\overline{\mu}})_{k,j} = \theta_{\overline{\mu}}(k-j)$ for $0 \le j \le k$ and $(\Psi_{\overline{\mu}})_{k,j} = \mathcal{O}$, $(\Theta_{\overline{\mu}})_{k,j} = \mathcal{O}$ for $0 \le k < j$. Then:*
(a) *the linear operator $\mathbf{P}_{\overline{\mu}}$ admits the factorization*

$$\mathbf{P}_{\overline{\mu}} = (\Psi_{\overline{\mu}})^{\top}\overline{\Psi}_{\overline{\mu}};$$

(b) *the inverse operator $(\mathbf{P}_{\overline{\mu}})^{-1}$ admits the factorization*

$$(\mathbf{P}_{\overline{\mu}})^{-1} = \overline{\Theta}_{\overline{\mu}}(\Theta_{\overline{\mu}})^{\top}.$$

Lemma 5.1.2. *Let factorizations* (5.17) *and* (5.18) *take place. Define by $\widetilde{\vec{e}}_{\overline{\mu}}(m) = (\Theta_{\overline{\mu}}^{\top}\mathbf{S}_{\overline{\mu}}\widetilde{\mathbf{a}}_{\overline{\mu}})_m$, $m \ge 0$, the mth entry of the vector $\widetilde{\mathbf{e}}_{\overline{\mu}} = \Theta_{\overline{\mu}}^{\top}\mathbf{S}_{\overline{\mu}}\widetilde{\mathbf{a}}_{\overline{\mu}}$. Then*

$$\widetilde{\vec{e}}_{\overline{\mu}}(m) = \sum_{j=-d(\gamma)}^{\infty} Z_{\overline{\mu}}(m+j+1)\vec{a}_{-\overline{\mu}}(j),$$

where $Z_{\overline{\mu}}(j)$, $j \in \mathbb{Z}$, are defined as

$$Z_{\overline{\mu}}(j) = \sum_{l=0}^{\infty} \overline{\psi}_{\overline{\mu}}(l)\overline{g}(l-j), \quad j \in \mathbb{Z}, \quad g(k) = \sum_{m=\max\{0,-k\}}^{\infty} \phi(m)\phi^*(k+m), \quad k \in \mathbb{Z}.$$

Proof. Factorizations (3.30) and (3.31) and Remark 3.1.2 imply

$$\sum_{k\in\mathbb{Z}} \vec{s}_{\overline{\mu}}(k)e^{i\lambda k} = \frac{|\beta^{(d)}(i\lambda)|^2}{|\chi_{\overline{\mu}}^{(d)}(e^{-i\lambda})|^2}(g(\lambda)(f(\lambda) + |\beta^{(d)}(i\lambda)|^2 g(\lambda))^{-1})^{\top}$$

$$
\begin{aligned}
&= (\Psi_{\overline{\mu}}(e^{-i\lambda}))^{\top}\overline{\Psi}_{\overline{\mu}}(e^{-i\lambda})\overline{g}(\lambda)\\
&= \sum_{l=0}^{\infty}\psi_{\overline{\mu}}^{\top}(l)e^{-i\lambda l}\sum_{j\in\mathbb{Z}}Z_{\overline{\mu}}(j)e^{i\lambda j}\\
&= \sum_{k\in\mathbb{Z}}\sum_{l=0}^{\infty}\psi_{\overline{\mu}}^{\top}(l)Z_{\overline{\mu}}(l+k)e^{i\lambda k}.
\end{aligned}
$$

Then

$$
\begin{aligned}
(\Theta_{\overline{\mu}}^{\top}\mathbf{S}_{\overline{\mu}}\tilde{\mathbf{a}}_{\overline{\mu}})_m &= \sum_{j=-d(\gamma)}^{\infty}\sum_{p=m}^{\infty}\theta_{\overline{\mu}}^{\top}(p-m)S_{\overline{\mu}}(p+j+1)\vec{a}_{-\overline{\mu}}(j)\\
&= \sum_{j=-d(\gamma)}^{\infty}\sum_{p=m}^{\infty}\sum_{l=0}^{\infty}\theta_{\overline{\mu}}^{\top}(p-m)\psi_{\overline{\mu}}^{\top}(l)Z_{\overline{\mu}}(l+p+j+1)\vec{a}_{-\overline{\mu}}(j)\\
&= \sum_{j=-d(\gamma)}^{\infty}\sum_{p=m}^{\infty}\sum_{k=p}^{\infty}(\psi_{\overline{\mu}}(k-p)\theta_{\overline{\mu}}(p-m))^{\top}Z_{\overline{\mu}}(k+j+1)\vec{a}_{-\overline{\mu}}(j)\\
&= \sum_{j=-d(\gamma)}^{\infty}\sum_{k=m}^{\infty}\mathrm{diag}_q(\delta_{k,m})Z_{\overline{\mu}}(k+j+1)\vec{a}_{-\overline{\mu}}(j)\\
&= \sum_{j=-d(\gamma)}^{\infty}Z_{\overline{\mu}}(m+j+1)\vec{a}_{-\overline{\mu}}(j).
\end{aligned}
$$

The representation for $Z_{\overline{\mu}}(j)$ follows from

$$
\sum_{j\in\mathbb{Z}}Z_{\overline{\mu}}(j)e^{i\lambda j} = \overline{\Psi}_{\overline{\mu}}(e^{-i\lambda})\overline{g}(\lambda) = \sum_{k\in\mathbb{Z}}\sum_{l=0}^{\infty}\overline{\Psi}_{\overline{\mu}}(l)\overline{g}(l-k)e^{i\lambda k}. \qquad \square
$$

Define the linear operators $\mathbf{G}^{-}$, $\mathbf{G}^{+}$, $\widetilde{\Phi}^{+}$ in the space ℓ_2 by matrices with the matrix entries $(\mathbf{G}^{-})_{l,k} = \overline{g}(l-k)$, $(\mathbf{G}^{+})_{l,k} = \overline{g}(l+k)$, and $(\widetilde{\Phi}^{+})_{l,k} = \phi^{\top}(k+j)$, $l,k \geq 0$, and the linear operator $\widetilde{\Phi}$ in the space ℓ_2 determined by a matrix with the matrix entries $(\widetilde{\Phi})_{k,j} = \phi^{\top}(k-j)$ for $0 \leq j \leq k$ and $(\widetilde{\Phi})_{k,j} = \mathcal{O}$ for $0 \leq k < j$.

Define also the coefficients $\{\vec{b}_{-\overline{\mu}}(k) : k \geq 0\}$ as follows: $\vec{b}_{-\overline{\mu}}(0) = \vec{0}$, $\vec{b}_{-\overline{\mu}}(k) = \vec{a}_{-\overline{\mu}}(-k)$ for $1 \leq k \leq d(\gamma)$ and $\vec{b}_{-\overline{\mu}}(k) = \vec{0}$ for $k > d(\gamma)$, where coefficients $\vec{a}_{-\overline{\mu}}(k)$ are calculated by formula (5.6) and the vectors

$$
\begin{aligned}
\mathbf{a}_{-\overline{\mu}} &= ((\vec{a}_{-\overline{\mu}}(0))^{\top}, (\vec{a}_{-\overline{\mu}}(1))^{\top}, (\vec{a}_{-\overline{\mu}}(2))^{\top}, \ldots)^{\top},\\
\mathbf{b}_{-\overline{\mu}} &= ((\vec{b}_{-\overline{\mu}}(0))^{\top}, (\vec{b}_{-\overline{\mu}}(1))^{\top}, (\vec{b}_{-\overline{\mu}}(2))^{\top}, \ldots)^{\top}.
\end{aligned}
$$

The following theorem describes a solution to the filtering problem in the case when the spectral densities $f(\lambda)$ and $g(\lambda)$ admit canonical factorizations (5.17)–(5.19).

Theorem 5.1.2. *A solution $\widehat{A}\vec{\xi}$ of the filtering problem for the linear functional $A\vec{\xi}$ which depends on the values of a vector-valued stochastic sequence $\vec{\xi}(m)$ with stationary GM increments under condition* (5.2) *and provided that the spectral densities $f(\lambda)$ and $g(\lambda)$ of the* *stochastic vector sequences* $\vec{\xi}(m)$ *and* $\vec{\eta}(m)$ *admit canonical factorizations* (5.17)–(5.19) *is calculated by formula* (5.5). *The spectral characteristic* $\vec{h}_{\overline{\mu}}(\lambda)$ *is calculated by the formulas*

$$\vec{h}_{\overline{\mu}}(\lambda) = \frac{\chi_{\overline{\mu}}^{(d)}(e^{-i\lambda})}{\beta^{(d)}(i\lambda)}\left(\sum_{k=0}^{\infty}\psi_{\overline{\mu}}^{\top}(k)e^{-i\lambda k}\right)\sum_{m=0}^{\infty}((\widetilde{\Psi}_{\overline{\mu}})^{*}\mathbf{G}^{-}\mathbf{a}_{-\overline{\mu}} + (\widetilde{\Psi}_{\overline{\mu}})^{*}\mathbf{G}^{+}\mathbf{b}_{-\overline{\mu}})_{m}e^{-i\lambda m}$$
$$= \frac{\chi_{\overline{\mu}}^{(d)}(e^{-i\lambda})}{\beta^{(d)}(i\lambda)}\left(\sum_{k=0}^{\infty}\psi_{\overline{\mu}}^{\top}(k)e^{-i\lambda k}\right)\sum_{m=0}^{\infty}(\overline{\psi}_{\overline{\mu}}\mathbf{C}_{\overline{\mu},g}^{-} + \overline{\psi}_{\overline{\mu}}\mathbf{C}_{\overline{\mu},g}^{+})_{m}e^{-i\lambda m}, \tag{5.20}$$

where $\overline{\psi}_{\overline{\mu}} = (\overline{\psi}_{\overline{\mu}}(0), \overline{\psi}_{\overline{\mu}}(1), \overline{\psi}_{\overline{\mu}}(2), \ldots)$,

$$(\overline{\psi}_{\overline{\mu}}\mathbf{C}_{\overline{\mu},g}^{\pm})_{m} = \sum_{k=0}^{\infty}\overline{\psi}_{\overline{\mu}}(k)\mathbf{c}_{\overline{\mu},g}^{\pm}(k+m),$$
$$\mathbf{c}_{\overline{\mu},g}^{-}(m) = \sum_{k=0}^{\infty}\overline{g}(m-k)\vec{a}_{-\overline{\mu}}(k) = \sum_{l=0}^{\infty}\overline{\phi}(l)\sum_{k=0}^{l+m}\phi^{\top}(l+m-k)\vec{a}_{-\overline{\mu}}(k)$$
$$= \sum_{l=0}^{\infty}\overline{\phi}(l)(\widetilde{\Phi}\mathbf{a}_{-\overline{\mu}})_{l+m} = (\widetilde{\Phi}^{*}\widetilde{\Phi}\mathbf{a}_{-\overline{\mu}})_{m},$$
$$\mathbf{c}_{\overline{\mu},g}^{+}(m) = \sum_{k=0}^{\infty}\overline{g}(m+k)\vec{b}_{-\overline{\mu}}(k) = \sum_{l=0}^{\infty}\overline{\phi}(l)\sum_{k=0}^{\infty}\phi^{\top}(l+m+k)\vec{b}_{-\overline{\mu}}(k)$$
$$= \sum_{l=0}^{\infty}\overline{\phi}(l)(\widetilde{\Phi}^{+}\mathbf{b}_{-\overline{\mu}})_{l+m} = (\widetilde{\Phi}^{*}\widetilde{\Phi}^{+}\mathbf{b}_{-\overline{\mu}})_{m}.$$

The MSE $\Delta(f,g;\widehat{A}\vec{\xi})$ *is calculated by the formula*

$$\Delta(f,g;\widehat{A}\xi) = \|\widetilde{\Phi}\mathbf{a}\|^{2} - \|\overline{\psi}_{\overline{\mu}}(\mathbf{C}_{\overline{\mu},g}^{-} + \mathbf{C}_{\overline{\mu},g}^{+})\|^{2}. \tag{5.21}$$

Proof. Under the conditions of Lemmas 5.1.1 and 5.1.2 on the spectral densities $f(\lambda)$ and $g(\lambda)$, formulas (5.7) and (5.8) can be rewritten. We make the following transformations:

$$\frac{|\beta^{(d)}(i\lambda)|^{2}}{\chi_{\overline{\mu}}^{(d)}(e^{-i\lambda})|^{2}}((f(\lambda) + |\beta^{(d)}(i\lambda)|^{2}g(\lambda))^{-1})^{\top}(g(\lambda))^{\top}A(e^{-i\lambda})\chi_{\overline{\mu}}^{(d)}(e^{i\lambda})$$
$$= \Psi_{\overline{\mu}}^{\top}(e^{-i\lambda})\overline{\Psi}_{\overline{\mu}}(e^{-i\lambda})\overline{g}(\lambda)A(e^{-i\lambda})\chi_{\overline{\mu}}^{(d)}(e^{i\lambda})$$
$$= \left(\sum_{k=0}^{\infty}\psi_{\overline{\mu}}^{\top}(k)e^{-i\lambda k}\right)\sum_{m\in\mathbb{Z}}\sum_{j=-d(\gamma)}^{\infty}Z_{\overline{\mu}}(m+j)\vec{a}_{-\overline{\mu}}(j)e^{i\lambda m}$$

and

$$\frac{|\beta^{(d)}(i\lambda)|^2}{|\chi_{\overline{\mu}}^{(d)}(e^{-i\lambda})|^2}((f(\lambda)+|\beta^{(d)}(i\lambda)|^2 g(\lambda))^{-1})^{\top}\left(\sum_{k=0}^{\infty}(\mathbf{P}_{\overline{\mu}}^{-1}\mathbf{S}_{\overline{\mu}}\mathbf{a}_{\overline{\mu}})_k e^{i\lambda(k+1)}\right)$$
$$=\left(\sum_{k=0}^{\infty}\psi_{\overline{\mu}}^{\top}(k)e^{-i\lambda k}\right)\sum_{j=0}^{\infty}\sum_{k=0}^{\infty}\overline{\psi}_{\overline{\mu}}(j)(\overline{\Theta}_{\overline{\mu}}\widetilde{\mathbf{e}}_{\overline{\mu}})_k e^{i\lambda(k+j+1)}$$
$$=\left(\sum_{k=0}^{\infty}\psi_{\overline{\mu}}^{\top}(k)e^{-i\lambda k}\right)\sum_{m=0}^{\infty}\sum_{p=0}^{m}\sum_{k=p}^{m}\overline{\psi}_{\overline{\mu}}(m-k)\overline{\theta}_{\overline{\mu}}(k-p)\vec{\widetilde{e}}_{\overline{\mu}}(p)e^{i\lambda(m+1)}$$
$$=\left(\sum_{k=0}^{\infty}\psi_{\overline{\mu}}^{\top}(k)e^{-i\lambda k}\right)\sum_{m=0}^{\infty}\sum_{p=0}^{m}\operatorname{diag}_q(\delta_{m,p})\vec{\widetilde{e}}_{\overline{\mu}}(m)e^{i\lambda(m+1)}$$
$$=\left(\sum_{k=0}^{\infty}\psi_{\overline{\mu}}^{\top}(k)e^{-i\lambda k}\right)\sum_{m=0}^{\infty}\vec{\widetilde{e}}_{\overline{\mu}}(m)e^{i\lambda(m+1)}.$$

Then we obtain

$$\vec{h}_{\overline{\mu}}(\lambda)=\frac{\chi_{\overline{\mu}}^{(d)}(e^{-i\lambda})}{\beta^{(d)}(i\lambda)}\left(\sum_{k=0}^{\infty}\psi_{\overline{\mu}}^{\top}(k)e^{-i\lambda k}\right)\sum_{m=0}^{\infty}\sum_{j=-d(\gamma)}^{\infty}Z_{\overline{\mu}}(j-m)\vec{a}_{-\overline{\mu}}(j)e^{-i\lambda m}$$
$$=\frac{\chi_{\overline{\mu}}^{(d)}(e^{-i\lambda})}{\beta^{(d)}(i\lambda)}\left(\sum_{k=0}^{\infty}\psi_{\overline{\mu}}^{\top}(k)e^{-i\lambda k}\right)$$
$$\times\sum_{m=0}^{\infty}\left(\sum_{j=0}^{\infty}Z_{\overline{\mu}}(j-m)\vec{a}_{-\overline{\mu}}(j)+\sum_{j=1}^{d(\gamma)}Z_{\overline{\mu}}(-j-m)\vec{b}_{-\overline{\mu}}(j)\right)e^{-i\lambda m}$$
$$=\frac{\chi_{\overline{\mu}}^{(d)}(e^{-i\lambda})}{\beta^{(d)}(i\lambda)}\left(\sum_{k=0}^{\infty}\psi_{\overline{\mu}}^{\top}(k)e^{-i\lambda k}\right)$$
$$\times\left(\sum_{m=0}^{\infty}\sum_{j=0}^{\infty}\sum_{p=m}^{\infty}\overline{\psi}_{\overline{\mu}}(p-m)\overline{g}(p-j)\vec{a}_{-\overline{\mu}}(j)e^{-i\lambda m}\right.$$
$$\left.+\sum_{m=0}^{\infty}\sum_{j=1}^{d(\gamma)}\sum_{p=m}^{\infty}\overline{\psi}_{\overline{\mu}}(p-m)\overline{g}(p+j)\vec{b}_{-\overline{\mu}}(j)e^{-i\lambda m}\right)$$
$$=\frac{\chi_{\overline{\mu}}^{(d)}(e^{-i\lambda})}{\beta^{(d)}(i\lambda)}\left(\sum_{k=0}^{\infty}\psi_{\overline{\mu}}^{\top}(k)e^{-i\lambda k}\right)\sum_{m=0}^{\infty}((\widetilde{\Psi}_{\overline{\mu}})^{*}\mathbf{G}^{-}\mathbf{a}_{-\overline{\mu}}+(\widetilde{\Psi}_{\overline{\mu}})^{*}\mathbf{G}^{+}\mathbf{b}_{-\overline{\mu}})_m e^{-i\lambda m}.$$

Finally,

$$\vec{h}_{\overline{\mu}}(\lambda)=\frac{\chi_{\overline{\mu}}^{(d)}(e^{-i\lambda})}{\beta^{(d)}(i\lambda)}\left(\sum_{k=0}^{\infty}\psi_{\overline{\mu}}^{\top}(k)e^{-i\lambda k}\right)\sum_{m=0}^{\infty}(\overline{\psi}_{\overline{\mu}}\mathbf{C}_{\overline{\mu},g}^{-}+\overline{\psi}_{\overline{\mu}}\mathbf{C}_{\overline{\mu},g}^{+})_m e^{-i\lambda m}.$$

The MSE $\Delta(f,g;\widehat{A}\xi)$ is calculated by the formula

$$\Delta(f,g;\widehat{A}\xi)=\Delta(f,g;\widehat{A}\eta)=\mathsf{E}|A\eta-\widehat{A}\eta|^2$$

$$= \frac{1}{2\pi}\int_{-\pi}^{\pi} (\vec{A}(e^{-i\lambda}))^\top g(\lambda)\overline{\vec{A}(e^{-i\lambda})}d\lambda$$
$$+ \frac{1}{2\pi}\int_{-\pi}^{\pi} (\vec{h}_{\overline{\mu}}(\lambda))^\top (f(\lambda) + |\beta^{(d)}(i\lambda)|^2 g(\lambda))\overline{\vec{h}_{\overline{\mu}}(\lambda)}d\lambda$$
$$- \frac{1}{2\pi}\int_{-\pi}^{\pi} (\vec{h}_{\overline{\mu}}(\lambda))^\top \beta^{(d)}(i\lambda) g(\lambda)\overline{A(e^{-i\lambda})}d\lambda$$
$$- \frac{1}{2\pi}\int_{-\pi}^{\pi} (A(e^{-i\lambda}))^\top \overline{\beta^{(d)}(i\lambda)} g(\lambda)\overline{\vec{h}_{\overline{\mu}}(\lambda)}d\lambda$$
$$= \|\widetilde{\Phi}\mathbf{a}\|^2 - \|\overline{\psi}_{\overline{\mu}}(\mathbf{C}^-_{\overline{\mu},g} + \mathbf{C}^+_{\overline{\mu},g})\|^2. \qquad \square$$

Remark 5.1.2. The following factorizations hold:

$$\mathbf{G}^- = \widetilde{\Phi}^*\widetilde{\Phi}, \quad \mathbf{G}^+ = \widetilde{\Phi}^*\widetilde{\Phi}^+.$$

The filtering problem for the functional $A_N\vec{\xi}$ is solved directly by Theorem 5.1.2 by putting $\vec{a}(k) = \vec{0}$ for $k > N$. To solve the filtering problem for the pth coordinate of the single vector $\vec{\xi}(-N)$, we put $\vec{a}(N) = \vec{\delta}_p$, $\vec{a}(k) = \vec{0}$ for $k \neq N$.

The following corollaries hold.

Corollary 5.1.4. *A solution $\widehat{A}_N\vec{\xi}$ of the filtering problem for the linear functional $A_N\vec{\xi}$ which depends on the values of a vector-valued stochastic sequence $\vec{\xi}(m)$ with stationary GM increments under condition* (5.2) *is calculated by the formula*

$$\widehat{A}_N\vec{\xi} = \sum_{k=0}^{N} (\vec{a}(k))^\top (\vec{\xi}(-k) + \vec{\eta}(-k)) - \int_{-\pi}^{\pi} (\vec{h}_{\overline{\mu},N}(\lambda))^\top d\vec{Z}_{\xi^{(d)}+\eta^{(d)}}(\lambda). \tag{5.22}$$

The spectral characteristic $\vec{h}_{\overline{\mu},N}(\lambda)$ and the MSE $\Delta(f,g;\widehat{A}_N\vec{\xi})$ of the optimal estimate $\widehat{A}_N\vec{\xi}$ are calculated by formulas (5.20) *and* (5.21) *for the vectors* $\mathbf{a}$, $\mathbf{a}_{-\overline{\mu}}$, *and* $\mathbf{b}_{-\overline{\mu}}$ *calculated as*

$$\mathbf{a}_N = ((\vec{a}(0))^\top, (\vec{a}(1))^\top, (\vec{a}(2))^\top, \dots, (\vec{a}(N))^\top, \vec{0}^\top, \dots)^\top,$$
$$\mathbf{a}_{-\overline{\mu},N} = ((\vec{a}_{-\overline{\mu},N}(0))^\top, (\vec{a}_{-\overline{\mu},N}(1))^\top, \dots, (\vec{a}_{-\overline{\mu},N}(N))^\top, \vec{0}^\top, \dots)^\top,$$
$$\mathbf{b}_{-\overline{\mu},N} = (0, (\vec{a}_{-\overline{\mu},N}(-1))^\top, (\vec{a}_{-\overline{\mu},N}(-2))^\top, \dots, (\vec{a}_{-\overline{\mu},N}(-d(\gamma)))^\top, \vec{0}^\top, \dots)^\top,$$

where the vectors $\vec{a}_{-\overline{\mu},N}(m)$, $-d(\gamma) \le m \le N$, are calculated by formula (5.12).

Corollary 5.1.5. *The optimal linear estimate $\widehat{\xi}_p(-N)$ of an unobserved value $\xi_p(-N)$, $N \geq 0$, of the stochastic vector sequence $\vec{\xi}(m)$ with GM stationary increments based on observations of the sequence $\vec{\xi}(m) + \vec{\eta}(m)$ at points $m = 0, -1, -2, \dots$, where the noise sequence $\vec{\eta}(m)$ is uncorrelated with $\vec{\xi}(m)$, is calculated by the formula*

$$\widehat{\xi}_p(-N) = (\xi_p(-N) + \eta_p(-N)) - \int_{-\pi}^{\pi} (\vec{h}_{\overline{\mu},N,p}(\lambda))^\top d\vec{Z}_{\xi^{(d)}+\eta^{(d)}}(\lambda). \tag{5.23}$$

Set

$$\vec{a}_{-\overline{\mu},N,p}(m) = e_\gamma(N-m)\vec{\delta}_p, \quad N - d(\gamma) \le m \le N.$$

If $N < d(\gamma)$, *the spectral characteristic* $\vec{h}_{\overline{\mu},N,p}(\lambda)$ *and the MSE* $\Delta(f,g;\widehat{\xi}_p(-N))$ *of the optimal estimate* $\widehat{\xi}_p(-N)$ *are calculated by formulas* (5.20) *and* (5.21) *for the vectors* $\mathbf{a}$, $\mathbf{a}_{-\overline{\mu}}$, *and* $\mathbf{b}_{-\overline{\mu}}$ *calculated as*

$$\begin{aligned}
\mathbf{a}_{N,p} &= (\vec{0}^\top,\dots,\vec{0}^\top,(\vec{\delta}_p)^\top,\vec{0}^\top,\dots)^\top,\\
\mathbf{a}_{-\overline{\mu},N,p} &= ((\vec{a}_{-\overline{\mu},N,p}(0))^\top,(\vec{a}_{-\overline{\mu},N,p}(1))^\top,\dots,(\vec{a}_{-\overline{\mu},N,p}(N))^\top,\vec{0}^\top,\dots)^\top,\\
\mathbf{b}_{-\mu,N,p} &= (\vec{0}^\top,(\vec{a}_{-\mu,N,p}(-1))^\top,(\vec{a}_{-\mu,N,p}(-2))^\top,\dots,(\vec{a}_{-\mu,N,p}(N-d(\gamma)))^\top,\vec{0}^\top,\dots)^\top.
\end{aligned}$$

If $N \ge d(\gamma)$, *the spectral characteristic* $\vec{h}_{\overline{\mu},N,p}(\lambda)$ *and the MSE* $\Delta(f,g;\widehat{\xi}_p(-N))$ *of the optimal estimate* $\widehat{\xi}_p(-N)$ *are calculated by formulas*

$$\vec{h}_{\overline{\mu},N,p}(\lambda) = \frac{\chi_{\overline{\mu}}^{(d)}(e^{-i\lambda})}{\beta^{(d)}(i\lambda)}\left(\sum_{k=0}^{\infty}\psi_{\overline{\mu}}^\top(k)e^{-i\lambda k}\right)\sum_{m=0}^{\infty}((\widetilde{\Psi}_{\overline{\mu}})^*\mathbf{G}^-\mathbf{a}_{-\overline{\mu},N,p})_m e^{-i\lambda m}$$

and

$$\Delta(f,g;\widehat{\xi}_p(-N)) = \langle \overline{g}(0)\vec{\delta}_p,\vec{\delta}_p\rangle - \|(\widetilde{\Psi}_{\overline{\mu}})^*\mathbf{G}^-\mathbf{a}_{-\overline{\mu},N,p}\|^2,$$

where

$$\mathbf{a}_{-\overline{\mu},N,p} = (\vec{0}^\top,\dots,\vec{0}^\top,(\vec{a}_{-\overline{\mu},N,p}(N-d(\gamma)))^\top,\dots,(\vec{a}_{-\overline{\mu},N,p}(N))^\top,\vec{0}^\top,\dots)^\top.$$

5.1.4 Filtering of stochastic sequences with periodically stationary GM increment

Consider the filtering problem for the functionals

$$A\xi = \sum_{k=0}^{\infty} a^{(\xi)}(k)\xi(-k), \quad A_M\xi = \sum_{k=0}^{N} a^{(\xi)}(k)\xi(-k)$$

which depend on unobserved values of a stochastic sequence $\xi(m)$ with periodically stationary GM increments. Estimates are based on observations of the sequence $\xi(m) + \eta(m)$ at points $m = 0,-1,-2,\dots$, where the periodically stationary noise sequence $\eta(m)$ is uncorrelated with $\xi(m)$.

The functional $A\xi$ can be represented in the form

$$A\xi = \sum_{k=0}^{\infty} a^{(\xi)}(k)\xi(-k) = \sum_{m=0}^{\infty}\sum_{p=1}^{T} a^{(\xi)}(mT+p-1)\xi(-mT-p+1)$$
$$= \sum_{m=0}^{\infty}\sum_{p=1}^{T} a_p(m)\xi_p(-m) = \sum_{m=0}^{\infty} (\vec{a}(m))^{\top}\vec{\xi}(-m) = A\vec{\xi},$$

where

$$\vec{\xi}(m) = (\xi_1(m), \xi_2(m), \dots, \xi_T(m))^{\top}, \quad \xi_p(m) = \xi(mT+p-1), \quad p = 1, 2, \dots, T, \tag{5.24}$$

and

$$\vec{a}(m) = (a_1(m), a_2(m), \dots, a_T(m))^{\top}, \quad a_p(m) = a^{(\xi)}(mT+p-1), \quad p = 1, 2, \dots, T. \tag{5.25}$$

In the same way, the functional $A\eta$ is represented as

$$A\eta = \sum_{k=0}^{\infty} a^{(\xi)}(k)\eta(-k) = \sum_{m=0}^{\infty} (\vec{a}(m))^{\top}\vec{\eta}(-m) = A\vec{\eta},$$

where

$$\vec{\eta}(m) = (\eta_1(m), \eta_2(m), \dots, \eta_T(m))^{\top}, \quad \eta_p(m) = \eta(mT+p-1), \quad p = 1, 2, \dots, T. \tag{5.26}$$

Making use of the introduced notations and statements of Theorem 2.1.1, we can claim that the following theorem holds.

Theorem 5.1.3. *Let a stochastic sequence $\xi(m)$ with periodically stationary GM increments and a stochastic periodically stationary sequence $\eta(m)$ generate by formulas* (5.24) *and* (5.26) *a multidimensional stochastic sequence $\vec{\xi}(m)$ with stationary GM increments and a multidimensional stationary stochastic sequence $\vec{\eta}(m)$ with the spectral densities matrices $f(\lambda) = \{f_{ij}(\lambda)\}_{i,j=1}^{T}$ and $g(\lambda) = \{g_{ij}(\lambda)\}_{i,j=1}^{T}$. A solution $\widehat{A}\xi$ to the filtering problem for the functional $A\xi = A\vec{\xi}$ under conditions* (5.2) *and* (5.3) *is calculated by formula* (5.5) *with the coefficients $\vec{a}(m)$, $m \geq 0$, defined in* (5.25). *The spectral characteristic $\vec{h}_{\overline{\mu}}(\lambda)$ and the value of the MSE $\Delta(f, g; \widehat{A}\xi)$ of the estimate $\widehat{A}\xi$ are calculated by formulas* (5.7) *and* (5.8), *respectively.*

The functional $A_M\xi$ can be represented in the form

$$A_M\xi = \sum_{k=0}^{M} a^{(\xi)}(k)\xi(-k) = \sum_{m=0}^{N}\sum_{p=1}^{T} a^{(\xi)}(mT+p-1)\xi(-mT-p+1)$$
$$= \sum_{m=0}^{N}\sum_{p=1}^{T} a_p(m)\xi_p(-m) = \sum_{m=0}^{N} (\vec{a}(m))^{\top}\vec{\xi}(-m) = A_N\vec{\xi},$$

where $N = [\frac{M}{T}]$, the sequence $\vec{\xi}(m)$ is determined by formula (5.24), and

$$\begin{aligned} \vec{a}(m) &= (a_1(m), a_2(m), \dots, a_T(m))^\top, \\ a_p(m) &= a^\zeta(mT + p - 1), \quad 0 \le m \le N,\ 1 \le p \le T,\ mT + p - 1 \le M, \\ a_p(N) &= 0, \quad M + 1 \le NT + p - 1 \le (N+1)T - 1,\ 1 \le p \le T. \end{aligned} \tag{5.27}$$

An estimate of a single unobserved value $\xi(-M)$, $M \ge 0$, of a stochastic sequence $\xi(m)$ with periodically stationary GM increments is obtained by making use of the notations $\xi(-M) = \xi_p(-N) = (\vec{\delta}_p)^\top \vec{\xi}(N)$, $N = [\frac{M}{T}]$, $p = M + 1 - NT$. We can conclude that the following corollaries hold.

Corollary 5.1.6. *Let the conditions of Theorem* 5.1.3 *be satisfied. A solution $\widehat{A}_M\xi$ to the filtering problem for the functional $A_M\xi = A_N\vec{\xi}$ is calculated by formula* (5.10) *for the coefficients $\vec{a}(m)$, $0 \le m \le N$, defined in* (5.27). *The spectral characteristic and the value of the MSE of the estimate $\widehat{A}_M\xi$ are calculated by formulas* (5.11) *and* (5.13), *respectively.*

Corollary 5.1.7. *Let the conditions of Theorem* 5.1.3 *be satisfied. A solution $\widehat{\xi}(-M)$ to the filtering problem for an unobserved value $\xi(-M) = \xi_p(-N) = (\vec{\delta}_p)^\top \vec{\xi}(-N)$, $N = [\frac{M}{T}]$, $p = M + 1 - NT$, is calculated by formula* (5.14). *The spectral characteristic and the value of the MSE of the estimate $\widehat{\xi}(-M)$ are calculated by formulas* (5.15) *and* (5.16), *respectively.*

When the spectral densities admit canonical factorizations, the following statements hold.

Theorem 5.1.4. *Let a stochastic sequence $\xi(m)$ with periodically stationary GM increments and a stochastic periodically stationary sequence $\eta(m)$ generate by formulas* (5.24) *and* (5.26) *a multidimensional stochastic sequence $\vec{\xi}(m)$ with stationary GM increments and a multidimensional stationary stochastic sequence $\vec{\eta}(m)$ with the spectral densities matrices $f(\lambda) = \{f_{ij}(\lambda)\}_{i,j=1}^T$ and $g(\lambda) = \{g_{ij}(\lambda)\}_{i,j=1}^T$ admitting canonical factorizations* (5.17)–(5.19). *A solution $\widehat{A}\xi$ to the filtering problem for the functional $A\xi = A\vec{\xi}$ under conditions* (5.2) *is calculated by formula* (5.5) *for the coefficients $\vec{a}(m)$, $m \ge 0$, defined in* (5.25). *The spectral characteristic $\vec{h}_{\overline{\mu}}(\lambda)$ and the MSE $\Delta(f, g; \widehat{A}\xi)$ of the estimate $\widehat{A}\xi$ are calculated by formulas* (5.20) *and* (5.21), *respectively.*

Corollary 5.1.8. *Let the conditions of Theorem* 5.1.4 *be satisfied. A solution $\widehat{A}_M\xi$ to the filtering problem for the functional $A_M\xi = A_N\vec{\xi}$ is calculated by formula* (5.22) *for the coefficients $\vec{a}(m)$, $0 \le m \le N$, defined in* (5.27). *The spectral characteristic and the MSE of the estimate $\widehat{A}_M\xi$ are calculated by formulas* (5.20) *and* (5.21), *respectively.*

Corollary 5.1.9. *Let the conditions of Theorem* 5.1.4 *be satisfied. A solution $\widehat{\xi}(-M)$ to the filtering problem for an unobserved value $\xi(-M) = \xi_p(-N) = (\vec{\delta}_p)^\top \vec{\xi}(-N)$, $N = [\frac{M}{T}]$, $p = M + 1 - NT$, is calculated by formula* (5.23). *The spectral characteristic and the MSE of the estimate $\widehat{\xi}(-M)$ are calculated by formulas* (5.20) *and* (5.21), *respectively.*

5.2 Minimax (robust) method of filtering

5.2.1 Least favorable spectral densities and minimax-robust spectral characteristic

Solutions of the problem of estimating the functionals $A\vec{\xi}$ and $A_N\vec{\xi}$ constructed from unobserved values of the stochastic sequence $\vec{\xi}(m)$ with stationary GM increments $\chi^{(d)}_{\overline{\mu},\overline{s}}(\vec{\xi}(m))$ having the spectral density matrix $f(\lambda)$ based on its observations with stationary noise $\vec{\xi}(m)+\vec{\eta}(m)$ at points $m=0,-1,-2,\dots$ are proposed in Theorem 5.1.1 and Corollary 5.1.1 in the case where the spectral density matrices $f(\lambda)$ and $g(\lambda)$ of the target sequence and the noise are exactly known.

In this section, we consider the case where the complete information about the spectral density matrices is not available, while some sets of admissible spectral densities $\mathcal{D}=\mathcal{D}_f\times\mathcal{D}_g$ are known. The minimax approach of estimation of the functionals from unobserved values of stochastic sequences is considered, which consists in finding an estimate that minimizes the maximal values of the MSEs for all spectral densities from a class $\mathcal{D}$ simultaneously. This method will be applied for the concrete classes of spectral densities.

To proceed with the stated problem, we recall the following definitions [134].

Definition 5.2.1. For a given class of spectral densities $\mathcal{D}=\mathcal{D}_f\times\mathcal{D}_g$, the spectral densities $f^0(\lambda)\in\mathcal{D}_f$ and $g^0(\lambda)\in\mathcal{D}_g$ are called the least favorable densities in the class $\mathcal{D}$ for optimal linear filtering of the functional $A\xi$ if the following relation holds:

$$\Delta(f^0,g^0)=\Delta(h(f^0,g^0);f^0,g^0)=\max_{(f,g)\in\mathcal{D}_f\times\mathcal{D}_g}\Delta(h(f,g);f,g).$$

Definition 5.2.2. For a given class of spectral densities $\mathcal{D}=\mathcal{D}_f\times\mathcal{D}_g$, the spectral characteristic $\vec{h}^0(\lambda)$ of the optimal estimate of the functional $A\xi$ is called minimax-robust if the following relations hold:

$$\vec{h}^0(\lambda)\in H_{\mathcal{D}}=\bigcap_{(f,g)\in\mathcal{D}_f\times\mathcal{D}_g}L_2^0(f(\lambda)+|\beta^{(d)}(i\lambda)|^2 g(\lambda)),$$

$$\min_{\vec{h}\in H_{\mathcal{D}}}\max_{(f,g)\in\mathcal{D}_f\times\mathcal{D}_g}\Delta(\vec{h};f,g)=\max_{(f,g)\in\mathcal{D}_f\times\mathcal{D}_g}\Delta(\vec{h}^0;f,g).$$

Taking into account the introduced definitions and the relations derived in the previous sections we can verify that the following lemma holds.

Lemma 5.2.1. *The spectral densities $f^0\in\mathcal{D}_f$ and $g^0\in\mathcal{D}_g$ which satisfy the minimality condition* (5.3) *are the least favorable spectral densities in the class $\mathcal{D}$ for the optimal linear filtering of the functional $A\xi$ if the matrices $\mathbf{P}^0_{\overline{\mu}}$, $\mathbf{S}^0_{\overline{\mu}}$, $\mathbf{Q}^0$ defined by the Fourier coefficients of the functions*

$$|\beta^{(d)}(i\lambda)|^2|\chi^{(d)}_{\overline{\mu}}(e^{-i\lambda})|^{-2}((f^0(\lambda)+|\beta^{(d)}(i\lambda)|^2 g^0(\lambda))^{-1})^{\top},$$

$$|\beta^{(d)}(i\lambda)|^2|\chi_{\overline{\mu}}^{(d)}(e^{-i\lambda})|^{-2}(g^0(\lambda)(f^0(\lambda)+|\beta^{(d)}(i\lambda)|^2g^0(\lambda))^{-1})^\top,$$
$$(f^0(\lambda)(f^0(\lambda)+|\beta^{(d)}(i\lambda)|^2g^0(\lambda))^{-1}g^0(\lambda))^\top$$

determine a solution of the constrained optimization problem

$$\max_{(f,g)\in\mathcal{D}_f\times\mathcal{D}_g}\left(\langle \mathbf{S}_{\overline{\mu}}\mathbf{a}_{\overline{\mu}},\mathbf{P}_{\overline{\mu}}^{-1}\mathbf{S}_{\overline{\mu}}\mathbf{a}_{\overline{\mu}}\rangle+\langle\mathbf{Q}\mathbf{a},\mathbf{a}\rangle\right)=\langle \mathbf{S}_{\overline{\mu}}^0\mathbf{a}_{\overline{\mu}},(\mathbf{P}_{\overline{\mu}}^0)^{-1}\mathbf{S}_{\overline{\mu}}^0\mathbf{a}_{\overline{\mu}}\rangle+\langle\mathbf{Q}^0\mathbf{a},\mathbf{a}\rangle. \tag{5.28}$$

The minimax spectral characteristic $h^0 = h_{\overline{\mu}}(f^0,g^0)$ is calculated by formula (5.7) *if $h_{\overline{\mu}}(f^0,g^0)\in H_{\mathcal{D}}$.*

Lemma 5.2.2. *The spectral densities $f^0\in\mathcal{D}_f$ and $g^0\in\mathcal{D}_g$ which admit canonical factorizations* (1.14), (5.17), *and* (5.18) *are least favorable densities in the class $\mathcal{D}$ for the optimal linear filtering of the functional $A\vec{\xi}$ based on observations of the sequence $\vec{\xi}(m)+\vec{\eta}(m)$ at points $m=0,-1,-2,\dots$ if the matrix coefficients of canonical factorizations* (5.17) *and* (5.18) *determine a solution to the constrained optimization problem*

$$\|\widetilde{\Phi}\mathbf{a}\|^2-\|\overline{\psi}_{\overline{\mu}}(\mathbf{C}_{\overline{\mu},g}^-+\mathbf{C}_{\overline{\mu},g}^+)\|^2\to\sup, \tag{5.29}$$
$$f(\lambda)=\frac{|\beta^{(d)}(i\lambda)|^2}{|\chi_{\overline{\mu}}^{(d)}(e^{-i\lambda})|^2}\Theta_{\overline{\mu}}(e^{-i\lambda})\Theta_{\overline{\mu}}^*(e^{-i\lambda})-|\beta^{(d)}(i\lambda)|^2\Phi(e^{-i\lambda})\Phi^*(e^{-i\lambda})\in\mathcal{D}_f,$$
$$g(\lambda)=\Phi(e^{-i\lambda})\Phi^*(e^{-i\lambda})\in\mathcal{D}_g.$$

The minimax spectral characteristic $h^0=\vec{h}_{\overline{\mu}}(f^0,g^0)$ is calculated by formula (5.20) *if $\vec{h}_{\overline{\mu}}(f^0,g^0)\in H_{\mathcal{D}}$.*

Lemma 5.2.3. *The spectral density $g^0\in\mathcal{D}_g$ which admits canonical factorizations* (5.17) *and* (5.18) *with the known spectral density $f(\lambda)$ is the least favorable in the class $\mathcal{D}_g$ for the optimal linear filtering of the functional $A\xi$ based on observations of the sequence $\vec{\xi}(m)+\vec{\eta}(m)$ at points $m=0,-1,-2,\dots$ if the matrix coefficients of the canonical factorizations*

$$f(\lambda)+|\beta^{(d)}(i\lambda)|^2g^0(\lambda)=\frac{|\beta^{(d)}(i\lambda)|^2}{|\chi_{\overline{\mu}}^{(d)}(e^{-i\lambda})|^2}\left(\sum_{k=0}^{\infty}\theta_{\overline{\mu}}^0(k)e^{-i\lambda k}\right)\left(\sum_{k=0}^{\infty}\theta_{\overline{\mu}}^0(k)e^{-i\lambda k}\right)^*, \tag{5.30}$$
$$g^0(\lambda)=\left(\sum_{k=0}^{\infty}\phi^0(k)e^{-i\lambda k}\right)\left(\sum_{k=0}^{\infty}\phi^0(k)e^{-i\lambda k}\right)^* \tag{5.31}$$

are determined by the equation $\Psi_{\overline{\mu}}^0(e^{-i\lambda})\Theta_{\overline{\mu}}^0(e^{-i\lambda})=E_q$ and a solution $\{\psi_{\overline{\mu}}^0(k),\phi^0(k):k\geq 0\}$ to the constrained optimization problem

$$\|\widetilde{\Phi}\mathbf{a}\|^2-\|\overline{\psi}_{\overline{\mu}}(\mathbf{C}_{\overline{\mu},g}^-+\mathbf{C}_{\overline{\mu},g}^+)\|^2\to\sup, \tag{5.32}$$
$$g(\lambda)=\Phi(e^{-i\lambda})\Phi^*(e^{-i\lambda})\in\mathcal{D}_g.$$

The minimax spectral characteristic $h^0 = \vec{h}_{\overline{\mu}}(f,g^0)$ *is calculated by formula* (5.20) *if* $\vec{h}_{\overline{\mu}}(f,g^0) \in H_{\mathcal{D}}$.

Lemma 5.2.4. *The spectral density* $f^0 \in \mathcal{D}_f$ *which admits canonical factorizations* (1.14) *and* (3.29) *with the known spectral density* $g(\lambda)$ *is the least favorable spectral density in the class* $\mathcal{D}_f$ *for the optimal linear filtering of the functional* $A\vec{\xi}$ *based on observations of the sequence* $\vec{\xi}(m) + \vec{\eta}(m)$ *at points* $m = 0, -1, -2, \ldots$ *if matrix coefficients of the canonical factorization*

$$f^0(\lambda) + |\beta^{(d)}(i\lambda)|^2 g(\lambda) = \frac{|\beta^{(d)}(i\lambda)|^2}{|\chi_{\overline{\mu}}^{(d)}(e^{-i\lambda})|^2} \left(\sum_{k=0}^{\infty} \theta_{\overline{\mu}}^0(k) e^{-i\lambda k}\right) \left(\sum_{k=0}^{\infty} \theta_{\overline{\mu}}^0(k) e^{-i\lambda k}\right)^*$$

are determined by the equation $\Psi_{\overline{\mu}}^0(e^{-i\lambda})\Theta_{\overline{\mu}}^0(e^{-i\lambda}) = E_q$ *and a solution* $\{\psi_{\overline{\mu}}^0(k) : k \geq 0\}$ *to the constrained optimization problem*

$$\|\overline{\psi}_{\overline{\mu}}(\mathbf{C}_{\overline{\mu},g}^- + \mathbf{C}_{\overline{\mu},g}^+)\|^2 \to \inf, \tag{5.33}$$

$$f(\lambda) = \frac{|\beta^{(d)}(i\lambda)|^2}{|\chi_{\overline{\mu}}^{(d)}(e^{-i\lambda})|^2} \Theta_{\overline{\mu}}(e^{-i\lambda})\Theta_{\overline{\mu}}^*(e^{-i\lambda}) - |\beta^{(d)}(i\lambda)|^2 \Phi(e^{-i\lambda})\Phi^*(e^{-i\lambda}) \in \mathcal{D}_f$$

for the fixed matrix coefficients $\{\phi(k) : k \geq 0\}$. *The minimax spectral characteristic* $h^0 = \vec{h}_{\overline{\mu}}(f^0,g)$ *is calculated by formula* (5.20) *if* $\vec{h}_{\overline{\mu}}(f^0,g) \in H_{\mathcal{D}}$.

More detailed analysis of the properties of the least favorable spectral densities and the minimax-robust spectral characteristics shows that the minimax spectral characteristic h^0 and the least favorable spectral densities f^0 and g^0 form a saddle point of the function $\Delta(h;f,g)$ on the set $H_{\mathcal{D}} \times \mathcal{D}$. The saddle point inequalities

$$\Delta(h;f^0,g^0) \geq \Delta(h^0;f^0,g^0) \geq \Delta(h^0;f,g) \quad \forall (f,g) \in \mathcal{D}, \forall h \in H_{\mathcal{D}}$$

hold if $h^0 = \vec{h}_{\overline{\mu}}(f^0,g^0)$, $\vec{h}_{\overline{\mu}}(f^0,g^0) \in H_{\mathcal{D}}$, and (f^0,g^0) is a solution of the constrained optimization problem

$$\widetilde{\Delta}(f,g) = -\Delta(\vec{h}_{\overline{\mu}}(f^0,g^0);f,g) \to \inf, \quad (f,g) \in \mathcal{D}, \tag{5.34}$$

where the functional $\Delta(\vec{h}_{\overline{\mu}}(f^0,g^0);f,g)$ is calculated by the formula

$$\begin{aligned}
&\Delta(\vec{h}_{\overline{\mu}}(f^0,g^0);f,g) \\
&= \frac{1}{2\pi} \int_{-\pi}^{\pi} \frac{|\beta^{(d)}(i\lambda)|^2}{|\chi_{\overline{\mu}}^{(d)}(e^{-i\lambda})|^2} (\chi_{\overline{\mu}}^{(d)}(e^{i\lambda})(\vec{A}(e^{-i\lambda}))^{\top} g^0(\lambda) - (\vec{C}_{\overline{\mu}}^0(e^{i\lambda}))^{\top}) \\
&\times (f^0(\lambda) + |\beta^{(d)}(i\lambda)|^2 g^0(\lambda))^{-1} f(\lambda)(f^0(\lambda) + |\beta^{(d)}(i\lambda)|^2 g^0(\lambda))^{-1} \\
&\times (\chi_{\overline{\mu}}^{(d)}(e^{-i\lambda}) g^0(\lambda)\overline{\vec{A}(e^{-i\lambda})} - \overline{\vec{C}_{\overline{\mu}}^0(e^{i\lambda})}) d\lambda
\end{aligned}$$

$$+\frac{1}{2\pi}\int_{-\pi}^{\pi}\frac{1}{|\chi_{\overline{\mu}}^{(d)}(e^{-i\lambda})|^2}(\chi_{\overline{\mu}}^{(d)}(e^{i\lambda})(\vec{A}(e^{-i\lambda}))^{\top}f^0(\lambda)+|\beta^{(d)}(i\lambda)|^2(\vec{C}_{\overline{\mu}}^0(e^{i\lambda}))^{\top})$$
$$\times(f^0(\lambda)+|\beta^{(d)}(i\lambda)|^2g^0(\lambda))^{-1}g(\lambda)(f^0(\lambda)+|\beta^{(d)}(i\lambda)|^2g^0(\lambda))^{-1}$$
$$\times(\chi_{\overline{\mu}}^{(d)}(e^{-i\lambda})f^0(\lambda)\overline{\vec{A}(e^{-i\lambda})}+|\beta^{(d)}(i\lambda)|^2\overline{\vec{C}_{\overline{\mu}}^0(e^{i\lambda})})d\lambda$$
$$=\langle \mathbf{S}_{\overline{\mu}}^0\mathbf{a}_{\overline{\mu}}),(\mathbf{P}_{\overline{\mu}}^0)^{-1}\mathbf{S}_{\overline{\mu}}^0\mathbf{a}_{\overline{\mu}}\rangle+\langle\mathbf{Q}^0\mathbf{a},\mathbf{a}\rangle,$$

with

$$\vec{C}_{\overline{\mu}}^0(e^{i\lambda})=\sum_{k=0}^{\infty}((\mathbf{P}_{\overline{\mu}}^0)^{-1}\mathbf{S}_{\overline{\mu}}^0\mathbf{a}_{\overline{\mu}})_k e^{i\lambda(k+1)},$$

or it is calculated by the formula

$$\Delta(\vec{h}_{\overline{\mu}}(f^0,g^0);f,g)$$
$$=\frac{1}{2\pi}\int_{-\pi}^{\pi}\frac{|\chi_{\overline{\mu}}^{(d)}(e^{-i\lambda})|^2}{|\beta^{(d)}(i\lambda)|^2}(\mathbf{h}_{\overline{\mu},f}^0(e^{-i\lambda}))^{\top}\Psi_{\overline{\mu}}^0(e^{-i\lambda})f(\lambda)(\Psi_{\overline{\mu}}^0(e^{-i\lambda}))^*\overline{\mathbf{h}_{\overline{\mu},f}^0(e^{-i\lambda})}d\lambda$$
$$+\frac{1}{2\pi}\int_{-\pi}^{\pi}(\mathbf{h}_{\overline{\mu},g}^0(e^{-i\lambda}))^{\top}\Psi_{\overline{\mu}}^0(e^{-i\lambda})g(\lambda)(\Psi_{\overline{\mu}}^0(e^{-i\lambda}))^*\overline{\mathbf{h}_{\overline{\mu},g}^0(e^{-i\lambda})}d\lambda,$$

where

$$\mathbf{h}_{\overline{\mu},f}^0(e^{-i\lambda})=\sum_{m=0}^{\infty}(\overline{\psi}_{\overline{\mu}}^0(\mathbf{C}_{\overline{\mu},g}^-)^0+\overline{\psi}_{\overline{\mu}}^0(\mathbf{C}_{\overline{\mu},g}^+)^0)_m e^{-i\lambda m},$$
$$\mathbf{h}_{\overline{\mu},g}^0(e^{-i\lambda})=\left(\sum_{k=0}^{\infty}(\theta_{\overline{\mu}}^0(k))^{\top}e^{-i\lambda k}\right)\left(\sum_{k=0}^{\infty}\vec{a}(k)e^{-i\lambda k}\right)$$
$$-\chi_{\overline{\mu}}^{(d)}(e^{-i\lambda})\left(\sum_{m=0}^{\infty}(\overline{\psi}_{\overline{\mu}}^0(\mathbf{C}_{\overline{\mu},g}^-)^0+\overline{\psi}_{\overline{\mu}}^0(\mathbf{C}_{\overline{\mu},g}^+)^0)_m e^{-i\lambda m}\right).$$

The constrained optimization problem (5.34) is equivalent to the unconstrained optimization problem

$$\Delta_{\mathcal{D}}(f,g)=\widetilde{\Delta}(f,g)+\delta(f,g|\mathcal{D})\to\inf, \tag{5.35}$$

where $\delta(f,g|\mathcal{D})$ is the indicator function of the set $\mathcal{D}$, namely, $\delta(f,g|\mathcal{D})=0$ if $(f,g)\in\mathcal{D}$ and $\delta(f|\mathcal{D})=+\infty$ if $(f,g)\notin\mathcal{D}$. The condition $0\in\partial\Delta_{\mathcal{D}}(f^0,g^0)$ characterizes a solution (f^0,g^0) of the stated unconstrained optimization problem. This condition is the necessary and sufficient condition that the point (f^0,g^0) belongs to the set of minimums of the convex functional $\Delta_{\mathcal{D}}(f,g)$ [146, 180]. Thus, it allows us to find the equalities for the least favorable spectral densities in some special classes of spectral densities $\mathcal{D}$.

The form of the functional $\Delta(\vec{h}_{\overline{\mu}}(f^0, g^0); f, g)$ is suitable for application of the Lagrange method of indefinite multipliers to the constrained optimization problem (5.34). Thus, the complexity of the problem is reduced to finding the subdifferential of the indicator function of the set of admissible spectral densities. We illustrate the solving of the problem (5.35) for concrete sets of admissible spectral densities in the following subsections. A semi-uncertain filtering problem, when the spectral density $f(\lambda)$ is known and the spectral density $g(\lambda)$ belongs to class $\mathcal{D}_g$, is considered as well.

5.2.2 Least favorable spectral densities in classes $\mathcal{D}_0 \times \mathcal{D}_V^U$

Consider the minimax filtering problem for the functional $A\vec{\xi}$ for sets of admissible spectral densities $\mathcal{D}_0^k$, $k = 1, 2, 3, 4$, of the sequence with GM increments $\vec{\xi}(m)$:

$$\mathcal{D}_0^1 = \left\{ f(\lambda) \left| \frac{1}{2\pi} \int_{-\pi}^{\pi} \frac{|\chi_{\overline{\mu}}^{(d)}(e^{-i\lambda})|^2}{|\beta^{(d)}(i\lambda)|^2} f(\lambda) d\lambda = P \right. \right\},$$

$$\mathcal{D}_0^2 = \left\{ f(\lambda) \left| \frac{1}{2\pi} \int_{-\pi}^{\pi} \frac{|\chi_{\overline{\mu}}^{(d)}(e^{-i\lambda})|^2}{|\beta^{(d)}(i\lambda)|^2} \mathrm{Tr}[f(\lambda)] d\lambda = p \right. \right\},$$

$$\mathcal{D}_0^3 = \left\{ f(\lambda) \left| \frac{1}{2\pi} \int_{-\pi}^{\pi} \frac{|\chi_{\overline{\mu}}^{(d)}(e^{-i\lambda})|^2}{|\beta^{(d)}(i\lambda)|^2} f_{kk}(\lambda) d\lambda = p_k, k = \overline{1, T} \right. \right\},$$

$$\mathcal{D}_0^4 = \left\{ f(\lambda) \left| \frac{1}{2\pi} \int_{-\pi}^{\pi} \frac{|\chi_{\overline{\mu}}^{(d)}(e^{-i\lambda})|^2}{|\beta^{(d)}(i\lambda)|^2} \langle B_1, f(\lambda) \rangle d\lambda = p \right. \right\},$$

where p, p_k, $k = \overline{1, T}$, are given numbers, P, B_1 are given positive-definite Hermitian matrices, and we have the following sets of admissible spectral densities $\mathcal{D}_V^U$, $k = 1, 2, 3, 4$, for the stationary noise sequence $\vec{\eta}(m)$:

$$\mathcal{D}_V^{U1} = \left\{ g(\lambda) \left| V(\lambda) \leq g(\lambda) \leq U(\lambda), \frac{1}{2\pi} \int_{-\pi}^{\pi} g(\lambda) d\lambda = Q \right. \right\},$$

$$\mathcal{D}_V^{U2} = \left\{ g(\lambda) \left| \mathrm{Tr}[V(\lambda)] \leq \mathrm{Tr}[g(\lambda)] \leq \mathrm{Tr}[U(\lambda)], \frac{1}{2\pi} \int_{-\pi}^{\pi} \mathrm{Tr}[g(\lambda)] d\lambda = q \right. \right\},$$

$$\mathcal{D}_V^{U3} = \left\{ g(\lambda) \left| v_{kk}(\lambda) \leq g_{kk}(\lambda) \leq u_{kk}(\lambda), \frac{1}{2\pi} \int_{-\pi}^{\pi} g_{kk}(\lambda) d\lambda = q_k, k = \overline{1, T} \right. \right\},$$

$$\mathcal{D}_V^{U4} = \left\{ g(\lambda) \left| \langle B_2, V(\lambda) \rangle \leq \langle B_2, g(\lambda) \rangle \leq \langle B_2, U(\lambda) \rangle, \frac{1}{2\pi} \int_{-\pi}^{\pi} \langle B_2, g(\lambda) \rangle d\lambda = q \right. \right\},$$

where the spectral densities $V(\lambda)$ and $U(\lambda)$ are known and fixed, $q, q_k, k = \overline{1,T}$, are given numbers, and Q, B_2 are given positive-definite Hermitian matrices.

From the condition $0 \in \partial\Delta_{\mathcal{D}}(f^0, g^0)$ we find the following equations which determine the least favorable spectral densities for these given sets of admissible spectral densities.

Define

$$(\mathbf{C}_{\overline{\mu}}^{f0}(e^{i\lambda}))^{\top} := \chi_{\overline{\mu}}^{(d)}(e^{i\lambda})(\vec{A}(e^{-i\lambda}))^{\top} g^0(\lambda) - (\vec{C}_{\overline{\mu}}^{0}(e^{i\lambda}))^{\top},$$

$$(\mathbf{C}_{\overline{\mu}}^{g0}(e^{i\lambda}))^{\top} := \frac{|\chi_{\overline{\mu}}^{(d)}(e^{-i\lambda})|^2}{|\beta^{(d)}(i\lambda)|^2}(\vec{A}(e^{-i\lambda}))^{\top} f^0(\lambda) + \chi_{\overline{\mu}}^{(d)}(e^{-i\lambda})(\vec{C}_{\overline{\mu}}^{0}(e^{i\lambda}))^{\top}.$$

For the first pair of sets of admissible spectral densities $\mathcal{D}_0^1 \times \mathcal{D}_V^{U1}$, we have

$$\begin{aligned}(\mathbf{C}_{\overline{\mu}}^{f0}(e^{i\lambda}))(\mathbf{C}_{\overline{\mu}}^{f0}(e^{i\lambda}))^* &= \left(\frac{|\chi_{\overline{\mu}}^{(d)}(e^{-i\lambda})|^2}{|\beta^{(d)}(i\lambda)|^2}(f^0(\lambda) + |\beta^{(d)}(i\lambda)|^2 g^0(\lambda))\right)\\ &\quad\times \vec{\alpha}\cdot\vec{\alpha}^*\left(\frac{|\chi_{\overline{\mu}}^{(d)}(e^{-i\lambda})|^2}{|\beta^{(d)}(i\lambda)|^2}(f^0(\lambda) + |\beta^{(d)}(i\lambda)|^2 g^0(\lambda))\right),\end{aligned} \tag{5.36}$$

$$\begin{aligned}(\mathbf{C}_{\overline{\mu}}^{g0}(e^{i\lambda}))(\mathbf{C}_{\overline{\mu}}^{g0}(e^{i\lambda}))^* &= \left(\frac{|\chi_{\overline{\mu}}^{(d)}(e^{-i\lambda})|^2}{|\beta^{(d)}(i\lambda)|^2}(f^0(\lambda) + |\beta^{(d)}(i\lambda)|^2 g^0(\lambda))\right)\\ &\quad\times (\vec{\beta}\cdot\vec{\beta}^* + \Gamma_1(\lambda) + \Gamma_2(\lambda))\left(\frac{|\chi_{\overline{\mu}}^{(d)}(e^{-i\lambda})|^2}{|\beta^{(d)}(i\lambda)|^2}(f^0(\lambda) + |\beta^{(d)}(i\lambda)|^2 g^0(\lambda))\right),\end{aligned} \tag{5.37}$$

where $\vec{\alpha}, \vec{\beta}$ are vectors of Lagrange multipliers, $\Gamma_1(\lambda) \leq 0$ and $\Gamma_1(\lambda) = 0$ if $g^0(\lambda) > V(\lambda)$, and $\Gamma_2(\lambda) \geq 0$ and $\Gamma_2(\lambda) = 0$ if $g^0(\lambda) < U(\lambda)$.

For the second pair of sets of admissible spectral densities $\mathcal{D}_0^2 \times \mathcal{D}_V^{U2}$, we have

$$(\mathbf{C}_{\overline{\mu}}^{f0}(e^{i\lambda}))(\mathbf{C}_{\overline{\mu}}^{f0}(e^{i\lambda}))^* = \alpha^2\left(\frac{|\chi_{\overline{\mu}}^{(d)}(e^{-i\lambda})|^2}{|\beta^{(d)}(i\lambda)|^2}(f^0(\lambda) + |\beta^{(d)}(i\lambda)|^2 g^0(\lambda))\right)^2, \tag{5.38}$$

$$\begin{aligned}(\mathbf{C}_{\overline{\mu}}^{g0}(e^{i\lambda}))(\mathbf{C}_{\overline{\mu}}^{g0}(e^{i\lambda}))^* &= (\beta^2 + \gamma_1(\lambda) + \gamma_2(\lambda))\\ &\quad\times\left(\frac{|\chi_{\overline{\mu}}^{(d)}(e^{-i\lambda})|^2}{|\beta^{(d)}(i\lambda)|^2}(f^0(\lambda) + |\beta^{(d)}(i\lambda)|^2 g^0(\lambda))\right)^2,\end{aligned} \tag{5.39}$$

where α^2, β^2 are Lagrange multipliers, $\gamma_1(\lambda) \leq 0$ and $\gamma_1(\lambda) = 0$ if $\mathrm{Tr}[g^0(\lambda)] > \mathrm{Tr}[V(\lambda)]$, and $\gamma_2(\lambda) \geq 0$ and $\gamma_2(\lambda) = 0$ if $\mathrm{Tr}[g^0(\lambda)] < \mathrm{Tr}[U(\lambda)]$.

For the third pair of sets of admissible spectral densities $\mathcal{D}_0^3 \times \mathcal{D}_V^{U^3}$, we have

$$(\mathbf{C}_{\overline{\mu}}^{f0}(e^{i\lambda}))(\mathbf{C}_{\overline{\mu}}^{f0}(e^{i\lambda}))^* = \left(\frac{|\chi_{\overline{\mu}}^{(d)}(e^{-i\lambda})|^2}{|\beta^{(d)}(i\lambda)|^2}(f^0(\lambda) + |\beta^{(d)}(i\lambda)|^2 g^0(\lambda))\right)$$
$$\times \{\alpha_k^2 \delta_{kl}\}_{k,l=1}^T \left(\frac{|\chi_{\overline{\mu}}^{(d)}(e^{-i\lambda})|^2}{|\beta^{(d)}(i\lambda)|^2}(f^0(\lambda) + |\beta^{(d)}(i\lambda)|^2 g^0(\lambda))\right), \quad (5.40)$$

$$(\mathbf{C}_{\overline{\mu}}^{g0}(e^{i\lambda}))(\mathbf{C}_{\overline{\mu}}^{g0}(e^{i\lambda}))^* = \left(\frac{|\chi_{\overline{\mu}}^{(d)}(e^{-i\lambda})|^2}{|\beta^{(d)}(i\lambda)|^2}(f^0(\lambda) + |\beta^{(d)}(i\lambda)|^2 g^0(\lambda))\right)$$
$$\times \{(\beta_k^2 + \gamma_{1k}(\lambda) + \gamma_{2k}(\lambda))\delta_{kl}\}_{k,l=1}^T$$
$$\times \left(\frac{|\chi_{\overline{\mu}}^{(d)}(e^{-i\lambda})|^2}{|\beta^{(d)}(i\lambda)|^2}(f^0(\lambda) + |\beta^{(d)}(i\lambda)|^2 g^0(\lambda))\right), \quad (5.41)$$

where α_k^2, β_k^2 are Lagrange multipliers, δ_{kl} are Kronecker symbols, $\gamma_{1k}(\lambda) \leq 0$ and $\gamma_{1k}(\lambda) = 0$ if $g_{kk}^0(\lambda) > v_{kk}(\lambda)$, and $\gamma_{2k}(\lambda) \geq 0$ and $\gamma_{2k}(\lambda) = 0$ if $g_{kk}^0(\lambda) < u_{kk}(\lambda)$.

For the fourth pair of sets of admissible spectral densities $\mathcal{D}_0^4 \times \mathcal{D}_V^{U^4}$, we have

$$(\mathbf{C}_{\overline{\mu}}^{f0}(e^{i\lambda}))(\mathbf{C}_{\overline{\mu}}^{f0}(e^{i\lambda}))^* = \alpha^2\left(\frac{|\chi_{\overline{\mu}}^{(d)}(e^{-i\lambda})|^2}{|\beta^{(d)}(i\lambda)|^2}(f^0(\lambda) + |\beta^{(d)}(i\lambda)|^2 g^0(\lambda))\right)$$
$$\times B_1^{\top}\left(\frac{|\chi_{\overline{\mu}}^{(d)}(e^{-i\lambda})|^2}{|\beta^{(d)}(i\lambda)|^2}(f^0(\lambda) + |\beta^{(d)}(i\lambda)|^2 g^0(\lambda))\right), \quad (5.42)$$

$$(\mathbf{C}_{\overline{\mu}}^{g0}(e^{i\lambda}))(\mathbf{C}_{\overline{\mu}}^{g0}(e^{i\lambda}))^* = (\beta^2 + \gamma_1'(\lambda) + \gamma_2'(\lambda))$$
$$\times \left(\frac{|\chi_{\overline{\mu}}^{(d)}(e^{-i\lambda})|^2}{|\beta^{(d)}(i\lambda)|^2}(f^0(\lambda) + |\beta^{(d)}(i\lambda)|^2 g^0(\lambda))\right)$$
$$\times B_2^{\top}\left(\frac{|\chi_{\overline{\mu}}^{(d)}(e^{-i\lambda})|^2}{|\beta^{(d)}(i\lambda)|^2}(f^0(\lambda) + |\beta^{(d)}(i\lambda)|^2 g^0(\lambda))\right), \quad (5.43)$$

where α^2, β^2 are Lagrange multipliers, $\gamma_1'(\lambda) \leq 0$ and $\gamma_1'(\lambda) = 0$ if $\langle B_2, g^0(\lambda)\rangle > \langle B_2, V(\lambda)\rangle$, and $\gamma_2'(\lambda) \geq 0$ and $\gamma_2'(\lambda) = 0$ if $\langle B_2, g^0(\lambda)\rangle < \langle B_2, U(\lambda)\rangle$.

The following theorem holds.

Theorem 5.2.1. *The least favorable spectral densities $f^0(\lambda)$ and $g^0(\lambda)$ in the classes $\mathcal{D}_0^k \times \mathcal{D}_V^{U^k}$, $k = 1, 2, 3, 4$, for the optimal linear filtering of the functional $A\vec{\xi}$ are determined by pairs of equations* (5.36)–(5.37), (5.38)–(5.39), (5.40)–(5.41), *and* (5.42)–(5.43), *the minimality condition* (5.3), *the constrained optimization problem* (5.28), *and restrictions on densities from the corresponding classes $\mathcal{D}_0^k \times \mathcal{D}_V^{U^k}$, $k = 1, 2, 3, 4$. The minimax-robust spectral characteristic $\vec{h}_{\overline{\mu}}(f^0, g^0)$ of the optimal estimate of the functional $A\vec{\xi}$ is determined by formula* (5.7).

5.2.3 Least favorable spectral densities in classes $\mathcal{D}_0 \times \mathcal{D}_{1\delta}$

Consider the minimax filtering problem for the functional $A\vec{\xi}$, when the spectral densities admit the canonical factorizations, for sets of admissible spectral densities $\mathcal{D}_0^k$, $k = 1,2,3,4$, of the sequence with GM increments $\vec{\xi}(m)$:

$$\mathcal{D}_0^1 = \left\{ f(\lambda) \left| \frac{1}{2\pi} \int_{-\pi}^{\pi} \frac{|\chi_{\overline{\mu}}^{(d)}(e^{-i\lambda})|^2}{|\beta^{(d)}(i\lambda)|^2} f(\lambda) d\lambda = P \right. \right\},$$

$$\mathcal{D}_0^2 = \left\{ f(\lambda) \left| \frac{1}{2\pi} \int_{-\pi}^{\pi} \frac{|\chi_{\overline{\mu}}^{(d)}(e^{-i\lambda})|^2}{|\beta^{(d)}(i\lambda)|^2} \operatorname{Tr}[f(\lambda)] d\lambda = p \right. \right\},$$

$$\mathcal{D}_0^3 = \left\{ f(\lambda) \left| \frac{1}{2\pi} \int_{-\pi}^{\pi} \frac{|\chi_{\overline{\mu}}^{(d)}(e^{-i\lambda})|^2}{|\beta^{(d)}(i\lambda)|^2} f_{kk}(\lambda) d\lambda = p_k, k = \overline{1,T} \right. \right\},$$

$$\mathcal{D}_0^4 = \left\{ f(\lambda) \left| \frac{1}{2\pi} \int_{-\pi}^{\pi} \frac{|\chi_{\overline{\mu}}^{(d)}(e^{-i\lambda})|^2}{|\beta^{(d)}(i\lambda)|^2} \langle B_1, f(\lambda) \rangle d\lambda = p \right. \right\},$$

where p, p_k, $k = \overline{1,T}$, are given numbers, P, B_1 are given positive-definite Hermitian matrices, and we have the following sets of admissible spectral densities $\mathcal{D}_{1\delta}^k$, $k = 1,2,3,4$, for the stationary noise sequence $\vec{\eta}(m)$:

$$\mathcal{D}_{1\delta}^1 = \left\{ g(\lambda) \left| \frac{1}{2\pi} \int_{-\pi}^{\pi} |g_{ij}(\lambda) - g_{ij}^1(\lambda)| d\lambda \le \delta_i^j, i,j = \overline{1,T} \right. \right\},$$

$$\mathcal{D}_{1\delta}^2 = \left\{ g(\lambda) \left| \frac{1}{2\pi} \int_{-\pi}^{\pi} |\operatorname{Tr}(g(\lambda) - g_1(\lambda))| d\lambda \le \delta \right. \right\},$$

$$\mathcal{D}_{1\delta}^3 = \left\{ g(\lambda) \left| \frac{1}{2\pi} \int_{-\pi}^{\pi} |g_{kk}(\lambda) - g_{kk}^1(\lambda)| d\lambda \le \delta_k, k = \overline{1,T} \right. \right\},$$

$$\mathcal{D}_{1\delta}^4 = \left\{ g(\lambda) \left| \frac{1}{2\pi} \int_{-\pi}^{\pi} |\langle B_2, g(\lambda) - g_1(\lambda) \rangle| d\lambda \le \delta \right. \right\},$$

where $g_1(\lambda) = \{g_{ij}^1(\lambda)\}_{i,j=1}^T$ is a fixed spectral density, B_2 is a given positive-definite Hermitian matrix, and δ, δ_k, $k = \overline{1,T}$, and δ_i^j, $i,j = \overline{1,T}$, are given numbers.

The condition $0 \in \partial\Delta_{\mathcal{D}}(f^0, g^0)$ implies the following equations which determine the least favorable spectral densities for these given sets of admissible spectral densities.

For the first set of admissible spectral densities $\mathcal{D}_0^1 \times \mathcal{D}_{1\delta}^1$, we have

$$(\mathbf{h}_{\overline{\mu},f}^0(e^{i\lambda}))(\mathbf{h}_{\overline{\mu},f}^0(e^{i\lambda}))^* = (\Theta_{\overline{\mu}}^0(e^{-i\lambda}))^\top \vec{\alpha} \cdot \vec{\alpha}^* \overline{\Theta_{\overline{\mu}}^0(e^{-i\lambda})}, \tag{5.44}$$

$$(\mathbf{h}_{\overline{\mu},g}^0(e^{i\lambda}))(\mathbf{h}_{\overline{\mu},g}^0(e^{i\lambda}))^* = (\Theta_{\overline{\mu}}^0(e^{-i\lambda}))^\top \{\beta_{ij}\gamma_{ij}(\lambda)\}_{i,j=1}^T \overline{\Theta_{\overline{\mu}}^0(e^{-i\lambda})}, \tag{5.45}$$

$$\frac{1}{2\pi}\int_{-\pi}^{\pi}|g_{ij}^{0}(\lambda)-g_{ij}^{1}(\lambda)|d\lambda=\delta_{i}^{j}, \tag{5.46}$$

where $\vec{\alpha}$, β_{ij} are Lagrange multipliers, functions $|\gamma_{ij}(\lambda)|\leq 1$, and

$$\gamma_{ij}(\lambda)=\frac{g_{ij}^{0}(\lambda)-g_{ij}^{1}(\lambda)}{|g_{ij}^{0}(\lambda)-g_{ij}^{1}(\lambda)|}\;:\;g_{ij}^{0}(\lambda)-g_{ij}^{1}(\lambda)\neq 0,\quad i,j=\overline{1,T}.$$

For the second set of admissible spectral densities $\mathcal{D}_0^2\times\mathcal{D}_{1\delta}^2$, we have

$$(\mathbf{h}_{\overline{\mu},f}^{0}(e^{i\lambda}))(\mathbf{h}_{\overline{\mu},f}^{0}(e^{i\lambda}))^{*}=\alpha^{2}(\Theta_{\overline{\mu}}^{0}(e^{-i\lambda}))^{\top}\overline{\Theta_{\overline{\mu}}^{0}(e^{-i\lambda})}, \tag{5.47}$$

$$(\mathbf{h}_{\overline{\mu},g}^{0}(e^{i\lambda}))(\mathbf{h}_{\overline{\mu},g}^{0}(e^{i\lambda}))^{*}=\beta^{2}\gamma_{2}(\lambda)(\Theta_{\overline{\mu}}^{0}(e^{-i\lambda}))^{\top}\overline{\Theta_{\overline{\mu}}^{0}(e^{-i\lambda})}, \tag{5.48}$$

$$\frac{1}{2\pi}\int_{-\pi}^{\pi}|\mathrm{Tr}(g^{0}(\lambda)-g_{1}(\lambda))|d\lambda=\delta, \tag{5.49}$$

where α^2, β^2 are Lagrange multipliers, function $|\gamma_2(\lambda)|\leq 1$, and

$$\gamma_{2}(\lambda)=\mathrm{sign}(\mathrm{Tr}(g^{0}(\lambda)-g_{1}(\lambda)))\;:\;\mathrm{Tr}(g^{0}(\lambda)-g_{1}(\lambda))\neq 0.$$

For the third set of admissible spectral densities $\mathcal{D}_0^3\times\mathcal{D}_{1\delta}^3$, we have

$$(\mathbf{h}_{\overline{\mu},f}^{0}(e^{i\lambda}))(\mathbf{h}_{\overline{\mu},f}^{0}(e^{i\lambda}))^{*}=(\Theta_{\overline{\mu}}^{0}(e^{-i\lambda}))^{\top}\{\alpha_{k}^{2}\delta_{kl}\}_{k,l=1}^{T}\overline{\Theta_{\overline{\mu}}^{0}(e^{-i\lambda})}, \tag{5.50}$$

$$(\mathbf{h}_{\overline{\mu},g}^{0}(e^{i\lambda}))(\mathbf{h}_{\overline{\mu},g}^{0}(e^{i\lambda}))^{*}=(\Theta_{\overline{\mu}}^{0}(e^{-i\lambda}))^{\top}\{\beta_{k}^{2}\gamma_{k}^{2}(\lambda)\delta_{kl}\}_{k,l=1}^{T}\overline{\Theta_{\overline{\mu}}^{0}(e^{-i\lambda})}, \tag{5.51}$$

$$\frac{1}{2\pi}\int_{-\pi}^{\pi}|g_{kk}^{0}(\lambda)-g_{kk}^{1}(\lambda)|d\lambda=\delta_{k}, \tag{5.52}$$

where α_k^2, β_k^2 are Lagrange multipliers, δ_{kl} are Kronecker symbols, functions $|\gamma_k^2(\lambda)|\leq 1$, and

$$\gamma_{k}^{2}(\lambda)=\mathrm{sign}(g_{kk}^{0}(\lambda)-g_{kk}^{1}(\lambda))\;:\;g_{kk}^{0}(\lambda)-g_{kk}^{1}(\lambda)\neq 0,\quad k=\overline{1,T}.$$

For the fourth set of admissible spectral densities $\mathcal{D}_0^4\times\mathcal{D}_{1\delta}^4$, we have

$$(\mathbf{h}_{\overline{\mu},f}^{0}(e^{i\lambda}))(\mathbf{h}_{\overline{\mu},f}^{0}(e^{i\lambda}))^{*}=\alpha^{2}(\Theta_{\overline{\mu}}^{0}(e^{-i\lambda}))^{\top}B_{1}\overline{\Theta_{\overline{\mu}}^{0}(e^{-i\lambda})}, \tag{5.53}$$

$$(\mathbf{h}_{\overline{\mu},g}^{0}(e^{i\lambda}))(\mathbf{h}_{\overline{\mu},g}^{0}(e^{i\lambda}))^{*}=\beta^{2}\gamma_{2}'(\lambda)(\Theta_{\overline{\mu}}^{0}(e^{-i\lambda}))^{\top}B_{2}\overline{\Theta_{\overline{\mu}}^{0}(e^{-i\lambda})}, \tag{5.54}$$

$$\frac{1}{2\pi}\int_{-\pi}^{\pi}|\langle B_{2},g^{0}(\lambda)-g_{1}(\lambda)\rangle|d\lambda=\delta, \tag{5.55}$$

where α^2, β^2 are Lagrange multipliers, function $|\gamma_2'(\lambda)| \le 1$, and

$$\gamma_2'(\lambda) = \operatorname{sign}\langle B_2, g^0(\lambda) - g_1(\lambda)\rangle : \ \langle B_2, g^0(\lambda) - g_1(\lambda)\rangle \neq 0.$$

The following theorem holds.

Theorem 5.2.2. *The least favorable spectral densities $f^0(\lambda)$ and $g^0(\lambda)$ in the classes $\mathcal{D}_0^k \times \mathcal{D}_{1\delta}^k$, $k = 1,2,3,4$, for the optimal linear filtering of the functional $A\vec{\xi}$ from observations of the sequence $\vec{\xi}(m) + \vec{\eta}(m)$ at points $m = 0,-1,-2,\dots$ are determined by canonical factorizations* (1.14), (5.17), *and* (5.18), *equations* (5.44)–(5.46), (5.47)–(5.49), (5.50)–(5.52), *and* (5.53)–(5.55), *respectively, the constrained optimization problem* (5.29), *and restrictions on densities from the corresponding classes $\mathcal{D}_0^k$, $\mathcal{D}_{1\delta}^k$, $k = 1,2,3,4$. The minimax-robust spectral characteristic of the optimal estimate of the functional $A\vec{\xi}$ is determined by formula* (5.20).

5.2.4 Semi-uncertain filtering problem in classes $\mathcal{D}_\varepsilon$ with known noise sequence density

Consider a semi-uncertain filtering problem for the functional $A\vec{\xi}$, where the spectral density $g(\lambda)$ of the noise sequence $\vec{\eta}(m)$ is known and the spectral density $f(\lambda)$ of the sequence $\vec{\eta}(m)$ with stationary GM increments belongs to the following sets of admissible spectral densities $\mathcal{D}_\varepsilon^k$, $k = 1,2,3,4$:

$$\mathcal{D}_\varepsilon^1 = \left\{ f(\lambda) \,\middle|\, f(\lambda) = (1-\varepsilon)f_1(\lambda) + \varepsilon W(\lambda), \frac{1}{2\pi}\int_{-\pi}^{\pi} \frac{|\chi_{\overline{\mu}}^{(d)}(e^{-i\lambda})|^2}{|\beta^{(d)}(i\lambda)|^2} f(\lambda)d\lambda = Q \right\},$$

$$\mathcal{D}_\varepsilon^2 = \left\{ f(\lambda) \,\middle|\, \operatorname{Tr}[f(\lambda)] = (1-\varepsilon)\operatorname{Tr}[f_1(\lambda)] + \varepsilon \operatorname{Tr}[W(\lambda)], \right.$$
$$\left. \frac{1}{2\pi}\int_{-\pi}^{\pi} \frac{|\chi_{\overline{\mu}}^{(d)}(e^{-i\lambda})|^2}{|\beta^{(d)}(i\lambda)|^2} \operatorname{Tr}[f(\lambda)]d\lambda = q \right\},$$

$$\mathcal{D}_\varepsilon^3 = \left\{ f(\lambda) \,\middle|\, f_{kk}(\lambda) = (1-\varepsilon)f_{kk}^1(\lambda) + \varepsilon w_{kk}(\lambda), \right.$$
$$\left. \frac{1}{2\pi}\int_{-\pi}^{\pi} \frac{|\chi_{\overline{\mu}}^{(d)}(e^{-i\lambda})|^2}{|\beta^{(d)}(i\lambda)|^2} f_{kk}(\lambda)d\lambda = q_k, k = \overline{1,T} \right\},$$

$$\mathcal{D}_\varepsilon^4 = \left\{ f(\lambda) \,\middle|\, \langle B_1, f(\lambda)\rangle = (1-\varepsilon)\langle B_1, f_1(\lambda)\rangle + \varepsilon\langle B_1, W(\lambda)\rangle, \right.$$
$$\left. \frac{1}{2\pi}\int_{-\pi}^{\pi} \frac{|\chi_{\overline{\mu}}^{(d)}(e^{-i\lambda})|^2}{|\beta^{(d)}(i\lambda)|^2} \langle B_2, f(\lambda)\rangle d\lambda = q \right\},$$

where $f_1(\lambda)$ is a fixed spectral density, $W(\lambda)$ is an unknown spectral density, $q, q_k, k = \overline{1, T}$, are given numbers, and Q, B_1 are given positive-definite Hermitian matrices.

The condition $0 \in \partial\Delta_{\mathcal{D}}(f^0, g)$ implies the equations which determine the least favorable spectral density of the sequence $\vec{\xi}(m)$.

Define

$$\mathbf{C}_{\overline{\mu}}^{f0,\text{semi}}(e^{i\lambda}) := \chi_{\overline{\mu}}^{(d)}(e^{i\lambda})(\vec{A}(e^{-i\lambda}))^{\top} g(\lambda) - (\vec{C}_{\overline{\mu}}^{0,\text{semi}}(e^{i\lambda}))^{\top},$$

where $\vec{C}_{\overline{\mu}}^{0,\text{semi}}(e^{i\lambda})$ is determined by the matrices $\mathbf{P}_{\overline{\mu}}^{0,\text{semi}}$ and $\mathbf{S}_{\overline{\mu}}^{0,\text{semi}}$ defined by the Fourier coefficients of the functions

$$|\beta^{(d)}(i\lambda)|^2 |\chi_{\overline{\mu}}^{(d)}(e^{-i\lambda})|^{-2}((f^0(\lambda) + |\beta^{(d)}(i\lambda)|^2 g(\lambda))^{-1})^{\top},$$
$$|\beta^{(d)}(i\lambda)|^2 |\chi_{\overline{\mu}}^{(d)}(e^{-i\lambda})|^{-2}(g(\lambda)(f^0(\lambda) + |\beta^{(d)}(i\lambda)|^2 g(\lambda))^{-1})^{\top}.$$

For the sets of admissible spectral densities $\mathcal{D}_{\varepsilon}^k$, $k = 1, 2, 3, 4$, we have

$$\begin{aligned}(\mathbf{C}_{\overline{\mu}}^{f0,\text{semi}}(e^{i\lambda}))(\mathbf{C}_{\overline{\mu}}^{f0,\text{semi}}(e^{i\lambda}))^* &= \frac{|\chi_{\overline{\mu}}^{(d)}(e^{-i\lambda})|^2}{|\beta^{(d)}(i\lambda)|^2}(f^0(\lambda) + |\beta^{(d)}(i\lambda)|^2 g(\lambda)) \\ &\quad \times (\vec{\beta} \cdot \vec{\beta}^* + \Gamma(\lambda)) \frac{|\chi_{\overline{\mu}}^{(d)}(e^{-i\lambda})|^2}{|\beta^{(d)}(i\lambda)|^2}(f^0(\lambda) + |\beta^{(d)}(i\lambda)|^2 g(\lambda)),\end{aligned} \tag{5.56}$$

where $\vec{\beta}$ is a vector of Lagrange multipliers and $\Gamma(\lambda) \leq 0$ and $\Gamma(\lambda) = 0$ if $f^0(\lambda) > (1-\varepsilon)f_1(\lambda)$,

$$(\mathbf{C}_{\overline{\mu}}^{f0,\text{semi}}(e^{i\lambda}))(\mathbf{C}_{\overline{\mu}}^{f0,\text{semi}}(e^{i\lambda}))^* = (\beta^2 + \gamma_2(\lambda))\left(\frac{|\chi_{\overline{\mu}}^{(d)}(e^{-i\lambda})|^2}{|\beta^{(d)}(i\lambda)|^2}(f^0(\lambda) + |\beta^{(d)}(i\lambda)|^2 g(\lambda))\right)^2, \tag{5.57}$$

where β^2 is a Lagrange multiplier and $\gamma_2(\lambda) \leq 0$ and $\gamma_2(\lambda) = 0$ if $\operatorname{Tr}[f^0(\lambda)] > (1 - \varepsilon)\operatorname{Tr}[f_1(\lambda)]$,

$$\begin{aligned}(\mathbf{C}_{\overline{\mu}}^{f0,\text{semi}}(e^{i\lambda}))(\mathbf{C}_{\overline{\mu}}^{f0,\text{semi}}(e^{i\lambda}))^* &= \frac{|\chi_{\overline{\mu}}^{(d)}(e^{-i\lambda})|^2}{|\beta^{(d)}(i\lambda)|^2}(f^0(\lambda) + |\beta^{(d)}(i\lambda)|^2 g(\lambda)) \\ &\quad \times \{(\beta_k^2 + \gamma_k^2(\lambda))\delta_{kl}\}_{k,l=1}^{T} \\ &\quad \times \frac{|\chi_{\overline{\mu}}^{(d)}(e^{-i\lambda})|^2}{|\beta^{(d)}(i\lambda)|^2}(f^0(\lambda) + |\beta^{(d)}(i\lambda)|^2 g(\lambda)),\end{aligned} \tag{5.58}$$

where β_k^2 are Lagrange multipliers and $\gamma_k^2(\lambda) \leq 0$ and $\gamma_k^2(\lambda) = 0$ if $f_{kk}^0(\lambda) > (1 - \varepsilon)f_{kk}^1(\lambda)$, and

$$(\mathbf{C}_{\overline{\mu}}^{f0,\mathrm{semi}}(e^{i\lambda}))(\mathbf{C}_{\overline{\mu}}^{f0,\mathrm{semi}}(e^{i\lambda}))^* = (\beta^2 + \gamma_2'(\lambda))\frac{|\chi_{\overline{\mu}}^{(d)}(e^{-i\lambda})|^2}{|\beta^{(d)}(i\lambda)|^2}(f^0(\lambda) + |\beta^{(d)}(i\lambda)|^2 g(\lambda)) \times B_1^{\top}\frac{|\chi_{\overline{\mu}}^{(d)}(e^{-i\lambda})|^2}{|\beta^{(d)}(i\lambda)|^2}(f^0(\lambda) + |\beta^{(d)}(i\lambda)|^2 g(\lambda)), \tag{5.59}$$

where β^2 is a Lagrange multiplier and $\gamma_2'(\lambda) \leq 0$ and $\gamma_2'(\lambda) = 0$ if $\langle B_1, f^0(\lambda)\rangle > (1-\varepsilon)\langle B_1, f_1(\lambda)\rangle$.

The following theorem holds.

Theorem 5.2.3. *The least favorable spectral density $f^0(\lambda)$ in the classes $\mathcal{D}_\varepsilon^k$, $k = 1,2,3,4$, for the optimal linear filtering of the functional $A\vec{\xi}$, where the spectral density $g(\lambda)$ of the stationary sequence $\vec{\eta}(m)$ is known, is determined by equations* (5.56), (5.57), (5.58), *and* (5.59), *the minimality condition* (5.3), *the constrained optimization problem* (5.28) *with fixed $g(\lambda)$, and the restrictions on densities from the corresponding classes $\mathcal{D}_\varepsilon^k$, $k = 1,2,3,4$. The minimax-robust spectral characteristic $\vec{h}_{\overline{\mu}}(f^0, g)$ of the optimal estimate of the functional $A\vec{\xi}$ is determined by formula* (5.7).

If the spectral densities $f^0(\lambda)$ and $g(\lambda)$ admit canonical factorizations (5.18) and (5.17), we can derive the following equation for the least favorable spectral densities, where $\mathbf{C}_{\overline{\mu},g}^+$ and $\mathbf{C}_{\overline{\mu},g}^-$ are known and defined by the coefficients of the canonical factorization of the spectral density $g(\lambda)$.

For the first set of admissible spectral densities $\mathcal{D}_\varepsilon^1$, we have

$$\left(\sum_{m=0}^{\infty}(\overline{\psi}_{\overline{\mu}}^0\mathbf{C}_{\overline{\mu},g}^- + \overline{\psi}_{\overline{\mu}}^0\mathbf{C}_{\overline{\mu},g}^+)_m e^{-i\lambda m}\right)\left(\sum_{m=0}^{\infty}(\overline{\psi}_{\overline{\mu}}^0\mathbf{C}_{\overline{\mu},g}^- + \overline{\psi}_{\overline{\mu}}^0\mathbf{C}_{\overline{\mu},g}^+)_m e^{-i\lambda m}\right)^* = \left(\sum_{k=0}^{\infty}\theta_{\overline{\mu}}^0(k)e^{-i\lambda k}\right)^{\top}(\vec{\alpha}\cdot\vec{\alpha}^* + \Gamma(\lambda))\left(\overline{\sum_{k=0}^{\infty}\theta_{\overline{\mu}}^0(k)e^{-i\lambda k}}\right), \tag{5.60}$$

where $\vec{\alpha}$ is a vector of Lagrange multipliers and matrix $\Gamma(\lambda) \leq 0$ and $\Gamma(\lambda) = 0$ if $f^0(\lambda) > (1-\varepsilon)f_1(\lambda)$.

For the second set of admissible spectral densities $\mathcal{D}_\varepsilon^2$, we have

$$\left(\sum_{m=0}^{\infty}(\overline{\psi}_{\overline{\mu}}^0\mathbf{C}_{\overline{\mu},g}^- + \overline{\psi}_{\overline{\mu}}^0\mathbf{C}_{\overline{\mu},g}^+)_m e^{-i\lambda m}\right)\left(\sum_{m=0}^{\infty}(\overline{\psi}_{\overline{\mu}}^0\mathbf{C}_{\overline{\mu},g}^- + \overline{\psi}_{\overline{\mu}}^0\mathbf{C}_{\overline{\mu},g}^+)_m e^{-i\lambda m}\right)^* = (\alpha^2 + \gamma(\lambda))\left(\sum_{k=0}^{\infty}\theta_{\overline{\mu}}^0(k)e^{-i\lambda k}\right)^{\top}\left(\overline{\sum_{k=0}^{\infty}\theta_{\overline{\mu}}^0(k)e^{-i\lambda k}}\right), \tag{5.61}$$

where α^2 is a Lagrange multiplier and function $\gamma(\lambda) \leq 0$ and $\gamma(\lambda) = 0$ if $\mathrm{Tr}[f^0(\lambda)] > (1-\varepsilon)\,\mathrm{Tr}[f_1(\lambda)]$.

For the third set of admissible spectral densities $\mathcal{D}_\varepsilon^3$, we have

$$\left(\sum_{m=0}^{\infty}(\overline{\psi}_{\overline{\mu}}^0\mathbf{C}_{\overline{\mu},g}^- + \overline{\psi}_{\overline{\mu}}^0\mathbf{C}_{\overline{\mu},g}^+)_m e^{-i\lambda m}\right)\left(\sum_{m=0}^{\infty}(\overline{\psi}_{\overline{\mu}}^0\mathbf{C}_{\overline{\mu},g}^- + \overline{\psi}_{\overline{\mu}}^0\mathbf{C}_{\overline{\mu},g}^+)_m e^{-i\lambda m}\right)^*$$
$$= \left(\sum_{k=0}^{\infty}\theta_{\overline{\mu}}^0(k)e^{-i\lambda k}\right)^{\top}\{(\alpha_k^2+\gamma_k(\lambda))\delta_{kl}\}_{k,l=1}^{T}\left(\overline{\sum_{k=0}^{\infty}\theta_{\overline{\mu}}^0(k)e^{-i\lambda k}}\right), \tag{5.62}$$

where α_k^2 are Lagrange multipliers, δ_{kl} are Kronecker symbols, and functions $\gamma_k(\lambda) \le 0$ and $\gamma_k(\lambda) = 0$ if $f_{kk}^0(\lambda) > (1-\varepsilon)f_{kk}^1(\lambda)$.

For the fourth set of admissible spectral densities $\mathcal{D}_\varepsilon^4$, we have

$$\left(\sum_{m=0}^{\infty}(\overline{\psi}_{\overline{\mu}}^0\mathbf{C}_{\overline{\mu},g}^- + \overline{\psi}_{\overline{\mu}}^0\mathbf{C}_{\overline{\mu},g}^+)_m e^{-i\lambda m}\right)\left(\sum_{m=0}^{\infty}(\overline{\psi}_{\overline{\mu}}^0\mathbf{C}_{\overline{\mu},g}^- + \overline{\psi}_{\overline{\mu}}^0\mathbf{C}_{\overline{\mu},g}^+)_m e^{-i\lambda m}\right)^*$$
$$= (\alpha^2+\gamma'(\lambda))\left(\sum_{k=0}^{\infty}\theta_{\overline{\mu}}^0(k)e^{-i\lambda k}\right)^{\top} B_1\left(\overline{\sum_{k=0}^{\infty}\theta_{\overline{\mu}}^0(k)e^{-i\lambda k}}\right), \tag{5.63}$$

where α^2 is a Lagrange multiplier and function $\gamma'(\lambda) \le 0$ and $\gamma'(\lambda) = 0$ if $\langle B_1, f^0(\lambda)\rangle > (1-\varepsilon)\langle B_1, f_1(\lambda)\rangle$.

The following theorems hold.

Theorem 5.2.4. *If the spectral density $g(\lambda)$ is known, the least favorable spectral density $f^0(\lambda)$ in the classes $\mathcal{D}_\varepsilon^k$, $k = 1,2,3,4$, for the optimal linear filtering of the functional $A\vec{\xi}$ from observations of the sequence $\vec{\xi}(m) + \vec{\eta}(m)$ at points $m = 0,-1,-2,\ldots$ is determined by canonical factorizations* (5.18) *and* (5.17), *equations* (5.60), (5.61), (5.62), *and* (5.63), *respectively, the constrained optimization problem* (5.33), *and restrictions on density from the corresponding classes $\mathcal{D}_\varepsilon^k$, $k = 1,2,3,4$. The minimax-robust spectral characteristic of the optimal estimate of the functional $A\vec{\xi}$ is determined by formula* (5.20).

Conclusions

In this chapter, we describe methods of solution of the filtering problem for linear functionals which are constructed from unobserved values of a sequence with periodically stationary long memory multi-seasonal increments. Estimates are based on observations of the sequence with a periodically stationary noise sequence. Estimates are obtained by representing the sequence under investigation as a multidimensional sequence with stationary GM increments.

The Hilbert space projection method is applied to derive formulas for calculation of the spectral characteristics and the MSEs of the optimal estimates of the functionals in the case of spectral certainty, where the spectral densities of the sequences are exactly known. In the case of spectral uncertainty, where the spectral density is not exactly known while, instead, a set of admissible spectral densities is specified, the minimax-robust method is applied. We propose a representation of the MSE in the form of a linear functional in L_1 space with respect to spectral densities, which allows us to solve the corresponding constrained optimization problem and describe the minimax (robust) estimates of the functionals.

Formulas that determine the least favorable spectral densities and minimax (robust) spectral characteristic of the optimal linear estimates of the functionals are derived for a collection of specific classes of admissible spectral densities. This method was illustrated with the help of sets of admissible spectral densities: $\mathcal{D}_0 \times \mathcal{D}_V^U$ and $\mathcal{D}_0 \times \mathcal{D}_{1\varepsilon}$. Here D_0 are classes of spectral densities with the moment restrictions, D_V^U are classes describing the "strip" models of spectral densities, and $D_{1\delta}$ are classes describing the "δ-neighborhood" models in the space L_1 of a fixed bounded spectral density.

The results described in this chapter have been published in [118, 126]. The corresponding results for minimax filtering of sequences with periodically stationary increments (a particular case of GM increments) are described in the paper by Luz and Moklyachuk [119]. The results presented in this chapter rely on and develop prior studies of the minimax filtering problem by Moklyachuk and Masyutka [149] for multidimensional stationary sequences, Golichenko (Dubovets'ka) and Moklyachuk [41] for periodically correlated sequences, and Luz and Moklyachuk [106, 113] for sequences with stationary increments.

6 Continuous time stochastic processes with periodically correlated increments

In this chapter, we deal with the continuous time stochastic processes with periodically correlated increments defined and studied in [122]. The spectral representations of stochastic processes with periodically correlated increments are presented. A brief review of properties of such stochastic processes is presented.

6.1 Periodically correlated processes and generated vector stationary sequences

In this section, we present a brief review of properties of periodically correlated processes and describe an approach to presenting these processes as stationary H-valued sequences [84, 147]. In the next section, this approach is applied to develop the spectral theory for periodically correlated increment processes.

Definition 6.1.1 (Gladyshev [59]). A mean square continuous stochastic process $\eta : \mathbb{R} \to H = L_2(\Omega, \mathcal{F}, \mathbb{P})$, with $\mathsf{E}\eta(t) = 0$, is called periodically correlated with period T if its correlation function $K(t,s) = \mathsf{E}\eta(t)\overline{\eta(s)}$ for all $t, s \in \mathbb{R}$ and some fixed $T > 0$ is such that

$$K(t,s) = \mathsf{E}\eta(t)\overline{\eta(s)} = \mathsf{E}\eta(t+T)\overline{\eta(s+T)} = K(t+T, s+T).$$

For a periodically correlated stochastic process $\eta(t)$, one can construct the following sequence of stochastic functions [42, 147]:

$$\{\eta_j(u) = \eta(u+jT), u \in [0,T), j \in \mathbb{Z}\}. \tag{6.1}$$

Sequence (6.1) forms an $L_2([0,T);H)$-valued stationary sequence $\{\eta_j, j \in \mathbb{Z}\}$ with the correlation function $B_\eta(l,j) = \langle \eta_l, \eta_j \rangle_H = B_\eta(l-j)$:

$$\begin{aligned} B_\eta(l,j) &= \int_0^T \mathsf{E}[\eta(u+lT)\overline{\eta(u+jT)}]du \\ &= \int_0^T K_\eta(u+(l-j)T, u)du \\ &= B_\eta(l-j), \end{aligned}$$

where $K_\eta(t,s) = \mathsf{E}\eta(t)\overline{\eta(s)}$ is the correlation function of the periodically correlated process $\eta(t)$. We choose the orthonormal basis

$$\left\{ \frac{1}{\sqrt{T}} e^{2\pi i k u/T}, k \in \mathbb{Z} \right\},$$

https://doi.org/10.1515/9783111325620-006

in the space $L_2([0,T);\mathbb{R})$, which can be represented in the form

$$\left\{\widetilde{e}_k = \frac{1}{\sqrt{T}}e^{2\pi i\{(-1)^k[\frac{k}{2}]\}u/T}, k = 1,2,3,\dots\right\}, \quad \langle\widetilde{e}_j,\widetilde{e}_k\rangle = \delta_{kj}. \tag{6.2}$$

Then the stationary sequence $\{\eta_j, j \in \mathbb{Z}\}$ can be represented in the form

$$\eta_j = \sum_{k=1}^{\infty}\eta_{kj}\widetilde{e}_k, \tag{6.3}$$

where

$$\eta_{kj} = \langle\eta_j,\widetilde{e}_k\rangle = \frac{1}{\sqrt{T}}\int_0^T \eta_j(v)e^{-2\pi i\{(-1)^k[\frac{k}{2}]\}v/T}dv.$$

The sequence $\{\eta_j, j \in \mathbb{Z}\}$, or the vector sequence corresponding to it,

$$\{\vec{\eta}_j = (\eta_{kj}, k = 1,2,\dots)^\top, j \in \mathbb{Z}\},$$

is called a vector stationary sequence generated by the process $\{\eta(t), t \in \mathbb{R}\}$. The components $\{\eta_{kj} : k = 1,2,\dots; j \in \mathbb{Z}\}$ of the generated stationary sequence $\{\eta_j, j \in \mathbb{Z}\}$ satisfy the relations [84, 136]

$$\mathsf{E}\eta_{kj} = 0, \quad \|\eta_j\|_H^2 = \sum_{k=1}^{\infty}\mathsf{E}|\eta_{kj}|^2 \le P_\eta = B_\eta(0), \quad \mathsf{E}\eta_{kl}\overline{\eta}_{nj} = \langle R_\eta(l-j)\widetilde{e}_k,\widetilde{e}_n\rangle.$$

The correlation function $R_\eta(j)$ of the generated stationary sequence $\{\eta_j, j \in \mathbb{Z}\}$ is a correlation operator function. The correlation operator $R_\eta(0) = R_\eta$ is a kernel operator and its kernel norm satisfies the following properties:

$$\|\eta_j\|_H^2 = \sum_{k=1}^{\infty}\langle R_\eta\widetilde{e}_k,\widetilde{e}_k\rangle \le P_\eta.$$

The generated stationary sequence $\{\eta_j, j \in \mathbb{Z}\}$ has the spectral density function $g(\lambda) = \{g_{kn}(\lambda)\}_{k,n=1}^{\infty}$, which is a positive-valued operator function of variable $\lambda \in [-\pi,\pi)$, if its correlation function $R_\eta(j)$ can be represented in the form

$$\langle R_\eta(j)\widetilde{e}_k,\widetilde{e}_n\rangle = \frac{1}{2\pi}\int_{-\pi}^{\pi}e^{i\lambda j}\langle g(\lambda)\widetilde{e}_k,\widetilde{e}_n\rangle d\lambda.$$

We finish our review by the statement that for almost all $\lambda \in [-\pi,\pi)$ the spectral density $g(\lambda)$ is a kernel operator with an integrable kernel norm

$$\sum_{k=1}^{\infty}\frac{1}{2\pi}\int_{-\pi}^{\pi}\langle g(\lambda)\widetilde{e}_k,\widetilde{e}_k\rangle d\lambda = \sum_{k=1}^{\infty}\langle R_\eta\widetilde{e}_k,\widetilde{e}_k\rangle = \|\eta_j\|_H^2 \le P_\eta.$$

6.2 Stochastic processes with periodically correlated *d*th increments

6.2.1 Definition and spectral representation of generated stationary increment

For a given stochastic process $\{\xi(t), t \in \mathbb{R}\}$, consider the stochastic dth increment process

$$\xi^{(d)}(t,\tau) = (1 - B_\tau)^d \xi(t) = \sum_{l=0}^{d} (-1)^l \binom{d}{l} \xi(t - l\tau), \tag{6.4}$$

with the step $\tau \in \mathbb{R}$, generated by the stochastic process $\xi(t)$. Here B_τ is the backward shift operator: $B_\tau \xi(t) = \xi(t-\tau)$, $\tau \in \mathbb{R}$.

We prefer to use the notation $\xi^{(d)}(t,\tau)$ instead of the widely used $\Delta_\tau^d \xi(t)$ to avoid duplication of the MSE notation.

Definition 6.2.1 (Luz and Moklyachuk [122]). A stochastic process $\{\xi(t), t \in \mathbb{R}\}$ is called a *stochastic process with periodically stationary (periodically correlated) increments* with the step $\tau \in \mathbb{Z}$ and the period $T > 0$ if the mathematical expectations exist and satisfy the relations

$$\begin{aligned} \mathsf{E}\xi^{(d)}(t+T,\tau T) &= \mathsf{E}\xi^{(d)}(t,\tau T) = c^{(d)}(t,\tau T), \\ \mathsf{E}\xi^{(d)}(t+T,\tau_1 T)\xi^{(d)}(s+T,\tau_2 T) &= D^{(d)}(t+T,s+T;\tau_1 T,v_2 T) \\ &= D^{(d)}(t,s;\tau_1 T,\tau_2 T) \end{aligned}$$

for every $t, s \in \mathbb{R}$, $\tau_1, \tau_2 \in \mathbb{Z}$ and for some fixed $T > 0$.

The functions $c^{(d)}(t,\tau T)$ and $D^{(d)}(t,s;\tau_1 T,\tau_2 T)$ from Definition 6.2.1 are called the *mean value* and the *structural function* of the stochastic process $\xi(t)$ with periodically stationary (periodically correlated) increments.

For the stochastic process $\{\xi(t), t \in \mathbb{R}\}$ with periodically correlated increments $\xi^{(d)}(t,\tau T)$ and the integer step τ, we follow the procedure described in Section 6.1 and construct a sequence of stochastic functions

$$\{\xi_j^{(d)}(u) := \xi_{j,\tau}^{(d)}(u) = \xi_j^{(d)}(u + jT, \tau T),\ u \in [0,T), j \in \mathbb{Z}\}. \tag{6.5}$$

The sequence (6.5) forms an $L_2([0,T);H)$-valued stationary increment sequence $\{\xi_j^{(d)}, j \in \mathbb{Z}\}$ with the structural function $B_{\xi^{(d)}}(l,j) = \langle \xi_l^{(d)}, \xi_j^{(d)} \rangle_H = B_{\xi^{(d)}}(l-j)$:

$$\begin{aligned} B_{\xi^{(d)}}(l,j) &= \int_0^T \mathsf{E}[\xi_j^{(d)}(u+lT,\tau_1 T)\overline{\xi_j^{(d)}(u+jT,\tau_2 T)}]du \\ &= \int_0^T D^{(d)}(u+(l-j)T,u;\tau_1 T,\tau_2 T)du \end{aligned}$$

$$= B_{\xi^{(d)}}(l-j).$$

Making use of the orthonormal basis (6.2), the stationary increment sequence $\{\xi_j^{(d)}, j \in \mathbb{Z}\}$ can be represented in the form

$$\xi_j^{(d)} = \sum_{k=1}^{\infty} \xi_{kj}^{(d)} \widetilde{e}_k, \tag{6.6}$$

where

$$\xi_{kj}^{(d)} = \langle \xi_j^{(d)}, \widetilde{e}_k \rangle = \frac{1}{\sqrt{T}} \int_0^T \xi_j^{(d)}(v) e^{-2\pi i\{(-1)^k [\frac{k}{2}]\} v/T} dv.$$

We call this sequence, $\{\xi_j^{(d)}, j \in \mathbb{Z}\}$, or the vector sequence corresponding to it,

$$\{\vec{\xi}^{(d)}(j,\tau) = \vec{\xi}_j^{(d)} = (\xi_{kj}^{(d)}, k = 1,2,\dots)^\top = (\xi_k^{(d)}(j,\tau), k = 1,2,\dots)^\top, j \in \mathbb{Z}\}, \tag{6.7}$$

an infinite-dimensional vector stationary increment sequence generated by the increment process $\{\xi^{(d)}(t,\tau T), t \in \mathbb{R}\}$, or *generated stationary increment sequence*. Further, we will omit the word vector in the notion generated vector stationary increment sequence.

The components $\{\xi_{kj}^{(d)}\}$: $k = 1,2,\dots, j \in \mathbb{Z}$ of the generated stationary increment sequence $\{\xi_j^{(d)}, j \in \mathbb{Z}\}$ are such that [84, 136]

$$\mathsf{E}\xi_{kj}^{(d)} = 0, \quad \|\xi_j^{(d)}\|_H^2 = \sum_{k=1}^{\infty} \mathsf{E}|\xi_{kj}^{(d)}|^2 \le P_{\xi^{(d)}} = B_{\xi^{(d)}}(0),$$

and

$$\mathsf{E}\xi_{kl}^{(d)}\overline{\xi_{nj}^{(d)}} = \langle R_{\xi^{(d)}}(l-j;\tau_1,\tau_2)\widetilde{e}_k, \widetilde{e}_n\rangle.$$

The *structural function* $R_{\xi^{(d)}}(j) := R_{\xi^{(d)}}(j;\tau_1,\tau_2)$ of the generated stationary increment sequence $\{\xi_j^{(d)}, j \in \mathbb{Z}\}$ is a correlation operator function. The correlation operator $R_{\xi^{(d)}}(0) = R_{\xi^{(d)}}$ is a kernel operator and its kernel norm satisfies the following properties:

$$\|\xi_j^{(d)}\|_H^2 = \sum_{k=1}^{\infty} \langle R_{\xi^{(d)}}\widetilde{e}_k, \widetilde{e}_k\rangle \le P_{\xi^{(d)}}.$$

We can represent the structural function $R_{\xi^{(d)}}(j)$ in the form

$$\langle R_{\xi^{(d)}}(j;\tau_1,\tau_2)\widetilde{e}_k, \widetilde{e}_n\rangle = \frac{1}{2\pi} \int_{-\pi}^{\pi} e^{i\lambda j} \frac{(1-e^{-i\tau_1\lambda})^d (1-e^{i\tau_2\lambda})^d}{\lambda^{2d}} \langle f(\lambda)\widetilde{e}_k, \widetilde{e}_n\rangle d\lambda,$$

where $f(\lambda) = \{f_{kn}(\lambda)\}_{k,n=1}^{\infty} = \langle f(\lambda)\widetilde{e}_k, \widetilde{e}_n\rangle$ is a *spectral density function* of the generated stationary increment sequence $\{\xi_j^{(d)}, j \in \mathbb{Z}\}$. It is a positive-valued operator functions of variable $\lambda \in [-\pi, \pi)$, and for almost all $\lambda \in [-\pi, \pi)$ it is a kernel operator with an integrable kernel norm

$$\sum_{k=1}^{\infty} \frac{1}{2\pi} \int_{-\pi}^{\pi} \frac{(1 - e^{-i\tau_1\lambda})^d (1 - e^{i\tau_2\lambda})^d}{\lambda^{2d}} \langle f(\lambda)\widetilde{e}_k, \widetilde{e}_k\rangle d\lambda = \sum_{k=1}^{\infty} \langle R_{\xi^{(d)}}\widetilde{e}_k, \widetilde{e}_k\rangle = \|\zeta_j\|_H^2 \leq P_{\xi^{(d)}}. \tag{6.8}$$

In this case the stationary dth increment sequence $\vec{\xi}_j^{(d)}$ admits the spectral representation

$$\vec{\xi}_j^{(d)} = \int_{-\pi}^{\pi} e^{i\lambda j} (1 - e^{-i\tau\lambda})^d \frac{1}{(i\lambda)^d} d\vec{Z}_{\xi^{(d)}}(\lambda), \tag{6.9}$$

where $\vec{Z}_{\xi^{(d)}}(\lambda) = \{Z_k(\lambda)\}_{k=1}^{\infty}$ is a vector-valued random process with uncorrelated increments on $[-\pi, \pi)$.

Consider the generated stationary stochastic sequence $\vec{\eta}_j$ defined in Section 6.1, which is uncorrelated with the increment sequence $\vec{\xi}_j^{(d)}$. It admits the spectral representation

$$\vec{\eta}_j = \int_{-\pi}^{\pi} e^{i\lambda j} d\vec{Z}_{\eta}(\lambda), \tag{6.10}$$

where $\vec{Z}_{\eta}(\lambda)$ is a vector-valued random process with uncorrelated increments on $[-\pi, \pi)$. The spectral representation of the sequence $\vec{\zeta}_j^{(d)}$, generated by the process $\zeta(t) = \xi(t) + \eta(t)$ with the spectral densities $f(\lambda)$ and $g(\lambda)$ of the corresponding generated sequences, admits the spectral representation

$$\vec{\zeta}_j^{(d)} = \int_{-\pi}^{\pi} e^{i\lambda j} (1 - e^{-i\tau\lambda})^d \frac{1}{(i\lambda)^d} d\vec{Z}_{\xi^{(d)} + \eta^{(d)}}(\lambda). \tag{6.11}$$

The random processes $\vec{Z}_{\eta}(\lambda)$ and $\vec{Z}_{\eta^{(d)}}(\lambda)$ are connected by the relation $d\vec{Z}_{\eta^{(d)}}(\lambda) = (i\lambda)^d d\vec{Z}_{\eta}(\lambda)$, $\lambda \in [-\pi, \pi)$; see [115]. The spectral density $p(\lambda) = \{p_{kn}(\lambda)\}_{k,n=1}^{\infty}$ of the sequence $\vec{\zeta}_j^{(d)}$ is determined by the spectral densities $f(\lambda)$ and $g(\lambda)$ by the relation

$$p(\lambda) = f(\lambda) + \lambda^{2d} g(\lambda).$$

6.2.2 Moving average representation of generated stationary increment

In the space $H = L_2(\Omega, \mathcal{F}, \mathsf{P})$, consider a closed linear subspace

$$H(\vec{\xi}^{(d)}) = \overline{\operatorname{span}}\{\xi_{kj}^{(d)} : k = 1, 2, \ldots; j \in \mathbb{Z}\}$$

generated by the components of the generated stationary increment sequence $\vec{\xi}^{(d)} = \{\xi_{kj}^{(d)} = \xi_k^{(d)}(j, \tau), \tau > 0\}$. For $q \in \mathbb{Z}$, consider also a closed linear subspace

$$H^q(\vec{\xi}^{(d)}) = \overline{\operatorname{span}}\{\xi_{kj}^{(d)} : k = 1, 2, \ldots; j \le q\}.$$

Define a subspace

$$S(\vec{\xi}^{(d)}) = \bigcap_{q \in \mathbb{Z}} H^q(\vec{\xi}^{(d)})$$

of the Hilbert space $H(\vec{\xi}^{(d)})$. The space $H(\vec{\xi}^{(d)})$ admits a decomposition $H(\vec{\xi}^{(d)}) = S(\vec{\xi}^{(d)}) \oplus R(\vec{\xi}^{(d)})$, where $R(\vec{\xi}^{(d)})$ is the orthogonal complement of the subspace $S(\vec{\xi}^{(d)})$ in the space $H(\vec{\xi}^{(d)})$.

Definition 6.2.2. A stationary (wide sense) increment sequence $\vec{\xi}_j^{(d)} = \{\xi_{kj}^{(d)}\}_{k=1}^{\infty}$ is called regular if $H(\vec{\xi}^{(d)}) = R(\vec{\xi}^{(d)})$, and it is called singular if $H(\vec{\xi}^{(d)}) = S(\vec{\xi}^{(d)})$.

Theorem 6.2.1. *A stationary increment sequence $\xi_j^{(d)}$ is uniquely represented in the form*

$$\xi_{kj}^{(d)} = \xi_{S,kj}^{(d)} + \xi_{R,kj}^{(d)}, \tag{6.12}$$

where $\xi_{R,kj}^{(d)}$, $k = 1, \ldots, \infty$, is a regular stationary increment sequence and $\xi_{S,kj}^{(d)}$, $k = 1, \ldots, \infty$, is a singular stationary increment sequence. The increment sequences $\xi_{R,kj}^{(d)}$ and $\xi_{S,kj}^{(d)}$ are orthogonal for all $j \in \mathbb{Z}$. They are defined by the formulas

$$\xi_{S,kj}^{(d)} = \mathsf{E}[\xi_{kj}^{(d)} | S(\vec{\xi}^{(d)})],$$
$$\xi_{R,kj}^{(d)} = \xi_{kj}^{(d)} - \xi_{S,kj}^{(d)}.$$

Consider an innovation sequence $\vec{\varepsilon}(u) = \{\varepsilon_m(u)\}_{m=1}^{M}$, $u \in \mathbb{Z}$, for a regular stationary increment, namely, the sequence of uncorrelated random variables such that $\mathsf{E}\varepsilon_m(u)\overline{\varepsilon}_j(v) = \delta_{mj}\delta_{uv}$, $\mathsf{E}|\varepsilon_m(u)|^2 = 1$, $m, j = 1, \ldots, M$, $u \in \mathbb{Z}$, and $H^r(\vec{\xi}^{(d)}) = H^r(\vec{\varepsilon})$ holds for all $r \in \mathbb{Z}$, where $H^r(\vec{\varepsilon})$ is the Hilbert space generated by elements $\{\varepsilon_m(u) : m = 1, \ldots, M; u \le r\}$; δ_{mj} and δ_{uv} are Kronecker symbols.

Theorem 6.2.2. *A stationary increment sequence $\vec{\xi}_j^{(d)}$ is regular if and only if there exists an innovation sequence $\vec{\varepsilon}(u) = \{\varepsilon_m(u)\}_{m=1}^{M}$, $u \in \mathbb{Z}$, and a sequence of matrix-valued functions $\varphi^{(d)}(l, \tau) = \{\varphi_{km}^{(d)}(l, \tau)\}_{k=\overline{1,\infty}}^{m=\overline{1,M}}$, $l \ge 0$, such that*

$$\sum_{l=0}^{\infty}\sum_{k=1}^{\infty}\sum_{m=1}^{M}|\varphi_{km}^{(d)}(l,\tau)|^2<\infty,\quad \xi_{kj}^{(d)}=\sum_{l=0}^{\infty}\sum_{m=0}^{M}\varphi_{km}^{(d)}(l,\tau)\vec{\varepsilon}_m(j-l). \tag{6.13}$$

Representation (6.13) *is called the canonical moving average representation of the generated stationary increment sequence* $\vec{\xi}_j^{(d)}$.

The spectral function $F(\lambda)$ of a stationary increment sequence $\vec{\xi}_j^{(d)}$ which admits canonical representation (6.13) has the spectral density $f(\lambda)=\{f_{ij}(\lambda)\}_{i,j=1}^{\infty}$ admitting the canonical factorization

$$f(\lambda)=\varphi(e^{-i\lambda})\varphi^*(e^{-i\lambda}), \tag{6.14}$$

where the function $\varphi(z)=\sum_{k=0}^{\infty}\varphi(k)z^k$ has analytic in the unit circle $\{z:|z|\le 1\}$ components $\varphi_{ij}(z)=\sum_{k=0}^{\infty}\varphi_{ij}(k)z^k$, $i=1,\dots,\infty$, $j=1,\dots,M$. Based on moving average representation (6.13), we define

$$\varphi_\tau(z)=\sum_{k=0}^{\infty}\varphi^{(d)}(k,\tau)z^k=\sum_{k=0}^{\infty}\varphi_\tau(k)z^k.$$

Then the following canonical factorization holds:

$$\frac{|1-e^{i\lambda\tau}|^{2d}}{\lambda^{2d}}f(\lambda)=\varphi_\tau(e^{-i\lambda})\varphi_\tau^*(e^{-i\lambda}),\quad \varphi_\tau(e^{-i\lambda})=\sum_{k=0}^{\infty}\varphi_\tau(k)e^{-i\lambda k}. \tag{6.15}$$

Conclusions

In this chapter we present a brief review of properties of periodically correlated processes and describe an approach to presenting these processes as stationary H-valued sequences [84, 147]. This approach is applied to develop the spectral theory for periodically stationary (periodically correlated) increment processes. Properties of the generated stationary increment sequences are described. The moving average representation of the generated stationary increment sequence is analyzed. A criterion of regularity is proposed.

7 Extrapolation of processes with periodically correlated increments

In this chapter, we deal with the extrapolation problem for stochastic processes with periodically correlated dth increments. More precisely, we consider the problem of the mean square optimal linear estimation of the functionals

$$A\xi = \int_0^\infty a(t)\xi(t)dt, \quad A_{NT}\xi = \int_0^{(N+1)T} a(t)\xi(t)dt,$$

which depend on the unknown values of a stochastic process $\xi(t)$ with periodically correlated dth increments. Estimates are based on observations of the process $\xi(t)$ at points $t < 0$.

7.1 Generated stationary increment sequence representation of the functional $A\xi$

Lemma 7.1.1 (Luz and Moklyachuk [115]). *Any linear functional*

$$A\xi = \int_0^\infty a(t)\xi(t)dt$$

allows a representation

$$A\xi = B\xi - V\xi,$$

where

$$B\xi = \int_0^\infty b^\tau(t)\xi^{(d)}(t,\tau T)dt, \quad V\xi = \int_{-\tau Td}^0 v^\tau(t)\xi(t)dt,$$

where

$$v^\tau(t) = \sum_{l=\lceil -\frac{t}{\tau T}\rceil}^{d} (-1)^l \binom{d}{l} b^\tau(t + l\tau T), \quad t \in [-\tau Td; 0),$$

$$b^\tau(t) = \sum_{k=0}^\infty a(t + \tau Tk)d(k) = D^{\tau T}\mathbf{a}(t), \quad t \ge 0,$$

where $\lceil x\rceil$ denotes the least integer greater than or equal to x, coefficients $\{d(k) : k \ge 0\}$ are determined by the relation

https://doi.org/10.1515/9783111325620-007

$$\sum_{k=0}^{\infty} d(k)x^k = \left(\sum_{j=0}^{\infty} x^j\right)^d,$$

and $D^{\tau T}$ *is the linear transformation acting on an arbitrary function* $x(t)$, $t \geq 0$, *as follows:*

$$D^{\tau T}\mathbf{x}(t) = \sum_{k=0}^{\infty} x(t + \tau Tk)d(k).$$

Corollary 7.1.1. *A linear functional*

$$A_{NT}\xi = \int_{0}^{(N+1)T} a(t)\xi(t)dt$$

allows the representation

$$A_{NT}\xi = B_{NT}\xi - V_{NT}\xi,$$

where

$$B_{NT}\xi = \int_{0}^{(N+1)T} b^{\tau,N}(t)\xi^{(d)}(t,\tau T)dt, \quad V_{NT}\xi = \int_{-\tau Td}^{0} v^{\tau,N}(t)\xi(t)dt,$$

where

$$v^{\tau,N}(t) = \sum_{l=\lceil -\frac{t}{\tau T}\rceil}^{\min\{[\frac{(N+1)T-t}{\tau T}],d\}} (-1)^l \binom{d}{l} b^{\tau,N}(t + l\tau T), \quad t \in [-\tau Td; 0), \tag{7.1}$$

$$b^{\tau,N}(t) = \sum_{k=0}^{[\frac{(N+1)T-t}{\tau T}]} a(t + \tau Tk)d(k) = D^{\tau T,N}\mathbf{a}(t), \quad t \in [0; (N+1)T], \tag{7.2}$$

where the linear transformation $D^{\tau T,N}$ *acts on an arbitrary function* $x(t)$, $t \in [0; (N+1)T]$, *as follows:*

$$D^{\tau T,N}\mathbf{x}(t) = \sum_{k=0}^{[\frac{(N+1)T-t}{\tau T}]} x(t + \tau Tk)d(k).$$

Let $\widehat{A}\xi$ denote the mean square optimal linear estimate of the functional $A\xi$ from observations of the process $\xi(t)$ at points $t < 0$ and let $\widehat{B}\xi$ denote the mean square optimal linear estimate of the functional $B\xi$ from observations of the stochastic dth increment process $\xi^{(d)}(t, \tau T)$ at points $t < 0$.

Let $\Delta(f;\widehat{A}\xi) = \mathsf{E}|A\xi - \widehat{A}\xi|^2$ and $\Delta(f;\widehat{B}\xi) = \mathsf{E}|B\xi - \widehat{B}\xi|^2$ denote the MSEs of the estimates $\widehat{A}\xi$ and $\widehat{B}\xi$, respectively. Since values $\xi(t)$ at points $t \in [-\tau Tn; 0)$ are known, the following equality comes from Lemma 7.1.1:

$$A\xi = B\xi - V\xi, \quad \widehat{A}\xi = \widehat{B}\xi - V\zeta, \tag{7.3}$$

and

$$\begin{aligned}\Delta(f;\widehat{A}\xi) &= \mathsf{E}|A\xi - \widehat{A}\xi|^2 = \mathsf{E}|B\xi - V\xi - \widehat{B}\xi + V\xi|^2\\ &= \mathsf{E}|B\xi - \widehat{B}\xi|^2 = \Delta(f;\widehat{B}\xi).\end{aligned}$$

Similar relations for the functional $A_{NT}\xi$ are valid:

$$A_{NT}\xi = B_{NT}\xi - V_{NT}\xi, \quad \widehat{A}_{NT}\xi = \widehat{B}_{NT}\xi - V_{NT}\zeta, \tag{7.4}$$

and

$$\begin{aligned}\Delta(f;\widehat{A}_{NT}\xi) &= \mathsf{E}|A_{NT}\xi - \widehat{A}_{NT}\xi|^2 = \mathsf{E}|B_{NT}\xi - V_{NT}\xi - \widehat{B}_{NT}\xi + V_{NT}\xi|^2\\ &= \mathsf{E}|B_{NT}\xi - \widehat{B}_{NT}\xi|^2 = \Delta(f;\widehat{B}_{NT}\xi).\end{aligned}$$

Thus, to find the mean square optimal linear estimate of the functionals $A\xi$, $A_{NT}\xi$, we have to find estimates of the functionals $B\xi$, $B_{NT}\xi$.

The functional $B_{NT}\xi$ can be represented in the form

$$B_{NT}\xi = \int_0^{(N+1)T} b^{\tau,N}(t)\xi^{(d)}(t,\tau T)dt = \sum_{j=0}^{N}\int_0^T b_j^{\tau,N}(u)\xi_j^{(d)}(u)du,$$

where

$$b_j^{\tau,N}(u) = b^{\tau,N}(u+jT), \quad \xi_j^{(d)}(u) = \xi_j^{(d)}(u+jT,\tau T), \quad u \in [0,T).$$

Making use of the decomposition (6.6) of the generated stationary increment sequence $\{\xi_j^{(d)}, j \in \mathbb{Z}\}$ and the solution of the equation

$$(-1)^k\left[\frac{k}{2}\right] + (-1)^m\left[\frac{m}{2}\right] = 0 \tag{7.5}$$

of two variables (k,m), which is given by pairs $(1,1)$, $(2l+1,2l)$, and $(2l,2l+1)$ for $l = 2,3,\dots$, the functional $B_{NT}\xi$ can be rewritten in the form [147]

$$B_{NT}\xi = \int_0^{(N+1)T} b^{\tau,N}(t)\xi^{(d)}(t,\tau T)dt = \sum_{j=0}^{N}\int_0^T b_j^{\tau,N}(u)\xi_j^{(d)}(u)du$$

$$= \sum_{j=0}^{N} \frac{1}{T} \int_0^T \left(\sum_{k=1}^{\infty} b_{kj}^{\tau,N} e^{2\pi i\{(-1)^k [\frac{k}{2}]\}u/T} \right) \left(\sum_{m=1}^{\infty} \xi_{mj}^{(d)} e^{2\pi i\{(-1)^m [\frac{m}{2}]\}u/T} \right) du$$

$$= \sum_{j=0}^{N} \sum_{k=1}^{\infty} \sum_{m=1}^{\infty} b_{kj}^{\tau,N} \xi_{mj}^{(d)} \frac{1}{T} \int_0^T e^{2\pi i\{(-1)^k [\frac{k}{2}]+(-1)^m [\frac{m}{2}]\}u/T} du$$

$$= \sum_{j=0}^{N} \sum_{k=1}^{\infty} b_{kj}^{\tau,N} \xi_{kj}^{(d)} = \sum_{j=0}^{N} (\vec{b}_j^{\tau,N})^\top \vec{\xi}_j^{(d)} = B_N \xi,$$

where the infinite-dimensional vectors $\vec{\xi}_j^{(d)}$ and $\vec{b}_j^{\tau,N}$ are defined as follows:

$$\vec{\xi}_j^{(d)} = (\xi_{kj}^{(d)}, k = 1, 2, \ldots)^\top$$

and

$$\vec{b}_j^{\tau,N} = (b_{kj}^{\tau,N}, k = 1, 2, \ldots)^\top = (b_{1j}^{\tau,N}, b_{3j}^{\tau,N}, b_{2j}^{\tau,N}, \ldots, b_{2k+1,j}^{\tau,N}, b_{2k,j}^{\tau,N}, \ldots)^\top.$$

Here

$$b_{kj}^{\tau,N} = \langle b_j^{\tau,N}, \widetilde{e}_k \rangle = \frac{1}{\sqrt{T}} \int_0^T b_j^{\tau,N}(v) e^{-2\pi i\{(-1)^k [\frac{k}{2}]\}v/T} dv,$$

where

$$b_j^{\tau,N}(u) = \sum_{l=0}^{[\frac{(N+1)T-u-jT}{\tau T}]} a(u + jT + \tau Tl) d(l) = D^{\tau T,N} a(u), \quad u \in [0; t), \ j = 0, 1, \ldots, N,$$

and

$$b_{kj}^{\tau,N} = \sum_{k=0}^{[\frac{N-j}{\tau}]} a_{kj+\tau l} d(k), \quad j = 0, 1, \ldots, N,$$

where

$$a_{kj} = \langle a_j, \widetilde{e}_k \rangle = \frac{1}{\sqrt{T}} \int_0^T a_j(v) e^{-2\pi i\{(-1)^k [\frac{k}{2}]\}v/T} dv, \quad k = 1, 2, \ldots, \ j = 0, 1, \ldots, N.$$

Assume that coefficients $\{\vec{b}_j^{\tau,N}, j = 0, 1, \ldots, N\}$, which determine the functional $B_{NT}\xi$, satisfy the conditions

$$\|\vec{b}_j^{\tau,N}\| < \infty, \quad \|\vec{b}_j^{\tau,N}\|^2 = \sum_{k=1}^{\infty} |b_{kj}^{\tau,N}|^2, \quad j = 0, 1, \ldots, N. \tag{7.6}$$

It follows from the condition (7.6) that the functional

$$B_{NT}\xi = \sum_{j=0}^{N}\sum_{k=1}^{\infty} b_{kj}^{\tau,N}\xi_{kj}^{(d)} = \sum_{j=0}^{N}\left(\vec{b}_j^{\tau,N}\right)^{\top}\vec{\xi}_j^{(d)} = B_N\vec{\xi}$$

has a finite second moment.

7.2 Extrapolation problem: the greatest value of the MSE

Let us calculate the greatest value of the MSE $\Delta(\xi, \hat{B}_{NT})$ of an estimate $\hat{B}_{NT}\xi$ of the functional $B_{NT}\xi$ in the case where the process $\xi(t)$ is not "fixed." Denote by Λ the set of all linear estimates of the functional $B_{NT}\xi$, based on observations of the process $\xi(t)$ at points $t < 0$.

Let $\mathbf{Y}$ denote the class of the mean square continuous periodically correlated increment processes $\xi^{(d)}(t, \tau T)$, such that $\mathsf{E}\xi^{(d)}(t, \tau T) = 0$, $\mathsf{E}|\xi^{(d)}(t, \tau T)|^2 \le P_{\xi^{(d)}}$. Let $\mathbf{Y}_R$ denote the class of all regular stationary sequences that satisfy the condition $\|\xi_j^{(d)}\|_H^2 \le P_{\xi^{(d)}}$. Then we can formulate the following theorem (see [65, 136, 145]).

Theorem 7.2.1. *Let the coefficients $\{\vec{b}_j^{\tau,N}, j = 0, 1, \dots, N\}$, which determine the functional $B_{NT}\xi$, satisfy condition (7.6). The function $\Delta(\xi, \hat{B}_{NT})$ has a saddle point on the set $\mathbf{Y} \times \Lambda$:*

$$\min_{\hat{B}_{NT}\in\Lambda}\max_{\xi\in\mathbf{Y}}\Delta(\xi, \hat{B}_{NT}) = \max_{\xi\in\mathbf{Y}}\min_{\hat{B}_{NT}\in\Lambda}\Delta(\xi, \hat{B}_{NT}) = P_{\xi^{(d)}}\cdot\nu_N^2,$$

where ν_N^2 is the greatest eigenvalue of the self-adjoint compact operator Q_N in the space ℓ_2, determined by the matrix $\{Q_N(p,q)\}_{p,q=0}^{N}$, which is constructed from the block-matrices $Q_N(p,q) = \{Q_{kn}^N(p,q)\}_{k,n=1}^{\infty}$ with the entries

$$Q_{kn}^N(p,q) = \sum_{s=0}^{\min(N-p,N-q)} b_{k,s+p}^{\tau,N}\cdot\overline{b_{n,s+q}^{\tau,N}}, \quad k,n = 1,2,\dots, \quad p,q = 0,1,\dots,N.$$

The least favorable stochastic sequence $\vec{\xi}_j^{(d)}$ in the class $\mathbf{Y}$ for the optimal estimation of the functional $B_{NT}\xi$ is a one-sided moving average sequence of order N of the form

$$\vec{\xi}_j^{(d)} = \sum_{s=j-N}^{j} g(j-s)\vec{\varepsilon}(s),$$

where $g_N = (g(0), g(1), \dots, g(N))^{\top}$ is the eigenvector, which corresponds to ν_N^2, which is constructed from matrices $g(p) = \{g_{km}(p)\}_{k=\overline{1,\infty}}^{m=\overline{1,M}}$ and is determined by the condition $\|g_N\|^2 = \sum_{p=0}^{N}\|g(p)\|^2 = P_{\xi^{(d)}}$, and $\vec{\varepsilon}(s) = \{\varepsilon_m(s)\}_{m=1}^{M}$ is a vector-valued stationary stochastic sequence with orthogonal values.

Similar properties can be obtained for the functional $B\xi$ [147]:

$$B\xi = \int_0^{\infty} b^{\tau}(t)\xi^{(d)}(t,\tau T)dt = \sum_{j=0}^{\infty} (\vec{b}_j^{\tau})^{\top} \vec{\xi}_j^{(d)} = B\vec{\xi},$$

where the infinite-dimensional vectors $\vec{\xi}_j^{(d)}$ and $\vec{b}_j^{\tau}$ are defined as follows:

$$\vec{\xi}_j^{(d)} = (\xi_{kj}^{(d)}, k = 1,2,\dots)^{\top},$$
$$\vec{b}_j^{\tau} = (b_{kj}^{\tau}, k = 1,2,\dots)^{\top} = (b_{1j}^{\tau}, b_{3j}^{\tau}, b_{2j}^{\tau}, \dots, b_{2k+1,j}^{\tau}, b_{2k,j}^{\tau}, \dots)^{\top}.$$

Here

$$b_{kj}^{\tau} = \langle b_j^{\tau}, \widetilde{e}_k \rangle = \frac{1}{\sqrt{T}} \int_0^T b_j^{\tau}(v) e^{-2\pi i \{(-1)^k [\frac{k}{2}]\} v/T} dv,$$

$$b_j^{\tau}(u) = \sum_{l=0}^{\infty} a(u + jT + \tau Tl) d(l) = D^{\tau} a_j(u), \quad u \in [0,T),\ j = 0,1,\dots,$$

and

$$b_{k,j}^{\tau} = \sum_{l=0}^{\infty} a_{k,j+\tau l} d(l), \quad j = 0,1,\dots,\ k = 1,2,\dots,$$

$$a_{kj} = \langle a_j, \widetilde{e}_k \rangle = \frac{1}{\sqrt{T}} \int_0^T a_j(v) e^{-2\pi i \{(-1)^k [\frac{k}{2}]\} v/T} dv.$$

Assume that coefficients $\{\vec{b}_j^{\tau}, j = 0,1,\dots\}$, which determine the functional $B\xi$, satisfy the conditions

$$\sum_{j=0}^{\infty} \|\vec{b}_j^{\tau}\| < \infty, \quad \sum_{j=0}^{\infty} (j+1) \|\vec{b}_j^{\tau}\|^2 < \infty, \quad \|\vec{b}_j^{\tau}\|^2 = \sum_{k=1}^{\infty} |b_{kj}^{\tau}|^2. \tag{7.7}$$

It follows from the condition (7.7) that the functional

$$B\xi = \sum_{j=0}^{\infty} \sum_{k=1}^{\infty} b_{kj}^{\tau} \xi_{kj}^{(d)} = \sum_{j=0}^{\infty} (\vec{b}_j^{\tau})^{\top} \vec{\xi}_j^{(n)}$$

has a finite second moment. The following theorem provides the greatest value of the MSE $\Delta(\xi, \hat{B}\xi)$ of the estimate $\hat{B}\xi$ of the functional $B\xi$ (see [136, 145]).

Theorem 7.2.2. *Let the coefficients $\{\vec{b}_j^{\tau}, j = 0,1,\dots\}$, which determine the functional $B\xi$, satisfy condition* (7.7). *The function $\Delta(\xi, \hat{B}\xi)$ has a saddle point on the set $\mathbf{Y} \times \Lambda$:*

$$\min_{\hat{B}\in\Lambda}\max_{\xi\in\mathbf{Y}}\Delta(\xi,\hat{B}\xi)=\max_{\xi\in\mathbf{Y}}\min_{\hat{B}\in\Lambda}\Delta(\xi,\hat{B})=P_{\xi^{(d)}}\cdot v^2,$$

where v^2 is the greatest eigenvalue of the self-adjoint compact operator Q in the space ℓ_2, determined by the matrix $\{Q(p,q)\}_{p,q=0}^{\infty}$, which is constructed from the block-matrices $Q(p,q)=\{Q_{kn}(p,q)\}_{k,n=1}^{\infty}$ with elements

$$Q_{kn}(p,q)=\sum_{s=0}^{\infty}b_{k,s+p}^{\tau}\cdot\overline{b_{n,s+q}^{\tau}},\quad k,n=1,2,\ldots,\quad p,q=0,1,\ldots.$$

The least favorable stochastic sequence $\vec{\xi}_j^{(d)}$ in the class $\mathbf{Y}$ for the optimal estimate of the functional $B\xi$ is a one-sided moving average sequence of the form

$$\vec{\xi}_j^{(d)}=\sum_{s=-\infty}^{j}g(j-s)\vec{\varepsilon}(s),$$

where $g=(g(0),g(1),\ldots)^{\top}$ is the eigenvector, which corresponds to v^2, which is constructed from matrices $g(p)=\{g_{km}(p)\}_{k=\overline{1,\infty}}^{m=\overline{1,M}}$ and is determined by the condition $\|g\|^2=\sum_{p=0}^{\infty}\|g(p)\|^2=P_{\xi^{(d)}}$, and $\vec{\varepsilon}(s)=\{\varepsilon_m(s)\}_{m=1}^{M}$ is a vector-valued stationary stochastic sequence with orthogonal values.

7.3 Hilbert space projection method of extrapolation

Let the spectral density $f(\lambda)$ satisfy the *minimality condition*

$$\int_{-\pi}^{\pi}\mathrm{Tr}\left[\frac{\lambda^{2d}}{|1-e^{i\lambda\tau}|^{2d}}f^{-1}(\lambda)\right]d\lambda<\infty. \tag{7.8}$$

This is the necessary and sufficient condition under which the MSEs of the estimates of the functionals $A\vec{\xi}$ and $A_N\vec{\xi}$ are not equal to 0.

Consider the Hilbert space $H=L_2(\Omega,\mathcal{F},\mathsf{P})$ of random variables γ with zero mean value, $\mathsf{E}\gamma=0$, and finite variance, $\mathsf{E}|\gamma|^2<\infty$, endowed with the inner product $(\gamma_1;\gamma_2)=\mathsf{E}\gamma_1\overline{\gamma_2}$. With the generated increment sequence $\{\xi_j^{(d)}\}$, we can associate the closed linear subspace

$$H^{-1}(\vec{\xi}^{(d)})=\overline{\mathrm{span}}\{\xi_{kj}^{(d)}:k=1,2,\ldots;j\le-1\}.$$

Consider also the closed linear subspace

$$L_2^{-1}(f)=\overline{\mathrm{span}}\{e^{i\lambda j}(1-e^{-i\lambda\tau})^{d}(i\lambda)^{-d}\vec{\delta}_k:k=1,2,\ldots,\infty,j\le-1\}$$

of the Hilbert space $L_2(f)$ of vector-valued functions, endowed with the inner product $\langle g_1; g_2\rangle = \int_{-\pi}^{\pi} (g_1(\lambda))^\top f(\lambda)\overline{g_2(\lambda)}d\lambda$. Here $\vec{\delta}_k = \{\delta_{lk}\}_{l=1}^{\infty}$, where δ_{lk} are Kronecker symbols. The relation

$$\xi_{kj}^{(d)} = \int_{-\pi}^{\pi} e^{i\lambda j}(1-e^{-i\lambda\tau})^d \frac{1}{(i\lambda)^d} dZ_k(\lambda), \quad k = 1, 2, \ldots, \tag{7.9}$$

implies a map between the subspaces $H^{-1}(\vec{\xi}^{(d)})$ and $L_2^{-1}(f)$. Here $Z_k(\lambda)$ are components of a vector-valued stochastic process $\vec{Z}_{\xi^{(d)}}(\lambda) = \{Z_k(\lambda)\}_{k=1}^{\infty}$ with uncorrelated increments on $[-\pi, \pi)$ connected with the spectral function $F(\lambda)$ by the relation

$$\begin{gathered} \mathsf{E}(Z_k(\lambda_2) - Z_k(\lambda_1))\overline{(Z_n(\lambda_2) - Z_n(\lambda_1))} = F_{kn}(\lambda_2) - F_{kn}(\lambda_1), \\ -\pi \le \lambda_1 < \lambda_2 < \pi, \quad k, n = 1, 2, \ldots, \infty. \end{gathered}$$

We define the Fourier coefficients of the function $\lambda^{2d}|1-e^{i\lambda\tau}|^{-2d}(f^{-1}(\lambda))^\top$:

$$F_{j,l}^{\tau} = \frac{1}{2\pi}\int_{-\pi}^{\pi} e^{-i\lambda(j-l)} \frac{\lambda^{2d}}{|1-e^{i\lambda\tau}|^{2d}} (f^{-1}(\lambda))^\top d\lambda, \quad j, l \ge 0.$$

Define the vector

$$\vec{a}_j = (a_{kj}, k = 1, 2, \ldots)^\top = (a_{1j}, a_{3j}, a_{2j}, \ldots, a_{2k+1,j}, a_{2k,j}, \ldots)^\top.$$

Lemma 7.1.1 implies

$$\vec{b}_j^{\tau} = \sum_{m=j}^{\infty} \mathrm{diag}_{\infty}(d_\tau(m-j))\vec{a}_m = (D^\tau \mathbf{a})_j, \quad j = 0, 1, 2, \ldots, \tag{7.10}$$

where $\mathbf{a} = ((\vec{a}_0)^\top, (\vec{a}_1)^\top, (\vec{a}_2)^\top, \ldots)^\top$, the coefficients $\{d_\tau(k) : k \ge 0\}$ are determined by the relationship

$$\sum_{k=0}^{\infty} d_\tau(k)x^k = \left(\sum_{j=0}^{\infty} x^{\tau j}\right)^d,$$

and D^τ is a linear transformation determined by a matrix with infinite-dimensional matrix entries $D^\tau(k,j)$, $k,j = 0,1,2,\ldots$, such that $D^\tau(k,j) = \mathrm{diag}_{\infty}(d_\tau(j-k))$ if $0 \le k \le j$ and $D^\tau(k,j) = \mathcal{O}$ for $0 \le j < k$; $\mathrm{diag}_{\infty}(x)$ denotes an infinite-dimensional diagonal matrix with the entry x on its diagonal, and $\mathcal{O}$ denotes an infinite-dimensional matrix with zero entries.

Theorem 7.3.1. *Let a stochastic process $x(t)$, $t \in \mathbb{R}$, with periodically stationary increments determine a generated stationary dth increment sequence $\vec{\xi}_m^{(d)}$ with the spectral* *density matrix* $f(\lambda) = \{f_{ij}(\lambda)\}_{i,j=1}^{\infty}$ *satisfying minimality condition* (7.8). *Let the coefficients* $\vec{a}_j$, $j \geqslant 0$, *generated by the function* $a(t)$, $t \geq 0$, *satisfy conditions* (7.7) *taking into account* (7.10). *The optimal linear estimate* $\widehat{A}\xi$ *of the functional* $A\xi$ *based on observations of the process* $\xi(t)$ *on the interval* $t < 0$ *is calculated by the formula*

$$\widehat{A}\xi = \int_{-\pi}^{\pi} (\vec{h}_\tau(\lambda))^\top d\vec{Z}_{\xi^{(d)}}(\lambda) - \int_{-\tau T d}^{0} v^\tau(t)\xi(t)dt, \tag{7.11}$$

where $\vec{h}_\tau(\lambda) = \{h_k(\lambda)\}_{k=1}^{\infty}$ *is the spectral characteristic of the estimate* $\widehat{A}\vec{\xi}$. *The spectral characteristic* $\vec{h}_\tau(\lambda)$ *and the value of the MSE* $\Delta(f;\widehat{A}\xi)$ *of the optimal estimate* $\widehat{A}\xi$ *are calculated by the formulas*

$$(\vec{h}_\tau(\lambda))^\top = (\vec{B}_\tau(e^{i\lambda}))^\top \frac{(1-e^{-i\lambda\tau})^d}{(i\lambda)^d} - \frac{(-i\lambda)^d}{(1-e^{i\lambda\tau})^d}(\vec{C}_\tau(e^{i\lambda}))^\top f^{-1}(\lambda) \tag{7.12}$$

and

$$\begin{aligned}\Delta(f;\widehat{A}\xi) &= \Delta(f;\widehat{B}\vec{\xi}) = \mathsf{E}|B\vec{\xi} - \widehat{B}\vec{\xi}|^2 \\ &= \frac{1}{2\pi}\int_{-\pi}^{\pi} \frac{(-i\lambda)^d}{(1-e^{i\lambda\tau})^d}(\vec{C}_\tau(e^{i\lambda}))^\top f(\lambda)\overline{\vec{C}_\tau(e^{i\lambda})}\frac{(i\lambda)^d}{(1-e^{-i\lambda\tau})^d}d\lambda \\ &= \langle D^\tau \mathbf{a}, \mathbf{F}_\tau^{-1} D^\tau \mathbf{a}\rangle,\end{aligned} \tag{7.13}$$

respectively, where

$$\vec{B}_\tau(e^{i\lambda}) = \sum_{j=0}^{\infty} \vec{b}_j^\tau e^{i\lambda j}, \quad \vec{C}_\tau(e^{i\lambda}) = \sum_{j=0}^{\infty} (\mathbf{F}_\tau^{-1} D^\tau \mathbf{a})_j e^{i\lambda j};$$

$\mathbf{F}_\tau$ *is a linear operator in the space* ℓ_2 *which is determined by a matrix with the infinite matrix entries* $(\mathbf{F}_\tau)_{j,l} = F_{j,l}^\tau$, $j, l \geq 0$.

Proof. Representation (7.11) for the estimate $\widehat{A}\xi$ comes from relation (7.3). The functional $B\vec{\xi}$ allows the spectral representation

$$B\vec{\xi} = \int_{-\pi}^{\pi} (\vec{B}_\tau(e^{i\lambda}))^\top \frac{(1-e^{-i\tau\lambda})^d}{(i\lambda)^d} d\vec{Z}_{\xi^{(d)}}(\lambda).$$

Thus, the spectral characteristic $\vec{h}_\tau(\lambda)$ is characterized by the following conditions, which determine an orthogonal projection of the element $\vec{B}_\tau(e^{i\lambda})\frac{(1-e^{-i\lambda\tau})^d}{(i\lambda)^d}$ on the subspace $L_2^{0-}(f)$:

$$\vec{h}_\tau(\lambda) \in L_2^{0-}(f), \tag{7.14}$$

$$\left(\vec{B}_\tau(e^{i\lambda})\frac{(1-e^{-i\lambda\tau})^d}{(i\lambda)^d} - \vec{h}_\tau(\lambda)\right) \perp L_2^{0-}(f). \tag{7.15}$$

From the condition (7.15), we obtain the following relation, which holds for all $j \leqslant -1$ and for all $k \geq 1$:

$$\int_{-\pi}^{\pi} \left(\vec{B}_\tau(e^{i\lambda})\frac{(1-e^{-i\tau\lambda})^d}{(i\lambda)^d} - \vec{h}_\tau(\lambda)\right)^\top f(\lambda)\vec{\delta}_k e^{-i\lambda j}\frac{(1-e^{i\tau\lambda})^d}{(-i\lambda)^d} d\lambda = 0. \tag{7.16}$$

Thus, the spectral characteristic of the estimate $\widehat{B\vec{\xi}}$ can be represented as

$$(\vec{h}_\tau(\lambda))^\top = (\vec{B}_\tau(e^{i\lambda}))^\top \frac{(1-e^{-i\lambda\tau})^d}{(i\lambda)^d} - \frac{(-i\lambda)^d}{(1-e^{i\lambda\tau})^d}(\vec{C}_\tau(e^{i\lambda}))^\top f^{-1}(\lambda),$$

where

$$\vec{C}_\tau(e^{i\lambda}) = \sum_{j=0}^{\infty} \vec{c}_j^{\,\tau} e^{ij\lambda},$$

where vectors $\vec{c}_j^{\,\tau} = \{c_{kj}^\tau\}_{k=1}^\infty$, $j \geqslant 0$, are unknown and are to be found.

Condition (7.14) implies the relations

$$\int_{-\pi}^{\pi} \left((\vec{B}_\tau(e^{i\lambda}))^\top - \frac{\lambda^{2d}(\vec{C}_\tau(e^{i\lambda}))^\top}{(1-e^{-i\lambda\tau})^d(1-e^{i\lambda\tau})^d} f^{-1}(\lambda)\right)e^{-i\lambda j} d\lambda = 0, \quad j \geqslant 0. \tag{7.17}$$

Making use of the introduced Fourier coefficients $F_{j,l}^\tau$, relation (7.17) can be presented as a system of linear equations

$$\vec{b}_l^\tau = \sum_{j=0}^{\infty} F_{l,j}^\tau \vec{c}_j^{\,\tau}, \quad l \geq 0,$$

allowing the matrix form representation

$$D^\tau \mathbf{a} = \mathbf{F}_\tau \mathbf{c}^\tau, \tag{7.18}$$

where $\mathbf{c}^\tau = ((\vec{c}_0^{\,\tau})^\top, (\vec{c}_1^{\,\tau})^\top, (\vec{c}_2^{\,\tau})^\top, \ldots)^\top$.

The derived expressions and the coefficients $\vec{c}_j^{\,\tau}, j \geq 0$, prove formulas (7.12) for the spectral characteristic $\vec{h}_\tau(\lambda)$ and (7.13) for the value of the MSE of the estimate $\hat{A}\xi$. □

Consider the extrapolation problem for the functional $A_{NT}\xi$. We can apply the results of Theorem 2.1.1 by putting $\vec{a}_j = \vec{0}$ for $j > N$. Corollary 7.1.1 implies

$$\vec{b}_j^\tau = \sum_{m=j}^{N} \mathrm{diag}_\infty(d_\tau(m-j))\vec{a}_m = (D_N^\tau \mathbf{a}_N)_j, \quad j = 0, 1, \dots, N,$$

where D_N^τ is the linear transformation determined by an infinite matrix with the entries $(D_N^\tau)(l,j) = \mathrm{diag}_\infty(d_\tau(j-l))$ if $0 \le l \le j \le N$ and $(D_N^\tau)(l,j) = \mathcal{O}$ if $j < l$ or $j, l > N$ and

$$\mathbf{a}_N = ((\vec{a}_0)^\top, (\vec{a}_1)^\top, \dots, (\vec{a}_N)^\top, \vec{0}^\top, \dots)^\top.$$

Theorem 7.3.2. *Let a stochastic process $\xi(t)$, $t \in \mathbb{R}$, with periodically stationary increments determine a generated stationary dth increment sequence $\vec{\xi}_m^{(d)}$ with the spectral density matrix $f(\lambda) = \{f_{ij}(\lambda)\}_{i,j=1}^{\infty}$ satisfying minimality condition (7.8). The optimal linear estimate $\widehat{A}_{NT}\xi$ of the functional $A_{NT}\xi$ based on observations of the process $\xi(t)$ on the interval $t < 0$ is calculated by the formula*

$$\widehat{A}_{NT}\xi = \int_{-\pi}^{\pi} (\vec{h}_{\tau,N}(\lambda))^\top d\vec{Z}_{\xi^{(d)}}(\lambda) - \int_{-\tau T d}^{0} v^{\tau,N}(t)\xi(t)dt. \tag{7.19}$$

The spectral characteristic $\vec{h}_{\tau,N}(\lambda) = \{h_{\tau,N,k}(\lambda)\}_{k=1}^{\infty}$ and the value of the MSE $\Delta(f; \widehat{A}_N\xi)$ are calculated by the formulas

$$(\vec{h}_{\tau,N}(\lambda))^\top = (\vec{B}_{\tau,N}(e^{i\lambda}))^\top \frac{(1-e^{-i\lambda\tau})^d}{(i\lambda)^d} - \frac{(-i\lambda)^d}{(1-e^{i\lambda\tau})^d}(\vec{C}_{\tau,N}(e^{i\lambda}))^\top f^{-1}(\lambda) \tag{7.20}$$

and

$$\begin{aligned}\Delta(f, \widehat{A}_{NT}\xi) &= \Delta(f, \widehat{B}_N\vec{\xi}) = \mathsf{E}|B_N\vec{\xi} - \widehat{B}_N\vec{\xi}|^2 \\ &= \frac{1}{2\pi}\int_{-\pi}^{\pi} \frac{(-i\lambda)^d}{(1-e^{i\lambda\tau})^d}(\vec{C}_{\tau,N}(e^{i\lambda}))^\top f(\lambda)\overline{\vec{C}_{\tau,N}(e^{i\lambda})}\frac{(i\lambda)^d}{(1-e^{-i\lambda\tau})^d}d\lambda \\ &= \langle D_N^\tau \mathbf{a}_N, \mathbf{F}_{\tau,N}^{-1} D_N^\tau \mathbf{a}_N\rangle,\end{aligned} \tag{7.21}$$

respectively, where

$$\vec{B}_{\tau,N}(e^{i\lambda}) = \sum_{j=0}^{N}(D_N^\tau \mathbf{a}_N)_j e^{i\lambda j}, \quad \vec{C}_{\tau,N}(e^{i\lambda}) = \sum_{j=0}^{\infty}(\mathbf{F}_{\tau,N}^{-1} D_N^\tau \mathbf{a}_N)_j e^{i\lambda j},$$

where $\mathbf{F}_{\tau,N}$ is a linear operator in ℓ_2 determined by a matrix with the infinite matrix entries $(\mathbf{F}_{\tau,N})_{l,m} = F_{l,m}^\tau$, $l \ge 0$, $0 \le m \le N$, and $(\mathbf{F}_{\tau,N})_{l,m} = \mathcal{O}$, $l \ge 0$, $m > N$.

7.4 Minimax (robust) method of extrapolation

7.4.1 Least favorable spectral density and minimax-robust spectral characteristic

The values of the MSEs and the spectral characteristics of the optimal estimates of the functionals $A\vec{\xi}$ and $A_{NT}\vec{\xi}$ depending on the unobserved values of a stochastic process $\xi(t)$, which determine the periodically stationary stochastic increment process $\xi^{(d)}(t,\tau T)$ or the corresponding infinite-dimensional vector stationary increment sequence $\{\vec{\xi}_j^{(d)} = (\xi_{kj}^{(d)}, k = 1,2,\dots)^\top, j \in \mathbb{Z}\}$, with the spectral density matrix $f(\lambda)$, can be calculated by formulas (7.12)–(7.13) and (7.20)–(7.21), respectively, provided the spectral density $f(\lambda)$ of the stochastic process $\xi(t)$ is exactly known. If the spectral density is not known while a set $\mathcal{D}$ of admissible spectral densities is defined, the minimax method of estimation of the functionals depending on unobserved values of stochastic sequences with stationary increments may be applied. This method consists in finding an estimate that minimizes the maximal values of the MSEs for all spectral densities from a given class $\mathcal{D}$ of admissible spectral densities simultaneously. The described method is formalized by the two following definitions [145].

Definition 7.4.1. For a given class of spectral densities $\mathcal{D}$, the spectral density $f^0(\lambda) \in \mathcal{D}$ is called the least favorable in $\mathcal{D}$ for the optimal linear estimation of the functional $A\vec{\xi}$ if the following relation holds:

$$\Delta(f^0) = \Delta(h(f^0); f^0) = \max_{f\in\mathcal{D}} \Delta(h(f); f).$$

Definition 7.4.2. For a given class of spectral densities $\mathcal{D}$, the spectral characteristic $h^0(\lambda)$ of the optimal linear estimate of the functional $A\xi$ is called minimax-robust if the following conditions are satisfied:

$$h^0(\lambda) \in H_{\mathcal{D}} = \bigcap_{f\in\mathcal{D}} L_2^{-1}(f),$$

$$\min_{h\in H_{\mathcal{D}}} \max_{f\in\mathcal{D}} \Delta(h; f) = \sup_{f\in\mathcal{D}} \Delta(h^0; f).$$

Taking into account the introduced definitions and the derived relations we can verify that the following lemmas hold.

Lemma 7.4.1. *The spectral density $f^0(\lambda) \in \mathcal{D}$ satisfying the minimality condition* (7.8) *is the least favorable density in the class $\mathcal{D}$ for the optimal linear extrapolation of the functional $A\xi$ based on observations of the process $\xi(t)$ on the interval $t < 0$ if the operator $\mathbf{F}_\tau^0$ defined by the Fourier coefficients of the function*

$$\lambda^{2d}|1 - e^{i\lambda\tau}|^{-2d}((f^0(\lambda))^{-1})^\top \tag{7.22}$$

determines a solution to the constrained optimization problem

$$\max_{f\in\mathcal{D}}(\langle D^{\tau}\mathbf{a}, \mathbf{F}_{\tau}^{-1}D^{\tau}\mathbf{a}\rangle) = \langle D^{\tau}\mathbf{a}, (\mathbf{F}_{\tau}^{0})^{-1}D^{\tau}\mathbf{a}\rangle. \tag{7.23}$$

The minimax spectral characteristic $h^0 = \vec{h}_\tau(f^0)$ *is calculated by formula* (7.12) *if* $\vec{h}_\tau(f^0) \in H_{\mathcal{D}}$.

For more detailed analysis of properties of the least favorable spectral densities and the minimax-robust spectral characteristics, we observe that the minimax spectral characteristic h^0 and the least favorable spectral density f_0 form a saddle point of the function $\Delta(h;f)$ on the set $H_{\mathcal{D}} \times \mathcal{D}$ [145]. The saddle point inequalities

$$\Delta(h;f_0) \geq \Delta(h^0;f^0) \geq \Delta(h^0;f) \quad \forall f \in \mathcal{D}, \forall h \in H_{\mathcal{D}}$$

hold if $h^0 = \vec{h}_\tau(f^0)$, $\vec{h}_\tau(f^0) \in H_{\mathcal{D}}$, and f^0 is a solution of the constrained optimization problem

$$\widetilde{\Delta}(f) = -\Delta(\vec{h}_\tau(f^0);f) \to \inf, \quad f \in \mathcal{D}. \tag{7.24}$$

The functional $\Delta(\vec{h}_\tau(f^0);f)$ is calculated by the formula

$$\Delta(\vec{h}_\tau(f^0);f) = \frac{1}{2\pi}\int_{-\pi}^{\pi} \frac{(-i\lambda)^d}{(1-e^{i\lambda\tau})^d}(\vec{C}_\tau^0(e^{i\lambda}))^{\top}(f^0(\lambda))^{-1}f(\lambda)(f^0(\lambda))^{-1}\overline{\vec{C}_\tau^0(e^{i\lambda})}\frac{(i\lambda)^d}{(1-e^{-i\lambda\tau})^d}d\lambda,$$

where

$$\vec{C}_\tau^0(e^{i\lambda}) = \sum_{j=0}^{\infty}((\mathbf{F}_\tau^0)^{-1}D^{\tau}\mathbf{a})_j e^{i\lambda j}.$$

The constrained optimization problem (7.24) is equivalent to the unconstrained optimization problem

$$\Delta_{\mathcal{D}}(f) = \widetilde{\Delta}(f) + \delta(f|\mathcal{D}) \to \inf,$$

where $\delta(f|\mathcal{D})$ is the indicator function of the set $\mathcal{D}$, namely, $\delta(f|\mathcal{D}) = 0$ if $f \in \mathcal{D}$ and $\delta(f|\mathcal{D}) = +\infty$ if $f \notin \mathcal{D}$. A solution f^0 of the unconstrained optimization problem is characterized by the condition $0 \in \partial\Delta_{\mathcal{D}}(f^0)$, which is the necessary and sufficient condition under which the point f^0 belongs to the set of minimums of the convex functional $\Delta_{\mathcal{D}}(f)$ [51, 145, 180]. This condition makes it possible to find the least favorable spectral densities in some special classes of spectral densities D.

The form of the functional $\widetilde{\Delta}(f)$ allows us to apply the Lagrange method of indefinite multipliers for investigating the constrained optimization problem (7.24).

7.4.2 Least favorable spectral density in classes $\mathcal{D}_0$

Consider the extrapolation problem for the functional $A\xi$ which depends on unobserved values of a process $\xi(t)$ with periodically stationary increments based on observations of this process at points $t < 0$ under the condition that the sets of admissible spectral densities $\mathcal{D}_0^k$, $k = 1, 2, 3, 4$, are defined as follows:

$$\mathcal{D}_0^1 = \left\{ f(\lambda) \left| \frac{1}{2\pi} \int_{-\pi}^{\pi} \frac{|1 - e^{i\lambda\tau}|^{2d}}{\lambda^{2d}} f(\lambda) d\lambda = P \right. \right\},$$

$$\mathcal{D}_0^2 = \left\{ f(\lambda) \left| \frac{1}{2\pi} \int_{-\pi}^{\pi} \frac{|1 - e^{i\lambda\tau}|^{2d}}{\lambda^{2d}} \operatorname{Tr}[f(\lambda)] d\lambda = p \right. \right\},$$

$$\mathcal{D}_0^3 = \left\{ f(\lambda) \left| \frac{1}{2\pi} \int_{-\pi}^{\pi} \frac{|1 - e^{i\lambda\tau}|^{2d}}{\lambda^{2d}} f_{kk}(\lambda) d\lambda = p_k, k = 1, 2, 3, \dots \right. \right\},$$

$$\mathcal{D}_0^4 = \left\{ f(\lambda) \left| \frac{1}{2\pi} \int_{-\pi}^{\pi} \frac{|1 - e^{i\lambda\tau}|^{2d}}{\lambda^{2d}} \langle B_1, f(\lambda) \rangle d\lambda = p \right. \right\},$$

where p, p_k, $k = 1, 2, 3, \dots$, are given numbers and P, B_1 are given positive-definite Hermitian matrices.

From the condition $0 \in \partial\Delta_{\mathcal{D}}(f^0)$ we find the following equations, which determine the least favorable spectral densities for these given sets of admissible spectral densities.

For the set $\mathcal{D}_0^1$ of admissible spectral densities, we have

$$\begin{aligned} &\left(\sum_{j=0}^{\infty} ((\mathbf{F}_\tau^0)^{-1} D^\tau \mathbf{a})_j e^{i\lambda j} \right) \left(\sum_{j=0}^{\infty} ((\mathbf{F}_\tau^0)^{-1} D^\tau \mathbf{a})_j e^{i\lambda j} \right)^* \\ &\quad = \left(\frac{|1 - e^{i\lambda\tau}|^{2d}}{\lambda^{2d}} f^0(\lambda) \right) \vec{\alpha} \cdot \vec{\alpha}^* \left(\frac{|1 - e^{i\lambda\tau}|^{2d}}{\lambda^{2d}} f^0(\lambda) \right), \end{aligned} \tag{7.25}$$

where $\vec{\alpha}$ is a vector of Lagrange multipliers.

For the set $\mathcal{D}_0^2$ of admissible spectral densities, we have

$$\begin{aligned} &\left(\sum_{j=0}^{\infty} ((\mathbf{F}_\tau^0)^{-1} D^\tau \mathbf{a})_j e^{i\lambda j} \right) \left(\sum_{j=0}^{\infty} ((\mathbf{F}_\tau^0)^{-1} D^\tau \mathbf{a})_j e^{i\lambda j} \right)^* \\ &\quad = \alpha^2 \left(\frac{|1 - e^{i\lambda\tau}|^{2d}}{\lambda^{2d}} f^0(\lambda) \right)^2, \end{aligned} \tag{7.26}$$

where α^2 is a Lagrange multiplier.

For the set $\mathcal{D}_0^3$ of admissible spectral densities, we have

$$\left(\sum_{j=0}^{\infty}((\mathbf{F}_\tau^0)^{-1}D^\tau\mathbf{a})_j e^{i\lambda j}\right)\left(\sum_{j=0}^{\infty}((\mathbf{F}_\tau^0)^{-1}D^\tau\mathbf{a})_j e^{i\lambda j}\right)^* = \left(\frac{|1-e^{i\lambda\tau}|^{2d}}{\lambda^{2d}}f^0(\lambda)\right)\{\alpha_k^2\delta_{kl}\}_{k,l=1}^{\infty}\left(\frac{|1-e^{i\lambda\tau}|^{2d}}{\lambda^{2d}}f^0(\lambda)\right), \tag{7.27}$$

where α_k^2 are Lagrange multipliers and δ_{kl} are Kronecker symbols.

For the set $\mathcal{D}_0^4$ of admissible spectral densities, we have

$$\left(\sum_{j=0}^{\infty}((\mathbf{F}_\tau^0)^{-1}D^\tau\mathbf{a})_j e^{i\lambda j}\right)\left(\sum_{j=0}^{\infty}((\mathbf{F}_\tau^0)^{-1}D^\tau\mathbf{a})_j e^{i\lambda j}\right)^* = \alpha^2\left(\frac{|1-e^{i\lambda\tau}|^{2d}}{\lambda^{2d}}f^0(\lambda)\right)B_1^\top\left(\frac{|1-e^{i\lambda\tau}|^{2d}}{\lambda^{2d}}f^0(\lambda)\right), \tag{7.28}$$

where α^2 is a Lagrange multiplier.

The following theorem holds.

Theorem 7.4.1. *The least favorable spectral density $f^0(\lambda)$ in the classes $\mathcal{D}_0^k$, $k = 1,2,3,4$, for the optimal linear extrapolation of the functional $A\xi$ from observations of the process $\xi(t)$ on the interval $t < 0$ is determined by equations* (7.25), (7.26), (7.27), *and* (7.28), *the minimality condition* (7.8), *the constrained optimization problem* (7.23), *and restrictions on densities from the corresponding classes $\mathcal{D}_0^k$, $k = 1,2,3,4$. The minimax-robust spectral characteristic of the optimal estimate of the functional $A\xi$ is determined by formula* (7.12).

7.4.3 Least favorable spectral density in classes $\mathcal{D}_{1\delta}$

Consider the extrapolation problem for the functional $A\xi$ which depends on unobserved values of the process $\xi(t)$ with periodically stationary increments based on its observations at points $t < 0$ under the condition that the sets of admissible spectral densities $D_{1\delta}^k$, $k = 1,2,3,4$, are defined as follows:

$$\mathcal{D}_{1\delta}^1 = \left\{f(\lambda)\,\middle|\,\frac{1}{2\pi}\int_{-\pi}^{\pi}\frac{|1-e^{i\lambda\tau}|^{2d}}{\lambda^{2d}}|f_{ij}(\lambda)-f_{ij}^1(\lambda)|d\lambda \le \delta_i^j,\ i,j=\overline{1,\infty}\right\},$$

$$\mathcal{D}_{1\delta}^2 = \left\{f(\lambda)\,\middle|\,\frac{1}{2\pi}\int_{-\pi}^{\pi}\frac{|1-e^{i\lambda\tau}|^{2d}}{\lambda^{2d}}|\operatorname{Tr}(f(\lambda)-f_1(\lambda))|d\lambda \le \delta\right\},$$

$$\mathcal{D}_{1\delta}^3 = \left\{f(\lambda)\,\middle|\,\frac{1}{2\pi}\int_{-\pi}^{\pi}\frac{|1-e^{i\lambda\tau}|^{2d}}{\lambda^{2d}}|f_{kk}(\lambda)-f_{kk}^1(\lambda)|d\lambda \le \delta_k,\ k=\overline{1,\infty}\right\},$$

$$\mathcal{D}_{1\delta}^4 = \left\{ f(\lambda) \left| \frac{1}{2\pi} \int_{-\pi}^{\pi} \frac{|1-e^{i\lambda\tau}|^{2d}}{\lambda^{2d}} |\langle B_2, f(\lambda) - f_1(\lambda) \rangle| d\lambda \leq \delta \right. \right\}.$$

Here $f_1(\lambda) = \{f_{ij}^1(\lambda)\}_{i,j=1}^{\infty}$ is a fixed spectral density, B_2 is a given positive-definite Hermitian matrix, and δ, δ_k, $k = \overline{1,\infty}$, and δ_i^j, $i,j = \overline{1,\infty}$, are given numbers.

From the condition $0 \in \partial\Delta_{\mathcal{D}}(f^0)$, we find the following equations, which determine the least favorable spectral densities for these given sets of admissible spectral densities.

For the set $\mathcal{D}_{1\delta}^1$ of admissible spectral densities, we have

$$\left(\sum_{j=0}^{\infty} ((\mathbf{F}_\tau^0)^{-1} D^\tau \mathbf{a})_j e^{i\lambda j} \right) \left(\sum_{j=0}^{\infty} ((\mathbf{F}_\tau^0)^{-1} D^\tau \mathbf{a})_j e^{i\lambda j} \right)^*$$
$$= \left(\frac{|1-e^{i\lambda\tau}|^{2d}}{\lambda^{2d}} f^0(\lambda) \right) \{\beta_{ij}\gamma_{ij}(\lambda)\}_{i,j=1}^{\infty} \left(\frac{|1-e^{i\lambda\tau}|^{2d}}{\lambda^{2d}} f^0(\lambda) \right), \tag{7.29}$$

$$\frac{1}{2\pi} \int_{-\pi}^{\pi} \frac{|1-e^{i\lambda\tau}|^{2d}}{\lambda^{2d}} |f_{ij}^0(\lambda) - f_{ij}^1(\lambda)| d\lambda = \delta_i^j, \tag{7.30}$$

where β_{ij} are Lagrange multipliers, functions $|\gamma_{ij}(\lambda)| \leq 1$, and

$$\gamma_{ij}(\lambda) = \frac{f_{ij}^0(\lambda) - f_{ij}^1(\lambda)}{|f_{ij}^0(\lambda) - f_{ij}^1(\lambda)|} : f_{ij}^0(\lambda) - f_{ij}^1(\lambda) \neq 0, \quad i,j = \overline{1,\infty}.$$

For the set $\mathcal{D}_{1\delta}^2$ of admissible spectral densities, we have

$$\left(\sum_{j=0}^{\infty} ((\mathbf{F}_\tau^0)^{-1} D^\tau \mathbf{a})_j e^{i\lambda j} \right) \left(\sum_{j=0}^{\infty} ((\mathbf{F}_\tau^0)^{-1} D^\tau \mathbf{a})_j e^{i\lambda j} \right)^*$$
$$= \beta^2 \gamma_2(\lambda) \left(\frac{|1-e^{i\lambda\tau}|^{2d}}{\lambda^{2d}} f^0(\lambda) \right)^2, \tag{7.31}$$

$$\frac{1}{2\pi} \int_{-\pi}^{\pi} \frac{|1-e^{i\lambda\tau}|^{2d}}{\lambda^{2d}} |\mathrm{Tr}(f^0(\lambda) - f_1(\lambda))| d\lambda = \delta, \tag{7.32}$$

where β^2 are Lagrange multipliers, the function $|\gamma_2(\lambda)| \leq 1$, and

$$\gamma_2(\lambda) = \mathrm{sign}(\mathrm{Tr}(f^0(\lambda) - f_1(\lambda))) : \mathrm{Tr}(f^0(\lambda) - f_1(\lambda)) \neq 0.$$

For the set $\mathcal{D}_{1\delta}^3$ of admissible spectral densities, we have

$$\left(\sum_{j=0}^{\infty} ((\mathbf{F}_\tau^0)^{-1} D^\tau \mathbf{a})_j e^{i\lambda j} \right) \left(\sum_{j=0}^{\infty} ((\mathbf{F}_\tau^0)^{-1} D^\tau \mathbf{a})_j e^{i\lambda j} \right)^*$$
$$= \left(\frac{|1-e^{i\lambda\tau}|^{2d}}{\lambda^{2d}} f^0(\lambda) \right) \{\beta_k^2 \gamma_k^2(\lambda) \delta_{kl}\}_{k,l=1}^{\infty} \left(\frac{|1-e^{i\lambda\tau}|^{2d}}{\lambda^{2d}} f^0(\lambda) \right), \tag{7.33}$$

$$\frac{1}{2\pi}\int_{-\pi}^{\pi}\frac{|1-e^{i\lambda\tau}|^{2d}}{\lambda^{2d}}|f_{kk}^{0}(\lambda)-f_{kk}^{1}(\lambda)|d\lambda=\delta_{k}, \tag{7.34}$$

where β_k^2 are Lagrange multipliers, δ_{kl} are Kronecker symbols, functions $|\gamma_k^2(\lambda)| \leq 1$, and

$$\gamma_k^2(\lambda)=\operatorname{sign}(f_{kk}^{0}(\lambda)-f_{kk}^{1}(\lambda)):\ f_{kk}^{0}(\lambda)-f_{kk}^{1}(\lambda)\neq 0,\quad k=\overline{1,\infty}.$$

For the set $\mathcal{D}_{1\delta}^{4}$ of admissible spectral densities, we have

$$\begin{aligned}&\left(\sum_{j=0}^{\infty}((\mathbf{F}_{\tau}^{0})^{-1}D^{\tau}\mathbf{a})_{j}e^{i\lambda j}\right)\left(\sum_{j=0}^{\infty}((\mathbf{F}_{\tau}^{0})^{-1}D^{\tau}\mathbf{a})_{j}e^{i\lambda j}\right)^{*}\\&\quad=\beta^{2}\gamma_{2}'(\lambda)\left(\frac{|1-e^{i\lambda\tau}|^{2d}}{\lambda^{2d}}f^{0}(\lambda)\right)B_{2}^{\top}\left(\frac{|1-e^{i\lambda\tau}|^{2d}}{\lambda^{2d}}f^{0}(\lambda)\right),\end{aligned} \tag{7.35}$$

$$\frac{1}{2\pi}\int_{-\pi}^{\pi}\frac{|1-e^{i\lambda\tau}|^{2d}}{\lambda^{2d}}|\langle B_{2},f^{0}(\lambda)-f_{1}(\lambda)\rangle|d\lambda=\delta, \tag{7.36}$$

where β^2 are Lagrange multipliers, function $|\gamma_2'(\lambda)| \leq 1$, and

$$\gamma_2'(\lambda)=\operatorname{sign}\langle B_{2},f^{0}(\lambda)-f_{1}(\lambda)\rangle:\ \langle B_{2},f^{0}(\lambda)-f_{1}(\lambda)\rangle\neq 0.$$

The derived results are summarized in the following theorem.

Theorem 7.4.2. *The least favorable spectral density* $f^0(\lambda) = \{f_{ij}^0(\lambda)\}_{i,j=1}^{\infty}$ *in the classes* $\mathcal{D}_{1\delta}^{k}$, $k = 1,2,3,4$, *for the optimal linear extrapolation of the functional* $A\xi$ *from observations of the process* $\xi(t)$ *at points* $t < 0$ *is determined by equations* (7.29)–(7.30), (7.31)–(7.32), (7.33)–(7.34), *and* (7.35)–(7.36), *respectively, the minimality condition* (7.8), *the constrained optimization problem* (7.23), *and restrictions on the densities from the corresponding classes* $\mathcal{D}_{1\delta}^{k}$, $k = 1,2,3,4$. *The minimax-robust spectral characteristic of the optimal estimate of the functional* $A\xi$ *is determined by formula* (7.12).

Conclusions

In this chapter, we describe methods of solution of the extrapolation problem for linear functionals which are constructed from unobserved values of stochastic processes with periodically stationary increments. Estimates of functionals are based on observations of the considered process without noise. The problem is solved by representing these processes as infinite-dimensional sequences with stationary increments.

The Hilbert space projection method is applied to derive formulas for calculating the spectral characteristics and the MSEs of the optimal estimates of the functionals in the case of spectral certainty, where the spectral densities of the generated sequences are exactly known. In the case of spectral uncertainty, where the spectral density is not exactly known while, instead, a set of admissible spectral densities is specified, the minimax-robust method is applied. We propose a representation of the MSE in the form of a linear functional in L_1 space with respect to spectral densities, which allows us to solve the corresponding constrained optimization problem and describe the minimax (robust) estimates of the functionals.

This method is illustrated with the help of concrete sets $\mathcal{D}_0$ and $\mathcal{D}_{1\delta}$ of admissible spectral densities. Here D_0 are classes of spectral densities with the moment restrictions and $D_{1\delta}$ are classes describing the "δ-neighborhood" models in the space L_1 of a fixed bounded spectral density.

The main results presented in this chapter have been published in [122]. These results rely on and develop prior studies of the minimax extrapolation problem by Moklyachuk and Masyutka [154] for multidimensional stationary processes, Golichenko (Dubovets'ka) and Moklyachuk [42] for periodically correlated processes, and Luz and Moklyachuk [112] for processes with stationary increments.

8 Extrapolation of processes with periodically correlated increments observed with noise

In this chapter, we deal with the extrapolation problem for stochastic processes with periodically stationary dth increments observed with noise.

By the *extrapolation problem* we understand the problem of the mean square optimal linear estimation of the functionals

$$A\xi = \int_0^{\infty} a(t)\xi(t)dt, \quad A_{NT}\xi = \int_0^{(N+1)T} a(t)\xi(t)dt,$$

which depend on the unknown values of a stochastic process $\xi(t)$ with periodically correlated dth increments. Estimates are based on observations of the process $\zeta(t) = \xi(t) + \eta(t)$ at points $t < 0$, where $\eta(t)$ is an uncorrelated with the process $\xi(t)$ periodically stationary stochastic process.

8.1 Hilbert space projection method of extrapolation

Let a periodically correlated increment process $\xi^{(d)}(t, \tau T)$, $t \in \mathbb{R}$, generate by formula (6.7) an infinite-dimensional vector stationary increment sequence $\{\vec{\xi}_j^{(d)}, j \in \mathbb{Z}\}$ which has the spectral density matrix $f(\lambda) = \{f_{ij}(\lambda)\}_{i,j=1}^{\infty}$. As a noise process, consider a periodically stationary stochastic process $\eta(t)$, $t \in \mathbb{R}$, uncorrelated with the process $\xi(t)$. Let the process $\eta(t)$ generate by formula (6.3) an infinite-dimensional vector stationary sequence $\{\vec{\eta}_j, j \in \mathbb{Z}\}$ with the spectral density matrix $g(\lambda) = \{g_{ij}(\lambda)\}_{i,j=1}^{\infty}$.

In this section, we consider the extrapolation problem under the condition that the spectral densities $f(\lambda)$ and $g(\lambda)$ are exactly known.

8.1.1 Extrapolation based on Fourier transformations of spectral densities

We need the following assumptions:

- the mean values of the increment sequence $\vec{\xi}_j^{(d)}$ and stationary sequence $\vec{\eta}_j$ equal 0; the increment step $\tau > 0$;
- the spectral densities $f(\lambda)$ and $g(\lambda)$ satisfy the *minimality condition*

$$\int_{-\pi}^{\pi} \mathrm{Tr}\left[\frac{\lambda^{2d}}{|1 - e^{i\lambda\tau}|^{2d}} \left(f(\lambda) + \lambda^{2d} g(\lambda)\right)^{-1}\right] d\lambda < \infty. \tag{8.1}$$

https://doi.org/10.1515/9783111325620-008

Remark 8.1.1. The latter assumption is the necessary and sufficient condition under which the MSEs of the optimal estimates of the functional $A\vec{\xi}$ to be defined below are not equal to 0.

The Hilbert space projection method of estimation may be applied under the condition that the element which is estimated belongs to the Hilbert space $H = L_2(\Omega, \mathcal{F}, \mathrm{P})$ of random variables with a zero mean value and finite variance. That is not the case for the functional $A\xi$. Thus we represent the functional $A\xi$ as the sum of a functional with finite second moment from the space H and a functional dependent on the observed values of the process $\zeta(t) = \xi(t) + \eta(t)$. This representation is described by the following two lemmas.

Lemma 8.1.1 ([122]). *The linear functional*

$$A\zeta = \int_0^\infty a(t)\zeta(t)dt$$

allows the representation

$$A\zeta = B\zeta - V\zeta,$$

where

$$B\zeta = \int_0^\infty b^\tau(t)\zeta^{(d)}(t, \tau T)dt, \quad V\zeta = \int_{-\tau Td}^0 v^\tau(t)\zeta(t)dt,$$

and

$$v^\tau(t) = \sum_{l=\lceil -\frac{t}{\tau T}\rceil}^{d} (-1)^l \binom{d}{l} b^\tau(t + l\tau T), \quad t \in [-\tau Td; 0), \tag{8.2}$$

$$b^\tau(t) = \sum_{k=0}^\infty a(t + \tau Tk)d(k) = D^{\tau T}\mathbf{a}(t), \quad t \geq 0. \tag{8.3}$$

Here $\lceil x \rceil$ denotes the least integer greater than or equal to x, $[x]$ denotes the integer part of x, coefficients $\{d(k) : k \geq 0\}$ are determined by the relation

$$\sum_{k=0}^\infty d(k)x^k = \left(\sum_{j=0}^\infty x^j\right)^d,$$

and $D^{\tau T}$ is the linear transformation acting on an arbitrary function $x(t)$, $t \geq 0$, as follows:

$$D^{\tau T}\mathbf{x}(t) = \sum_{k=0}^\infty x(t + \tau Tk)d(k).$$

From Lemma 8.1.1, we obtain the following representation of the functional $A\xi$:

$$A\xi = A\zeta - A\eta = B\zeta - A\eta - V\zeta = H\xi - V\zeta,$$

where

$$H\xi := B\zeta - A\eta$$

and

$$A\zeta = \int_0^\infty a(t)\zeta(t)dt, \quad A\eta = \int_0^\infty a(t)\eta(t)dt,$$

$$B\zeta = \int_0^\infty b^\tau(t)\zeta^{(d)}(t,\tau T)dt, \quad V\zeta = \int_{-\tau Td}^0 v^\tau(t)\zeta(t)dt,$$

where the functions $b^\tau(t)$, $t \in [0;\infty)$, and $v^\tau(t)$, $t \in [-\tau Td; 0)$, are calculated by formulas (8.3) and (8.2), respectively. The functional $H\xi$ allows a representation in terms of the sequences $\vec{\eta}_j = (\eta_{kj}, k = 1,2,\dots)^\top$ and $\vec{\zeta}_j^{(d)} = \vec{\xi}_j^{(d)} + \vec{\eta}_j^{(d)} = (\zeta_{kj}^{(d)}, k = 1,2,\dots)^\top, j \in \mathbb{Z}$, which is described in the following lemma.

Lemma 8.1.2. *The functional $H\xi = B\zeta - A\eta$ can be represented in the form*

$$H\xi = \sum_{j=0}^\infty (\vec{b}_j^\tau)^\top \vec{\zeta}_j^{(d)} - \sum_{j=0}^\infty (\vec{a}_j)^\top \vec{\eta}_j = B\vec{\zeta} - V\vec{\eta} =: H\vec{\xi},$$

where for $k = 1,2,\dots$ and $j = 0,1,\dots,\infty$ a vector

$$\vec{b}_j^\tau = (b_{kj}^\tau, k = 1,2,\dots)^\top = (b_{1j}^\tau, b_{3j}^\tau, b_{2j}^\tau, \dots, b_{2k+1,j}^\tau, b_{2k,j}^\tau, \dots)^\top$$

has the entries

$$b_{kj}^\tau = \langle b_j^\tau, \widetilde{e}_k \rangle = \frac{1}{\sqrt{T}} \int_0^T b_j^\tau(v) e^{-2\pi i \{(-1)^k [\frac{k}{2}]\} v/T} dv$$

and a vector

$$\vec{a}_j = (a_{kj}, k = 1,2,\dots)^\top = (a_{1j}, a_{3j}, a_{2j}, \dots, a_{2k+1,j}, a_{2k,j}, \dots)^\top$$

has the entries

$$a_{kj} = \langle a_j, \widetilde{e}_k \rangle = \frac{1}{\sqrt{T}} \int_0^T a_j(v) e^{-2\pi i \{(-1)^k [\frac{k}{2}]\} v/T} dv.$$

The coefficients $\{\vec{a}_j, j = 0, 1, \ldots, \infty\}$ *and* $\{\vec{b}_j^\tau, j = 0, 1, \ldots, \infty\}$ *are related as*

$$\vec{b}_j^\tau = \sum_{m=j}^{\infty} \mathrm{diag}_\infty(d_\tau(m-j))\vec{a}_m = (D^\tau \mathbf{a})_j, \quad j = 0, 1, \ldots, \infty, \tag{8.4}$$

where $\mathbf{a} = ((\vec{a}_0)^\top, (\vec{a}_1)^\top, \ldots)^\top$, *the coefficients* $\{d_\tau(k) : k \geq 0\}$ *are determined by the relationship*

$$\sum_{k=0}^{\infty} d_\tau(k)x^k = \left(\sum_{j=0}^{\infty} x^{\tau j}\right)^d,$$

and D^τ *is a linear transformation determined by a matrix with the infinite-dimensional matrix entries* $D^\tau(k,j)$, $k, j = 0, 1, \ldots$, *such that* $D^\tau(k,j) = \mathrm{diag}_\infty(d_\tau(j-k))$ *if* $0 \leq k \leq j \leq \infty$ *and* $D^\tau(k,j) = \mathcal{O}$ *if* $0 \leq j < k \leq \infty$; $\mathrm{diag}_\infty(x)$ *denotes an infinite-dimensional diagonal matrix with the entry* x *on its diagonal.*

Proof. Define $b_j^\tau(u) = b^\tau(u+jT)$, $\zeta_j^{(d)}(u) = \zeta_j^{(d)}(u+jT, \tau T)$, $u \in [0, T)$, and $a_j(u) = a(u+jT)$, $\eta_j(u) = \eta_j(u + jT)$, $u \in [0, T)$.

Making use of decomposition (6.6) for the increment sequence $\{\zeta_j^{(d)}, j \in \mathbb{Z}\}$ and the solution of equation

$$(-1)^k[k/2] + (-1)^m[m/2] = 0 \tag{8.5}$$

of two variables (k, m), which is given by pairs $(1, 1)$, $(2l + 1, 2l)$, and $(2l, 2l + 1)$ for $l = 2, 3, \ldots$, rewrite the functional $B\zeta$ as [147]

$$\begin{aligned}
B\zeta &= \int_0^\infty b^\tau(t)\zeta^{(d)}(t, \tau T)dt = \sum_{j=0}^{\infty} \int_0^T b_j^\tau(u)\zeta_j^{(d)}(u)du \\
&= \sum_{j=0}^{\infty} \frac{1}{T} \int_0^T \left(\sum_{k=1}^{\infty} b_{kj}^\tau e^{2\pi i\{(-1)^k[\frac{k}{2}]\}u/T}\right)\left(\sum_{m=1}^{\infty} \zeta_{mj}^{(d)} e^{2\pi i\{(-1)^m[\frac{m}{2}]\}u/T}\right)du \\
&= \sum_{j=0}^{\infty}\sum_{k=1}^{\infty}\sum_{m=1}^{\infty} b_{kj}^\tau \zeta_{mj}^{(d)} \frac{1}{T} \int_0^T e^{2\pi i\{(-1)^k[\frac{k}{2}]+(-1)^m[\frac{m}{2}]\}u/T} du \\
&= \sum_{j=0}^{\infty}\sum_{k=1}^{\infty} b_{kj}^\tau \zeta_{kj}^{(d)} = \sum_{j=0}^{\infty} (\vec{b}_j^\tau)^\top \vec{\zeta}_j^{(d)} = B\vec{\zeta}.
\end{aligned}$$

The representation of the functional $V\eta$ is obtained in the same way:

$$V\eta = \int_0^\infty a(t)\eta(t)dt = \sum_{j=0}^{\infty} \int_0^T a_j(u)\eta_j(u)du$$

$$= \sum_{j=0}^{\infty}\sum_{k=1}^{\infty} a_{kj}\eta_{kj} = \sum_{j=0}^{\infty} (\vec{a}_j)^{\top}\vec{\eta}_j = V\vec{\eta}.$$

From Lemma 8.1.1 we obtain

$$b_j^{\tau}(u) = \sum_{l=0}^{\infty} a(u + jT + \tau Tl)d(l) = D^{\tau T}a(u), \quad u \in [0;T), \ j = 0,1,\dots,$$

and $b_{kj}^{\tau} = \sum_{l=0}^{\infty} a_{kj+\tau l}d(l), j = 0,1,\dots$, which finalizes the proof. □

Assume that coefficients $\{\vec{a}_j, j = 0,1,\dots\}$ and $\{\vec{b}_j^{\tau}, j = 0,1,\dots\}$, which determine the functional $H\vec{\xi}$, satisfy the conditions

$$\sum_{j=0}^{\infty} \|\vec{a}_j\| < \infty, \quad \sum_{j=0}^{\infty}(j+1)\|\vec{a}_j\|^2 < \infty, \quad \|\vec{a}_j\|^2 = \sum_{k=1}^{\infty} |a_{kj}|^2, \tag{8.6}$$

$$\sum_{j=0}^{\infty} \|\vec{b}_j^{\tau}\| < \infty, \quad \sum_{j=0}^{\infty}(j+1)\|\vec{b}_j^{\tau}\|^2 < \infty, \quad \|\vec{b}_j^{\tau}\|^2 = \sum_{k=1}^{\infty} |b_{kj}^{\tau}|^2. \tag{8.7}$$

Under conditions (8.6)–(8.7) the functional $H\vec{\xi}$ has finite second moment. Since the functional $V\zeta$ depends on the observations $\{\xi(t) + \eta(t) : t \in [\tau Td; 0)\}$, the estimates $\widehat{A}\xi$ and $\widehat{H}\vec{\xi}$ of the functionals $A\xi$ and $H\vec{\xi}$ and the MSEs $\Delta(f,g;\widehat{A}\xi) = \mathsf{E}|A\xi - \widehat{A}\xi|^2$ and $\Delta(f,g;\widehat{H}\vec{\xi}) = \Delta(f,g;\widehat{H}\vec{\xi}) = \mathsf{E}|H\vec{\xi} - \widehat{H}\vec{\xi}|^2$ satisfy

$$\begin{aligned} \widehat{A}\xi &= \widehat{H}\vec{\xi} - V\zeta, \\ \Delta(f,g;\widehat{A}\xi) &= \mathsf{E}|A\xi - \widehat{A}\xi|^2 = \mathsf{E}|H\vec{\xi} - V\zeta - \widehat{H}\vec{\xi} + V\zeta|^2 \\ &= \mathsf{E}|H\vec{\xi} - \widehat{H}\vec{\xi}|^2 = \Delta(f,g;\widehat{H}\vec{\xi}). \end{aligned} \tag{8.8}$$

Thus, the functional $H\vec{\xi}$ is a target element to be estimated. Making use of representations (6.11) and (6.10), we obtain its spectral representation:

$$H\vec{\xi} = \int_{-\pi}^{\pi} (\vec{B}_{\tau}(e^{i\lambda}))^{\top} \frac{(1-e^{-i\lambda\tau})^d}{(i\lambda)^d} d\vec{Z}_{\xi^{(d)}+\eta^{(d)}}(\lambda) - \int_{-\pi}^{\pi} (\vec{A}(e^{i\lambda}))^{\top} d\vec{Z}_{\eta}(\lambda),$$

where

$$\vec{B}_{\tau}(e^{i\lambda}) = \sum_{j=0}^{\infty} \vec{b}_j^{\tau} e^{i\lambda j} = \sum_{j=0}^{\infty} (D^{\tau}\mathbf{a})_j e^{i\lambda j}, \quad \vec{A}(e^{i\lambda}) = \sum_{j=0}^{\infty} \vec{a}_j e^{i\lambda j}.$$

The classical approach of estimation consists in finding a projection of the element $H\vec{\xi}$ on the closed linear subspace of $H = L_2(\Omega, \mathcal{F}, \mathsf{P})$ generated by the observations. Let us define this subspace as

$$H^{-1}(\vec{\xi}^{(d)} + \vec{\eta}^{(d)}) = \overline{\text{span}}\{\vec{\xi}_{kj}^{(d)} + \vec{\eta}_{kj}^{(d)} : k = 1,\dots,\infty; j = -1,-2,-3,\dots\}.$$

Define also the closed linear subspaces of the Hilbert space $L_2(f(\lambda)+\lambda^{2d}g(\lambda))$ of vector-valued functions endowed with the inner product

$$\langle g_1; g_2\rangle = \int_{-\pi}^{\pi} (g_1(\lambda))^\top (f(\lambda)+\lambda^{2d}g(\lambda))\overline{g_2(\lambda)}d\lambda$$

as

$$L_2^{-1}(f(\lambda)+\lambda^{2d}g(\lambda)) = \overline{\text{span}}\left\{e^{i\lambda j}(1-e^{-i\lambda\tau})^d \frac{1}{(i\lambda)^d}\vec{\delta}_k, k=1,2,3,\dots; j=-1,-2,-3,\dots\right\},$$

where $\vec{\delta}_k = \{\delta_{kl}\}_{l=1}^{\infty}$, where δ_{kl} are Kronecker symbols.

Remark 8.1.2. Representation (6.11) yields a map between the elements $e^{i\lambda j}(1-e^{-i\lambda\tau})^d(i\lambda)^{-d}\vec{\delta}_k$ of the space $L_2^{-1}(f(\lambda)+\lambda^{2d}g(\lambda))$ and the elements $\vec{\xi}_{kj}^{(d)}+\vec{\eta}_{kj}^{(d)}$ of the space $H^{-1}(\vec{\xi}^{(d)}+\vec{\eta}^{(d)})$.

The mean square optimal estimate $\widehat{H}\vec{\xi}$ is found as an orthogonal projection of the element $H\vec{\xi}$ on the subspace $H^{-1}(\vec{\xi}^{(d)}+\vec{\eta}^{(d)})$:

$$\widehat{H}\vec{\xi} = \text{Proj}_{H^{-1}(\vec{\xi}^{(d)}+\vec{\eta}^{(d)})} H\vec{\xi}.$$

Relation (8.8) lets us write the optimal estimate $\widehat{A}\xi$ in the form

$$\widehat{A}\xi = \text{Proj}_{H^{-1}(\vec{\xi}^{(d)}+\vec{\eta}^{(d)})} H\vec{\xi} - V\zeta$$

or in the form

$$\widehat{A}\xi = \int_{-\pi}^{\pi} (\vec{h}_\tau(\lambda))^\top d\vec{Z}_{\xi^{(d)}+\eta^{(d)}}(\lambda) - \int_{-\tau Td}^{0} v^\tau(t)\zeta(t)dt, \tag{8.9}$$

where $\vec{h}_\tau(\lambda) = \{h_k^\tau(\lambda)\}_{k=1}^{\infty}$ is the spectral characteristic of the estimate $\widehat{H}\vec{\xi}$.

Denote

$$\vec{A}_\tau(e^{i\lambda}) = (1-e^{i\lambda\tau})^d \vec{A}(e^{i\lambda}) = \sum_{j=0}^{\infty} \vec{a}_j^\tau e^{i\lambda j},$$

where

$$\vec{a}_j^\tau = \sum_{l=0}^{\min\{[j/\tau],d\}} (-1)^l \binom{d}{l} \vec{a}(j-\tau l), \quad j \geq 0. \tag{8.10}$$

Define the vector

$$\mathbf{a}^\tau = ((\vec{a}_0^\tau)^\top, (\vec{a}_1^\tau)^\top, (\vec{a}_2^\tau)^\top, \dots)^\top.$$

With the help of the Fourier coefficients

$$T^{\tau}_{l,j} = \frac{1}{2\pi}\int_{-\pi}^{\pi} e^{i\lambda(j-l)} \frac{\lambda^{2d}}{|1-e^{i\lambda\tau}|^{2d}} \left(g(\lambda)(f(\lambda)+\lambda^{2d}g(\lambda))^{-1}\right)^{\top} d\lambda, \quad l,j \geq 0,$$

$$P^{\tau}_{l,j} = \frac{1}{2\pi}\int_{-\pi}^{\pi} e^{i\lambda(j-l)} \frac{\lambda^{2d}}{|1-e^{i\lambda\tau}|^{2d}} \left((f(\lambda)+\lambda^{2d}g(\lambda))^{-1}\right)^{\top} d\lambda, \quad l,j \geq 0,$$

$$Q_{l,j} = \frac{1}{2\pi}\int_{-\pi}^{\pi} e^{i\lambda(j-l)} \left(f(\lambda)(f(\lambda)+\lambda^{2d}g(\lambda))^{-1}g(\lambda)\right)^{\top} d\lambda, \quad l,j \geq 0,$$

of the corresponding matrix functions, define the linear operators $\mathbf{P}_{\tau}$, $\mathbf{T}_{\tau}$, and $\mathbf{Q}$ in the space ℓ_2 by matrices with the infinite-dimensional matrix entries $(\mathbf{P}_{\tau})_{l,j} = P^{\tau}_{l,j}$, $(\mathbf{T}_{\tau})_{l,j} = T^{\tau}_{l,j}$, and $(\mathbf{Q})_{l,j} = Q_{l,j}$, $l,j \geq 0$.

We use the following notation: $\langle \mathbf{x},\mathbf{y}\rangle = \sum_{j=0}^{\infty}(\vec{x}_j)^{\top}\overline{\vec{y}}_j$ for vectors $\mathbf{x} = ((\vec{x}_0)^{\top},(\vec{x}_1)^{\top}, (\vec{x}_2)^{\top},\dots)^{\top}$ and $\mathbf{y} = ((\vec{y}_0)^{\top},(\vec{y}_1)^{\top},(\vec{y}_2)^{\top},\dots)^{\top}$.

Theorem 8.1.1. *Consider two uncorrelated processes: a stochastic process $\xi(t)$, $t \in \mathbb{R}$, with periodically stationary increments, which determines a generated stationary dth increment sequence $\vec{\xi}^{(d)}_j$ with the spectral density matrix $f(\lambda) = \{f_{kn}(\lambda)\}_{k,n=1}^{\infty}$, and a periodically stationary stochastic process $\eta(t)$, $t \in \mathbb{R}$, which determines a generated stationary sequence $\vec{\eta}_j$ with the spectral density matrix $g(\lambda) = \{g_{kn}(\lambda)\}_{k,n=1}^{\infty}$. Let the coefficients $\vec{a}_j$, $\vec{b}^{\tau}_j$, $j = 0,1,\dots$, generated by the function $a(t)$, $t \geq 0$, satisfy conditions* (8.6)–(8.7). *Let the minimality condition* (8.1) *be satisfied. The optimal linear estimate $\widehat{A}\xi$ of the functional $A\xi$ based on observations of the process $\xi(t)+\eta(t)$ at points $t < 0$ is calculated by formula* (8.9). *The spectral characteristic $\vec{h}_{\tau}(\lambda)$ is calculated by formula*

$$\begin{aligned}(\vec{h}_{\tau}(\lambda))^{\top} &= (\vec{B}_{\tau}(e^{i\lambda}))^{\top}\frac{(1-e^{-i\lambda\tau})^d}{(i\lambda)^d} \\ &\quad - ((\vec{A}_{\tau}(e^{i\lambda}))^{\top}g(\lambda) + (\vec{C}_{\tau}(e^{i\lambda}))^{\top})\frac{(-i\lambda)^d}{(1-e^{i\lambda\tau})^d}(f(\lambda)+\lambda^{2d}g(\lambda))^{-1}, \end{aligned} \tag{8.11}$$

where

$$\vec{C}_{\tau}(e^{i\lambda}) = \sum_{j=0}^{\infty}(\mathbf{P}_{\tau}^{-1}D^{\tau}\mathbf{a} - \mathbf{P}_{\tau}^{-1}\mathbf{T}_{\tau}\mathbf{a}^{\tau})_j e^{i\lambda j}.$$

The value of the MSE is calculated by the formula

$$\Delta(f,g;\widehat{A}\xi) = \langle D^{\tau}\mathbf{a} - \mathbf{T}_{\tau}\mathbf{a}^{\tau}, \mathbf{P}_{\tau}^{-1}D^{\tau}\mathbf{a} - \mathbf{P}_{\tau}^{-1}\mathbf{T}_{\tau}\mathbf{a}^{\tau}\rangle + \langle \mathbf{Q}\mathbf{a},\mathbf{a}\rangle. \tag{8.12}$$

Proof. A projection $\widehat{H}\vec{\xi}$ of the element $H\vec{\xi}$ on the subspace $H^{0-}(\xi^{(d)}_{\tau} + \eta^{(d)}_{\tau})$ is characterized by two conditions:

(1) $\widehat{H}\vec{\xi} \in H^{-1}(\vec{\xi}^{(d)} + \vec{\eta}^{(d)})$;
(2) $(H\vec{\xi} - \widehat{H}\vec{\xi}) \perp H^{-1}(\vec{\xi}^{(d)} + \vec{\eta}^{(d)})$.

The second condition implies the following relation, which holds for all $j \leq -1$ and for all $k \geq 1$:

$$\int_{-\pi}^{\pi} \left((\vec{B}_\tau(e^{i\lambda}))^\top \frac{(1-e^{-i\lambda\tau})^d}{(i\lambda)^d} (f(\lambda) + \lambda^{2d} g(\lambda)) - (\vec{h}_\tau(\lambda))^\top (f(\lambda) + \lambda^{2d} g(\lambda)) \right.$$
$$\left. - (\vec{A}(e^{i\lambda}))^\top g(\lambda)(-i\lambda)^d \right) \vec{\delta}_k \frac{(1-e^{i\lambda\tau})^d}{(-i\lambda)^d} e^{-i\lambda j} d\lambda = 0.$$

From these relations, we conclude that the spectral characteristic $\vec{h}_\tau(\lambda)$ of the estimate $\widehat{H}\vec{\xi}$ allows a representation in the form (8.11), where

$$\vec{C}_\tau(e^{i\lambda}) = \sum_{j=0}^{\infty} \vec{c}_j^{\,\tau} e^{i\lambda j},$$

where $\vec{c}_j^{\,\tau} = \{c_{kj}^\tau\}_{k=1}^\infty, j = 0, 1, 2, \ldots$, are unknown coefficients to be found.

Condition (1) implies $(i\lambda)^d(1-e^{-i\lambda\tau})^{-d}\vec{h}_\tau(\lambda) \in L_2^{0-}$, and thus,

$$\int_{-\pi}^{\pi} \left((\vec{B}_\tau(e^{i\lambda}))^\top - (\vec{A}_\tau(e^{i\lambda}))^\top g(\lambda) \frac{\lambda^{2d}}{|1-e^{i\lambda\tau}|^{2d}} (f(\lambda) + \lambda^{2d} g(\lambda))^{-1} \right.$$
$$\left. - (\vec{C}_\tau(e^{i\lambda}))^\top \frac{\lambda^{2d}}{|1-e^{i\lambda\tau}|^{2d}} (f(\lambda) + \lambda^{2d} g(\lambda))^{-1} \right) e^{-i\lambda j} d\lambda = \vec{0}, \quad j \geq 0,$$

which can be presented as a system of linear equations

$$\vec{b}_l^\tau - \sum_{j=0}^{\infty} T_{l,j}^\tau \vec{a}_j^\tau = \sum_{j=0}^{\infty} P_{l,j}^\tau \vec{c}_j^{\,\tau}, \quad l \geq 0, \tag{8.13}$$

determining the unknown coefficients $\vec{c}_j^{\,\tau}, j \geq 0$. We rewrite the system of equations (8.13) in matrix form as

$$D^\tau \mathbf{a} - \mathbf{T}_\tau \mathbf{a}^\tau = \mathbf{P}_\tau \mathbf{c}^\tau,$$

where $\mathbf{c}^\tau = ((\vec{c}_0^{\,\tau})^\top, (\vec{c}_1^{\,\tau})^\top, (\vec{c}_2^{\,\tau})^\top, \ldots)^\top$. Consequently, the unknown coefficients $\vec{c}_j^{\,\tau}, j \geq 0$, determining the spectral characteristic $\vec{h}_\tau(\lambda)$ are as follows:

$$\vec{c}_j^{\,\tau} = (\mathbf{P}_\tau^{-1} D^\tau \mathbf{a} - \mathbf{P}_\tau^{-1} \mathbf{T}_\tau \mathbf{a}^\tau)_j, \quad j \geq 0,$$

where $(\mathbf{P}_\tau^{-1}D^\tau\mathbf{a}-\mathbf{P}_\tau^{-1}\mathbf{T}_\tau\mathbf{a}^\tau)_j, j \geq 0$, is the jth infinite-dimensional vector entry of the vector $\mathbf{P}_\tau^{-1}D^\tau\mathbf{a} - \mathbf{P}_\tau^{-1}\mathbf{T}_\tau\mathbf{a}^\tau$. The existence of the inverse matrix $(\mathbf{P}^\tau)^{-1}$ is justified in [15] under condition (8.1). Thus,

$$\vec{C}_\tau(e^{i\lambda}) = \sum_{j=0}^{\infty}(\mathbf{P}_\tau^{-1}D^\tau\mathbf{a} - \mathbf{P}_\tau^{-1}\mathbf{T}_\tau\mathbf{a}^\tau)_j e^{i\lambda j}$$

and the spectral characteristic $\vec{h}_\tau(\lambda)$ is calculated by the formula

$$\begin{aligned}(\vec{h}_\tau(\lambda))^\top &= (\vec{B}_\tau(e^{i\lambda}))^\top \frac{(1-e^{-i\lambda\tau})^d}{(i\lambda)^d} \\ &\quad - (\vec{A}_\tau(e^{i\lambda}))^\top g(\lambda)\frac{(-i\lambda)^d}{(1-e^{i\lambda\tau})^d}(f(\lambda)+\lambda^{2d}g(\lambda))^{-1} \\ &\quad - \frac{(-i\lambda)^d}{(1-e^{i\lambda\tau})^d}\left(\sum_{k=0}^{\infty}(\mathbf{P}_\tau^{-1}D^\tau\mathbf{a} - \mathbf{P}_\tau^{-1}\mathbf{T}_\tau\mathbf{a}^\tau)_k e^{i\lambda j}\right)^\top (f(\lambda)+\lambda^{2d}g(\lambda))^{-1}.\end{aligned}$$

The MSE of the estimate $\widehat{A}\vec{\xi}$ is calculated by the formula

$$\begin{aligned}\Delta(f,g;\widehat{A}\vec{\xi}) &= \Delta(f,g;\widehat{H}\vec{\xi}) = \mathsf{E}|H\vec{\xi}-\widehat{H}\vec{\xi}|^2 \\ &= \frac{1}{2\pi}\int_{-\pi}^{\pi}\frac{\lambda^{2d}}{|1-e^{i\lambda\tau}|^{2d}}((\vec{A}_\tau(e^{i\lambda}))^\top g(\lambda) + (\vec{C}_\tau(e^{i\lambda}))^\top) \\ &\quad\times (f(\lambda)+\lambda^{2d}g(\lambda))^{-1} f(\lambda)\,(f(\lambda)+\lambda^{2d}g(\lambda))^{-1} \\ &\quad\times (g(\lambda)\overline{\vec{A}_\tau(e^{i\lambda})} + \overline{\vec{C}_\tau(e^{i\lambda})})d\lambda \\ &\quad+ \frac{1}{2\pi}\int_{-\pi}^{\pi}\frac{\lambda^{4d}}{|1-e^{i\lambda\tau}|^{4d}}\left(\frac{|1-e^{i\lambda\tau}|^{2d}}{\lambda^{2d}}(\vec{A}(e^{i\lambda}))^\top f(\lambda) - (1-e^{-i\lambda\tau})^d(\vec{C}_\tau(e^{i\lambda}))^\top\right) \\ &\quad\times (f(\lambda)+\lambda^{2d}g(\lambda))^{-1} g(\lambda)\,(f(\lambda)+\lambda^{2d}g(\lambda))^{-1} \\ &\quad\times \left(\frac{|1-e^{i\lambda\tau}|^{2d}}{\lambda^{2d}}f(\lambda)\overline{\vec{A}(e^{i\lambda})} - (1-e^{i\lambda\tau})^d\overline{\vec{C}_\tau(e^{i\lambda})}\right)d\lambda \\ &= \langle D^\tau\mathbf{a} - \mathbf{T}_\tau\mathbf{a}^\tau, \mathbf{P}_\tau^{-1}D^\tau\mathbf{a} - \mathbf{P}_\tau^{-1}\mathbf{T}_\tau\mathbf{a}^\tau\rangle + \langle\mathbf{Q}\mathbf{a},\mathbf{a}\rangle.\end{aligned}$$

□

Corollary 8.1.1. *The spectral characteristics $\vec{h}_\tau^1(\lambda)$ and $\vec{h}_\tau^2(\lambda)$ of the optimal estimates $\widehat{B}\vec{\zeta}$ and $\widehat{A}\vec{\eta}$ of the functionals $B\vec{\zeta}$ and $A\vec{\eta}$ based on observations $\xi(t)+\eta(t)$ at points $t<0$ are calculated by the formulas*

$$\begin{aligned}(\vec{h}_\tau^1(\lambda))^\top &= (\vec{B}_\tau(e^{i\lambda}))^\top \frac{(1-e^{-i\lambda\tau})^d}{(i\lambda)^d} \\ &\quad - \frac{(-i\lambda)^d}{(1-e^{i\lambda\tau})^d}\left(\sum_{j=0}^{\infty}(\mathbf{P}_\tau^{-1}D^\tau\mathbf{a})_j e^{i\lambda j}\right)^\top (f(\lambda)+\lambda^{2d}g(\lambda))^{-1},\end{aligned}$$

$$(\vec{h}_\tau^2(\lambda))^\top = (\vec{A}_\tau(e^{i\lambda}))^\top g(\lambda)\frac{(-i\lambda)^d}{(1-e^{i\lambda\tau})^d}(f(\lambda)+\lambda^{2d}g(\lambda))^{-1}$$
$$-\frac{(-i\lambda)^d}{(1-e^{i\lambda\tau})^d}\left(\sum_{j=0}^{\infty}(\mathbf{P}_\tau^{-1}\mathbf{T}_\tau\mathbf{a}^\tau)_j e^{i\lambda j}\right)^\top (f(\lambda)+\lambda^{2d}g(\lambda))^{-1}.$$

The optimal estimate $\widehat{A}_{NT}\xi$ of the functional $A_{NT}\xi$, which depends on the unknown values of the process $\xi(t)$ at points $t \in [0, T(N+1)]$, based on observations of the process $\xi(t)+\eta(t)$ at points $t<0$ can be obtained by using Theorem 8.1.1.

We first formulate the following corollaries from Lemmas 8.1.1 and 8.1.2.

Corollary 8.1.2. *The linear functional*

$$A_{NT}\xi = \int_0^{(N+1)T} a(t)\xi(t)dt$$

allows the representation

$$A_{NT}\xi = B_{NT}\xi - V_{NT}\xi,$$

where

$$B_{NT}\xi = \int_0^{(N+1)T} b^{\tau,N}(t)\xi^{(d)}(t,\tau T)dt, \quad V_{NT}\xi = \int_{-\tau Td}^{0} v^{\tau,N}(t)\xi(t)dt,$$

and

$$v^{\tau,N}(t) = \sum_{l=\lceil -\frac{t}{\tau T}\rceil}^{\min\{[\frac{(N+1)T-t}{\tau T}],d\}} (-1)^l\binom{d}{l}b^{\tau,N}(t+l\tau T), \quad t\in[-\tau Td;0), \tag{8.14}$$

$$b^{\tau,N}(t) = \sum_{l=0}^{[\frac{(N+1)T-t}{\tau T}]} a(t+\tau Tl)d(l) = D^{\tau T,N}\mathbf{a}(t), \quad t\in[0;(N+1)T]. \tag{8.15}$$

The linear transformation $D^{\tau T,N}$ acts on an arbitrary function $x(t)$, $t\in[0;(N+1)T]$, as follows:

$$D^{\tau T,N}\mathbf{x}(t) = \sum_{k=0}^{[\frac{(N+1)T-t}{\tau T}]} x(t+\tau Tk)d(k).$$

Corollary 8.1.3. *The functional*

$$H_{NT}\xi = B_{NT}\zeta - A_{NT}\eta,$$

where

$$B_{NT}\zeta = \int_0^{T(N+1)} b^{\tau,N}(t)\zeta^{(d)}(t,\tau T)dt, \quad A_{NT}\eta = \int_0^{T(N+1)} a(t)\eta(t)dt,$$

can be represented in the form

$$H_{NT}\xi = \sum_{j=0}^{N} (\vec{b}_j^{\tau,N})^\top \vec{\zeta}_j^{(d)} - \sum_{j=0}^{N} (\vec{a}_j)^\top \vec{\eta}_j = B_N\vec{\zeta} - V_N\vec{\eta} =: H_N\vec{\xi},$$

where the vector

$$\vec{b}_j^{\tau,N} = (b_{kj}^{\tau,N}, k = 1,2,\dots)^\top = (b_{1j}^{\tau,N}, b_{3j}^{\tau,N}, b_{2j}^{\tau,N}, \dots, b_{2k+1,j}^{\tau,N}, b_{2k,j}^{\tau,N}, \dots)^\top,$$

with the entries

$$b_{kj}^{\tau,N} = \langle b_j^{\tau,N}, \widetilde{e}_k\rangle = \frac{1}{\sqrt{T}}\int_0^T b_j^{\tau,N}(v)e^{-2\pi i\{(-1)^k[\frac{k}{2}]\}v/T}dv.$$

The coefficients $\{\vec{a}_j, j = 0,1,\dots,N\}$ *and* $\{\vec{b}_j^{\tau,N}, j = 0,1,\dots,N\}$ *are related as*

$$\vec{b}_j^{\tau,N} = \sum_{m=j}^{N} \mathrm{diag}_\infty(d_\tau(m-j))\vec{a}_m = (D_N^\tau \mathbf{a}_N)_j, \quad j = 0,1,\dots,N, \tag{8.16}$$

where $\mathbf{a}_N = ((\vec{a}_0)^\top, (\vec{a}_1)^\top, \dots, (\vec{a}_N)^\top, \vec{0}^\top, \dots)^\top$ *and* D_N^τ *is a linear transformation determined by a matrix with the infinite-dimensional matrix entries* $D_N^\tau(l,j)$, $l,j \geq 0$, *such that* $D_N^\tau(l,j) = \mathrm{diag}_\infty(d_\tau(j-l))$ *if* $0 \leq l \leq j \leq N$ *and* $D_N^\tau(l,j) = \mathcal{O}$ *for* $0 \leq j < l$ *or* $j,l > N$.

Set $a(t) = 0$ for $t > T(N+1)$. Define vector coefficients $\{\vec{a}_j^{\tau,N} : 0 \leq j \leq N + \tau d\}$ by the formula

$$\vec{a}_j^{\tau,N} = \sum_{l=\max\{\lceil\frac{j-N}{\tau}\rceil,0\}}^{\min\{[\frac{j}{\tau}],d\}} (-1)^l \binom{d}{l}\vec{a}(j-\tau l), \quad 0 \leq j \leq N + \tau d,$$

and a vector

$$\mathbf{a}_N^\tau = ((\vec{a}_0^{\tau,N})^\top, (\vec{a}_1^{\tau,N})^\top, (\vec{a}_2^{\tau,N})^\top, \dots, (\vec{a}_{N+\tau d}^{\tau,N})^\top, \vec{0}^\top, \dots)^\top.$$

The following theorem holds.

Theorem 8.1.2. *Consider two uncorrelated processes: a stochastic process* $\xi(t)$, $t \in \mathbb{R}$, *with periodically stationary increments, which determines a generated stationary dth increment sequence* $\vec{\xi}_j^{(d)}$ *with the spectral density matrix* $f(\lambda) = \{f_{kn}(\lambda)\}_{k,n=1}^\infty$, *and a period-*

ically stationary stochastic process $\eta(t)$, $t \in \mathbb{R}$, which determines a generated stationary sequence $\vec{\eta}_j$ with the spectral density matrix $g(\lambda) = \{g_{kn}(\lambda)\}_{k,n=1}^{\infty}$. Let the coefficients $\vec{a}_j$, $\vec{b}_j^{\tau}$, $j = 0, 1, \ldots, N$, generated by the function $a(t)$, $0 \le t \le T(N+1)$, satisfy conditions

$$\|\vec{a}_j\| < \infty, \quad \|\vec{a}_j\|^2 = \sum_{k=1}^{\infty} |a_{kj}|^2, \quad j = 0, 1, \ldots, N,$$

and

$$\|\vec{b}_j^{\tau,N}\| < \infty, \quad \|\vec{b}_j^{\tau,N}\|^2 = \sum_{k=1}^{\infty} |b_{kj}^{\tau,N}|^2, \quad j = 0, 1, \ldots, N.$$

Let the minimality condition (8.1) *be satisfied. The optimal linear estimate $\widehat{A}_{NT}\xi$ of the functional $A_{NT}\xi$ based on observations of the process $\xi(t)+\eta(t)$ at points $t < 0$ is calculated by formula*

$$\widehat{A}_{NT}\xi = \int_{-\pi}^{\pi} (\vec{h}_{\tau,N}(\lambda))^{\top} d\vec{Z}_{\xi^{(d)}+\eta^{(d)}}(\lambda) - \int_{-\tau Td}^{0} v_{\tau,N}(t)\zeta(t)dt,$$

where the spectral characteristic $\vec{h}_{\tau,N}(\lambda)$ of the optimal estimate $\widehat{A}_{NT}\xi$ is calculated by formula

$$\begin{aligned}
(\vec{h}_{\tau,N}(\lambda))^{\top} &= \frac{(1-e^{-i\lambda\tau})^d}{(i\lambda)^d}\left(\sum_{j=0}^{N}(D_N^{\tau}\mathbf{a}_N)_j e^{i\lambda j}\right)^{\top} \\
&\quad - \frac{(-i\lambda)^d}{(1-e^{i\lambda\tau})^d}\left(\sum_{j=0}^{N+\tau d}\vec{a}_j^{\tau,N}e^{i\lambda j}\right)^{\top} g(\lambda)(f(\lambda)+\lambda^{2d}g(\lambda))^{-1} \\
&\quad - \frac{(-i\lambda)^d}{(1-e^{i\lambda\tau})^d}\left(\sum_{j=0}^{\infty}(\mathbf{P}_{\tau}^{-1}D_N^{\tau}\mathbf{a}_N - \mathbf{P}_{\tau}^{-1}\mathbf{T}_{\tau}\mathbf{a}_{\tau,N})_j e^{i\lambda j}\right)^{\top}(f(\lambda)+\lambda^{2d}g(\lambda))^{-1}.
\end{aligned}$$

The MSE is calculated by the formula

$$\begin{aligned}
\Delta(f,g;\widehat{A}_{NT}\xi) &= \Delta(f,g;\widehat{H}_N\vec{\xi}) = \mathsf{E}|H_N\vec{\xi} - \widehat{H}_N\vec{\xi}|^2 \\
&= \langle D_N^{\tau}\mathbf{a}_N - \mathbf{T}_{\tau}\mathbf{a}_{\tau,N}, \mathbf{P}_{\tau}^{-1}D_N^{\tau}\mathbf{a}_N - \mathbf{P}_{\tau}^{-1}\mathbf{T}_{\tau}\mathbf{a}_{\tau,N}\rangle + \langle \mathbf{Q}_N\mathbf{a}_N, \mathbf{a}_N\rangle,
\end{aligned}$$

where $\mathbf{Q}_N$ is a linear operator in the space ℓ_2 defined by the matrix with the infinite-dimensional matrix entries $(\mathbf{Q}_N)_{l,j} = Q_{l,j}$, $0 \le l, j \le N$, and $(\mathbf{Q}_N)_{l,j} = \mathcal{O}$ otherwise.

8.1.2 Extrapolation based on factorizations of spectral densities

Assume that the following canonical factorizations take place:

$$\frac{|1-e^{i\lambda\tau}|^{2d}}{\lambda^{2d}}(f(\lambda)+\lambda^{2d}g(\lambda)) = \Theta_\tau(e^{-i\lambda})\Theta_\tau^*(e^{-i\lambda}), \quad \Theta_\tau(e^{-i\lambda}) = \sum_{k=0}^{\infty}\theta_\tau(k)e^{-i\lambda k}, \tag{8.17}$$

$$g(\lambda) = \sum_{k=-\infty}^{\infty} g(k)e^{i\lambda k} = \Phi(e^{-i\lambda})\Phi^*(e^{-i\lambda}), \quad \Phi(e^{-i\lambda}) = \sum_{k=0}^{\infty}\phi(k)e^{-i\lambda k}. \tag{8.18}$$

Define the matrix-valued function $\Psi_\tau(e^{-i\lambda}) = \{\Psi_{ij}(e^{-i\lambda})\}_{i=\overline{1,M}}^{j=\overline{1,\infty}}$ by the equation

$$\Psi_\tau(e^{-i\lambda})\Theta_\tau(e^{-i\lambda}) = E_M,$$

where E_M is an identity $M\times M$ matrix. Then the following factorization takes place:

$$\frac{\lambda^{2d}}{|1-e^{i\lambda\tau}|^{2d}}(f(\lambda)+\lambda^{2d}g(\lambda))^{-1} = \Psi_\tau^*(e^{-i\lambda})\Psi_\tau(e^{-i\lambda}), \quad \Psi_\tau(e^{-i\lambda}) = \sum_{k=0}^{\infty}\psi_\tau(k)e^{-i\lambda k} \tag{8.19}$$

Remark 8.1.3. Any spectral density matrix $f(\lambda)$ is self-adjoint: $f(\lambda) = f^*(\lambda)$. Thus, $(f(\lambda))^\top = \overline{f(\lambda)}$. One can check that an inverse spectral density $f^{-1}(\lambda)$ is also self-adjoint: $f^{-1}(\lambda) = (f^{-1}(\lambda))^*$ and $(f^{-1}(\lambda))^\top = \overline{f^{-1}(\lambda)}$.

Let $\mathcal{O}$ denote an $M\times\infty$ or $\infty\times M$ matrix with zero entries. The following lemmas provide factorizations of the operators $\mathbf{P}_\tau$ and $\mathbf{T}_\tau$, which contain coefficients of factorizations (8.17)–(8.19).

Lemma 8.1.3. *Let factorization* (8.17) *take place and let $M\times\infty$ matrix function $\Psi_\tau(e^{-i\lambda})$ satisfy equation $\Psi_\tau(e^{-i\lambda})\Theta_\tau(e^{-i\lambda}) = E_M$. Define the linear operators Ψ_τ and Θ_τ in the space ℓ_2 by the matrices with the matrix entries $(\Psi_\tau)_{k,j} = \psi_\tau(k-j)$, $(\Theta_\tau)_{k,j} = \theta_\tau(k-j)$ for $0\le j\le k$ and $(\Psi_\tau)_{k,j} = \mathcal{O}$, $(\Theta_\tau)_{k,j} = \mathcal{O}$ for $0\le k<j$. Then:*

(a) *the linear operator $\mathbf{P}_\tau$ admits the factorization*

$$\mathbf{P}_\tau = (\Psi_\tau)^\top\overline{\Psi}_\tau;$$

(b) *the inverse operator $(\mathbf{P}_\tau)^{-1}$ admits the factorization*

$$(\mathbf{P}_\tau)^{-1} = \overline{\Theta}_\tau(\Theta_\tau)^\top.$$

Lemma 8.1.4. *Let factorizations* (8.17) *and* (8.18) *take place. Then the operator $\mathbf{T}_\tau$ admits the representation*

$$\mathbf{T}_\tau = (\Psi_\tau)^\top\mathbf{Z}_\tau,$$

where $\mathbf{Z}_\tau$ is a linear operator in the space ℓ_2 defined by a matrix with the entries

$$(\mathbf{Z}_\tau)_{k,j} = \sum_{l=j}^{\infty} \overline{\psi}_\tau(l-j)\overline{g}(l-k), \quad k,j \geq 0,$$

$$g(k) = \sum_{m=\max\{0,-k\}}^{\infty} \phi(m)\phi^*(k+m), \quad k \in \mathbb{Z}.$$

The proofs of Lemmas 8.1.3 and 8.1.4, as well as the justification of the following representations of the spectral characteristics $\vec{h}^1_\tau(\lambda)$ and $\vec{h}^2_\tau(\lambda)$, correspond to the ones in Section 3.1.3 for the finite-dimensional vector stationary GM increment sequences or in [124] for the finite-dimensional vector stationary increment sequences.

Remark 8.1.4. Lemmas 8.1.3 and 8.1.4 imply the factorization

$$(\mathbf{P}_\tau)^{-1}\mathbf{T}_\tau\mathbf{a}^\tau = \overline{\Theta}_\tau(\Theta_\tau)^\top(\Psi_\tau)^\top\mathbf{Z}_\tau\mathbf{a}^\tau = \overline{\Theta}_\tau\mathbf{Z}_\tau\mathbf{a}^\tau = \overline{\Theta}_\tau\mathbf{e}_\tau,$$

where $\mathbf{e}_\tau := \mathbf{Z}_\tau\mathbf{a}^\tau$.

The spectral characteristic $\vec{h}^2_\tau(\lambda)$ of the optimal estimate $\widehat{A}\vec{\eta}$ from Corollary 8.1.1 can be presented as

$$\vec{h}^2_\tau(\lambda) = \frac{(1-e^{-i\lambda\tau})^d}{(i\lambda)^d}\Psi^\top_\tau(e^{-i\lambda})\vec{C}_{\tau,g}(e^{-i\lambda}),$$

where $\overline{\psi}_\tau = (\overline{\psi}_\tau(0), \overline{\psi}_\tau(1), \overline{\psi}_\tau(2), \dots)$,

$$\vec{C}_{\tau,g}(e^{-i\lambda}) = \sum_{m=1}^{\infty}(\overline{\psi}_\tau\mathbf{C}_{\tau,g})_m e^{-i\lambda m},$$

$$(\overline{\psi}_\tau\mathbf{C}_{\tau,g})_m = \sum_{k=0}^{\infty}\overline{\psi}_\tau(k)\mathbf{c}_{\tau,g}(k+m),$$

$$\begin{aligned}\mathbf{c}_{\tau,g}(m) &= \sum_{k=0}^{\infty}\overline{g}(m+k)\vec{a}^\tau_k = \sum_{l=0}^{\infty}\overline{\phi}(l)\sum_{k=0}^{\infty}\phi^\top(l+m+k)\vec{a}^\tau_k \\ &= \sum_{l=0}^{\infty}\overline{\phi}(l)(\widetilde{\Phi}\mathbf{a}^\tau)_{l+m},\end{aligned}$$

$$(\widetilde{\Phi}\mathbf{a}^\tau)_m = \sum_{k=0}^{\infty}\phi^\top(m+k)\vec{a}^\tau_k.$$

The spectral characteristic $\vec{h}^1_\tau(\lambda)$ of the optimal estimate $\widehat{B}\vec{\xi}$ from Corollary 8.1.1 can be presented as

$$\begin{aligned}\vec{h}^1_\tau(\lambda) &= \frac{(1-e^{-i\lambda\tau})^d}{(i\lambda)^d}(\vec{B}_\tau(e^{i\lambda}) - \Psi^\top_\tau(e^{-i\lambda})\vec{r}_\tau(e^{i\lambda})) \\ &= \frac{(1-e^{-i\lambda\tau})^d}{(i\lambda)^d}\Psi^\top_\tau(e^{-i\lambda})\vec{C}_{\tau,b}(e^{-i\lambda}),\end{aligned}$$

where

$$\vec{r}_\tau(e^{i\lambda}) = \sum_{m=0}^{\infty} (\theta_\tau^\top D^\tau \mathbf{A})_m e^{i\lambda m} = \sum_{m=0}^{\infty} \sum_{p=0}^{\infty} \theta_\tau^\top(p) \vec{b}_{p+m}^\tau e^{i\lambda m},$$

$$\vec{C}_{\tau,b}(e^{-i\lambda}) = \sum_{m=1}^{\infty} (\theta_\tau^\top \widetilde{\mathbf{B}}_\tau)_m e^{-i\lambda m} = \sum_{m=1}^{\infty} \sum_{p=m}^{\infty} \theta_\tau^\top(p) \vec{b}_{p-m}^\tau e^{-i\lambda m}$$

$$= \sum_{m=1}^{\infty} \sum_{p=0}^{\infty} \theta_\tau^\top(m+p) \vec{b}_p^\tau e^{-i\lambda m},$$

where the vector

$$\theta_\tau^\top = ((\theta_\tau(0))^\top, (\theta_\tau(1))^\top, (\theta_\tau(2))^\top, \ldots),$$

$\mathbf{A}$ is a linear symmetric operator determined by the matrix with the vector entries $(\mathbf{A})_{k,j} = \vec{a}_{k+j}$, $k, j \geq 0$, and $\widetilde{\mathbf{B}}_\tau$ is a linear operator which is determined by a matrix with the vector entries $(\widetilde{\mathbf{B}}_\tau)_{k,j} = \vec{b}_{k-j}^\tau$ for $0 \leq j \leq k$ and $(\widetilde{\mathbf{B}}_\tau)_{k,j} = \vec{0}$ for $0 \leq k < j$.

Then the spectral characteristic $\vec{h}_\tau(\lambda)$ of the estimate $\widehat{A}\xi$ can be calculated by the formula

$$\vec{h}_\tau(\lambda) = \frac{(1-e^{-i\lambda\tau})^d}{(i\lambda)^d} \left(\sum_{k=0}^{\infty} \psi_\tau^\top(k) e^{-i\lambda k} \right) \sum_{m=1}^{\infty} (\theta_\tau^\top \widetilde{\mathbf{B}}_\tau - \overline{\psi}_\tau \mathbf{C}_{\tau,g})_m e^{-i\lambda m}$$

$$= \frac{(1-e^{-i\lambda\tau})^d}{(i\lambda)^d} \Psi_\tau^\top(e^{-i\lambda}) (\vec{C}_{\tau,b}(e^{-i\lambda}) - \vec{C}_{\tau,g}(e^{-i\lambda}))$$

$$= \vec{B}_\tau(e^{i\lambda}) \frac{(1-e^{-i\lambda\tau})^d}{(i\lambda)^d} - \widetilde{\vec{h}}_\tau(\lambda), \tag{8.20}$$

where

$$\widetilde{\vec{h}}_\tau(\lambda) = \frac{(1-e^{-i\lambda\tau})^d}{(i\lambda)^d} \left(\sum_{k=0}^{\infty} \psi_\tau^\top(k) e^{-i\lambda k} \right) \left(\sum_{m=0}^{\infty} (\theta_\tau^\top D^\tau \mathbf{A})_m e^{i\lambda m} + \sum_{m=1}^{\infty} (\overline{\psi}_\tau \mathbf{C}_{\tau,g})_m e^{-i\lambda m} \right)$$

$$= \frac{(1-e^{-i\lambda\tau})^d}{(i\lambda)^d} \Psi_\tau^\top(e^{-i\lambda}) (\vec{r}_\tau(e^{i\lambda}) + \vec{C}_{\tau,g}(e^{-i\lambda})).$$

The value of the MSE of the estimate $\widehat{A}\xi$ is calculated by the formula

$$\Delta(f, g; \widehat{A}\xi) = \Delta(f, g; \widehat{H}\vec{\xi}) = \mathsf{E}|H\vec{\xi} - \widehat{H}\vec{\xi}|^2$$

$$= \frac{1}{2\pi} \int_{-\pi}^{\pi} (\vec{A}(e^{i\lambda}))^\top g(\lambda) \overline{\vec{A}(e^{i\lambda})} d\lambda$$

$$+ \frac{1}{2\pi} \int_{-\pi}^{\pi} (\widetilde{\vec{h}}_\tau(e^{i\lambda}))^\top (f(\lambda) + \lambda^{2d} g(\lambda)) \overline{\widetilde{\vec{h}}_\tau(e^{i\lambda})} d\lambda$$

$$
\begin{aligned}
&-\frac{1}{2\pi}\int_{-\pi}^{\pi}\frac{(i\lambda)^d}{(1-e^{-i\lambda\tau})^d}(\vec{h}_\tau(e^{i\lambda}))^\top g(\lambda)\overline{\vec{A}^\tau(e^{i\lambda})}d\lambda \\
&-\frac{1}{2\pi}\int_{-\pi}^{\pi}\frac{(-i\lambda)^d}{(1-e^{i\lambda\tau})^d}(\vec{A}^\tau(e^{i\lambda}))^\top g(\lambda)\overline{\bar{\vec{h}}_\tau(e^{i\lambda})}d\lambda \\
&=\|\Phi^\top\mathbf{a}^\tau\|^2+\|\widetilde{\Phi}\mathbf{a}^\tau\|_1^2+\langle\theta_\tau^\top D^\tau\mathbf{A}-\overline{\psi}_\tau\mathbf{C}_{\tau,g},\theta_\tau^\top D^\tau\mathbf{A}\rangle \\
&-\langle\theta_\tau^\top D^\tau\mathbf{A},\mathbf{Z}_\tau\mathbf{a}^\tau\rangle-\langle\mathbf{Z}_\tau\mathbf{a}^\tau,\overline{\psi}_\tau\mathbf{C}_{\tau,g}\rangle_1,
\end{aligned}
\tag{8.21}
$$

where $\|\mathbf{x}\|_1^2=\langle\mathbf{x},\mathbf{x}\rangle_1$ and $\langle\mathbf{x},\mathbf{y}\rangle_1=\sum_{j=1}^{\infty}(\vec{x}_j)^\top\overline{\vec{y}}_j$ for the vectors

$$
\mathbf{x}=((\vec{x}_0)^\top,(\vec{x}_1)^\top,(\vec{x}_2)^\top,\dots)^\top,\quad \mathbf{y}=((\vec{y}_0)^\top,(\vec{y}_1)^\top,(\vec{y}_2)^\top,\dots)^\top.
$$

The obtained results are summarized in the form of the following theorem.

Theorem 8.1.3. *Let the conditions of Theorem* 8.1.1 *be fulfilled and let the spectral densities $f(\lambda)$ and $g(\lambda)$ of the stochastic processes $\xi(t)$ and $\eta(t)$ admit canonical factorizations* (8.17)–(8.19). *Then the spectral characteristic $\vec{h}_\tau(\lambda)$ and the value of the MSE $\Delta(f,g;\widehat{A}\xi)$ of the optimal estimate $\widehat{A}\xi$ of the functional $A\xi$ based on observations of the processes $\xi(t)+\eta(t)$ at points $t<0$ can be calculated by formulas* (8.20) *and* (8.21), *respectively.*

8.2 Minimax (robust) method of extrapolation

8.2.1 Least favorable spectral densities and minimax-robust spectral characteristic

Consider the estimation problem for the functional $A\xi$ based on the observations $\xi(t)+\eta(t)$ at points $t<0$ when the spectral densities of sequences are not exactly known while a set $\mathcal{D}=\mathcal{D}_f\times\mathcal{D}_g$ of admissible spectral densities is defined. The minimax (robust) approach of estimation is applied. It is formalized by the following two definitions.

Definition 8.2.1. For a given class of spectral densities $\mathcal{D}=\mathcal{D}_f\times\mathcal{D}_g$ the spectral densities $f^0(\lambda)\in\mathcal{D}_f$ and $g^0(\lambda)\in\mathcal{D}_g$ are called least favorable in the class $\mathcal{D}$ for the optimal linear extrapolation of the functional $A\xi$ if the following relation holds:

$$
\Delta(f^0,g^0)=\Delta(h(f^0,g^0);f^0,g^0)=\max_{(f,g)\in\mathcal{D}_f\times\mathcal{D}_g}\Delta(h(f,g);f,g).
$$

Definition 8.2.2. For a given class of spectral densities $\mathcal{D}=\mathcal{D}_f\times\mathcal{D}_g$ the spectral characteristic $h^0(\lambda)$ of the optimal linear estimate of the functional $A\xi$ is called minimax-robust if the following conditions are satisfied:

$$
h^0(\lambda)\in H_{\mathcal{D}}=\bigcap_{(f,g)\in\mathcal{D}_f\times\mathcal{D}_g}L_2^{-1}(f(\lambda)+\lambda^{2d}g(\lambda))
$$

and

$$\min_{h \in H_{\mathcal{D}}} \max_{(f,g) \in \mathcal{D}_f \times \mathcal{D}_g} \Delta(h; f, g) = \max_{(f,g) \in \mathcal{D}_f \times \mathcal{D}_g} \Delta(h^0; f, g).$$

Taking into account the introduced definitions and the derived relations we can verify that the following lemmas hold.

Lemma 8.2.1. *Spectral densities $f^0 \in \mathcal{D}_f$ and $g^0 \in \mathcal{D}_g$ which satisfy condition* (8.1) *are least favorable in the class $\mathcal{D} = \mathcal{D}_f \times \mathcal{D}_g$ for the optimal linear extrapolation of the functional $A\vec{\xi}$ if operators $\mathbf{P}_\tau^0$, $\mathbf{T}_\tau^0$, $\mathbf{Q}^0$ defined by the Fourier coefficients of the functions*

$$\frac{\lambda^{2d}}{|1 - e^{i\lambda\tau}|^{2d}} \left(g^0(\lambda)(f^0(\lambda) + \lambda^{2d} g^0(\lambda))^{-1}\right)^{\top},$$
$$\frac{\lambda^{2d}}{|1 - e^{i\lambda\tau}|^{2d}} \left((f^0(\lambda) + \lambda^{2d} g^0(\lambda))^{-1}\right)^{\top},$$
$$(f^0(\lambda)(f^0(\lambda) + \lambda^{2d} g^0(\lambda))^{-1} g^0(\lambda))^{\top}$$

determine a solution of the constrained optimization problem

$$\begin{aligned} &\max_{(f,g) \in \mathcal{D}_f \times \mathcal{D}_g} \left(\langle D^\tau \mathbf{a} - \mathbf{T}_\tau \mathbf{a}^\tau, \mathbf{P}_\tau^{-1} D^\tau \mathbf{a} - \mathbf{P}_\tau^{-1} \mathbf{T}_\tau \mathbf{a}^\tau \rangle + \langle \mathbf{Q}\mathbf{a}, \mathbf{a} \rangle \right) \\ &= \langle D^\tau \mathbf{a} - \mathbf{T}_\tau^0 \mathbf{a}^\tau, (\mathbf{P}_\tau^0)^{-1} D^\tau \mathbf{a} - (\mathbf{P}_\tau^0)^{-1} \mathbf{T}_\tau^0 \mathbf{a}^\tau \rangle + \langle \mathbf{Q}^0 \mathbf{a}, \mathbf{a} \rangle. \end{aligned} \tag{8.22}$$

The minimax spectral characteristic $h^0 = \vec{h}_\tau(f^0, g^0)$ is calculated by formula (8.11) *if $\vec{h}_\tau(f^0, g^0) \in H_{\mathcal{D}}$.*

Lemma 8.2.2. *The spectral densities $f^0 \in \mathcal{D}_f$ and $g^0 \in \mathcal{D}_g$ which admit canonical factorizations* (6.15), (8.17), *and* (8.18) *are least favorable densities in the class $\mathcal{D}$ for the optimal linear prediction of the functional $A\xi$ based on observations of the process $\xi(t) + \eta(t)$ at points $t < 0$ if the matrix coefficients of canonical factorizations* (8.17) *and* (8.18) *determine a solution to the constrained optimization problem*

$$\begin{aligned} &\|\Phi^\top \mathbf{a}^\tau\|^2 + \|\widetilde{\Phi} \mathbf{a}^\tau\|_1^2 + \langle \theta_\tau^\top D^\tau \mathbf{A} - \overline{\psi}_\tau \mathbf{C}_{\tau,g}, \theta_\tau^\top D^\tau \mathbf{A} \rangle \\ &- \langle \theta_\tau^\top D^\tau \mathbf{A}, \mathbf{Z}_\tau \mathbf{a}^\tau \rangle - \langle \mathbf{Z}_\tau \mathbf{a}^\tau, \overline{\psi}_\tau \mathbf{C}_{\tau,g} \rangle_1 \to \sup, \end{aligned} \tag{8.23}$$

for

$$f(\lambda) = \frac{\lambda^{2d}}{|1 - e^{i\lambda\tau}|^{2d}} \Theta_\tau(e^{-i\lambda}) \Theta_\tau^*(e^{-i\lambda}) - \lambda^{2d} \Phi(e^{-i\lambda}) \Phi^*(e^{-i\lambda}) \in \mathcal{D}_f,$$
$$g(\lambda) = \Phi(e^{-i\lambda}) \Phi^*(e^{-i\lambda}) \in \mathcal{D}_g.$$

The minimax spectral characteristic $h^0 = \vec{h}_\tau(f^0, g^0)$ is calculated by formula (8.20) *if $\vec{h}_\tau(f^0, g^0) \in H_{\mathcal{D}}$.*

Lemma 8.2.3. *The spectral density $g^0 \in \mathcal{D}_g$ which admits canonical factorizations* (8.17) *and* (8.18) *with the known spectral density $f(\lambda)$ is the least favorable in the class $\mathcal{D}_g$ for the optimal linear prediction of the functional $A\xi$ based on observations of the process $\xi(t) + \eta(t)$ at points $t < 0$ if the matrix coefficients of the canonical factorizations*

$$f(\lambda) + \lambda^{2d} g^0(\lambda) = \frac{\lambda^{2d}}{|1 - e^{i\lambda\tau}|^{2d}} \left(\sum_{k=0}^{\infty} \theta_\tau^0(k) e^{-i\lambda k} \right) \left(\sum_{k=0}^{\infty} \theta_\tau^0(k) e^{-i\lambda k} \right)^*, \tag{8.24}$$

$$g^0(\lambda) = \left(\sum_{k=0}^{\infty} \phi^0(k) e^{-i\lambda k} \right) \left(\sum_{k=0}^{\infty} \phi^0(k) e^{-i\lambda k} \right)^* \tag{8.25}$$

and the equation $\Psi_\tau^0(e^{-i\lambda})\Theta_\tau^0(e^{-i\lambda}) = E_M$ determine a solution to the constrained optimization problem

$$\begin{aligned} &\|\Phi^\top \mathbf{a}^\tau\|^2 + \|\widetilde{\Phi}\mathbf{a}^\tau\|_1^2 + \langle \theta_\tau^\top D^\tau \mathbf{A} - \overline{\psi}_\tau \mathbf{C}_{\tau,g}, \theta_\tau^\top D^\tau \mathbf{A} \rangle \\ &\quad - \langle \theta_\tau^\top D^\tau \mathbf{A}, \mathbf{Z}_\tau \mathbf{a}^\tau \rangle - \langle \mathbf{Z}_\tau \mathbf{a}^\tau, \overline{\psi}_\tau \mathbf{C}_{\tau,g} \rangle_1 \to \sup, \end{aligned} \tag{8.26}$$

for

$$g(\lambda) = \Phi(e^{-i\lambda})\Phi^*(e^{-i\lambda}) \in \mathcal{D}_g.$$

The minimax spectral characteristic $h^0 = \vec{h}_\tau(f, g^0)$ is calculated by formula (8.20) *if $\vec{h}_\tau(f, g^0) \in H_{\mathcal{D}}$.*

Lemma 8.2.4. *The spectral density $f^0 \in \mathcal{D}_f$ which admits canonical factorizations* (6.15) *and* (8.17) *with the known spectral density $g(\lambda)$ is the least favorable spectral density in the class $\mathcal{D}_f$ for the optimal linear extrapolation of the functional $A\xi$ based on observations of the process $\xi(t) + \eta(t)$ at points $t < 0$ if matrix coefficients of the canonical factorization*

$$f^0(\lambda) + \lambda^{2d} g(\lambda) = \frac{\lambda^{2d}}{|1 - e^{i\lambda\tau}|^{2d}} \left(\sum_{k=0}^{\infty} \theta_\tau^0(k) e^{-i\lambda k} \right) \left(\sum_{k=0}^{\infty} \theta_\tau^0(k) e^{-i\lambda k} \right)^* \tag{8.27}$$

and the equation $\Psi_\tau^0(e^{-i\lambda})\Theta_\tau^0(e^{-i\lambda}) = E_M$ determine a solution to the constrained optimization problem

$$\langle \theta_\tau^\top D^\tau \mathbf{A} - \overline{\psi}_\tau \mathbf{C}_{\tau,g}, \theta_\tau^\top D^\tau \mathbf{A} \rangle - \langle \theta_\tau^\top D^\tau \mathbf{A}, \mathbf{Z}_\tau \mathbf{a}^\tau \rangle - \langle \mathbf{Z}_\tau \mathbf{a}^\tau, \overline{\psi}_\tau \mathbf{C}_{\tau,g} \rangle_1 \to \sup, \tag{8.28}$$

for

$$f(\lambda) = \frac{\lambda^{2d}}{|1 - e^{i\lambda\tau}|^{2d}} \Theta_\tau(e^{-i\lambda})\Theta_\tau^*(e^{-i\lambda}) - \lambda^{2d}\Phi(e^{-i\lambda})\Phi^*(e^{-i\lambda}) \in \mathcal{D}_f$$

for the fixed matrix coefficients $\{\phi(k) : k \geq 0\}$. The minimax spectral characteristic $\vec{h}^0 = \vec{h}_\tau(f^0, g)$ is calculated by formula (8.20) *if $h_\tau(f^0, g) \in H_{\mathcal{D}}$.*

For more detailed analysis of properties of the least favorable spectral densities and minimax-robust spectral characteristics we observe that the minimax spectral characteristic h^0 and the least favorable spectral densities (f^0, g^0) form a saddle point of the function $\Delta(h; f, g)$ on the set $H_{\mathcal{D}} \times \mathcal{D}$.

The saddle point inequalities

$$\Delta(h; f^0, g^0) \geq \Delta(h^0; f^0, g^0) \geq \Delta(h^0; f, g) \quad \forall f \in \mathcal{D}_f,\ \forall g \in \mathcal{D}_g,\ \forall h \in H_{\mathcal{D}}$$

hold if $h^0 = \vec{h}_\tau(f^0, g^0)$ and $\vec{h}_\tau(f^0, g^0) \in H_{\mathcal{D}}$, where (f^0, g^0) is a solution of the constrained optimization problem

$$\widetilde{\Delta}(f, g) = -\Delta(\vec{h}_\tau(f^0, g^0); f, g) \to \inf, \quad (f, g) \in \mathcal{D}, \tag{8.29}$$

where the functional $\Delta(\vec{h}_\tau(f^0, g^0); f, g)$ is calculated by the formula

$$\begin{aligned}
&\Delta(\vec{h}_\tau(f^0, g^0); f, g) \\
&= \frac{1}{2\pi} \int_{-\pi}^{\pi} \frac{\lambda^{2d}}{|1 - e^{i\lambda\tau}|^{2d}} (\mathbf{C}_\tau^{f0}(e^{i\lambda}))^\top (f^0(\lambda) + \lambda^{2d} g^0(\lambda))^{-1} f(\lambda) \\
&\times (f^0(\lambda) + \lambda^{2d} g^0(\lambda))^{-1} \overline{\mathbf{C}_\tau^{f0}(e^{i\lambda})} d\lambda \\
&+ \frac{1}{2\pi} \int_{-\pi}^{\pi} \frac{\lambda^{4d}}{|1 - e^{i\lambda\tau}|^{4d}} (\mathbf{C}_\tau^{g0}(e^{i\lambda}))^\top (f^0(\lambda) + \lambda^{2d} g^0(\lambda))^{-1} g(\lambda) \\
&\times (f^0(\lambda) + \lambda^{2d} g^0(\lambda))^{-1} \overline{\mathbf{C}_\tau^{g0}(e^{i\lambda})} d\lambda,
\end{aligned}$$

where

$$\mathbf{C}_\tau^{f0}(e^{i\lambda}) = \overline{g^0(\lambda)} \vec{A}_\tau(e^{i\lambda}) + \sum_{j=0}^{\infty} ((\mathbf{P}_\tau^0)^{-1} D^\tau \mathbf{a} - (\mathbf{P}_\tau^0)^{-1} \mathbf{T}_\tau^0 \mathbf{a}^\tau)_j e^{i\lambda j},$$

$$\mathbf{C}_\tau^{g0}(e^{i\lambda}) = |1 - e^{i\lambda\tau}|^{2d} \lambda^{-2d} \overline{f^0(\lambda)} \vec{A}(e^{i\lambda}) - (1 - e^{-i\lambda\tau})^d \sum_{j=0}^{\infty} ((\mathbf{P}_\tau^0)^{-1} D^\tau \mathbf{a} - (\mathbf{P}_\tau^0)^{-1} \mathbf{T}_\tau^0 \mathbf{a}^\tau)_j e^{i\lambda j},$$

or it is calculated by the formula

$$\begin{aligned}
&\Delta(\vec{h}_\tau(f^0, g^0); f, g) \\
&= \frac{1}{2\pi} \int_{-\pi}^{\pi} \frac{|1 - e^{i\lambda\tau}|^{2d}}{\lambda^{2d}} (\mathbf{r}_{\tau,f}^0(e^{-i\lambda}))^\top \Psi_\tau^0(e^{-i\lambda}) f(\lambda) (\Psi_\tau^0(e^{-i\lambda}))^* \overline{\mathbf{r}_{\tau,f}^0(e^{-i\lambda})} d\lambda \\
&+ \frac{1}{2\pi} \int_{-\pi}^{\pi} (\mathbf{r}_{\tau,g}^0(e^{-i\lambda}))^\top \Psi_\tau^0(e^{-i\lambda}) g(\lambda) (\Psi_\tau^0(e^{-i\lambda}))^* \overline{\mathbf{r}_{\tau,g}^0(e^{-i\lambda})} d\lambda,
\end{aligned}$$

where

$$\mathbf{r}^0_{\tau,f}(e^{-i\lambda}) = \sum_{m=0}^{\infty}((\theta^0_\tau)^\top D^\tau \mathbf{A})_m e^{i\lambda m} + \sum_{m=1}^{\infty}(\overline{\psi}^0_\tau \mathbf{C}^0_{\tau,g})_m e^{-i\lambda m},$$

$$\mathbf{r}^0_{\tau,g}(e^{-i\lambda}) = (1-e^{-i\lambda\tau})^d\left(\sum_{m=0}^{\infty}((\theta^0_\tau)^\top D^\tau \mathbf{A})_m e^{i\lambda m} + \sum_{m=1}^{\infty}(\overline{\psi}^0_\tau \mathbf{C}^0_{\tau,g})_m e^{-i\lambda m}\right)$$
$$- (\Theta^0_\tau(e^{-i\lambda}))^\top A(e^{i\lambda}).$$

The constrained optimization problem (8.29) is equivalent to the unconstrained optimization problem

$$\Delta_{\mathcal{D}}(f,g) = \widetilde{\Delta}(f,g) + \delta(f,g|\mathcal{D}_f \times \mathcal{D}_g) \to \inf, \tag{8.30}$$

where $\delta(f,g|\mathcal{D}_f \times \mathcal{D}_g)$ is the indicator function of the set $\mathcal{D} = \mathcal{D}_f \times \mathcal{D}_g$. Solution (f^0,g^0) of this unconstrained optimization problem is characterized by the condition $0 \in \partial\Delta_{\mathcal{D}}(f^0,g^0)$, where $\partial\Delta_{\mathcal{D}}(f^0,g^0)$ is the subdifferential of the functional $\Delta_{\mathcal{D}}(f,g)$ at point $(f^0,g^0) \in \mathcal{D} = \mathcal{D}_f \times \mathcal{D}_g$, that is, the set of all continuous linear functionals Λ on $L_1 \times L_1$ which satisfy the inequality $\Delta_{\mathcal{D}}(f,g) - \Delta_{\mathcal{D}}(f^0,g^0) \geq \Lambda((f,g)-(f^0,g^0))$, $(f,g) \in \mathcal{D}$ (see [146, 180] for more details). This condition makes it possible to find the least favorable spectral densities in some special classes of spectral densities $\mathcal{D} = \mathcal{D}_f \times \mathcal{D}_g$.

The form of the functional $\Delta(\vec{h}_\tau(f^0,g^0);f,g)$ is convenient for application of the Lagrange method of indefinite multipliers for finding a solution to the problem (8.30). Making use of the method of Lagrange multipliers and the form of the subdifferential of the indicator functions $\delta(f,g|\mathcal{D}_f \times \mathcal{D}_g)$ of the set $\mathcal{D}_f \times \mathcal{D}_g$ of spectral densities, we describe relations that determine least favorable spectral densities in some special classes of spectral densities (see [115, 146] for additional details).

8.2.2 Least favorable spectral densities in classes $\mathcal{D}_0 \times \mathcal{D}_{1\delta}$

Consider the extrapolation problem for the functional $A\xi$ which depends on unobserved values of a process $\xi(t)$ with periodically stationary increments based on observations of the process $\xi(t) + \eta(t)$ at points $t < 0$ under the condition that the sets of admissible spectral densities $\mathcal{D}_0^k, D_{1\delta}^k$, $k = 1,2,3,4$, are defined as follows:

$$\mathcal{D}_0^1 = \left\{ f(\lambda) \,\middle|\, \frac{1}{2\pi}\int_{-\pi}^{\pi} \frac{|1-e^{i\lambda\tau}|^{2d}}{\lambda^{2d}} f(\lambda)d\lambda = P \right\},$$

$$\mathcal{D}_0^2 = \left\{ f(\lambda) \,\middle|\, \frac{1}{2\pi}\int_{-\pi}^{\pi} \frac{|1-e^{i\lambda\tau}|^{2d}}{\lambda^{2d}} \operatorname{Tr}[f(\lambda)]d\lambda = p \right\},$$

$$\mathcal{D}_0^3 = \left\{ f(\lambda) \,\middle|\, \frac{1}{2\pi}\int_{-\pi}^{\pi} \frac{|1-e^{i\lambda\tau}|^{2d}}{\lambda^{2d}} f_{kk}(\lambda)d\lambda = p_k, k = \overline{1,\infty} \right\},$$

$$\mathcal{D}_0^4 = \left\{ f(\lambda) \left| \frac{1}{2\pi} \int_{-\pi}^{\pi} \frac{|1-e^{i\lambda\tau}|^{2d}}{\lambda^{2d}} \langle B_1, f(\lambda) \rangle d\lambda = p \right. \right\}$$

and

$$\mathcal{D}_{1\delta}^1 = \left\{ g(\lambda) \left| \frac{1}{2\pi} \int_{-\pi}^{\pi} |g_{ij}(\lambda) - g_{ij}^1(\lambda)| d\lambda \le \delta_i^j, i,j = \overline{1,\infty} \right. \right\},$$

$$\mathcal{D}_{1\delta}^2 = \left\{ g(\lambda) \left| \frac{1}{2\pi} \int_{-\pi}^{\pi} |\mathrm{Tr}(g(\lambda) - g_1(\lambda))| d\lambda \le \delta \right. \right\},$$

$$\mathcal{D}_{1\delta}^3 = \left\{ g(\lambda) \left| \frac{1}{2\pi} \int_{-\pi}^{\pi} |g_{kk}(\lambda) - g_{kk}^1(\lambda)| d\lambda \le \delta_k, k = \overline{1,\infty} \right. \right\},$$

$$\mathcal{D}_{1\delta}^4 = \left\{ g(\lambda) \left| \frac{1}{2\pi} \int_{-\pi}^{\pi} |\langle B_2, g(\lambda) - g_1(\lambda) \rangle| d\lambda \le \delta \right. \right\}.$$

Here $g_1(\lambda) = \{g_{ij}^1(\lambda)\}_{i,j=1}^{\infty}$ is a fixed spectral density, $p, p_k, k = \overline{1,\infty}$, are given numbers, P, B_2 are given positive-definite Hermitian matrices, and $\delta, \delta_k, k = \overline{1,\infty}$, and δ_i^j, $i, j = \overline{1,\infty}$, are given numbers.

From the condition $0 \in \partial\Delta_{\mathcal{D}}(f^0, g^0)$ we find the following equations, which determine the least favorable spectral densities for these given sets of admissible spectral densities.

For the first set of admissible spectral densities $\mathcal{D}_0^1 \times \mathcal{D}_{1\delta}^1$, we have

$$(\mathbf{C}_\tau^{f0}(e^{i\lambda}))(\mathbf{C}_\tau^{f0}(e^{i\lambda}))^* = \left(\frac{|1-e^{i\lambda\tau}|^{2d}}{\lambda^{2d}} (f^0(\lambda) + \lambda^{2d} g^0(\lambda)) \right) \vec{\alpha} \cdot \vec{\alpha}^* \times \left(\frac{|1-e^{i\lambda\tau}|^{2d}}{\lambda^{2d}} (f^0(\lambda) + \lambda^{2d} g^0(\lambda)) \right), \tag{8.31}$$

$$(\mathbf{C}_\tau^{g0}(e^{i\lambda}))(\mathbf{C}_\tau^{g0}(e^{i\lambda}))^* = \left(\frac{|1-e^{i\lambda\tau}|^{2d}}{\lambda^{2d}} (f^0(\lambda) + \lambda^{2d} g^0(\lambda)) \right) \{\beta_{ij}\gamma_{ij}(\lambda)\}_{i,j=1}^{\infty} \times \left(\frac{|1-e^{i\lambda\tau}|^{2d}}{\lambda^{2d}} (f^0(\lambda) + \lambda^{2d} g^0(\lambda)) \right), \tag{8.32}$$

$$\frac{1}{2\pi} \int_{-\pi}^{\pi} |g_{ij}^0(\lambda) - g_{ij}^1(\lambda)| d\lambda = \delta_i^j, \tag{8.33}$$

where $\vec{\alpha}$, β_{ij} are Lagrange multipliers, functions $|\gamma_{ij}(\lambda)| \le 1$, and

$$\gamma_{ij}(\lambda) = \frac{g_{ij}^0(\lambda) - g_{ij}^1(\lambda)}{|g_{ij}^0(\lambda) - g_{ij}^1(\lambda)|} : \ g_{ij}^0(\lambda) - g_{ij}^1(\lambda) \ne 0, \quad i,j = \overline{1,\infty}.$$

For the second set of admissible spectral densities $\mathcal{D}_0^2 \times \mathcal{D}_{1\delta}^2$, we have

$$(\mathbf{C}_\tau^{f0}(e^{i\lambda}))(\mathbf{C}_\tau^{f0}(e^{i\lambda}))^* = \alpha^2\left(\frac{|1-e^{i\lambda\tau}|^{2d}}{\lambda^{2d}}(f^0(\lambda)+\lambda^{2d}g^0(\lambda))\right)^2, \tag{8.34}$$

$$(\mathbf{C}_\tau^{g0}(e^{i\lambda}))(\mathbf{C}_\tau^{g0}(e^{i\lambda}))^* = \beta^2\gamma_2(\lambda)\left(\frac{|1-e^{i\lambda\tau}|^{2d}}{\lambda^{2d}}(f^0(\lambda)+\lambda^{2d}g^0(\lambda))\right)^2, \tag{8.35}$$

$$\frac{1}{2\pi}\int_{-\pi}^{\pi}|\mathrm{Tr}(g^0(\lambda)-g_1(\lambda))|d\lambda = \delta, \tag{8.36}$$

where α^2, β^2 are Lagrange multipliers, the function $|\gamma_2(\lambda)| \le 1$, and

$$\gamma_2(\lambda) = \mathrm{sign}(\mathrm{Tr}(g^0(\lambda)-g_1(\lambda))) : \mathrm{Tr}(g^0(\lambda)-g_1(\lambda)) \ne 0.$$

For the third set of admissible spectral densities $\mathcal{D}_0^3 \times \mathcal{D}_{1\delta}^3$, we have

$$\begin{aligned}(\mathbf{C}_\tau^{f0}(e^{i\lambda}))(\mathbf{C}_\tau^{f0}(e^{i\lambda}))^* &= \left(\frac{|1-e^{i\lambda\tau}|^{2d}}{\lambda^{2d}}(f^0(\lambda)+\lambda^{2d}g^0(\lambda))\right)\{\alpha_k^2\delta_{kl}\}_{k,l=1}^{\infty}\\ &\times\left(\frac{|1-e^{i\lambda\tau}|^{2d}}{\lambda^{2d}}(f^0(\lambda)+\lambda^{2d}g^0(\lambda))\right),\end{aligned} \tag{8.37}$$

$$\begin{aligned}(\mathbf{C}_\tau^{g0}(e^{i\lambda}))(\mathbf{C}_\tau^{g0}(e^{i\lambda}))^* &= \left(\frac{|1-e^{i\lambda\tau}|^{2d}}{\lambda^{2d}}(f^0(\lambda)+\lambda^{2d}g^0(\lambda))\right)\{\beta_k^2\gamma_k^2(\lambda)\delta_{kl}\}_{k,l=1}^{\infty}\\ &\times\left(\frac{|1-e^{i\lambda\tau}|^{2d}}{\lambda^{2d}}(f^0(\lambda)+\lambda^{2d}g^0(\lambda))\right),\end{aligned} \tag{8.38}$$

$$\frac{1}{2\pi}\int_{-\pi}^{\pi}|g_{kk}^0(\lambda)-g_{kk}^1(\lambda)|d\lambda = \delta_k, \tag{8.39}$$

where α_k^2, β_k^2 are Lagrange multipliers, δ_{kl} are Kronecker symbols, functions $|\gamma_k^2(\lambda)| \le 1$, and

$$\gamma_k^2(\lambda) = \mathrm{sign}(g_{kk}^0(\lambda)-g_{kk}^1(\lambda)) : g_{kk}^0(\lambda)-g_{kk}^1(\lambda) \ne 0, \quad k=\overline{1,\infty}.$$

For the fourth set of admissible spectral densities $\mathcal{D}_0^4 \times \mathcal{D}_{1\delta}^4$, we have

$$\begin{aligned}(\mathbf{C}_\tau^{f0}(e^{i\lambda}))(\mathbf{C}_\tau^{f0}(e^{i\lambda}))^* &= \alpha^2\left(\frac{|1-e^{i\lambda\tau}|^{2d}}{\lambda^{2d}}(f^0(\lambda)+\lambda^{2d}g^0(\lambda))\right)\\ &\times B_1^{\top}\left(\frac{|1-e^{i\lambda\tau}|^{2d}}{\lambda^{2d}}(f^0(\lambda)+\lambda^{2d}g^0(\lambda))\right),\end{aligned} \tag{8.40}$$

$$\begin{aligned}(\mathbf{C}_\tau^{g0}(e^{i\lambda}))(\mathbf{C}_\tau^{g0}(e^{i\lambda}))^* &= \beta^2\gamma_2'(\lambda)\left(\frac{|1-e^{i\lambda\tau}|^{2d}}{\lambda^{2d}}(f^0(\lambda)+\lambda^{2d}g^0(\lambda))\right)\\ &\times B_2^{\top}\left(\frac{|1-e^{i\lambda\tau}|^{2d}}{\lambda^{2d}}(f^0(\lambda)+\lambda^{2d}g^0(\lambda))\right),\end{aligned} \tag{8.41}$$

$$\frac{1}{2\pi}\int_{-\pi}^{\pi}\left|\left\langle B_2, g^0(\lambda)-g_1(\lambda)\right\rangle\right|d\lambda=\delta, \tag{8.42}$$

where α^2, β^2 are Lagrange multipliers, function $|\gamma_2'(\lambda)| \le 1$, and

$$\gamma_2'(\lambda)=\operatorname{sign}\left\langle B_2, g^0(\lambda)-g_1(\lambda)\right\rangle : \left\langle B_2, g^0(\lambda)-g_1(\lambda)\right\rangle \neq 0.$$

The derived results are summarized in the following theorem.

Theorem 8.2.1. *The least favorable spectral densities $f^0(\lambda)$ and $g^0(\lambda)$, in the classes $\mathcal{D}_0^k \times \mathcal{D}_{1\delta}^k$, $k = 1,2,3,4$, for the optimal linear extrapolation of the functional $A\xi$ from observations of the process $\xi(t)+\eta(t)$ at points $t<0$ are determined by equations* (8.31)–(8.33), (8.34)–(8.36), (8.37)–(8.39), *and* (8.40)–(8.42), *respectively, the minimality condition* (8.1), *the constrained optimization problem* (8.22), *and restrictions on densities from the corresponding classes $\mathcal{D}_0^k$, $\mathcal{D}_{1\delta}^k$, $k = 1,2,3,4$. The minimax-robust spectral characteristic of the optimal estimate of the functional $A\xi$ is determined by formula* (8.11).

In the case where the spectral densities $f(\lambda)$ and $g(\lambda)$ admit canonical factorizations (8.17) and (8.18), the equations for the least favorable spectral densities are described below.

For the first set of admissible spectral densities $\mathcal{D}_0^1 \times \mathcal{D}_{1\delta}^1$, we have

$$(\mathbf{r}_{\tau,f}^0(e^{i\lambda}))(\mathbf{r}_{\tau,f}^0(e^{i\lambda}))^* = (\Theta_\tau^0(e^{-i\lambda}))^\top \vec{\alpha}\cdot\vec{\alpha}^*\overline{\Theta_\tau^0(e^{-i\lambda})}, \tag{8.43}$$

$$(\mathbf{r}_{\tau,g}^0(e^{i\lambda}))(\mathbf{r}_{\tau,g}^0(e^{i\lambda}))^* = (\Theta_\tau^0(e^{-i\lambda}))^\top \{\beta_{ij}\gamma_{ij}(\lambda)\}_{i,j=1}^{\infty}\overline{\Theta_\tau^0(e^{-i\lambda})}, \tag{8.44}$$

$$\frac{1}{2\pi}\int_{-\pi}^{\pi}|g_{ij}^0(\lambda)-g_{ij}^1(\lambda)|d\lambda=\delta_i^j, \tag{8.45}$$

where $\vec{\alpha}, \beta_{ij}$ are Lagrange multipliers, functions $|\gamma_{ij}(\lambda)| \le 1$, and

$$\gamma_{ij}(\lambda)=\frac{g_{ij}^0(\lambda)-g_{ij}^1(\lambda)}{|g_{ij}^0(\lambda)-g_{ij}^1(\lambda)|} : g_{ij}^0(\lambda)-g_{ij}^1(\lambda)\neq 0, \quad i,j=\overline{1,\infty}.$$

For the second set of admissible spectral densities $\mathcal{D}_0^2 \times \mathcal{D}_{1\delta}^2$, we have

$$(\mathbf{r}_{\tau,f}^0(e^{i\lambda}))(\mathbf{r}_{\tau,f}^0(e^{i\lambda}))^* = \alpha^2(\Theta_\tau^0(e^{-i\lambda}))^\top\overline{\Theta_\tau^0(e^{-i\lambda})}, \tag{8.46}$$

$$(\mathbf{r}_{\tau,g}^0(e^{i\lambda}))(\mathbf{r}_{\tau,g}^0(e^{i\lambda}))^* = \beta^2\gamma_2(\lambda)(\Theta_\tau^0(e^{-i\lambda}))^\top\overline{\Theta_\tau^0(e^{-i\lambda})}, \tag{8.47}$$

$$\frac{1}{2\pi}\int_{-\pi}^{\pi}|\mathrm{Tr}(g^0(\lambda)-g_1(\lambda))|d\lambda=\delta, \tag{8.48}$$

where α^2, β^2 are Lagrange multipliers, function $|\gamma_2(\lambda)| \le 1$, and

$$\gamma_2(\lambda) = \mathrm{sign}(\mathrm{Tr}(g^0(\lambda) - g_1(\lambda))) : \ \mathrm{Tr}(g^0(\lambda) - g_1(\lambda)) \neq 0.$$

For the third set of admissible spectral densities $\mathcal{D}_0^3 \times \mathcal{D}_{1\delta}^3$, we have

$$(\mathbf{r}^0_{\tau,f}(e^{i\lambda}))(\mathbf{r}^0_{\tau,f}(e^{i\lambda}))^* = (\Theta^0_\tau(e^{-i\lambda}))^\top \{\alpha_k^2 \delta_{kl}\}_{k,l=1}^\infty \overline{\Theta^0_\tau(e^{-i\lambda})}, \tag{8.49}$$

$$(\mathbf{r}^0_{\tau,g}(e^{i\lambda}))(\mathbf{r}^0_{\tau,g}(e^{i\lambda}))^* = (\Theta^0_\tau(e^{-i\lambda}))^\top \{\beta_k^2 \gamma_k^2(\lambda) \delta_{kl}\}_{k,l=1}^\infty \overline{\Theta^0_\tau(e^{-i\lambda})}, \tag{8.50}$$

$$\frac{1}{2\pi} \int_{-\pi}^{\pi} |g^0_{kk}(\lambda) - g^1_{kk}(\lambda)| d\lambda = \delta_k, \tag{8.51}$$

where α_k^2, β_k^2 are Lagrange multipliers, δ_{kl} are Kronecker symbols, functions $|\gamma_k^2(\lambda)| \leq 1$, and

$$\gamma_k^2(\lambda) = \mathrm{sign}(g^0_{kk}(\lambda) - g^1_{kk}(\lambda)) : \ g^0_{kk}(\lambda) - g^1_{kk}(\lambda) \neq 0, \quad k = \overline{1, \infty}.$$

For the fourth set of admissible spectral densities $\mathcal{D}_0^4 \times \mathcal{D}_{1\delta}^4$, we have

$$(\mathbf{r}^0_{\tau,f}(e^{i\lambda}))(\mathbf{r}^0_{\tau,f}(e^{i\lambda}))^* = \alpha^2 (\Theta^0_\tau(e^{-i\lambda}))^\top B_1 \overline{\Theta^0_\tau(e^{-i\lambda})}, \tag{8.52}$$

$$(\mathbf{r}^0_{\tau,g}(e^{i\lambda}))(\mathbf{r}^0_{\tau,g}(e^{i\lambda}))^* = \beta^2 \gamma_2'(\lambda) (\Theta^0_\tau(e^{-i\lambda}))^\top B_2 \overline{\Theta^0_\tau(e^{-i\lambda})}, \tag{8.53}$$

$$\frac{1}{2\pi} \int_{-\pi}^{\pi} |\langle B_2, g^0(\lambda) - g_1(\lambda)\rangle| d\lambda = \delta, \tag{8.54}$$

where α^2, β^2 are Lagrange multipliers, function $|\gamma_2'(\lambda)| \leq 1$, and

$$\gamma_2'(\lambda) = \mathrm{sign}\langle B_2, g^0(\lambda) - g_1(\lambda)\rangle : \ \langle B_2, g^0(\lambda) - g_1(\lambda)\rangle \neq 0.$$

The following theorems hold.

Theorem 8.2.2. *The least favorable spectral densities $f^0(\lambda)$ and $g^0(\lambda)$ in the classes $\mathcal{D}_0^k \times \mathcal{D}_{1\delta}^k$, $k = 1, 2, 3, 4$, for the optimal linear extrapolation of the functional $A\xi$ from observations of the process $\xi(t) + \eta(t)$ at points $t < 0$ are determined by canonical factorizations* (8.17) *and* (8.18), *equations* (8.43)–(8.45), (8.46)–(8.48), (8.49)–(8.51), *and* (8.52)–(8.54), *respectively, the constrained optimization problem* (8.23), *and restrictions on densities from the corresponding classes $\mathcal{D}_0^k$, $\mathcal{D}_{1\delta}^k$, $k = 1, 2, 3, 4$. The minimax-robust spectral characteristic of the optimal estimate of the functional $A\vec{\xi}$ is determined by formula* (8.20).

Theorem 8.2.3. *In the case where the spectral density $g(\lambda)$ is known, the least favorable spectral density $f^0(\lambda)$ in the classes $\mathcal{D}_0^k$, $k = 1, 2, 3, 4$, for the optimal linear extrapolation of the functional $A\xi$ from observations of the process $\xi(t) + \eta(t)$ at points $t < 0$ is determined by canonical factorizations* (8.18) *and* (8.17), *equations* (8.43), (8.46), (8.49), *and* (8.52), *respectively, the constrained optimization problem* (8.28), *and restrictions on density from the corresponding classes $\mathcal{D}_0^k$, $k = 1, 2, 3, 4$. The minimax-robust spectral characteristic of the optimal estimate of the functional $A\xi$ is determined by formula* (8.20).

8.2.3 Least favorable spectral densities in classes $\mathcal{D}_\varepsilon \times \mathcal{D}_V^U$

Consider the prediction problem for the functional $A\xi$ depending on unobserved values of the process $\xi(t)$ with periodically stationary increments based on observations of the process $\xi(t) + \eta(t)$ at points $t < 0$ under the condition that the sets of admissible spectral densities $\mathcal{D}_\varepsilon^k, \mathcal{D}_V^{Uk}$, $k = 1, 2, 3, 4$, are defined as follows:

$$\mathcal{D}_\varepsilon^1 = \left\{ f(\lambda) \,\middle|\, f(\lambda) = (1-\varepsilon) f_1(\lambda) + \varepsilon W(\lambda), \frac{1}{2\pi} \int_{-\pi}^{\pi} \frac{|1 - e^{i\lambda\tau}|^{2d}}{\lambda^{2d}} f(\lambda) d\lambda = P \right\},$$

$$\mathcal{D}_\varepsilon^2 = \left\{ f(\lambda) \,\middle|\, \operatorname{Tr}[f(\lambda)] = (1-\varepsilon) \operatorname{Tr}[f_1(\lambda)] + \varepsilon \operatorname{Tr}[W(\lambda)], \right.$$
$$\left. \frac{1}{2\pi} \int_{-\pi}^{\pi} \frac{|1 - e^{i\lambda\tau}|^{2d}}{\lambda^{2d}} \operatorname{Tr}[f(\lambda)] d\lambda = p \right\},$$

$$\mathcal{D}_\varepsilon^3 = \left\{ f(\lambda) \,\middle|\, f_{kk}(\lambda) = (1-\varepsilon) f_{kk}^1(\lambda) + \varepsilon w_{kk}(\lambda), \right.$$
$$\left. \frac{1}{2\pi} \int_{-\pi}^{\pi} \frac{|1 - e^{i\lambda\tau}|^{2d}}{\lambda^{2d}} f_{kk}(\lambda) d\lambda = p_k, k = \overline{1, \infty} \right\},$$

$$\mathcal{D}_\varepsilon^4 = \left\{ f(\lambda) \,\middle|\, \langle B_1, f(\lambda) \rangle = (1-\varepsilon) \langle B_1, f_1(\lambda) \rangle + \varepsilon \langle B_1, W(\lambda) \rangle, \right.$$
$$\left. \frac{1}{2\pi} \int_{-\pi}^{\pi} \frac{|1 - e^{i\lambda\tau}|^{2d}}{\lambda^{2d}} \langle B_1, f(\lambda) \rangle d\lambda = p \right\}$$

and

$$\mathcal{D}_V^{U1} = \left\{ g(\lambda) \,\middle|\, V(\lambda) \le g(\lambda) \le U(\lambda), \frac{1}{2\pi} \int_{-\pi}^{\pi} g(\lambda) d\lambda = Q \right\},$$

$$\mathcal{D}_V^{U2} = \left\{ g(\lambda) \,\middle|\, \operatorname{Tr}[V(\lambda)] \le \operatorname{Tr}[g(\lambda)] \le \operatorname{Tr}[U(\lambda)], \frac{1}{2\pi} \int_{-\pi}^{\pi} \operatorname{Tr}[g(\lambda)] d\lambda = q \right\},$$

$$\mathcal{D}_V^{U3} = \left\{ g(\lambda) \,\middle|\, v_{kk}(\lambda) \le g_{kk}(\lambda) \le u_{kk}(\lambda), \frac{1}{2\pi} \int_{-\pi}^{\pi} g_{kk}(\lambda) d\lambda = q_k, k = \overline{1, \infty} \right\},$$

$$\mathcal{D}_V^{U4} = \left\{ g(\lambda) \,\middle|\, \langle B_2, V(\lambda) \rangle \le \langle B_2, g(\lambda) \rangle \le \langle B_2, U(\lambda) \rangle, \frac{1}{2\pi} \int_{-\pi}^{\pi} \langle B_2, g(\lambda) \rangle d\lambda = q \right\}.$$

Here $f_1(\lambda)$, $V(\lambda)$, and $U(\lambda)$ are known and fixed spectral densities, $W(\lambda)$ is an unknown spectral density, p, p_k, q, q_k, $k = \overline{1,\infty}$, are given numbers, and P, B_1, Q, B_2 are given positive-definite Hermitian matrices.

The condition $0 \in \partial\Delta_{\mathcal{D}}(f^0, g^0)$ implies the following equations determining the least favorable spectral densities for these given sets of admissible spectral densities.

For the first set of admissible spectral densities $\mathcal{D}_\varepsilon^1 \times \mathcal{D}_V^{U1}$, we have

$$
\begin{aligned}
(\mathbf{C}_\tau^{f0}(e^{i\lambda}))(\mathbf{C}_\tau^{f0}(e^{i\lambda}))^* &= \left(\frac{|1-e^{i\lambda\tau}|^{2d}}{\lambda^{2d}}(f^0(\lambda)+\lambda^{2d}g^0(\lambda))\right)(\vec{\alpha}\cdot\vec{\alpha}^* + \Gamma(\lambda)) \\
&\quad \times \left(\frac{|1-e^{i\lambda\tau}|^{2d}}{\lambda^{2d}}(f^0(\lambda)+\lambda^{2d}g^0(\lambda))\right), \qquad (8.55)
\end{aligned}
$$

$$
\begin{aligned}
(\mathbf{C}_\tau^{g0}(e^{i\lambda}))(\mathbf{C}_\tau^{g0}(e^{i\lambda}))^* &= \left(\frac{|1-e^{i\lambda\tau}|^{2d}}{\lambda^{2d}}(f^0(\lambda)+\lambda^{2d}g^0(\lambda))\right)(\vec{\beta}\cdot\vec{\beta}^* + \Gamma_1(\lambda)+\Gamma_2(\lambda)) \\
&\quad \times \left(\frac{|1-e^{i\lambda\tau}|^{2d}}{\lambda^{2d}}(f^0(\lambda)+\lambda^{2d}g^0(\lambda))\right), \qquad (8.56)
\end{aligned}
$$

where $\vec{\alpha}$ and $\vec{\beta}$ are vectors of Lagrange multipliers, function $\Gamma(\lambda) \le 0$ and $\Gamma(\lambda) = 0$ if $f^0(\lambda) > (1-\varepsilon)f_1(\lambda)$, the matrix $\Gamma_1(\lambda) \le 0$ and $\Gamma_1(\lambda) = 0$ if $g^0(\lambda) > V(\lambda)$, and the matrix $\Gamma_2(\lambda) \ge 0$ and $\Gamma_2(\lambda) = 0$ if $g^0(\lambda) < U(\lambda)$.

For the second set of admissible spectral densities $\mathcal{D}_\varepsilon^2 \times \mathcal{D}_V^{U2}$, we have

$$
(\mathbf{C}_\tau^{f0}(e^{i\lambda}))(\mathbf{C}_\tau^{f0}(e^{i\lambda}))^* = (\alpha^2+\gamma(\lambda))\left(\frac{|1-e^{i\lambda\tau}|^{2d}}{\lambda^{2d}}(f^0(\lambda)+\lambda^{2d}g^0(\lambda))\right)^2, \qquad (8.57)
$$

$$
(\mathbf{C}_\tau^{g0}(e^{i\lambda}))(\mathbf{C}_\tau^{g0}(e^{i\lambda}))^* = (\beta^2+\gamma_1(\lambda)+\gamma_2(\lambda))\left(\frac{|1-e^{i\lambda\tau}|^{2d}}{\lambda^{2d}}(f^0(\lambda)+\lambda^{2d}g^0(\lambda))\right)^2, \qquad (8.58)
$$

where α^2, β^2 are Lagrange multipliers, function $\gamma(\lambda) \le 0$ and $\gamma(\lambda) = 0$ if $\mathrm{Tr}[f^0(\lambda)] > (1-\varepsilon)\,\mathrm{Tr}[f_1(\lambda)]$, function $\gamma_1(\lambda) \le 0$ and $\gamma_1(\lambda) = 0$ if $\mathrm{Tr}[g^0(\lambda)] > \mathrm{Tr}[V(\lambda)]$, and the function $\gamma_2(\lambda) \ge 0$ and $\gamma_2(\lambda) = 0$ if $\mathrm{Tr}[g^0(\lambda)] < \mathrm{Tr}[U(\lambda)]$.

For the third set of admissible spectral densities $\mathcal{D}_\varepsilon^3 \times \mathcal{D}_V^{U3}$, we have

$$
\begin{aligned}
(\mathbf{C}_\tau^{f0}(e^{i\lambda}))(\mathbf{C}_\tau^{f0}(e^{i\lambda}))^* &= \left(\frac{|1-e^{i\lambda\tau}|^{2d}}{\lambda^{2d}}(f^0(\lambda)+\lambda^{2d}g^0(\lambda))\right)\{(\alpha_k^2+\gamma_k(\lambda))\delta_{kl}\}_{k,l=1}^\infty \\
&\quad \times \left(\frac{|1-e^{i\lambda\tau}|^{2d}}{\lambda^{2d}}(f^0(\lambda)+\lambda^{2d}g^0(\lambda))\right), \qquad (8.59)
\end{aligned}
$$

$$
\begin{aligned}
(\mathbf{C}_\tau^{g0}(e^{i\lambda}))(\mathbf{C}_\tau^{g0}(e^{i\lambda}))^* &= \left(\frac{|1-e^{i\lambda\tau}|^{2d}}{\lambda^{2d}}(f^0(\lambda)+\lambda^{2d}g^0(\lambda))\right)\{(\beta_k^2+\gamma_{1k}(\lambda)+\gamma_{2k}(\lambda))\delta_{kl}\}_{k,l=1}^\infty \\
&\quad \times \left(\frac{|1-e^{i\lambda\tau}|^{2d}}{\lambda^{2d}}(f^0(\lambda)+\lambda^{2d}g^0(\lambda))\right), \qquad (8.60)
\end{aligned}
$$

where α_k^2, β_k^2 are Lagrange multipliers, δ_{kl} are Kronecker symbols, functions $\gamma_k(\lambda) \leq 0$ and $\gamma_k(\lambda) = 0$ if $f_{kk}^0(\lambda) > (1-\varepsilon)f_{kk}^1(\lambda)$, functions $\gamma_{1k}(\lambda) \leq 0$ and $\gamma_{1k}(\lambda) = 0$ if $g_{kk}^0(\lambda) > v_{kk}(\lambda)$, and functions $\gamma_{2k}(\lambda) \geq 0$ and $\gamma_{2k}(\lambda) = 0$ if $g_{kk}^0(\lambda) < u_{kk}(\lambda)$.

For the fourth set of admissible spectral densities $\mathcal{D}_\varepsilon^4 \times \mathcal{D}_V^{U4}$, we have

$$(\mathbf{C}_\tau^{f0}(e^{i\lambda}))(\mathbf{C}_\tau^{f0}(e^{i\lambda}))^* = (\alpha^2 + \gamma'(\lambda))\left(\frac{|1-e^{i\lambda\tau}|^{2d}}{\lambda^{2d}}(f^0(\lambda) + \lambda^{2d}g^0(\lambda))\right) \times B_1^\top\left(\frac{|1-e^{i\lambda\tau}|^{2d}}{\lambda^{2d}}(f^0(\lambda) + \lambda^{2d}g^0(\lambda))\right), \tag{8.61}$$

$$(\mathbf{C}_\tau^{g0}(e^{i\lambda}))(\mathbf{C}_\tau^{g0}(e^{i\lambda}))^* = (\beta^2 + \gamma_1'(\lambda) + \gamma_2'(\lambda))\left(\frac{|1-e^{i\lambda\tau}|^{2d}}{\lambda^{2d}}(f^0(\lambda) + \lambda^{2d}g^0(\lambda))\right) \times B_2^\top\left(\frac{|1-e^{i\lambda\tau}|^{2d}}{\lambda^{2d}}(f^0(\lambda) + \lambda^{2d}g^0(\lambda))\right), \tag{8.62}$$

where α^2, β^2, are Lagrange multipliers, function $\gamma'(\lambda) \leq 0$ and $\gamma'(\lambda) = 0$ if $\langle B_1, f^0(\lambda)\rangle > (1-\varepsilon)\langle B_1, f_1(\lambda)\rangle$, functions $\gamma_1'(\lambda) \leq 0$ and $\gamma_1'(\lambda) = 0$ if $\langle B_2, g^0(\lambda)\rangle > \langle B_2, V(\lambda)\rangle$, and functions $\gamma_2'(\lambda) \geq 0$ and $\gamma_2'(\lambda) = 0$ if $\langle B_2, g^0(\lambda)\rangle < \langle B_2, U(\lambda)\rangle$.

The following theorem holds.

Theorem 8.2.4. *The least favorable spectral densities $f^0(\lambda)$ and $g^0(\lambda)$ in classes $\mathcal{D}_\varepsilon^k \times \mathcal{D}_V^{Uk}$, $k = 1,2,3,4$, for the optimal linear extrapolation of the functional $A\xi$ from observations of the process $\xi(t) + \eta(t)$ at points $t < 0$ are determined by equations* (8.55)–(8.56), (8.57)–(8.58), (8.59)–(8.60), *and* (8.61)–(8.62), *respectively, the minimality condition* (8.1), *the constrained optimization problem* (8.22), *and restrictions on densities from the corresponding classes $\mathcal{D}_\varepsilon^k$, $\mathcal{D}_V^{Uk}$, $k = 1,2,3,4$. The minimax-robust spectral characteristic of the optimal estimate of the functional $A\xi$ is determined by formula* (8.11).

Let the spectral densities $f(\lambda)$ and $g(\lambda)$ admit the canonical factorizations (8.17) and (8.18). The equations for the least favorable spectral densities are described below.

For the first set of admissible spectral densities $\mathcal{D}_\varepsilon^1 \times \mathcal{D}_V^{U1}$, we have

$$(\mathbf{r}_{\tau,f}^0(e^{i\lambda}))(\mathbf{r}_{\tau,f}^0(e^{i\lambda}))^* = (\Theta_\tau^0(e^{-i\lambda}))^\top(\vec{\alpha}\cdot\vec{\alpha}^* + \Gamma(\lambda))\overline{\Theta_\tau^0(e^{-i\lambda})}, \tag{8.63}$$

$$(\mathbf{r}_{\tau,g}^0(e^{i\lambda}))(\mathbf{r}_{\tau,g}^0(e^{i\lambda}))^* = (\Theta_\tau^0(e^{-i\lambda}))^\top(\vec{\beta}\cdot\vec{\beta}^* + \Gamma_1(\lambda) + \Gamma_2(\lambda))\overline{\Theta_\tau^0(e^{-i\lambda})}, \tag{8.64}$$

where $\vec{\alpha}$ and $\vec{\beta}$ are vectors of Lagrange multipliers, matrix $\Gamma(\lambda) \leq 0$ and $\Gamma(\lambda) = 0$ if $f^0(\lambda) > (1-\varepsilon)f_1(\lambda)$, matrix $\Gamma_1(\lambda) \leq 0$ and $\Gamma_1(\lambda) = 0$ if $g^0(\lambda) > V(\lambda)$, and matrix $\Gamma_2(\lambda) \geq 0$ and $\Gamma_2(\lambda) = 0$ if $g^0(\lambda) < U(\lambda)$.

For the second set of admissible spectral densities $\mathcal{D}_\varepsilon^2 \times \mathcal{D}_V^{U2}$, we have

$$(\mathbf{r}_{\tau,f}^0(e^{i\lambda}))(\mathbf{r}_{\tau,f}^0(e^{i\lambda}))^* = (\alpha^2 + \gamma(\lambda))(\Theta_\tau^0(e^{-i\lambda}))^\top\overline{\Theta_\tau^0(e^{-i\lambda})}, \tag{8.65}$$

$$(\mathbf{r}_{\tau,g}^0(e^{i\lambda}))(\mathbf{r}_{\tau,g}^0(e^{i\lambda}))^* = (\beta^2 + \gamma_1(\lambda) + \gamma_2(\lambda))(\Theta_\tau^0(e^{-i\lambda}))^\top\overline{\Theta_\tau^0(e^{-i\lambda})}, \tag{8.66}$$

where α^2, β^2 are Lagrange multipliers, function $\gamma(\lambda) \leq 0$ and $\gamma(\lambda) = 0$ if $\mathrm{Tr}[f^0(\lambda)] > (1-\varepsilon)\,\mathrm{Tr}[f_1(\lambda)]$, function $\gamma_1(\lambda) \leq 0$ and $\gamma_1(\lambda) = 0$ if $\mathrm{Tr}[g^0(\lambda)] > \mathrm{Tr}[V(\lambda)]$, and the function $\gamma_2(\lambda) \geq 0$ and $\gamma_2(\lambda) = 0$ if $\mathrm{Tr}[g^0(\lambda)] < \mathrm{Tr}[U(\lambda)]$.

For the third set of admissible spectral densities $\mathcal{D}_\varepsilon^3 \times \mathcal{D}_V^{U3}$, we have

$$(\mathbf{r}_{\tau,f}^0(e^{i\lambda}))(\mathbf{r}_{\tau,f}^0(e^{i\lambda}))^* = (\Theta_\tau^0(e^{-i\lambda}))^\top\{(\alpha_k^2 + \gamma_k(\lambda))\delta_{kl}\}_{k,l=1}^\infty\overline{\Theta_\tau^0(e^{-i\lambda})}, \tag{8.67}$$

$$(\mathbf{r}_{\tau,g}^0(e^{i\lambda}))(\mathbf{r}_{\tau,g}^0(e^{i\lambda}))^* = (\Theta_\tau^0(e^{-i\lambda}))^\top\{(\beta_k^2 + \gamma_{1k}(\lambda) + \gamma_{2k}(\lambda))\delta_{kl}\}_{k,l=1}^\infty\overline{\Theta_\tau^0(e^{-i\lambda})}, \tag{8.68}$$

where α_k^2, β_k^2 are Lagrange multipliers, δ_{kl} are Kronecker symbols, functions $\gamma_k(\lambda) \leq 0$ and $\gamma_k(\lambda) = 0$ if $f_{kk}^0(\lambda) > (1-\varepsilon)f_{kk}^1(\lambda)$, functions $\gamma_{1k}(\lambda) \leq 0$ and $\gamma_{1k}(\lambda) = 0$ if $g_{kk}^0(\lambda) > v_{kk}(\lambda)$, and functions $\gamma_{2k}(\lambda) \geq 0$ and $\gamma_{2k}(\lambda) = 0$ if $g_{kk}^0(\lambda) < u_{kk}(\lambda)$.

For the fourth set of admissible spectral densities $\mathcal{D}_\varepsilon^4 \times \mathcal{D}_V^{U4}$, we have

$$(\mathbf{r}_{\tau,f}^0(e^{i\lambda}))(\mathbf{r}_{\tau,f}^0(e^{i\lambda}))^* = (\alpha^2 + \gamma'(\lambda))(\Theta_\tau^0(e^{-i\lambda}))^\top B_1\overline{\Theta_\tau^0(e^{-i\lambda})}, \tag{8.69}$$

$$(\mathbf{r}_{\tau,g}^0(e^{i\lambda}))(\mathbf{r}_{\tau,g}^0(e^{i\lambda}))^* = (\beta^2 + \gamma_1'(\lambda) + \gamma_2'(\lambda))(\Theta_\tau^0(e^{-i\lambda}))^\top B_2\overline{\Theta_\tau^0(e^{-i\lambda})}, \tag{8.70}$$

where α^2, β^2, are Lagrange multipliers, function $\gamma'(\lambda) \leq 0$ and $\gamma'(\lambda) = 0$ if $\langle B_1, f^0(\lambda)\rangle > (1-\varepsilon)\langle B_1, f_1(\lambda)\rangle$, functions $\gamma_1'(\lambda) \leq 0$ and $\gamma_1'(\lambda) = 0$ if $\langle B_2, g^0(\lambda)\rangle > \langle B_2, V(\lambda)\rangle$, and functions $\gamma_2'(\lambda) \geq 0$ and $\gamma_2'(\lambda) = 0$ if $\langle B_2, g^0(\lambda)\rangle < \langle B_2, U(\lambda)\rangle$.

The following theorems hold.

Theorem 8.2.5. *The least favorable spectral densities $f^0(\lambda)$ and $g^0(\lambda)$ in the classes $\mathcal{D}_\varepsilon^k \times \mathcal{D}_V^{Uk}$, $k = 1,2,3,4$, for the optimal linear extrapolation of the functional $A\xi$ from observations of the process $\xi(t) + \eta(t)$ at points $t < 0$ are determined by canonical factorizations* (8.17) *and* (8.18), *equations* (8.63)–(8.64), (8.65)–(8.66), (8.67)–(8.68), *and* (8.69)–(8.70), *respectively, the constrained optimization problem* (8.23), *and restrictions on densities from the corresponding classes $\mathcal{D}_\varepsilon^k$, $\mathcal{D}_V^{Uk}$, $k = 1,2,3,4$. The minimax-robust spectral characteristic of the optimal estimate of the functional $A\xi$ is determined by formula* (8.20).

Theorem 8.2.6. *If the spectral density $g(\lambda)$ is known, the least favorable spectral density $f^0(\lambda)$ in the classes $\mathcal{D}_\varepsilon^k$, $k = 1,2,3,4$, for the optimal linear extrapolation of the functional $A\xi$ from observations of the process $\xi(t) + \eta(t)$ at points $t < 0$ is determined by canonical factorizations* (8.18) *and* (8.17), *equations* (8.63), (8.65), (8.67), *and* (8.69), *respectively, the constrained optimization problem* (8.28), *and restrictions on density from the corresponding classes $\mathcal{D}_\varepsilon^k$, $k = 1,2,3,4$. The minimax-robust spectral characteristic of the optimal estimate of the functional $A\xi$ is determined by formula* (8.20).

Conclusions

In this chapter, we describe methods of solution of the extrapolation problem for linear functionals which are constructed from unobserved values of stochastic processes with periodically stationary increments. Estimates of functionals are based on observations of the process $\xi(t)$ with an uncorrelated with the process $\xi(t)$ periodically stationary stochastic noise process $\eta(t)$. The problem is solved by representing these processes as infinite-dimensional sequences with stationary increments.

The Hilbert space projection method is applied to derive formulas for calculating the spectral characteristics and the MSEs of the optimal estimates of the functionals in the case of spectral certainty, where the spectral densities of the generated sequences are exactly known. In the case of spectral uncertainty, where the spectral densities are not exactly known while, instead, a set of admissible spectral densities is specified, the minimax-robust method is applied. We propose a representation of the MSE in the form of a linear functional in L_1 space with respect to spectral densities, which allows us to solve the corresponding constrained optimization problem and describe the minimax (robust) estimates of the functionals.

This method is illustrated with the help of concrete sets $\mathcal{D}_0 \times \mathcal{D}_{1\delta}$ and $\mathcal{D}_\varepsilon \times \mathcal{D}_V^U$ of admissible spectral densities. Here D_0 are classes of spectral densities with the moment restrictions, $D_{1\delta}$ are classes describing the “δ-neighborhood” models in the space L_1 of a fixed bounded spectral density, D_ε are classes describing the “ε-contaminated” models of a fixed bounded spectral density, and D_V^U are classes describing the “strip” models of spectral densities.

The main results presented in this chapter have been published in [127]. These results rely on and develop prior studies of the minimax extrapolation problem by Moklyachuk and Masyutka [148] for multidimensional stationary processes, Golichenko (Dubovets’ka) and Moklyachuk [43] for periodically correlated processes, and Luz and Moklyachuk [112] for processes with stationary increments.

9 Interpolation of processes with periodically correlated increments observed with or without noise

In this chapter, we deal with the interpolation problem for stochastic processes with periodically stationary dth increments observed with a periodically stationary stochastic noise process.

By the *interpolation problem* we understand the problem of the mean square optimal linear estimation of the functional

$$A_{NT}\xi = \int_{0}^{(N+1)T} a(t)\xi(t)dt$$

which depends on the unknown values of the stochastic process $\xi(t)$ with periodically correlated dth increments. Estimates are based on observations of the process $\zeta(t) = \xi(t) + \eta(t)$ at points $t \in \mathbb{R} \setminus [0;(N+1)T]$, where $\eta(t)$ is an uncorrelated with the process $\xi(t)$ periodically stationary stochastic process.

9.1 Hilbert space projection method of interpolation

Consider an infinite-dimensional vector stationary increment sequence $\{\vec{\xi}_j^{(d)}, j \in \mathbb{Z}\}$ (6.7) generated by the periodically stationary increment process $\xi^{(d)}(t, \tau T)$, $t \in \mathbb{R}$, which has the spectral density matrix $f(\lambda) = \{f_{ij}(\lambda)\}_{i,j=1}^{\infty}$.

Let $\eta(t)$, $t \in \mathbb{R}$, be an uncorrelated with the process $\xi(t)$ periodically stationary stochastic process, which generates an infinite-dimensional vector stationary sequence $\{\vec{\eta}_j, j \in \mathbb{Z}\}$ (6.3) with the spectral density matrix $g(\lambda) = \{g_{ij}(\lambda)\}_{i,j=1}^{\infty}$.

We will assume that the mean values of the increment sequence $\vec{\xi}_j^{(d)}$ and stationary sequence $\vec{\eta}_j$ equal 0. We will also consider the increment step $\tau > 0$.

In this section, we consider the interpolation problem under the condition that the spectral densities $f(\lambda)$ and $g(\lambda)$ are exactly known.

9.1.1 Interpolation of stochastic processes with periodically stationary *d*th increments observed with periodically stationary noise

Assume that spectral densities $f(\lambda)$ and $g(\lambda)$ satisfy the *minimality condition*

$$\int_{-\pi}^{\pi} \mathrm{Tr}\left[\frac{\lambda^{2d}}{|1-e^{i\lambda\tau}|^{2d}}\left(f(\lambda) + \lambda^{2d}g(\lambda)\right)^{-1}\right]d\lambda < \infty. \tag{9.1}$$

https://doi.org/10.1515/9783111325620-009

This is the necessary and sufficient condition under which the MSEs of the optimal estimates of the functional $A_N\vec{\xi}$ to be defined below are not equal to 0.

The functional $A_{NT}\xi$ does not belong to the Hilbert space $H = L_2(\Omega, \mathcal{F}, P)$ of random variables with zero mean value and finite variance. To apply the Hilbert space projection method of estimation, we represent the functional $A_{NT}\xi$ as the sum of a functional with finite second moment belonging to H and a functional depending on the observed values of the process $\zeta(t)$ ("initial values"). This representation is described by the following corollaries from Section 8.1.

Corollary 9.1.1. *The linear functional*

$$A_{NT}\zeta = \int_0^{(N+1)T} a(t)\zeta(t)dt$$

allows the representation

$$A_{NT}\zeta = B_{NT}\zeta - V_{NT}\zeta,$$

where

$$B_{NT}\zeta = \int_0^{(N+1)T} b^{\tau,N}(t)\zeta^{(d)}(t,\tau T)dt, \quad V_{NT}\zeta = \int_{-\tau Td}^{0} v^{\tau,N}(t)\zeta(t)dt,$$

and

$$v^{\tau,N}(t) = \sum_{l=\lceil -\frac{t}{\tau T}\rceil}^{\min\{[\frac{(N+1)T-t}{\tau T}],d\}} (-1)^l \binom{d}{l} b^{\tau,N}(t+l\tau T), \quad t \in [-\tau Td; 0), \tag{9.2}$$

$$b^{\tau,N}(t) = \sum_{k=0}^{[\frac{(N+1)T-t}{\tau T}]} a(t+\tau Tk)d(k) = D^{\tau T,N}\mathbf{a}(t), \quad t \in [0; (N+1)T]. \tag{9.3}$$

Here $\lceil x\rceil$ denotes the least integer greater than or equal to x, $[x]$ denotes the integer part of x, coefficients $\{d(k) : k \geq 0\}$ are determined by the relation

$$\sum_{k=0}^{\infty} d(k)x^k = \left(\sum_{j=0}^{\infty} x^j\right)^d,$$

and the linear transformation $D^{\tau T,N}$ acts on an arbitrary function $x(t)$, $t \in [0; (N+1)T]$, as follows:

$$D^{\tau T,N}\mathbf{x}(t) = \sum_{k=0}^{[\frac{(N+1)T-t}{\tau T}]} x(t+\tau Tk)d(k).$$

From Corollary 9.1.1 we obtain the following representation of the functional $A_{NT}\xi$:

$$A_{NT}\xi = A_{NT}\zeta - A_{NT}\eta = B_{NT}\zeta - A_{NT}\eta - V_{NT}\zeta = H_{NT}\xi - V_{NT}\zeta,$$

where

$$H_{NT}\xi := B_{NT}\zeta - A_{NT}\eta$$

and

$$A_{NT}\zeta = \int_0^{(N+1)T} a(t)\zeta(t)dt, \quad A_{NT}\eta = \int_0^{(N+1)T} a(t)\eta(t)dt,$$

$$B_{NT}\zeta = \int_0^{(N+1)T} b_{\tau,N}(t)\zeta^{(d)}(t,\tau T)dt, \quad V_{NT}\zeta = \int_{-\tau Td}^{0} v_{\tau,N}(t)\zeta(t)dt,$$

and the functions $b_{\tau,N}(t)$, $t \in [0;(N+1)T]$, and $v_{\tau,N}(t)$, $t \in [-\tau Td;0)$, are calculated by formulas (9.3) and (9.2), respectively.

The following corollary represents the functional $H_{NT}\xi$ in terms of the sequences $\vec{\eta}_j = (\eta_{kj}, k=1,2,\dots)^\top$ and $\vec{\zeta}_j^{(d)} = \vec{\xi}_j^{(d)} + \vec{\eta}_j^{(d)} = (\zeta_{kj}^{(d)}, k=1,2,\dots)^\top, j \in \mathbb{Z}$.

Corollary 9.1.2. *The functional $H_{NT}\xi$ allows a representation in the form*

$$H_{NT}\xi = \sum_{j=0}^{N} (\vec{b}_j^{\tau,N})^\top \vec{\zeta}_j^{(d)} - \sum_{j=0}^{N} (\vec{a}_j)^\top \vec{\eta}_j = B_N\vec{\zeta} - V_N\vec{\eta} =: H_N\vec{\xi},$$

where

$$\vec{b}_j^{\tau,N} = (b_{kj}^{\tau,N}, k=1,2,\dots)^\top = (b_{1j}^{\tau,N}, b_{3j}^{\tau,N}, b_{2j}^{\tau,N}, \dots, b_{2k+1,j}^{\tau,N}, b_{2k,j}^{\tau,N}, \dots)^\top,$$

$$b_{kj}^{\tau,N} = \langle b_j^{\tau,N}, \widetilde{e}_k\rangle = \frac{1}{\sqrt{T}}\int_0^T b_j^{\tau,N}(v)e^{-2\pi i\{(-1)^k[\frac{k}{2}]\}v/T}dv,$$

and

$$\vec{a}_j = (a_{kj}, k=1,2,\dots)^\top = (a_{1j}, a_{3j}, a_{2j}, \dots, a_{2k+1,j}, a_{2k,j}, \dots)^\top,$$

$$a_{kj} = \langle a_j, \widetilde{e}_k\rangle = \frac{1}{\sqrt{T}}\int_0^T a_j(v)e^{-2\pi i\{(-1)^k[\frac{k}{2}]\}v/T}dv.$$

The vector coefficients $\{\vec{a}_j, j=0,1,\dots,N\}$ and $\{\vec{b}_j^{\tau,N}, j=0,1,\dots,N\}$ are related as

$$\vec{b}_j^{\tau,N} = \sum_{m=j}^{N} \mathrm{diag}_\infty(d_\tau(m-j))\vec{a}_m = (D_N^\tau \mathbf{a}_N)_j, \quad j=0,1,\dots,N, \tag{9.4}$$

where $\mathbf{a}_N = ((\vec{a}_0)^\top, (\vec{a}_1)^\top, \dots, (\vec{a}_N)^\top)^\top$, *the coefficients* $\{d_\tau(k) : k \geq 0\}$ *are determined by the relationship*

$$\sum_{k=0}^{\infty} d_\tau(k)x^k = \left(\sum_{j=0}^{\infty} x^{\tau j}\right)^d,$$

and D_N^τ *is a linear transformation determined by an* $N \times N$ *matrix with the infinite-dimensional matrix entries* $D_N^\tau(k,j)$, $k,j = 0,1,\dots,N$, *such that* $D_N^\tau(k,j) = \operatorname{diag}_\infty(d_\tau(j-k))$ *if* $0 \leq k \leq j \leq N$ *and* $D_N^\tau(k,j) = \mathcal{O}$ *if* $0 \leq j < k \leq N$; $\operatorname{diag}_\infty(x)$ *denotes an infinite-dimensional diagonal matrix with the entry* x *on its diagonal.*

Assume that the coefficients $\{\vec{a}_j, j = 0,1,\dots,N\}$ and $\{\vec{b}_j^{\tau,N}, j = 0,1,\dots,N\}$ satisfy the conditions

$$\|\vec{a}_j\| < \infty, \quad \|\vec{a}_j\|^2 = \sum_{k=1}^{\infty} |a_{kj}|^2, \quad j = 0,1,\dots,N, \tag{9.5}$$

and

$$\|\vec{b}_j^{\tau,N}\| < \infty, \quad \|\vec{b}_j^{\tau,N}\|^2 = \sum_{k=1}^{\infty} |b_{kj}^{\tau,N}|^2, \quad j = 0,1,\dots,N. \tag{9.6}$$

It follows from conditions (9.5)–(9.6) that the functional $H_{NT}\xi = H_N\vec{\xi}$ has a finite second moment and the functional $V_{NT}\zeta$ depends on the observations of the process $\xi(t) + \eta(t)$ at points $t \in [\tau T d; 0)$. Therefore, the estimates $\widehat{A}_{NT}\xi$ and $\widehat{H}_N\vec{\xi}$ of the functionals $A_{NT}\xi$ and $H_N\vec{\xi}$, and the MSEs $\Delta(f,g;\widehat{A}_{NT}\xi) = \Delta(f,g;\widehat{A}_N\vec{\xi}) = \mathsf{E}|A_N\vec{\xi} - \widehat{A}_N\vec{\xi}|^2$ and $\Delta(f,g;\widehat{H}_{NT}\xi) = \Delta(f,g;\widehat{H}_N\vec{\xi}) = \mathsf{E}|H_N\vec{\xi} - \widehat{H}_N\vec{\xi}|^2$ of the estimates $\widehat{A}_{NT}\xi$ and $\widehat{H}_N\vec{\xi}$ satisfy the following relations:

$$\begin{aligned}
\widehat{A}_{NT}\xi &= \widehat{H}_N\vec{\xi} - V_{NT}\zeta, \\
\Delta(f,g;\widehat{A}_{NT}\xi) &= \mathsf{E}|A_{NT}\xi - \widehat{A}_{NT}\xi|^2 = \mathsf{E}|H_N\vec{\xi} - V_{NT}\zeta - \widehat{H}_N\vec{\xi} + V_{NT}\zeta|^2 \\
&= \mathsf{E}|H_N\vec{\xi} - \widehat{H}_N\vec{\xi}|^2 = \Delta(f,g;\widehat{H}_N\vec{\xi}).
\end{aligned} \tag{9.7}$$

Thus, it is enough to find the estimate of the functional $H_N\vec{\xi}$.

Making use of spectral representations (6.10) and (6.11), one can obtain the following spectral representation of the functional $H_N\vec{\xi}$:

$$H_N\vec{\xi} = \int_{-\pi}^{\pi} (\vec{B}_{\tau,N}(e^{i\lambda}))^\top \frac{(1-e^{-i\lambda\tau})^d}{(i\lambda)^d} d\vec{Z}_{\xi^{(d)}+\eta^{(d)}}(\lambda) - \int_{-\pi}^{\pi} (\vec{A}_N(e^{i\lambda}))^\top d\vec{Z}_\eta(\lambda),$$

where

$$\vec{B}_{\tau,N}(e^{i\lambda}) = \sum_{j=0}^{N} \vec{b}_j^{\tau,N} e^{i\lambda j} = \sum_{j=0}^{N} (D_N^{\tau}\mathbf{a}_N)_j e^{i\lambda j}, \quad \vec{A}_N(e^{i\lambda}) = \sum_{j=0}^{N} \vec{a}_j e^{i\lambda j}.$$

The next step is to define the subspace of $H = L_2(\Omega, \mathcal{F}, \mathsf{P})$ generated by the observations. Consider the closed linear subspaces

$$H^{0-}(\vec{\xi}_{\tau}^{(d)} + \vec{\eta}_{\tau}^{(d)}) = \overline{\operatorname{span}}\{\xi_k^{(d)}(j,\tau) + \eta_k^{(d)}(j,\tau) : k = 1,\dots,\infty; j \le -1\}$$

and

$$H^{N+}(\vec{\xi}_{-\tau}^{(d)} + \vec{\eta}_{-\tau}^{(d)}) = \overline{\operatorname{span}}\{\xi_k^{(d)}(j,-\tau) + \eta_k^{(d)}(j,-\tau) : k = 1,\dots,\infty; j \ge N+1\}.$$

The equality $\xi_k^{(d)}(j,-\tau) = (-1)^d \xi_k^{(d)}(j + \tau d, \tau)$ implies

$$H^{N+}(\vec{\xi}_{-\tau}^{(d)} + \vec{\eta}_{-\tau}^{(d)}) = H^{(N+\tau d)+}(\vec{\xi}_{\tau}^{(d)} + \vec{\eta}_{\tau}^{(d)}).$$

Define the following closed linear subspaces of the Hilbert space $L_2(f(\lambda) + \lambda^{2d} g(\lambda))$ of vector-valued functions endowed with the inner product

$$\langle g_1; g_2 \rangle = \int_{-\pi}^{\pi} (g_1(\lambda))^{\top} (f(\lambda) + \lambda^{2d} g(\lambda)) \overline{g_2(\lambda)} d\lambda,$$

which is generated by the functions

$$L_2^{0-}(f(\lambda) + \lambda^{2d} g(\lambda)) = \overline{\operatorname{span}}\left\{ e^{i\lambda j}(1 - e^{-i\lambda\tau})^d \frac{1}{(i\lambda)^d} \vec{\delta}_k, k = 1,\dots,\infty; j \le -1 \right\}$$

and

$$L_2^{N+}(f(\lambda) + \lambda^{2d} g(\lambda)) = \overline{\operatorname{span}}\left\{ e^{i\lambda j}(1 - e^{-i\lambda\tau})^d \frac{1}{(i\lambda)^d} \vec{\delta}_k, k = 1,\dots,\infty; j \ge N+1 \right\},$$

where $\vec{\delta}_k = \{\delta_{kl}\}_{l=1}^{\infty}$, where δ_{kl} are Kronecker symbols.

Remark 9.1.1. Representation (6.11) yields a map between the elements

$$e^{i\lambda j}(1 - e^{-i\lambda\tau})^d (i\lambda)^{-d} \vec{\delta}_k$$

of the space

$$L_2^{0-}(f(\lambda) + \lambda^{2d} g(\lambda)) \oplus L_2^{(N+\tau d)+}(f(\lambda) + \lambda^{2d} g(\lambda))$$

and the elements $\xi_{kj}^{(d)} + \eta_{kj}^{(d)}$ of the space

$$H^{0-}(\vec{\xi}_{\tau}^{(d)} + \vec{\eta}_{\tau}^{(d)}) \oplus H^{N+}(\vec{\xi}_{-\tau}^{(d)} + \vec{\eta}_{-\tau}^{(d)}) = H^{0-}(\vec{\xi}_{\tau}^{(d)} + \vec{\eta}_{\tau}^{(d)}) \oplus H^{(N+\tau d)+}(\vec{\xi}_{\tau}^{(d)} + \vec{\eta}_{\tau}^{(d)}).$$

Now we can apply the Hilbert space projection approach. The mean square optimal estimate $\widehat{H}_N\vec{\xi}$ is found as an orthogonal projection

$$\widehat{H}_N\vec{\xi} = \text{Proj}_{H^{0-}(\vec{\xi}_{\tau}^{(d)}+\vec{\eta}_{\tau}^{(d)})\oplus H^{(N+\tau d)+}(\vec{\xi}_{\tau}^{(d)}+\vec{\eta}_{\tau}^{(d)})} H_N\vec{\xi}. \tag{9.8}$$

Relation (9.7) implies that an optimal linear estimate $\widehat{A}_{NT}\xi$ of the functional $A_{NT}\xi$ can be represented as

$$\widehat{A}_{NT}\xi = \text{Proj}_{H^{0-}(\vec{\xi}_{\tau}^{(d)}+\vec{\eta}_{\tau}^{(d)})\oplus H^{(N+\tau d)+}(\vec{\xi}_{\tau}^{(d)}+\vec{\eta}_{\tau}^{(d)})} H_N\vec{\xi} - V_N\zeta$$

or as

$$\widehat{A}_{NT}\xi = \int_{-\pi}^{\pi} (\vec{h}_{\tau,N}(\lambda))^{\top} d\vec{Z}_{\xi^{(d)}+\eta^{(d)}}(\lambda) - \int_{-\tau Td}^{0} v_{\tau,N}(t)\zeta(t)dt, \tag{9.9}$$

where $\vec{h}_{\tau,N}(\lambda) = \{h_k^{\tau,N}(\lambda)\}_{k=1}^{\infty}$ is the spectral characteristic of the estimate $\widehat{H}_N\vec{\xi}$.

Define for $0 \le l, j \le N + \tau d$ the Fourier coefficients of the corresponding functions

$$T_{l,j}^{\tau} = \frac{1}{2\pi}\int_{-\pi}^{\pi} e^{i\lambda(j-l)} \frac{\lambda^{2d}}{|1-e^{i\lambda\tau}|^{2d}} (g(\lambda)(f(\lambda) + \lambda^{2d}g(\lambda))^{-1})^{\top} d\lambda,$$

$$P_{l,j}^{\tau} = \frac{1}{2\pi}\int_{-\pi}^{\pi} e^{i\lambda(j-l)} \frac{\lambda^{2d}}{|1-e^{i\lambda\tau}|^{2d}} ((f(\lambda) + \lambda^{2d}g(\lambda))^{-1})^{\top} d\lambda,$$

$$Q_{l,j} = \frac{1}{2\pi}\int_{-\pi}^{\pi} e^{i\lambda(j-l)} (f(\lambda)(f(\lambda) + \lambda^{2d}g(\lambda))^{-1}g(\lambda))^{\top} d\lambda.$$

Denote

$$\vec{A}_{\tau,N}(e^{i\lambda}) = (1-e^{i\lambda\tau})^d \vec{A}_N(e^{i\lambda}) = \sum_{j=0}^{N+\tau d} \vec{a}_j^{\tau,N} e^{i\lambda j},$$

where the coefficients $\{\vec{a}_j^{\tau,N} : 0 \le j \le N + \tau d\}$ are calculated by the formula

$$\vec{a}_j^{\tau,N} = \sum_{l=\max\{\lceil (j-N)/\tau \rceil, 0\}}^{\min\{[j/\tau], d\}} (-1)^l \binom{d}{l} \vec{a}_{j-\tau l}, \quad 0 \le j \le N + \tau d. \tag{9.10}$$

Define a vector

$$\mathbf{a}_N^{\tau} = ((\vec{a}_0^{\tau,N})^{\top}, (\vec{a}_1^{\tau,N})^{\top}, (\vec{a}_2^{\tau,N})^{\top}, \ldots, (\vec{a}_{N+\tau d}^{\tau,N})^{\top})^{\top}$$

of dimension $N+\tau d+1$ with infinite-dimensional entries, matrices $\mathbf{P}_N^{\tau}$ and $\mathbf{T}_N^{\tau}$ of dimension $(N+\tau d+1)\times(N+\tau d+1)$ with the infinite-dimensional matrix entries $(\mathbf{P}_N^{\tau})_{l,j} = P_{j,l}^{\tau}$ and $(\mathbf{T}_N^{\tau})_{l,j} = T_{j,l}^{\tau}$, $0 \le l, j \le N+\tau d$, and a matrix $\mathbf{Q}_N$ of dimension $(N+1)\times(N+1)$ with the infinite-dimensional matrix entries $(\mathbf{Q}_N)_{j,l} = Q_{j,l}$, $0 \le j, l \le N$.

Finally, denote by $[D_N^{\tau}\mathbf{a}_N]_{+\tau d}$ a vector of dimension $N+\tau d+1$ with infinite-dimensional entries, which is constructed by adding τd infinite-dimensional zero vectors to the vector $D_N^{\tau}\mathbf{a}_N$ of dimension $N+1$.

The following theorem holds.

Theorem 9.1.1. *Consider two uncorrelated processes: a stochastic process $\xi(t)$, $t \in \mathbb{R}$, with periodically stationary increments and a periodically stationary stochastic process $\eta(t)$, $t \in \mathbb{R}$, which determine the generated stationary dth increment sequence $\vec{\xi}_j^{(d)}$ with the spectral density matrix $f(\lambda) = \{f_{kn}(\lambda)\}_{k,n=1}^{\infty}$ and the generated stationary sequence $\vec{\eta}_j$ with the spectral density matrix $g(\lambda) = \{g_{kn}(\lambda)\}_{k,n=1}^{\infty}$, respectively. Let the coefficients $\vec{a}_j$, $\vec{b}_j^{\tau,N}$, $j = 0, 1, \ldots, N$, generated by the function $a(t)$, $t \ge 0$, satisfy conditions* (9.5)–(9.6). *Let the minimality condition* (9.1) *be satisfied. The optimal linear estimate $\widehat{A}_{NT}\xi$ of the functional $A_{NT}\xi$ based on observations of the process $\xi(t)+\eta(t)$ at points $t \in \mathbb{R}\backslash[0;(N+1)T]$ is calculated by formula* (9.9). *The spectral characteristic $\vec{h}_{\tau,N}(\lambda)$ of the estimate $\widehat{A}_{NT}\xi$ is calculated by the formula*

$$\begin{aligned}(\vec{h}_{\tau,N}(\lambda))^{\top} &= (\vec{B}_{\tau,N}(e^{i\lambda}))^{\top}\frac{(1-e^{-i\lambda\tau})^d}{(i\lambda)^d} \\ &\quad - (\vec{A}_{\tau,N}(e^{i\lambda}))^{\top} g(\lambda)\frac{(-i\lambda)^d}{(1-e^{i\lambda\tau})^d}(f(\lambda)+\lambda^{2d}g(\lambda))^{-1} \\ &\quad - (\vec{C}_{\tau,N}(e^{i\lambda}))^{\top}\frac{(-i\lambda)^d}{(1-e^{i\lambda\tau})^d}(f(\lambda)+\lambda^{2d}g(\lambda))^{-1}, \end{aligned} \tag{9.11}$$

where

$$\vec{C}_{\tau,N}(e^{i\lambda}) = \sum_{j=0}^{N+\tau d}((\mathbf{P}_N^{\tau})^{-1}[D_N^{\tau}\mathbf{a}_N]_{+\tau d} - (\mathbf{P}_N^{\tau})^{-1}\mathbf{T}_N^{\tau}\mathbf{a}_N^{\tau})_j e^{i\lambda j}. \tag{9.12}$$

The MSE of the estimate $\widehat{A}_{NT}\xi$ is calculated by the formula

$$\begin{aligned}\Delta(f,g;\widehat{A}_{NT}\xi) &= \Delta(f,g;\widehat{H}_N\vec{\xi}) = \mathsf{E}|H_N\vec{\xi} - \widehat{H}_N\vec{\xi}|^2 \\ &= \frac{1}{2\pi}\int_{-\pi}^{\pi}\frac{\lambda^{2d}}{|1-e^{i\lambda\tau}|^{2d}}(\mathbf{C}_{\tau,N}^f(e^{i\lambda}))^{\top}(f(\lambda)+\lambda^{2d}g(\lambda))^{-1}f(\lambda) \\ &\quad \times (f(\lambda)+\lambda^{2d}g(\lambda))^{-1}\overline{\mathbf{C}_{\tau,N}^f(e^{i\lambda})}d\lambda\end{aligned}$$

$$+\frac{1}{2\pi}\int_{-\pi}^{\pi}\frac{\lambda^{4d}}{|1-e^{i\lambda\tau}|^{4d}}(\mathbf{C}^g_{\tau,N}(e^{i\lambda}))^{\top}(f(\lambda)+\lambda^{2d}g(\lambda))^{-1}g(\lambda)$$
$$\times(f(\lambda)+\lambda^{2d}g(\lambda))^{-1}\overline{\mathbf{C}^g_{\tau,N}(e^{i\lambda})}d\lambda$$
$$=\langle[D^{\tau}_N\mathbf{a}_N]_{+\tau d}-\mathbf{T}^{\tau}_N\mathbf{a}^{\tau}_N,(\mathbf{P}^{\tau}_N)^{-1}[D^{\tau}_N\mathbf{a}_N]_{+\tau d}-(\mathbf{P}^{\tau}_N)^{-1}\mathbf{T}^{\tau}_N\mathbf{a}^{\tau}_N\rangle$$
$$+\langle\mathbf{Q}_N\mathbf{a}_N,\mathbf{a}_N\rangle, \tag{9.13}$$

where

$$\mathbf{C}^f_{\tau,N}(e^{i\lambda})=\overline{g(\lambda)\vec{A}_{\tau,N}(e^{i\lambda})}+\vec{C}_{\tau,N}(e^{i\lambda}),$$
$$\mathbf{C}^g_{\tau,N}(e^{i\lambda})=|1-e^{i\lambda\tau}|^{2d}\lambda^{-2d}\overline{f(\lambda)\vec{A}_N(e^{i\lambda})}-(1-e^{-i\lambda\tau})^d\vec{C}_{\tau,N}(e^{i\lambda}).$$

Proof. Projection (9.8) is characterized by two conditions:
(1) $\widehat{H}_N\vec{\xi}\in H^{0-}(\vec{\xi}^{(d)}_{\tau}+\vec{\eta}^{(d)}_{\tau})\oplus H^{(N+\tau d)+}(\vec{\xi}^{(d)}_{\tau}+\vec{\eta}^{(d)}_{\tau})$;
(2) $(H_N\vec{\xi}-\widehat{H}_N\vec{\xi})\perp H^{0-}(\vec{\xi}^{(d)}_{\tau}+\vec{\eta}^{(d)}_{\tau})\oplus H^{(N+\tau d)+}(\vec{\xi}^{(d)}_{\tau}+\vec{\eta}^{(d)}_{\tau})$.

The following relation follows from condition (2) and holds for all $j\le -1$, $j\ge N+\tau d+1$, and $k\ge 1$:

$$\int_{-\pi}^{\pi}\left(\left(\vec{B}_{\tau,N}(e^{i\lambda})\frac{(1-e^{-i\lambda\tau})^d}{(i\lambda)^d}-\vec{h}_{\tau,N}(\lambda)\right)^{\top}(f(\lambda)+\lambda^{2d}g(\lambda))\right.$$
$$\left.-(\vec{A}_N(e^{i\lambda}))^{\top}g(\lambda)(-i\lambda)^d\right)\vec{\delta}_k\frac{(1-e^{i\lambda\tau})^d}{(-i\lambda)^d}e^{-i\lambda j}d\lambda=0.$$

This implies representation (9.11) of the spectral characteristic $\vec{h}_{\tau,N}(\lambda)$ of the estimate $\widehat{H}_N\vec{\xi}$, where

$$\vec{C}_{\tau,N}(e^{i\lambda})=\sum_{j=0}^{N+\tau d}\vec{c}^{\,\tau,N}_j e^{i\lambda j},$$

where the vectors $\vec{c}^{\,\tau,N}_j=\{c^{\tau,N}_{kj}\}_{k=1}^{\infty}$, $j=0,1,\dots,N+\tau d$, are unknown vectors to be found.
Condition (1) implies that the following equations should be satisfied:

$$\int_{-\pi}^{\pi}\left((\vec{B}_{\tau,N}(e^{i\lambda}))^{\top}-(\vec{A}_{\tau,N}(e^{i\lambda}))^{\top}g(\lambda)\frac{\lambda^{2d}}{|1-e^{i\lambda\tau}|^{2d}}(f(\lambda)+\lambda^{2d}g(\lambda))^{-1}\right.$$
$$\left.-\frac{\lambda^{2d}}{|1-e^{i\lambda\tau}|^{2d}}(\vec{C}_{\tau,N}(e^{i\lambda}))^{\top}(f(\lambda)+\lambda^{2d}g(\lambda))^{-1}\right)e^{-i\lambda j}d\lambda=0,\quad 0\le j\le N+\tau d. \tag{9.14}$$

Making use of the defined Fourier coefficients, relation (9.14) can be presented as a system of $N+\tau d+1$ linear equations determining the unknown coefficients $\vec{c}^{\,\tau,N}_j$, $0\le j\le N+\tau d$:

$$\begin{cases} \vec{b}_l^{\tau,N} - \sum_{j=0}^{N+\tau d} T_{l,j}^{\tau} \vec{a}_j^{\tau,N} = \sum_{j=0}^{N+\tau d} P_{l,j}^{\tau} \vec{c}_j^{\tau,N}, & 0 \le l \le N, \\ - \sum_{j=0}^{N+\tau d} T_{l,j}^{\tau} \vec{a}_j^{\tau,N} = \sum_{j=0}^{N+\tau d} P_{l,j}^{\tau} \vec{c}_j^{\tau,N}, & N+1 \le l \le N+\tau d. \end{cases} \tag{9.15}$$

Denote a vector

$$\mathbf{c}_N^{\tau} = ((\vec{c}_0^{\tau,N})^{\top}, (\vec{c}_1^{\tau,N})^{\top}, (\vec{c}_2^{\tau,N})^{\top}, \ldots, (\vec{c}_{N+\tau d}^{\tau,N})^{\top})^{\top}$$

of dimension $N + \tau d + 1$ with infinite-dimensional entries. Then the system (9.15) can be represented in matrix form as

$$[D_N^{\tau} \mathbf{a}_N]_{+\tau d} - \mathbf{T}_N^{\tau} \mathbf{a}_N^{\tau} = \mathbf{P}_N^{\tau} \mathbf{c}_N^{\tau}.$$

Thus, the coefficients $\vec{c}_j^{\tau,N}$, $0 \le j \le N + \tau d$, are determined by the formula

$$\vec{c}_j^{\tau,N} = ((\mathbf{P}_N^{\tau})^{-1} [D_N^{\tau} \mathbf{a}_N]_{+\tau d} - (\mathbf{P}_N^{\tau})^{-1} \mathbf{T}_N^{\tau} \mathbf{a}_N^{\tau})_j, \quad 0 \le j \le N + \tau d,$$

where $((\mathbf{P}_N^{\tau})^{-1}[D_N^{\tau}\mathbf{a}_N]_{+\tau d} - (\mathbf{P}_N^{\tau})^{-1}\mathbf{T}_N^{\tau}\mathbf{a}_N^{\tau})_j$, $0 \le j \le N + \tau d$, is the jth entry of the vector $(\mathbf{P}_N^{\tau})^{-1}[D_N^{\tau}\mathbf{a}_N]_{+\tau d} - (\mathbf{P}_N^{\tau})^{-1}\mathbf{T}_N^{\tau}\mathbf{a}_N^{\tau}$.

The existence of the inverse matrix $(\mathbf{P}_N^{\tau})^{-1}$ was justified in [115] under condition (9.1).

The derived expressions justify the use of formulas (9.11) and (9.13) for calculating the spectral characteristic $\vec{h}_{\tau}(\lambda)$ and the MSE $\Delta(f, g; \widehat{A}_{NT}\xi)$ of the estimate $\widehat{A}_{NT}\xi$ of the functional $A_{NT}\xi$. □

Corollary 9.1.3. *The spectral characteristic $\vec{h}_{\tau,N}(\lambda)$ (9.11) admits the representation $\vec{h}_{\tau,N}(\lambda) = \vec{h}_{\tau,N}^1(\lambda) - \vec{h}_{\tau,N}^2(\lambda)$, where*

$$\begin{aligned} (\vec{h}_{\tau,N}^1(\lambda))^{\top} &= (\vec{B}_{\tau,N}(e^{i\lambda}))^{\top} \frac{(1 - e^{-i\lambda\tau})^d}{(i\lambda)^d} \\ &\quad - \frac{(-i\lambda)^d}{(1 - e^{i\lambda\tau})^d} \left(\sum_{j=0}^{N+\tau d} ((\mathbf{P}_N^{\tau})^{-1}[D_N^{\tau}\mathbf{a}_N]_{+\tau d})_j e^{i\lambda j} \right)^{\top} (f(\lambda) + \lambda^{2d} g(\lambda))^{-1}, \\ (\vec{h}_{\tau,N}^2(\lambda))^{\top} &= (\vec{A}_N(e^{i\lambda}))^{\top} (-i\lambda)^d g(\lambda)(f(\lambda) + \lambda^{2d} g(\lambda))^{-1} \\ &\quad - \frac{(-i\lambda)^d}{(1 - e^{i\lambda\tau})^d} \left(\sum_{j=0}^{N+\tau d} ((\mathbf{P}_N^{\tau})^{-1}\mathbf{T}_N^{\tau}\mathbf{a}_N^{\tau})_j e^{i\lambda j} \right)^{\top} (f(\lambda) + \lambda^{2d} g(\lambda))^{-1}. \end{aligned}$$

Here $\vec{h}_{\tau,N}^1(\lambda)$ and $\vec{h}_{\tau,N}^2(\lambda)$ are spectral characteristics of the optimal estimates $\widehat{B}_{NT}\zeta$ and $\widehat{A}_{NT}\eta$ of the functionals $B_{NT}\zeta$ and $A_{NT}\eta$, respectively, based on observations $\xi(t) + \eta(t)$ at points of the set $t \in \mathbb{R} \setminus [0; (N+1)T]$.

9.1.2 Interpolation of stochastic processes with periodically stationary dth increments observed without noise

In this subsection, we describe a corollary from the interpolation problem solutions for the case of observations without noise. Let an infinite-dimensional vector stationary increment sequence $\{\vec{\xi}_j^{(d)}, j \in \mathbb{Z}\}$ (6.7) generated by the periodically stationary increment process $\xi^{(d)}(t, \tau T)$, $t \in \mathbb{R}$, have the spectral density matrix $f(\lambda) = \{f_{ij}(\lambda)\}_{i,j=1}^{\infty}$ satisfying the minimality condition

$$\int_{-\pi}^{\pi} \operatorname{Tr}\left[\frac{\lambda^{2d}}{|1 - e^{i\lambda\tau}|^{2d}} f^{-1}(\lambda)\right] d\lambda < \infty. \tag{9.16}$$

By *interpolation problem without noise* we understand the problem of mean square optimal linear estimation of the functional

$$A_{NT}\xi = \int_0^{(N+1)T} a(t)\xi(t)dt,$$

which depends on the unobserved values of the stochastic process $\xi(t)$ with periodically correlated dth increments based on observations of the process $\xi(t)$ at points $t \in \mathbb{R} \setminus [0; (N+1)T]$.

Corollary 7.1.1 lets us exhibit the functional $A_{NT}\xi$ as the sum of two functionals: a functional $B_{NT}\xi$ from the space $H = L_2(\Omega, \mathcal{F}, \mathsf{P})$ and a functional $V_{NT}\xi$ of observed values $\xi(t)$ at points $t \in [0; (N+1)T]$. The following equality holds:

$$\widehat{A}_{NT}\xi = \widehat{B}_{NT}\xi - V_{NT}\xi. \tag{9.17}$$

The functional $B_{NT}\xi$ can be rewritten in the form

$$B_{NT}\xi = \int_0^{(N+1)T} b^{\tau,N}(t)\xi^{(d)}(t, \tau T)dt = \sum_{j=0}^{N} (\vec{b}_j^{\tau,N})^{\top} \vec{\xi}_j^{(d)} = B_N\vec{\xi}, \tag{9.18}$$

where the vector

$$\vec{b}_j^{\tau,N} = (b_{kj}^{\tau,N}, k = 1, 2, \dots)^{\top} = (b_{1j}^{\tau,N}, b_{3j}^{\tau,N}, b_{2j}^{\tau,N}, \dots, b_{2k+1,j}^{\tau,N}, b_{2k,j}^{\tau,N}, \dots)^{\top}$$

and the coefficients $b_{kj}^{\tau,N}$ are defined in Corollary 9.1.2.

Denote by $\Delta(f, \widehat{A}_{NT}\xi) := \mathsf{E}|A_{NT}\xi - \widehat{A}_{NT}\xi|^2$ and $\Delta(f, \widehat{B}_N\vec{\xi}) := \mathsf{E}|B_N\vec{\xi} - \widehat{B}_N\vec{\xi}|^2$ the MSEs of the optimal estimate $\widehat{A}_{NT}\xi$ of the functional $A_{NT}\xi$ and the optimal estimate $\widehat{B}_N\vec{\xi}$ of the functional $B_N\vec{\xi}$, respectively. Then

$$\Delta(f; \widehat{A}_{NT}\xi) = \mathsf{E}|A_{NT}\xi - \widehat{A}_{NT}\xi|^2 = \mathsf{E}|B_N\vec{\xi} - V\vec{\xi} - \widehat{B}_N\vec{\xi} + V_{NT}\xi|^2$$

$$= \mathsf{E}|B_N\vec{\xi} - \widehat{B}_N\vec{\xi}|^2 = \Delta(f;\widehat{B}_N\vec{\xi}).$$

The spectral representation of the functional $B_N\vec{\xi}$ is

$$B_N\vec{\xi} = \int_{-\pi}^{\pi} (\vec{B}_{\tau,N}(e^{i\lambda}))^\top \frac{(1-e^{-i\lambda\tau})^d}{(i\lambda)^d} d\vec{Z}_{\xi^{(d)}}(\lambda),$$

where

$$\vec{B}_{\tau,N}(e^{i\lambda}) = \sum_{j=0}^{N} \vec{b}_j^{\tau,N} e^{i\lambda j} = \sum_{j=0}^{N} (D_N^\tau \mathbf{a}_N)_j e^{i\lambda j}.$$

Equalities (9.17) and (9.18) imply that the linear estimate $\widehat{A}_{NT}\xi$ of the functional $A_{NT}\xi$ can be represented as

$$\widehat{A}_{NT}\xi = \int_{-\pi}^{\pi} (\vec{h}^{\xi}_{\tau,N}(\lambda))^\top d\vec{Z}_{\xi^{(d)}}(\lambda) - \int_{-\tau Td}^{0} v^{\tau,N}(t)\xi(t)dt, \tag{9.19}$$

where $\vec{h}^{\xi}_{\tau,N}(\lambda) = \{h^{\xi}_k(\lambda)\}_{k=1}^{\infty}$ is the spectral characteristic of the optimal estimate $\widehat{B}_N\vec{\xi}$.

Define

$$H^{0-}(\vec{\xi}_\tau^{(d)}) \oplus H^{(N+\tau d)+}(\vec{\xi}_\tau^{(d)}) = \overline{\text{span}}\{\xi_k^{(d)}(j,\tau) : j \in \mathbb{Z} \setminus \{0,1,\dots,N+\tau d\};\ k=1,\dots,\infty\}.$$

Then the estimate $\widehat{B}_N\vec{\xi}$ can be found as

$$\widehat{B}_N\vec{\xi} = \mathrm{Proj}_{H^{0-}(\vec{\xi}_\tau^{(d)})\oplus H^{(N+\tau d)+}(\vec{\xi}_\tau^{(d)})} B_N\vec{\xi}.$$

Or, equivalently, the spectral characteristic $\vec{h}^{\xi}_{\tau,N}(\lambda)$ can be found as

$$\vec{h}^{\xi}_{\tau,N}(\lambda) = \mathrm{Proj}_{L_2^{0-}(f)\oplus L_2^{(N+\tau d)+}(f)} \vec{B}_{\tau,N}(e^{i\lambda}),$$

where

$$L_2^{0-}(f) \oplus L_2^{(N+\tau d)+}(f) \\ = \overline{\text{span}}\left\{e^{i\lambda j}(1-e^{-i\lambda\tau})^d \frac{1}{(i\lambda)^d}\vec{\delta}_k : j \in \mathbb{Z} \setminus \{0,1,\dots,N+\tau d\};\ k=1,\dots,\infty\right\}$$

is a closed linear subspace of the Hilbert space $L_2(f)$ of vector-valued functions endowed with the inner product $\langle g_1;g_2\rangle = \int_{-\pi}^{\pi}(g_1(\lambda))^\top f(\lambda)\overline{g_2(\lambda)}d\lambda$. Then equality (9.17), taking into account (9.18), allows a representation

$$\widehat{A}_{NT}\xi = \mathrm{Proj}_{H^{0-}(\vec{\xi}_\tau^{(d)})\oplus H^{(N+\tau d)+}(\vec{\xi}_\tau^{(d)})} B_N\vec{\xi} - V_{NT}\xi.$$

Define for $0 \le l, j \le N + \tau d$ the Fourier coefficients

$$F^{\tau}_{l,j} = \frac{1}{2\pi} \int_{-\pi}^{\pi} e^{-i\lambda(l-j)} \frac{\lambda^{2d}}{|1 - e^{i\lambda\tau}|^{2d}} (f^{-1}(\lambda))^{\top} d\lambda$$

and a corresponding matrix $\mathbf{F}^{\tau}_{N}$ of dimension $(N + \tau d + 1) \times (N + \tau d + 1)$ with the infinite-dimensional matrix entries $(\mathbf{F}^{\tau}_{N})_{l,j} = F^{\tau}_{l,j}$, $0 \le l, j \le N + \tau d$.

The following corollary from Theorem 9.1.1 holds.

Corollary 9.1.4. *Let $\xi(t)$, $t \in \mathbb{R}$, be a stochastic process with periodically stationary increments which determines a generated stationary dth increment sequence $\vec{\xi}^{(d)}_j$ with the spectral density matrix $f(\lambda) = \{f_{kn}(\lambda)\}_{k,n=1}^{\infty}$. Let the minimality condition (9.16) be satisfied. The optimal linear estimate $\widehat{A}_{NT}\xi$ of the functional $A_{NT}\xi$ which depends on the unknown values of the process $\xi(t)$ at points $t \in [0, (N+1)T]$ from its observations without noise at points $\mathbb{R} \setminus [0, (N+1)T]$ is calculated by formula (9.19). The spectral characteristic $\vec{h}^{\xi}_{\tau,N}(\lambda)$ of the optimal estimate $\widehat{A}_{NT}\xi$ is calculated by the formula*

$$(\vec{h}^{\xi}_{\tau,N}(\lambda))^{\top} = (\vec{B}_{\tau,N}(e^{i\lambda}))^{\top} \frac{(1 - e^{-i\lambda\tau})^{d}}{(i\lambda)^{d}} - (\vec{C}^{\xi}_{\tau,N}(e^{i\lambda}))^{\top} \frac{(-i\lambda)^{d}}{(1 - e^{i\lambda\tau})^{d}} f^{-1}(\lambda), \tag{9.20}$$

where

$$\vec{C}^{\xi}_{\tau,N}(e^{i\lambda}) = \sum_{j=0}^{N+\tau d} ((\mathbf{F}^{\tau}_{N})^{-1} [D^{\tau}_{N}\mathbf{a}_{N}]_{+\tau d})_{j} e^{i\lambda j}.$$

The MSE of the estimate $\widehat{A}_{NT}\xi$ is calculated by the formula

$$\begin{aligned}
\Delta(f; \widehat{A}_{NT}\xi) &= \Delta(f; \widehat{B}_{N}\vec{\xi}) = \mathsf{E}|B_{N}\vec{\xi} - \widehat{B}_{N}\vec{\xi}|^{2} \\
&= \frac{1}{2\pi} \int_{-\pi}^{\pi} \frac{\lambda^{2d}}{|1 - e^{i\lambda\tau}|^{2d}} \left(\sum_{j=0}^{N+\tau d} ((\mathbf{F}^{\tau}_{N})^{-1} [D^{\tau}_{N}\mathbf{a}_{N}]_{+\tau d})_{j} e^{i\lambda j} \right)^{\top} \\
&\quad \times f(\lambda) \overline{\sum_{j=0}^{N+\tau d} ((\mathbf{F}^{\tau}_{N})^{-1} [D^{\tau}_{N}\mathbf{a}_{N}]_{+\tau d})_{j} e^{i\lambda j}} d\lambda \\
&= \langle [D^{\tau}_{N}\mathbf{a}_{N}]_{+\tau d}, (\mathbf{F}^{\tau}_{N})^{-1} [D^{\tau}_{N}\mathbf{a}_{N}]_{+\tau d} \rangle.
\end{aligned}$$

9.2 Minimax-robust method of interpolation

9.2.1 Least favorable spectral densities and minimax-robust spectral characteristic

The values of the MSEs and the spectral characteristics of the optimal estimate of the functional $A_{NT}\xi$ depending on the unobserved values of a stochastic process $\xi(t)$ which determine a generated stationary stochastic dth increment sequence $\vec{\xi}_j^{(d)}$ with the spectral density matrix $f(\lambda)$ based on observations of the process $\xi(t) + \eta(t)$ at points $\mathbb{R} \setminus [0;(N+1)T]$ can be calculated by formulas (9.11), (9.12), and (9.13), respectively, under the condition that spectral densities $f(\lambda)$ and $g(\lambda)$ of stochastic sequences $\vec{\xi}_j$ and $\vec{\eta}_j$ are exactly known.

In practical cases, however, the spectral densities usually are not exactly known. If in such cases a set $\mathcal{D} = \mathcal{D}_f \times \mathcal{D}_g$ of admissible spectral densities is defined, the minimax-robust approach to estimation of linear functionals depending on unobserved values of stochastic sequences with stationary increments may be applied. This method consists in finding an estimate that minimizes the maximal values of the MSEs for all spectral densities from a given class $\mathcal{D} = \mathcal{D}_f \times \mathcal{D}_g$ of admissible spectral densities simultaneously.

To formalize this approach we present the following definitions.

Definition 9.2.1. For a given class of spectral densities $\mathcal{D} = \mathcal{D}_f \times \mathcal{D}_g$ the spectral densities $f^0(\lambda) \in \mathcal{D}_f$ and $g^0(\lambda) \in \mathcal{D}_g$ are called least favorable in the class $\mathcal{D}$ for the optimal linear estimation of the functional $A_N\vec{\xi}$ if the following relation holds:

$$\Delta(f^0, g^0) = \Delta(h(f^0, g^0); f^0, g^0) = \max_{(f,g)\in\mathcal{D}_f\times\mathcal{D}_g} \Delta(h(f,g); f, g).$$

Definition 9.2.2. For a given class of spectral densities $\mathcal{D} = \mathcal{D}_f \times \mathcal{D}_g$ the spectral characteristic $h^0(\lambda)$ of the optimal linear estimate of the functional $A_N\vec{\xi}$ is called minimax-robust if the following conditions are satisfied:

$$h^0(\lambda) \in H_{\mathcal{D}} = \bigcap_{(f,g)\in\mathcal{D}_f\times\mathcal{D}_g} L_2^{0-}(f(\lambda) + \lambda^{2d}g(\lambda)) \oplus L_2^{(N+\tau d)+}(f(\lambda) + \lambda^{2d}g(\lambda))$$

and

$$\min_{h\in H_{\mathcal{D}}} \max_{(f,g)\in\mathcal{D}_f\times\mathcal{D}_g} \Delta(h; f, g) = \max_{(f,g)\in\mathcal{D}_f\times\mathcal{D}_g} \Delta(h^0; f, g).$$

Taking into account the introduced definitions and the derived relations we can verify that the following lemma holds.

Lemma 9.2.1. *The spectral densities $f^0 \in \mathcal{D}_f$ and $g^0 \in \mathcal{D}_g$ which satisfy minimality condition* (9.1) *are least favorable in the class $\mathcal{D} = \mathcal{D}_f \times \mathcal{D}_g$ for the optimal linear interpolation of the functional $A_{NT}\xi$ based on observations of the process $\xi(t) + \eta(t)$ at points $t \in \mathbb{R} \setminus [0;(N+1)T]$ if the matrices $(\mathbf{P}_N^\tau)^0$, $(\mathbf{T}_N^\tau)^0$, $(\mathbf{Q}_N)^0$ whose entries are defined by the Fourier coefficients of the functions*

$$\frac{\lambda^{2d}}{|1 - e^{i\lambda\tau}|^{2d}} (g^0(\lambda)(f^0(\lambda) + \lambda^{2d}g^0(\lambda))^{-1})^\top,$$

$$\frac{\lambda^{2d}}{|1 - e^{i\lambda\tau}|^{2d}} ((f^0(\lambda) + \lambda^{2d}g^0(\lambda))^{-1})^\top,$$

$$(f^0(\lambda)(f^0(\lambda)+\lambda^{2d}g^0(\lambda))^{-1}g^0(\lambda))^{\top}$$

determine a solution of the constrained optimization problem

$$\max_{(f,g)\in\mathcal{D}_f\times\mathcal{D}_g}(\langle\mathbf{P}_N^{\tau}\mathbf{c}_N^{\tau},\mathbf{c}_N^{\tau}\rangle+\langle\mathbf{Q}_N\mathbf{a}_N,\mathbf{a}_N\rangle)=\langle(\mathbf{P}_N^{\tau})^0(\mathbf{c}_N^{\tau})^0,(\mathbf{c}_N^{\tau})^0\rangle+\langle\mathbf{Q}_N^0\mathbf{a}_N,\mathbf{a}_N\rangle, \tag{9.21}$$

where

$$\mathbf{c}_N^{\tau}:=(\mathbf{P}_N^{\tau})^{-1}([D_N^{\tau}\mathbf{a}_N]_{+\tau d}-\mathbf{T}_N^{\tau}\mathbf{a}_N^{\tau}),$$
$$(\mathbf{c}_N^{\tau})^0:=((\mathbf{P}_N^{\tau})^0)^{-1}([D_N^{\tau}\mathbf{a}_N]_{+\tau d}-(\mathbf{T}_N^{\tau})^0\mathbf{a}_N^{\tau}).$$

The minimax spectral characteristic $h^0=\vec{h}_{\tau,N}(f^0,g^0)$ is calculated by formula (9.11) *if $\vec{h}_{\tau,N}(f^0,g^0)\in H_{\mathcal{D}}$.*

For the interpolation problem without noise, the corresponding definitions formalizing the minimax approach and lemmas concluded from them are the following [134].

Definition 9.2.3. For a given class of spectral densities $\mathcal{D}$ the spectral density $f^0(\lambda)\in\mathcal{D}$ is called least favorable in the class $\mathcal{D}$ for the optimal linear interpolation of the functional $A_{NT}\xi$ if the following relation holds:

$$\Delta(f^0)=\Delta(h(f^0);f^0)=\max_{f\in\mathcal{D}}\Delta(h(f);f).$$

Definition 9.2.4. For a given class of spectral densities $\mathcal{D}$ the spectral characteristic $h^0(\lambda)$ of the optimal linear estimate of the functional $A_{NT}\xi$ is called minimax-robust if the following conditions are satisfied:

$$h^0(\lambda)\in H_{\mathcal{D}}=\bigcap_{f\in\mathcal{D}_f}L_2^{0-}(f)\oplus L_2^{(N+\tau d)+}(f),$$
$$\min_{h\in H_{\mathcal{D}}}\max_{f\in\mathcal{D}}\Delta(h;f)=\max_{f\in\mathcal{D}}\Delta(h^0;f).$$

Lemma 9.2.2. *The spectral density $f^0\in\mathcal{D}$ which satisfies the minimality condition* (9.16) *is the least favorable one in the class $\mathcal{D}$ for the optimal linear interpolation of the functional $A_{NT}\xi$ based on observations of the process $\xi(t)$ at points $t\in\mathbb{R}\setminus[0,(N-1)T]$ if the matrix $(\mathbf{F}_N^{\tau})^0$ whose elements are defined by the Fourier coefficients of the function*

$$\frac{\lambda^{2d}}{|1-e^{i\lambda\tau}|^{2d}}((f^0(\lambda))^{-1})^{\top}$$

determine a solution of the constrained optimization problem

$$\max_{f\in\mathcal{D}}(\langle[D_N^{\tau}\mathbf{a}_N]_{+\tau d},(\mathbf{F}_N^{\tau})^{-1}[D_N^{\tau}\mathbf{a}_N]_{+\tau d}\rangle)=\langle[D_N^{\tau}\mathbf{a}_N]_{+\tau d},((\mathbf{F}_N^{\tau})^0)^{-1}[D_N^{\tau}\mathbf{a}_N]_{+\tau d}\rangle. \tag{9.22}$$

The minimax spectral characteristic $h^0 = \vec{h}^{\xi}_{\tau,N}(f^0)$ *is calculated by formula* (9.20) *if* $\vec{h}^{\xi}_{\tau,N}(f^0) \in H_{\mathcal{D}}$.

More detailed analysis of properties of the least favorable spectral densities and minimax-robust spectral characteristics shows that the minimax spectral characteristic h^0 and the pair of the least favorable spectral densities (f^0, g^0) (or the least favorable spectral density f^0 in the case of observations without noise) form a saddle point of the function $\Delta(h; f, g)$ (or of the function $\Delta(h; f)$) on the set $H_{\mathcal{D}} \times \mathcal{D}$. The saddle point inequalities

$$\Delta(h; f^0, g^0) \geq \Delta(h^0; f^0, g^0) \geq \Delta(h^0; f, g) \quad \forall (f, g) \in \mathcal{D}, \forall h \in H_{\mathcal{D}}$$

or

$$\Delta(h; f^0) \geq \Delta(h^0; f^0) \geq \Delta(h^0; f) \quad \forall f \in \mathcal{D}, \forall h \in H_{\mathcal{D}}$$

hold if $h^0 = \vec{h}_{\tau,N}(f^0, g^0)$, $\vec{h}_{\tau,N}(f^0, g^0) \in H_{\mathcal{D}}$ (or $h^0 = \vec{h}^{\xi}_{\tau,N}(f^0)$, $\vec{h}^{\xi}_{\tau,N}(f^0) \in H_{\mathcal{D}}$), and (f^0, g^0) (or f^0) is a solution of the constrained optimization problem

$$\widetilde{\Delta}(f, g) = -\Delta(\vec{h}_{\tau,N}(f^0, g^0); f, g) \to \inf, \quad (f, g) \in \mathcal{D}, \tag{9.23}$$

or

$$\widetilde{\Delta}(f) = -\Delta(\vec{h}^{\xi}_{\tau,N}(f^0); f) \to \inf, \quad f \in \mathcal{D}, \tag{9.24}$$

where the functional $\Delta(\vec{h}_{\tau,N}(f^0, g^0); f, g)$ is calculated by the formula

$$\begin{aligned}\Delta(\vec{h}_{\tau,N}(f^0, g^0); f, g) = {} & \frac{1}{2\pi} \int_{-\pi}^{\pi} \frac{\lambda^{2d}}{|1 - e^{i\lambda\tau}|^{2d}} (\mathbf{C}^{f0}_{\tau,N}(e^{i\lambda}))^{\top} (f^0(\lambda) + \lambda^{2d} g^0(\lambda))^{-1} f(\lambda) \\ & \times (f^0(\lambda) + \lambda^{2d} g^0(\lambda))^{-1} \overline{\mathbf{C}^{f0}_{\tau,N}(e^{i\lambda})} d\lambda \\ & + \frac{1}{2\pi} \int_{-\pi}^{\pi} \frac{\lambda^{4d}}{|1 - e^{i\lambda\tau}|^{4d}} (\mathbf{C}^{g0}_{\tau,N}(e^{i\lambda}))^{\top} (f^0(\lambda) + \lambda^{2d} g^0(\lambda))^{-1} g(\lambda) \\ & \times (f^0(\lambda) + \lambda^{2d} g^0(\lambda))^{-1} \overline{\mathbf{C}^{g0}_{\tau,N}(e^{i\lambda})} d\lambda,\end{aligned}$$

where

$$\begin{aligned}\mathbf{C}^{f0}_{\tau,N}(e^{i\lambda}) &= \overline{g^0(\lambda)} \vec{A}_{\tau,N}(e^{i\lambda}) + \vec{C}^{0}_{\tau,N}(e^{i\lambda}), \\ \mathbf{C}^{g0}_{\tau,N}(e^{i\lambda}) &= |1 - e^{i\lambda\tau}|^{2d} \lambda^{-2d} \overline{f^0(\lambda)} \vec{A}_{N}(e^{i\lambda}) - (1 - e^{-i\lambda\tau})^{d} \vec{C}^{0}_{\tau,N}(e^{i\lambda}),\end{aligned}$$

and

$$\vec{C}^0_{\tau,N}(e^{i\lambda}) = \sum_{j=0}^{N+\tau d} (((\mathbf{P}^\tau_N)^0)^{-1}[D^\tau_N \mathbf{a}_N]_{+\tau d} - ((\mathbf{P}^\tau_N)^0)^{-1}(\mathbf{T}^\tau_N)^0 \mathbf{a}^\tau_N)_j e^{i\lambda j},$$

or the functional $\Delta(\vec{h}^\xi_{\tau,N}(f^0);f)$, in the case of observations without noise, is calculated by the formula

$$\Delta(\vec{h}^\xi_{\tau,N}(f^0);f) = \frac{1}{2\pi}\int_{-\pi}^{\pi} \frac{\lambda^{2d}}{|1-e^{i\lambda\tau}|^{2d}} \left(\sum_{j=0}^{N+\tau d} (((\mathbf{F}^\tau_N)^0)^{-1}[D^\tau_N \mathbf{a}_N]_{+\tau d})_j e^{i\lambda j}\right)^\top$$
$$\times (f^0(\lambda))^{-1} f(\lambda) (f^0(\lambda))^{-1} \overline{\sum_{j=0}^{N+\tau d} (((\mathbf{F}^\tau_N)^0)^{-1}[D^\tau_N \mathbf{a}_N]_{+\tau d})_j e^{i\lambda j}} d\lambda.$$

The constrained optimization problems (9.23) and (9.24) are equivalent to the unconstrained optimization problems

$$\Delta_{\mathcal{D}}(f,g) = \widetilde{\Delta}(f,g) + \delta(f,g|\mathcal{D}_f \times \mathcal{D}_g) \to \inf, \tag{9.25}$$
$$\Delta_{\mathcal{D}}(f) = \widetilde{\Delta}(f) + \delta(f|\mathcal{D}) \to \inf, \tag{9.26}$$

where $\delta(f,g|\mathcal{D}_f \times \mathcal{D}_g)$ and $\delta(f|\mathcal{D})$ are the indicator functions of the sets $\mathcal{D} = \mathcal{D}_f \times \mathcal{D}_g$ and $\mathcal{D} = \mathcal{D}_f$. Solutions (f^0, g^0) and f^0 to these unconstrained optimization problems are characterized by the conditions $0 \in \partial\Delta_{\mathcal{D}}(f^0,g^0)$ and $0 \in \partial\Delta_{\mathcal{D}}(f^0)$, where $\partial\Delta_{\mathcal{D}}(f^0,g^0)$ and $\partial\Delta_{\mathcal{D}}(f^0)$ are the subdifferentials of the functionals $\Delta_{\mathcal{D}}(f,g)$ and $\Delta_{\mathcal{D}}(f)$ at points $(f^0,g^0) \in \mathcal{D} = \mathcal{D}_f \times \mathcal{D}_g$ and $f^0 \in \mathcal{D} = \mathcal{D}_f$. These conditions allow us to find the least favorable spectral densities in some special classes of spectral densities $\mathcal{D} = \mathcal{D}_f \times \mathcal{D}_g$ and $\mathcal{D} = \mathcal{D}_f$, respectively.

The forms of the functionals $\Delta(\vec{h}_{\tau,N}(f^0,g^0);f,g)$ and $\Delta(\vec{h}_{\tau,N}(f^0);f)$ are convenient for application of the Lagrange method of indefinite multipliers for finding solutions to the problems (9.25) and (9.26). Making use of the method of Lagrange multipliers and the form of subdifferentials of the defined indicator functions we describe relations that determine the least favorable spectral densities in some special classes of spectral densities (see [115, 146] for additional details).

9.2.2 Least favorable spectral densities in classes $\mathcal{D}_0 \times \mathcal{D}_V^U$

Consider the problem of optimal linear estimation of the functional $A_{NT}\xi$ which depends on unobserved values of the process $\xi(t)$ with periodically stationary increments based on observations of the process $\xi(t)+\eta(t)$ at points $\mathbb{R} \setminus [0;(N+1)T]$ under the condition that the sets of admissible spectral densities $\mathcal{D}_0^k, \mathcal{D}_V^{Uk}$, $k = 1,2,3,4$, are defined as follows:

$$\mathcal{D}_0^1 = \left\{ f(\lambda) \left| \frac{1}{2\pi}\int_{-\pi}^{\pi} \frac{|1-e^{i\lambda\tau}|^{2d}}{\lambda^{2d}} f(\lambda) d\lambda = P \right.\right\},$$

$$\mathcal{D}_0^2 = \left\{ f(\lambda) \left| \frac{1}{2\pi} \int_{-\pi}^{\pi} \frac{|1-e^{i\lambda\tau}|^{2d}}{\lambda^{2d}} \operatorname{Tr}[f(\lambda)] d\lambda = p \right. \right\},$$

$$\mathcal{D}_0^3 = \left\{ f(\lambda) \left| \frac{1}{2\pi} \int_{-\pi}^{\pi} \frac{|1-e^{i\lambda\tau}|^{2d}}{\lambda^{2d}} f_{kk}(\lambda) d\lambda = p_k, k = \overline{1,\infty} \right. \right\},$$

$$\mathcal{D}_0^4 = \left\{ f(\lambda) \left| \frac{1}{2\pi} \int_{-\pi}^{\pi} \frac{|1-e^{i\lambda\tau}|^{2d}}{\lambda^{2d}} \langle B_1, f(\lambda) \rangle d\lambda = p \right. \right\}$$

and

$$\mathcal{D}_V^{U1} = \left\{ g(\lambda) \left| V(\lambda) \le g(\lambda) \le U(\lambda), \frac{1}{2\pi} \int_{-\pi}^{\pi} g(\lambda) d\lambda = Q \right. \right\},$$

$$\mathcal{D}_V^{U2} = \left\{ g(\lambda) \left| \operatorname{Tr}[V(\lambda)] \le \operatorname{Tr}[g(\lambda)] \le \operatorname{Tr}[U(\lambda)], \frac{1}{2\pi} \int_{-\pi}^{\pi} \operatorname{Tr}[g(\lambda)] d\lambda = q \right. \right\},$$

$$\mathcal{D}_V^{U3} = \left\{ g(\lambda) \left| v_{kk}(\lambda) \le g_{kk}(\lambda) \le u_{kk}(\lambda), \frac{1}{2\pi} \int_{-\pi}^{\pi} g_{kk}(\lambda) d\lambda = q_k, k = \overline{1,\infty} \right. \right\},$$

$$\mathcal{D}_V^{U4} = \left\{ g(\lambda) \left| \langle B_2, V(\lambda) \rangle \le \langle B_2, g(\lambda) \rangle \le \langle B_2, U(\lambda) \rangle, \frac{1}{2\pi} \int_{-\pi}^{\pi} \langle B_2, g(\lambda) \rangle d\lambda = q \right. \right\}.$$

Here $V(\lambda)$ and $U(\lambda)$ are known and fixed spectral densities, $W(\lambda)$ is an unknown spectral density, $p, p_k, k = \overline{1,\infty}$, and $q, q_k, k = \overline{1,\infty}$, are given numbers, and P, Q, B_1, B_2 are given positive-definite Hermitian matrices.

From the condition $0 \in \partial\Delta_{\mathcal{D}}(f^0, g^0)$ we find the following equations, which determine the least favorable spectral densities for these given sets of admissible spectral densities.

For the first set of admissible spectral densities $\mathcal{D}_0^1 \times \mathcal{D}_V^{U1}$, we have

$$\begin{aligned}(\mathbf{C}_{\tau,N}^{f0}(e^{i\lambda}))(\mathbf{C}_{\tau,N}^{f0}(e^{i\lambda}))^* &= \left(\frac{|1-e^{i\lambda\tau}|^{2d}}{\lambda^{2d}} (f^0(\lambda) + \lambda^{2d} g^0(\lambda)) \right) \\ &\quad \times \vec{\alpha} \cdot \vec{\alpha}^* \left(\frac{|1-e^{i\lambda\tau}|^{2d}}{\lambda^{2d}} (f^0(\lambda) + \lambda^{2d} g^0(\lambda)) \right),\end{aligned} \tag{9.27}$$

$$\begin{aligned}(\mathbf{C}_{\tau,N}^{g0}(e^{i\lambda}))(\mathbf{C}_{\tau,N}^{g0}(e^{i\lambda}))^* &= \left(\frac{|1-e^{i\lambda\tau}|^{2d}}{\lambda^{2d}} (f^0(\lambda) + \lambda^{2d} g^0(\lambda)) \right) \\ &\quad \times (\vec{\beta} \cdot \vec{\beta}^* + \Gamma_1(\lambda) + \Gamma_2(\lambda)) \\ &\quad \times \left(\frac{|1-e^{i\lambda\tau}|^{2d}}{\lambda^{2d}} (f^0(\lambda) + \lambda^{2d} g^0(\lambda)) \right),\end{aligned} \tag{9.28}$$

where $\vec{\alpha}$ and $\vec{\beta}$ are vectors of Lagrange multipliers, the matrix $\Gamma_1(\lambda) \le 0$ and $\Gamma_1(\lambda) = 0$ if $g^0(\lambda) > V(\lambda)$, and the matrix $\Gamma_2(\lambda) \ge 0$ and $\Gamma_2(\lambda) = 0$ if $g^0(\lambda) < U(\lambda)$.

For the second set of admissible spectral densities $\mathcal{D}_0^2 \times \mathcal{D}_V^{U2}$, we have

$$(\mathbf{C}^{f0}_{\tau,N}(e^{i\lambda}))(\mathbf{C}^{f0}_{\tau,N}(e^{i\lambda}))^* = \alpha^2\left(\frac{|1-e^{i\lambda\tau}|^{2d}}{\lambda^{2d}}(f^0(\lambda)+\lambda^{2d}g^0(\lambda))\right)^2, \tag{9.29}$$

$$\begin{aligned}(\mathbf{C}^{g0}_{\tau,N}(e^{i\lambda}))(\mathbf{C}^{g0}_{\tau,N}(e^{i\lambda}))^* &= (\beta^2+\gamma_1(\lambda)+\gamma_2(\lambda))\\ &\times\left(\frac{|1-e^{i\lambda\tau}|^{2d}}{\lambda^{2d}}(f^0(\lambda)+\lambda^{2d}g^0(\lambda))\right)^2,\end{aligned} \tag{9.30}$$

where α^2, β^2 are Lagrange multipliers, the function $\gamma_1(\lambda) \le 0$ and $\gamma_1(\lambda) = 0$ if $\mathrm{Tr}[g^0(\lambda)] > \mathrm{Tr}[V(\lambda)]$, and the function $\gamma_2(\lambda) \ge 0$ and $\gamma_2(\lambda) = 0$ if $\mathrm{Tr}[g^0(\lambda)] < \mathrm{Tr}[U(\lambda)]$.

For the third set of admissible spectral densities $\mathcal{D}_0^3 \times \mathcal{D}_V^{U3}$, we have

$$\begin{aligned}(\mathbf{C}^{f0}_{\tau,N}(e^{i\lambda}))(\mathbf{C}^{f0}_{\tau,N}(e^{i\lambda}))^* &= \left(\frac{|1-e^{i\lambda\tau}|^{2d}}{\lambda^{2d}}(f^0(\lambda)+\lambda^{2d}g^0(\lambda))\right)\\ &\times\{\alpha_k^2\delta_{kl}\}_{k,l=1}^{\infty}\left(\frac{|1-e^{i\lambda\tau}|^{2d}}{\lambda^{2d}}(f^0(\lambda)+\lambda^{2d}g^0(\lambda))\right),\end{aligned} \tag{9.31}$$

$$\begin{aligned}(\mathbf{C}^{g0}_{\tau,N}(e^{i\lambda}))(\mathbf{C}^{g0}_{\tau,N}(e^{i\lambda}))^* &= \left(\frac{|1-e^{i\lambda\tau}|^{2d}}{\lambda^{2d}}(f^0(\lambda)+\lambda^{2d}g^0(\lambda))\right)\\ &\times\{(\beta_k^2+\gamma_{1k}(\lambda)+\gamma_{2k}(\lambda))\delta_{kl}\}_{k,l=1}^{\infty}\\ &\times\left(\frac{|1-e^{i\lambda\tau}|^{2d}}{\lambda^{2d}}(f^0(\lambda)+\lambda^{2d}g^0(\lambda))\right),\end{aligned} \tag{9.32}$$

where α_k^2, β_k^2 are Lagrange multipliers, δ_{kl} are Kronecker symbols, the functions $\gamma_{1k}(\lambda) \le 0$ and $\gamma_{1k}(\lambda) = 0$ if $g^0_{kk}(\lambda) > v_{kk}(\lambda)$, and the functions $\gamma_{2k}(\lambda) \ge 0$ and $\gamma_{2k}(\lambda) = 0$ if $g^0_{kk}(\lambda) < u_{kk}(\lambda)$.

For the fourth set of admissible spectral densities $\mathcal{D}_0^4 \times \mathcal{D}_V^{U4}$, we have

$$\begin{aligned}(\mathbf{C}^{f0}_{\tau,N}(e^{i\lambda}))(\mathbf{C}^{f0}_{\tau,N}(e^{i\lambda}))^* &= \alpha^2\left(\frac{|1-e^{i\lambda\tau}|^{2d}}{\lambda^{2d}}(f^0(\lambda)+\lambda^{2d}g^0(\lambda))\right)\\ &\times B_1^{\top}\left(\frac{|1-e^{i\lambda\tau}|^{2d}}{\lambda^{2d}}(f^0(\lambda)+\lambda^{2d}g^0(\lambda))\right),\end{aligned} \tag{9.33}$$

$$\begin{aligned}(\mathbf{C}^{g0}_{\tau,N}(e^{i\lambda}))(\mathbf{C}^{g0}_{\tau,N}(e^{i\lambda}))^* &= (\beta^2+\gamma_1'(\lambda)+\gamma_2'(\lambda))\\ &\times\left(\frac{|1-e^{i\lambda\tau}|^{2d}}{\lambda^{2d}}(f^0(\lambda)+\lambda^{2d}g^0(\lambda))\right)\\ &\times B_2^{\top}\left(\frac{|1-e^{i\lambda\tau}|^{2d}}{\lambda^{2d}}(f^0(\lambda)+\lambda^{2d}g^0(\lambda))\right),\end{aligned} \tag{9.34}$$

where α^2, β^2 are Lagrange multipliers, the functions $\gamma_1'(\lambda) \le 0$ and $\gamma_1'(\lambda) = 0$ if $\langle B_2, g^0(\lambda)\rangle > \langle B_2, V(\lambda)\rangle$, and functions $\gamma_2'(\lambda) \ge 0$ and $\gamma_2'(\lambda) = 0$ if $\langle B_2, g^0(\lambda)\rangle < \langle B_2, U(\lambda)\rangle$.

The following theorem holds.

Theorem 9.2.1. *Let minimality condition* (9.1) *hold. The least favorable spectral densities* $f^0(\lambda)$ *and* $g^0(\lambda)$ *in the classes* $\mathcal{D}_0^k \times \mathcal{D}_V^{Uk}$, $k = 1, 2, 3, 4$, *for the optimal linear estimation of the functional* $A_{NT}\xi$ *from observations of the process* $\xi(t) + \eta(t)$ *at points* $\mathbb{R} \setminus [0; (N+1)T]$ *are determined by equations* (9.27)–(9.28), (9.29)–(9.30), (9.31)–(9.32), *and* (9.33)–(9.34), *respectively, the minimality condition* (9.1), *the constrained optimization problem* (9.21), *and restrictions on densities from the corresponding classes* $\mathcal{D}_0^k$, $\mathcal{D}_V^{Uk}$, $k = 1, 2, 3, 4$. *The minimax-robust spectral characteristic of the optimal estimate of the functional* $A_{NT}\xi$ *is determined by formula* (9.11).

9.2.3 Example of the least favorable spectral density in the class $\mathcal{D}_0^-$

Consider the minimax interpolation problem for the functional $A_{NT}\xi$ depending on the unobserved values of the stochastic process $\xi(t)$ with periodically stationary increments from observations without noise for the set of admissible spectral densities

$$\mathcal{D}_0^- = \left\{ f(\lambda) \left| \frac{1}{2\pi} \int_{-\pi}^{\pi} \frac{\lambda^{2d}}{|1 - e^{i\lambda\tau}|^{2d}} f^{-1}(\lambda) d\lambda = P \right. \right\},$$

where P is a given positive-definite Hermitian matrix.

The condition $0 \in \partial\Delta_{\mathcal{D}}(f^0)$ implies the following relation determining the least favorable spectral density:

$$((f^0(\lambda))^{-1})^{\top} (\vec{C}^{\xi}_{\tau,N}(e^{i\lambda}))(\vec{C}^{\xi}_{\tau,N}(e^{i\lambda}))^* ((f^0(\lambda))^{-1})^{\top} = ((f^0(\lambda))^{-1})^{\top} \vec{\alpha} \cdot \vec{\alpha}^* ((f^0(\lambda))^{-1})^{\top},$$

where $\vec{\alpha} = \{\alpha_k\}_{k=1}^{\infty} = (\alpha_1, \alpha_2, \ldots)^{\top}$ is a vector of Lagrange multipliers. Thus, the Fourier coefficients

$$F_{l,j}^{0\tau} = \frac{1}{2\pi} \int_{-\pi}^{\pi} e^{-i\lambda(l-j)} \frac{\lambda^{2d}}{|1 - e^{i\lambda\tau}|^{2d}} ((f^0(\lambda))^{-1})^{\top} d\lambda$$

satisfy the following two equations:

$$\left(\sum_{j=0}^{N+\tau d} \vec{c}_j^{\,\xi,\tau} e^{i\lambda j} \right) \left(\sum_{j=0}^{N+\tau d} \vec{c}_j^{\,\xi,\tau} e^{i\lambda j} \right)^* = \vec{\alpha} \cdot \vec{\alpha}^* \tag{9.35}$$

and

$$(\mathbf{F}_N^{\tau})^0 \mathbf{c}_N^{\xi,\tau} = [D_N^{\tau} \mathbf{a}_N]_{+\tau d},$$

where

$$\mathbf{c}_N^{\xi,\tau} = ((\vec{c}_0^{\,\xi,\tau})^{\top}, (\vec{c}_1^{\,\xi,\tau})^{\top}, (\vec{c}_2^{\,\xi,\tau})^{\top}, \ldots, (\vec{c}_{N+\tau d}^{\,\xi,\tau})^{\top})^{\top}$$

is an $(N+\tau d+1)$-dimensional vector with the entries $\vec{c}_j^{\,\xi,\tau} = \{c_{kj}^{\xi,\tau}\}_{k=1}^{\infty}, j = 0,1,\dots,N+\tau d$. Equation (9.35) implies

$$\sum_{j=0}^{N+\tau d} \vec{c}_j^{\,\xi,\tau} \cdot (\vec{c}_j^{\,\xi,\tau})^* = \vec{a} \cdot \vec{a}^*.$$

As a particular case, equation (9.35) is satisfied by the $(N+\tau d+1)$-dimensional vector

$$\mathbf{c}_N^{\xi,\tau} = \boldsymbol{a}^{j^*,\tau} = (\vec{0}^\top, \dots, \vec{0}^\top, \vec{a}^\top, \vec{0}^\top, \dots, \vec{0}^\top)^\top$$

with infinite-dimensional entries, where the entry $\vec{a}^\top$ occupies the j^*th place. Thus, the Fourier coefficients $F_{l,j}^{0\tau}$ satisfy the equation

$$(\mathbf{F}_N^\tau)^0 \boldsymbol{a}^{j^*,\tau} = [D_N^\tau \mathbf{a}_N]_{+\tau d}.$$

Note that $[D_N^\tau \mathbf{a}_N]_{+\tau d}$ has zero vectors as the entries for $j = N+1,\dots,N+\tau d$. Since we require $F_{-j}^{0\tau} = (F_j^{0\tau})^*$, the appropriate values of j^* are $j^* = 0$ and $j^* = N$.

Let $j^* = 0$. Then we obtain a system of $N+1$ equations,

$$(F_j^{0\tau})^\top \vec{a} = \vec{b}_j^{\,\tau,N},$$

where $\vec{b}_j^{\,\tau,N} = ([D_N^\tau \mathbf{a}_N]_{+\tau d})_j$ is the jth entry of the vector $[D_N^\tau \mathbf{a}_N]_{+\tau d}$ and $(F_{j-l}^{0\tau})^\top = F_{j,l}^{0\tau}$, $j,l = 0,1,\dots,N$. From the definition of the set $\mathcal{D}_0^-$ we obtain $F_0^{0\tau} = P$, and therefore $P^\top \vec{a} = \vec{b}_0^{\,\tau,N}$, which implies

$$(F_j^{0\tau})^\top (P^\top)^{-1} \vec{b}_0^{\,\tau,N} = \vec{b}_j^{\,\tau,N}, \quad j = 0,1,\dots,N. \tag{9.36}$$

Define a vector $(\vec{b}_j^{\,\tau,N})^{-1}$ such that

$$((\vec{b}_0^{\,\tau,N})^{-1})^\top \vec{b}_0^{\,\tau,N} = 1.$$

Then equation (9.36) is satisfied by a solution of equations,

$$(F_j^{0\tau})^\top = \vec{b}_j^{\,\tau,N} ((\vec{b}_0^{\,\tau,N})^{-1})^\top P^\top, \quad j = 0,1,\dots,N.$$

Note that $\vec{b}_j^{\,\tau,N} = \vec{0}$ for $j = N+1,\dots,N+\tau d$. Suppose that the vectors $\vec{b}_j^{\,\tau,N}$ for $j = 0,\dots,N$ are such that the matrix function

$$\frac{\lambda^{2d}}{|1-e^{i\lambda\tau}|^{2d}} (f^0(\lambda))^{-1} = \sum_{j=-N}^{N} F_j^{0\tau} e^{i\lambda j}$$

is positive-definite and has a determinant which does not equal zero identically. Under such conditions the function $\frac{|1-e^{i\lambda\tau}|^{2d}}{\lambda^{2d}} f^0(\lambda)$ allows a canonical factorization.

The minimax spectral characteristic is calculated by the formula

$$\begin{aligned}\vec{h}^{0\xi}_{\tau,N}(\lambda) &= \frac{(1-e^{-i\lambda\tau})^d}{(i\lambda)^d}\left(\vec{B}_{\tau,N}(e^{i\lambda}) - \frac{\lambda^{2d}}{|1-e^{i\lambda\tau}|^{2d}}((f^0(\lambda))^{-1})^\top \vec{C}^{0\xi}_{\tau,N}(e^{i\lambda})\right)\\ &= \frac{(1-e^{-i\lambda\tau})^d}{(i\lambda)^d}\left(\sum_{j=0}^{N}\vec{b}_j^{\tau,N}e^{i\lambda j} - \sum_{j=-N}^{N}(F_j^{0\tau})^\top (P^\top)^{-1}\vec{b}_0^{\tau,N}e^{i\lambda j}\right)\\ &= \frac{(1-e^{-i\lambda\tau})^d}{(i\lambda)^d}\left(-\sum_{j=1}^{N}\overline{P(\vec{b}_0^{\tau,N})^{-1}(\vec{b}_j^{\tau,N})^*(P^\top)^{-1}\vec{b}_0^{\tau,N}}e^{-i\lambda j}\right),\end{aligned} \tag{9.37}$$

where $\vec{b}_j^{\tau,N} = ([D_N^\tau \mathbf{a}_N]_{+\tau d})_j$.

9.2.4 Least favorable spectral density in the class $\mathcal{D}_V^U$

Consider the minimax (robust) interpolation problem without noise for the functional $A_{NT}\xi$ for the set of admissible spectral densities

$$\mathcal{D}_V^U = \left\{f(\lambda)\,\middle|\, V(\lambda) \le f(\lambda) \le U(\lambda), \frac{1}{2\pi}\int_{-\pi}^{\pi}\frac{\lambda^{2d}}{|1-e^{i\lambda\tau}|^{2d}}f^{-1}(\lambda)d\lambda = P\right\},$$

where the spectral densities $V(\lambda)$ and $U(\lambda)$ from the definitions of the set $\mathcal{D}_V^U$ are known and fixed and P is a given positive-definite Hermitian matrix.

The condition $0 \in \partial\Delta_{\mathcal{D}}(f^0)$ implies the following equation determining the least favorable spectral density:

$$\begin{aligned}&((f^0(\lambda))^{-1})^\top(\vec{C}^{\xi}_{\tau,N}(e^{i\lambda}))(\vec{C}^{\xi}_{\tau,N}(e^{i\lambda}))^*((f^0(\lambda))^{-1})^\top\\ &\quad= ((f^0(\lambda))^{-1})^\top(\vec{\alpha}\cdot\vec{\alpha}^* + \Gamma_1(\lambda) + \Gamma_2(\lambda))((f^0(\lambda))^{-1})^\top,\end{aligned} \tag{9.38}$$

where $\vec{\alpha}$ is a vector of Lagrange multipliers, the matrix $\Gamma_1(\lambda) \le 0$ and $\Gamma_1(\lambda) = 0$ if $f^0(\lambda) > V(\lambda)$, and the matrix $\Gamma_2(\lambda) \ge 0$ and $\Gamma_2(\lambda) = 0$ if $f^0(\lambda) < U(\lambda)$. Thus, the Fourier coefficients $(F^{0\tau}_{l-j})^\top = F^{0\tau}_{l,j}$ of the function $\frac{\lambda^{2d}}{|1-e^{i\lambda\tau}|^{2d}}((f^0(\lambda))^{-1})^\top$ satisfy the following two equations:

$$\left(\sum_{j=0}^{N+\tau d}\vec{c}_j^{\,\xi,\tau}e^{i\lambda j}\right)\left(\sum_{j=0}^{N+\tau d}\vec{c}_j^{\,\xi,\tau}e^{i\lambda j}\right)^* = (\vec{\alpha}\cdot\vec{\alpha}^* + \Gamma_1(\lambda) + \Gamma_2(\lambda)) \tag{9.39}$$

and

$$(\mathbf{F}_N^\tau)^0\mathbf{c}_N^{\xi,\tau} = [D_N^\tau\mathbf{a}_N]_{+\tau d},$$

where the vector $\mathbf{c}_N^{\xi,\tau}$ is defined in the example considered in Section 9.2.3.

When

$$V(\lambda) \le f(\lambda) \le U(\lambda),$$

we have a case of the example from Section 9.2.3. Namely, the matrix function

$$\frac{\lambda^{2d}}{|1-e^{i\lambda\tau}|^{2d}}(f^0(\lambda))^{-1} = \sum_{j=-N}^{N} F_j^{0\tau} e^{i\lambda j} \tag{9.40}$$

is positive-definite and has a determinant which does not equal zero identically. It allows a representation

$$\frac{\lambda^{2d}}{|1-e^{i\lambda\tau}|^{2d}}(f^0(\lambda))^{-1} = \left(\sum_{j=0}^{N} \varphi^{0\tau}(j)e^{-i\lambda j}\right)\left(\sum_{j=0}^{N} \varphi^{0\tau}(j)e^{-i\lambda j}\right)^{*}.$$

The coefficients $F_j^{0\tau}$ are the following:

$$F_j^{0\tau} = P(\vec{b}_0^{\tau,N})^{-1}(\vec{b}_j^{\tau,N})^{\top}, \quad j = 0,1,\dots,N, \tag{9.41}$$

$$F_{-j}^{0\tau} = (F_j^{0\tau})^{*}, \quad j = 1,\dots,N. \tag{9.42}$$

The least favorable spectral density for the optimal interpolation without noise of the functional $A_N\vec{\xi}$ for the set $\mathcal{D}_V^U$ of admissible spectral densities is a spectral density of a vector-valued stochastic regression sequence of order N, determined by the matrix coefficients $\{\varphi^{0\tau}(j) : j = 0,1,\dots,N\}$, if the following inequality holds:

$$\frac{|1-e^{i\lambda\tau}|^{2d}}{\lambda^{2d}}V(\lambda) \le \left(\sum_{j=-N}^{N} F_j^{0\tau} e^{i\lambda j}\right)^{-1} \le \frac{|1-e^{i\lambda\tau}|^{2d}}{\lambda^{2d}}U(\lambda). \tag{9.43}$$

When equalities (9.43) are not satisfied, the derived relations imply

$$\frac{|1-e^{i\lambda\tau}|^{2d}}{\lambda^{2d}}f^0(\lambda) = \max\left\{\frac{|1-e^{i\lambda\tau}|^{2d}}{\lambda^{2d}}V(\lambda), \min\left\{\frac{|1-e^{i\lambda\tau}|^{2d}}{\lambda^{2d}}U(\lambda), \left(\sum_{j=-N}^{N} F_j^{0\tau} e^{i\lambda j}\right)^{-1}\right\}\right\}. \tag{9.44}$$

Theorem 9.2.2. *Let the function $a(t)$, $t \in [0,(N+1)T]$, determine the infinite-dimensional vectors $\vec{a}_j$, $j = 0,1,\dots,N$, such that matrix function* (9.40) *determined by matrix coefficients* (9.41)–(9.42) *is positive-definite and its determinant does not equal zero identically. If inequality* (9.43) *is satisfied, then the least favorable spectral density in the class $\mathcal{D}_V^U$ for the optimal interpolation of the functional $A_N\vec{\xi}$ is a spectral density of a vector autoregression sequence of order N, and the minimax spectral characteristic can be calculated by formula* (9.37). *If inequality* (9.43) *is not satisfied, then the least favorable spectral density is determined by relation* (9.44) *and the constrained optimization problem* (9.22), *and the minimax spectral characteristic is calculated by formula* (9.20).

Conclusions

In this chapter, we describe methods of solution of the interpolation problem for linear functionals which are constructed from unobserved values of stochastic processes with periodically stationary increments. Estimates of functionals are based on observations of the considered process $\xi(t)$ with an uncorrelated with the process $\xi(t)$ periodically stationary stochastic noise process $\eta(t)$. The problem is solved by representing these processes as infinite-dimensional sequences with stationary increments.

The Hilbert space projection method is applied to derive formulas for calculating the spectral characteristics and the MSEs of the optimal estimates of the functionals in the case of spectral certainty, where the spectral densities of the generated sequences are exactly known. In the case of spectral uncertainty, where the spectral densities are not exactly known while, instead, a set of admissible spectral densities is specified, the minimax-robust method is applied. We propose a representation of the MSE in the form of a linear functional in L_1 space with respect to spectral densities, which allows us to solve the corresponding constrained optimization problem and describe the minimax (robust) estimates of the functionals.

This method is illustrated with the help of concrete sets $\mathcal{D}_0 \times \mathcal{D}_V^U$ of admissible spectral densities. Here D_0 are classes of spectral densities with the moment restrictions and D_V^U are classes describing the "strip" models of spectral densities.

The main results presented in this chapter have been published in [123]. These results rely on and develop prior studies of the minimax interpolation problem by Masyutka, Moklyachuk, and Sidei [131] for multidimensional stationary processes, Golichenko (Dubovets'ka) and Moklyachuk [44] for periodically correlated processes, and Luz and Moklyachuk [109, 114] for processes with stationary increments.

10 Filtering of processes with periodically correlated increments

In this chapter, we deal with the filtering problem for stochastic processes with periodically correlated dth increments observed with a periodically stationary stochastic noise process.

By the *filtering problem* we understand the problem of the mean square optimal linear estimation of the functionals

$$A\xi = \int_0^\infty a(t)\xi(-t)dt, \quad A_{NT}\xi = \int_0^{(N+1)T} a(t)\xi(-t)dt,$$

which depend on the unknown values of the stochastic process $\xi(t)$ with periodically correlated dth increments. Estimates are based on observations of the process $\zeta(t) = \xi(t)+\eta(t)$ at points $t \leq 0$, where $\eta(t)$ is an uncorrelated with the process $\xi(t)$ periodically stationary stochastic process.

10.1 Hilbert space projection method of filtering

Let an increment process $\xi^{(d)}(t, \tau T)$, $t \in \mathbb{R}$, generate an infinite-dimensional vector stationary increment sequence $\{\vec{\xi}_j^{(d)}, j \in \mathbb{Z}\}$ (6.7) with the spectral density matrix $f(\lambda) = \{f_{ij}(\lambda)\}_{i,j=1}^\infty$. Let $\eta(t)$, $t \in \mathbb{R}$, be an uncorrelated with the increment process $\xi^{(d)}(t, \tau T)$ periodically stationary stochastic process, which generates an infinite-dimensional vector stationary sequence $\{\vec{\eta}_j, j \in \mathbb{Z}\}$ (6.3) with the spectral density matrix $g(\lambda) = \{g_{ij}(\lambda)\}_{i,j=1}^\infty$.

As in the previous chapters, assume that the mean values of the sequences $\vec{\xi}_j^{(d)}$ and $\vec{\eta}_j$ equal 0, and consider the increment step $\tau > 0$.

In this section, we consider the filtering problem in the case where the spectral densities $f(\lambda)$ and $g(\lambda)$ are exactly known.

Assume that spectral densities $f(\lambda)$ and $g(\lambda)$ satisfy the *minimality condition*

$$\int_{-\pi}^{\pi} \mathrm{Tr}\left[\frac{\lambda^{2d}}{|1 - e^{i\lambda\tau}|^{2d}}(f(\lambda) + \lambda^{2d} g(\lambda))^{-1}\right] d\lambda < \infty. \tag{10.1}$$

This is the necessary and sufficient condition under which the MSEs of the optimal estimates of the functionals $A\vec{\xi}$ and $A_{NT}\vec{\xi}$ to be defined below are not equal to 0.

The functional $A\xi$ does not belong to the Hilbert space $H = L_2(\Omega, \mathcal{F}, \mathrm{P})$. To apply a Hilbert space projection method of estimation, let us formulate the following lemma, which will allow to represent the functional A_ξ as the sum of a functional with finite second moment from the space H and *an initial value functional* from the observed values of the process $\zeta(t)$.

https://doi.org/10.1515/9783111325620-010

Lemma 10.1.1. *The functional $A\xi$ can be presented as*

$$A\xi = A\zeta - A\vec{\eta},$$

$$A\zeta = \int_0^\infty a(t)\zeta(-t)dt, \quad A\vec{\eta} = \sum_{j=0}^\infty (\vec{a}_j)^\top \vec{\eta}_{-j},$$

where for $k = 1, 2, \dots$ and $j = 0, 1, \dots$ the vector

$$\vec{a}_j = (a_{kj}, k = 1, 2, \dots)^\top = (a_{1j}, a_{3j}, a_{2j}, \dots, a_{2k+1,j}, a_{2k,j}, \dots)^\top$$

has the entries

$$a_{kj} = \langle a_j, \widetilde{e}_k \rangle = \frac{1}{\sqrt{T}} \int_0^T a_j(v) e^{-2\pi i \{(-1)^k [\frac{k}{2}]\} v/T} dv.$$

Proof. The representation of the functional $A\vec{\eta}$ comes from Lemma 8.1.2. □

The conditions

$$\sum_{j=0}^\infty \|\vec{a}_j\| < \infty, \quad \sum_{j=0}^\infty (j+1)\|\vec{a}_j\|^2 < \infty, \quad \|\vec{a}_j\|^2 = \sum_{k=1}^\infty |a_{kj}|^2, \tag{10.2}$$

imply that the functional $A\vec{\eta}$ has finite second moment and belongs to the space $H = L_2(\Omega, \mathcal{F}, \mathsf{P})$.

Denote by $\Delta(f, g; \widehat{A}\xi) = \mathsf{E}|A\xi - \widehat{A}\xi|^2$ and $\Delta(f, g; \widehat{A}\vec{\eta}) = \mathsf{E}|A\vec{\eta} - \widehat{A}\vec{\eta}|^2$ the MSEs of the optimal estimates $\widehat{A}\xi$ and $\widehat{A}\vec{\eta}$ of the functionals $A\xi$ and $A\vec{\zeta}$. Since the functional $A\zeta$ depends on the observed values of the stochastic sequence $\zeta(t)$, we have the relation

$$\widehat{A}\xi = A\zeta - \widehat{A}\vec{\eta} \tag{10.3}$$

and

$$\begin{aligned}\Delta(f, g; \widehat{A}\xi) &= \mathsf{E}|A\xi - \widehat{A}\xi|^2 = \mathsf{E}|A\zeta - A\vec{\eta} - A\zeta + \widehat{A}\vec{\eta}|^2 \\ &= \mathsf{E}|A\vec{\eta} - \widehat{A}\vec{\eta}|^2 = \Delta(f, g; \widehat{A}\vec{\eta}).\end{aligned}$$

Therefore, it is sufficient to find the optimal linear estimate of the functional $A\vec{\eta}$ in order to find the optimal linear estimate of the functional $A\xi$.

From spectral representation (6.10) of the stationary stochastic sequence $\vec{\eta}_j$ we obtain the spectral representation

$$\vec{\eta}_j^{(d)} = \int_{-\pi}^{\pi} e^{i\lambda j}(1 - e^{-i\lambda\tau})^d d\vec{Z}_\eta(\lambda)$$

of the dth increment sequence $\vec{\eta}_j^{(d)}$. The random processes $\vec{Z}_\eta(\lambda)$ and $\vec{Z}_{\eta^{(d)}}(\lambda)$ are connected by the relation $d\vec{Z}_{\eta^{(d)}}(\lambda) = (i\lambda)^d d\vec{Z}_\eta(\lambda)$. The stationary increment sequence $\vec{\zeta}_j^{(d)}$ admits the spectral representation

$$\vec{\zeta}_j^{(d)} = \int_{-\pi}^{\pi} e^{i\lambda j}(1-e^{-i\lambda\tau})^d \frac{1}{(i\lambda)^d} d\vec{Z}_{\xi^{(d)}}(\lambda) + \int_{-\pi}^{\pi} e^{i\lambda j}(1-e^{-i\lambda\tau})^d d\vec{Z}_\eta(\lambda).$$

The spectral density $p(\lambda)$ of the stochastic increment sequence $\vec{\zeta}_j^{(d)}$ is

$$p(\lambda) = f(\lambda) + \lambda^{2d} g(\lambda).$$

Let us define the subspace

$$H^0(\vec{\xi}^{(d)} + \vec{\eta}^{(d)}) = \overline{\operatorname{span}}\{\xi_k^{(d)}(j,\tau) + \eta_k^{(d)}(j,\tau) : k = 1,\dots,\infty; j \le 0\}$$

of $H = L_2(\Omega, \mathcal{F}, \mathsf{P})$ generated by the observations $\{\zeta(t) : t \le 0\}$ and the subspace

$$L_2^0(f(\lambda) + \lambda^{2d} g(\lambda)) = \overline{\operatorname{span}}\left\{e^{i\lambda j}(1-e^{-i\lambda\tau})^d \frac{1}{(i\lambda)^d}\vec{\delta}_k : k = 1,\dots,\infty;\ j \le 0\right\}$$

of the Hilbert space $L_2(f(\lambda)+\lambda^{2d}g(\lambda))$ of vector-valued functions endowed with the inner product $\langle g_1; g_2\rangle = \int_{-\pi}^{\pi}(g_1(\lambda))^\top(f(\lambda) + \lambda^{2d}g(\lambda))\overline{g_2(\lambda)}d\lambda$. Here $\vec{\delta}_k = \{\delta_{kl}\}_{l=1}^{\infty}$, where δ_{kl} are Kronecker symbols.

Remark 10.1.1. Representation (6.11) yields a map between the elements

$$e^{i\lambda j}(1-e^{-i\lambda\tau})^d (i\lambda)^{-d}\vec{\delta}_k$$

from the subspace $L_2^0(f(\lambda) + \lambda^{2d}g(\lambda))$ and the elements $\vec{\xi}_{kj}^{(d)} + \vec{\eta}_{kj}^{(d)}$ from the subspace $H^0(\vec{\xi}^{(d)} + \vec{\eta}^{(d)})$.

The final step is to find the mean square optimal estimate $\widehat{A}\vec{\eta}$ as an orthogonal projection of the element $A\vec{\eta}$ on the subspace $H^0(\vec{\xi}^{(d)} + \vec{\eta}^{(d)})$:

$$\widehat{A}\vec{\eta} = \mathsf{Proj}_{H^0(\vec{\xi}^{(d)}+\vec{\eta}^{(d)})} A\vec{\eta}.$$

Relation (10.3) implies that an optimal linear estimate $\widehat{A}\xi$ allows a representation

$$\widehat{A}\xi = A\zeta - \mathsf{Proj}_{H^0(\vec{\xi}^{(d)}+\vec{\eta}^{(d)})} A\vec{\eta}$$

or a representation

$$\widehat{A}\xi = \int_0^{\infty} a(t)(\xi(-t) + \eta(-t))dt - \int_{-\pi}^{\pi} (\vec{h}_\tau(\lambda))^\top d\vec{Z}_{\xi^{(d)}+\eta^{(d)}}(\lambda), \tag{10.4}$$

where $\vec{h}_\tau(\lambda) = \{h_p(\lambda)\}_{p=1}^{\infty}$ is the spectral characteristic of the estimate $\widehat{A}\vec{\eta}$.

Define the Fourier coefficients of the corresponding functions

$$S^{\tau}_{l,j} = \frac{1}{2\pi}\int_{-\pi}^{\pi} e^{-i\lambda(l+j)}\frac{\lambda^{2d}}{|1-e^{i\lambda\tau}|^{2d}}\left(g(\lambda)(f(\lambda)+\lambda^{2d}g(\lambda))^{-1}\right)^{\top}d\lambda, \quad l \geq 0, j \geq -\tau d,$$

$$P^{\tau}_{l,j} = \frac{1}{2\pi}\int_{-\pi}^{\pi} e^{-i\lambda(l-j)}\frac{\lambda^{2d}}{|1-e^{i\lambda\tau}|^{2d}}\left((f(\lambda)+\lambda^{2d}g(\lambda))^{-1}\right)^{\top}d\lambda, \quad l,j \geq 0,$$

$$Q_{l,j} = \frac{1}{2\pi}\int_{-\pi}^{\pi} e^{-i\lambda(l-j)}\left(f(\lambda)(f(\lambda)+\lambda^{2d}g(\lambda))^{-1}g(\lambda)\right)^{\top}d\lambda, \quad l,j \geq 0.$$

Denote

$$\vec{A}(e^{-i\lambda}) = \sum_{j=0}^{\infty}\vec{a}_j e^{-i\lambda j}, \quad \vec{A}_{\tau}(e^{-i\lambda}) = (1-e^{i\lambda\tau})^d\vec{A}(e^{-i\lambda}) = \sum_{j=-\tau d}^{\infty}\vec{a}_j^{-\tau}e^{-i\lambda j},$$

where coefficients $\vec{a}_j^{-\tau}, j \geq 0$, are calculated as

$$\vec{a}_j^{-\tau} = \sum_{l=\max\{\lceil -\frac{j}{\tau}\rceil,0\}}^{d}(-1)^l\binom{d}{l}\vec{a}_{j+\tau l}, \quad j \geq -\tau d. \tag{10.5}$$

Here $\lceil x\rceil$ denotes the least integer greater than or equal to x.

Define vectors

$$\mathbf{a} = ((\vec{a}_0)^{\top}, (\vec{a}_1)^{\top}, (\vec{a}_2)^{\top}, \dots)^{\top},$$
$$\mathbf{a}^{\tau} = ((\vec{a}_0^{\tau})^{\top}, (\vec{a}_1^{\tau})^{\top}, (\vec{a}_2^{\tau})^{\top}, \dots)^{\top}, \quad \vec{a}_j^{\tau} = \vec{a}_{j-\tau d}^{-\tau}, \; j \geq 0,$$

and linear operators $\mathbf{P}_{\tau}$, $\mathbf{S}_{\tau}$, $\mathbf{Q}$ in the space ℓ_2 defined by the matrix with the infinite-dimensional matrix entries $(\mathbf{P}_{\tau})_{l,j} = P^{\tau}_{l,j}$, $(\mathbf{S}_{\tau})_{l,j} = S^{\tau}_{l+1,j-\tau d}$, and $(\mathbf{Q})_{l,j} = Q_{l,j}$, $l,j \geq 0$.

Theorem 10.1.1. *Consider two uncorrelated processes: a stochastic process $\xi(t)$, $t \in \mathbb{R}$, with periodically stationary increments, which determines a generated stationary dth increment sequence $\vec{\xi}_j^{(d)}$ with the spectral density matrix $f(\lambda) = \{f_{kn}(\lambda)\}_{k,n=1}^{\infty}$, and a periodically stationary stochastic process $\eta(t)$, $t \in \mathbb{R}$, which determines a generated stationary sequence $\vec{\eta}_j$ with the spectral density matrix $g(\lambda) = \{g_{kn}(\lambda)\}_{k,n=1}^{\infty}$. Let the coefficients $\vec{a}_j$, $j = 0,1,\dots$, generated by the function $a(t)$, $t \geq 0$, satisfy conditions* (10.2). *Let the minimality condition* (10.1) *be satisfied. The optimal linear estimate $\widehat{A}\xi$ of the functional $A\xi$ based on observations of the process $\xi(t) + \eta(t)$ at points $t \leq 0$ is calculated by formula* (10.4). *The spectral characteristic $\vec{h}^{\tau}(\lambda)$ of the optimal estimate $\widehat{A}\xi$ is calculated by the formula*

$$(\vec{h}_{\tau}(\lambda))^{\top} = (\vec{A}(e^{-i\lambda}))^{\top}g(\lambda)(-i\lambda)^d(f(\lambda)+\lambda^{2d}g(\lambda))^{-1}$$
$$- \frac{(-i\lambda)^d}{(1-e^{i\lambda\tau})^d}(\vec{C}_{\tau}(e^{i\lambda}))^{\top}(f(\lambda)+\lambda^{2d}g(\lambda))^{-1}, \tag{10.6}$$

where

$$\vec{C}_\tau(e^{i\lambda}) = \sum_{j=0}^{\infty} (\mathbf{P}_\tau^{-1}\mathbf{S}_\tau \mathbf{a}^\tau)_j e^{i\lambda(j+1)}.$$

The MSE of the estimate $\widehat{A}\xi$ is calculated by the formula

$$\begin{aligned}
\Delta(f,g;\widehat{A}\xi) &= \Delta(f,g;\widehat{A}\vec{\eta}) = \mathsf{E}|A\vec{\eta} - \widehat{A}\vec{\eta}|^2 \\
&= \frac{1}{2\pi}\int_{-\pi}^{\pi} \frac{\lambda^{2d}}{|1-e^{i\lambda\tau}|^{2d}} ((\vec{A}_\tau(e^{-i\lambda}))^\top g(\lambda) + (\vec{C}_\tau(e^{i\lambda}))^\top) \\
&\quad \times (f(\lambda) + \lambda^{2d} g(\lambda))^{-1} f(\lambda)\,(f(\lambda) + \lambda^{2d} g(\lambda))^{-1} \\
&\quad \times (g(\lambda)\overline{\vec{A}_\tau(e^{-i\lambda})} + \overline{\vec{C}_\tau(e^{i\lambda})})d\lambda \\
&\quad + \frac{1}{2\pi}\int_{-\pi}^{\pi} \frac{\lambda^{4d}}{|1-e^{i\lambda\tau}|^{4d}} \left(\frac{|1-e^{i\lambda\tau}|^{2d}}{\lambda^{2d}} (\vec{A}(e^{-i\lambda}))^\top f(\lambda) - (1-e^{-i\lambda\tau})^d (\vec{C}_\tau(e^{i\lambda}))^\top\right) \\
&\quad \times (f(\lambda) + \lambda^{2d} g(\lambda))^{-1} g(\lambda)\,(f(\lambda) + \lambda^{2d} g(\lambda))^{-1} \\
&\quad \times \left(\frac{|1-e^{i\lambda\tau}|^{2d}}{\lambda^{2d}} f(\lambda)\overline{\vec{A}(e^{-i\lambda})} - (1-e^{i\lambda\tau})^d \overline{\vec{C}_\tau(e^{i\lambda})}\right)d\lambda \\
&= \langle \mathbf{S}_\tau \mathbf{a}^\tau, \mathbf{P}_\tau^{-1}\mathbf{S}_\tau \mathbf{a}^\tau\rangle + \langle \mathbf{Q}\mathbf{a}, \mathbf{a}\rangle.
\end{aligned} \tag{10.7}$$

Proof. The estimate $\widehat{A}\vec{\eta}$ is determined by two conditions:
(1) $\widehat{A}\vec{\eta} \in H^{0-}(\vec{\xi}^{(d)} + \vec{\eta}^{(d)})$;
(2) $(A\vec{\eta} - \widehat{A}\vec{\eta}) \perp H^{0-}(\vec{\xi}^{(d)} + \vec{\eta}^{(d)})$.

Condition (2) implies the following relation, which holds for all $j \le 0$ and for all $k \ge 1$:

$$\int_{-\pi}^{\pi} ((\vec{A}(e^{-i\lambda}))^\top g(\lambda)(-i\lambda)^d - (\vec{h}_\tau(\lambda))^\top (f(\lambda) + \lambda^{2d} g(\lambda))) \frac{(1-e^{i\lambda\tau})^d}{(-i\lambda)^d} \vec{\delta}_k e^{-i\lambda j} d\lambda = 0.$$

This relation allows us to derive the spectral characteristic $\vec{h}_\tau(\lambda)$ of the estimate $\widehat{A}\vec{\eta}$ in the form

$$\begin{aligned}
(\vec{h}_\tau(\lambda))^\top &= (\vec{A}(e^{-i\lambda}))^\top g(\lambda)(-i\lambda)^d (f(\lambda) + \lambda^{2d} g(\lambda))^{-1} \\
&\quad - \frac{(-i\lambda)^d}{(1-e^{i\lambda\tau})^d} (\vec{C}_\tau(e^{i\lambda}))^\top (f(\lambda) + \lambda^{2d} g(\lambda))^{-1},
\end{aligned}$$

where

$$\vec{C}_\tau(e^{i\lambda}) = \sum_{j=0}^{\infty} \vec{c}_j^{\,\tau} e^{i\lambda(j+1)},$$

where $\vec{c}_j^{\,\tau} = \{c_{kj}^\tau\}_{k=1}^\infty$, $j \ge 0$, are unknown coefficients to be found.

Condition (1) implies that the spectral characteristic $\vec{h}_\tau(\lambda)$ admits the representation

$$\vec{h}_\tau(\lambda) = \vec{h}(\lambda)(1 - e^{-i\lambda\tau})^d \frac{1}{(i\lambda)^d}, \quad \vec{h}(\lambda) = \sum_{j=0}^{\infty} \vec{s}_j e^{-i\lambda j},$$

where

$$\int_{-\pi}^{\pi} (\vec{h}(\lambda))^\top (f(\lambda) + \lambda^{2d} g(\lambda))\overline{\vec{h}(\lambda)} \frac{|1 - e^{i\lambda\tau}|^{2d}}{\lambda^{2d}} d\lambda < \infty, \qquad \frac{(i\lambda)^d}{(1 - e^{-i\lambda\tau})^d} \vec{h}_\tau(\lambda) \in L_2^0.$$

Thus, we obtain the following relation:

$$\int_{-\pi}^{\pi} ((\vec{A}(e^{-i\lambda}))^\top (1 - e^{i\lambda\tau})^d g(\lambda) - (\vec{C}_\tau(e^{i\lambda}))^\top) \frac{\lambda^{2d}}{|1 - e^{-i\lambda\tau}|^{2d}}$$
$$\times (f(\lambda) + \lambda^{2d} g(\lambda))^{-1} e^{-i\lambda j} d\lambda = 0, \quad j \geq 1. \tag{10.8}$$

Then relation (10.8) can be presented as a system of linear equations:

$$\sum_{j=-\tau d}^{\infty} S_{l+1,j}^\tau \vec{a}_j^{-\tau} = \sum_{j=0}^{\infty} P_{l+1,j+1}^\tau \vec{c}_j^{\,\tau}, \quad l \geq 0,$$

determining the unknown coefficients $\vec{c}_j^{\,\tau}, j \geq 0$. This system can be written in matrix form as

$$\mathbf{S}_\tau \mathbf{a}^\tau = \mathbf{P}_\tau \mathbf{c}_\tau,$$

where

$$\mathbf{c}^\tau = ((\vec{c}_1^{\,\tau})^\top, (\vec{c}_2^{\,\tau})^\top, (\vec{c}_3^{\,\tau})^\top, \ldots)^\top.$$

Finally, the coefficients $\vec{c}_j^{\,\tau}, j \geq 0$, determining the spectral characteristic $\vec{h}_\tau(\lambda)$ are calculated by the formula

$$\vec{c}_j^{\,\tau} = (\mathbf{P}_\tau^{-1} \mathbf{S}_\tau \mathbf{a}^\tau)_j, \quad j \geq 0,$$

where $(\mathbf{P}_\tau^{-1} \mathbf{S}_\tau \mathbf{a}^\tau)_j, j \geq 0$, is the jth infinite-dimensional entry of the vector $\mathbf{P}_\tau^{-1} \mathbf{S}_\tau \mathbf{a}^\tau$.

The derived expressions justify the use of formulas (10.6) and (10.7) for calculating the spectral characteristic $\vec{h}_\tau(\lambda)$ and the value of the MSE $\Delta(f, g; \widehat{A}\xi)$ of the estimate $\widehat{A}\xi$ of the functional $A\xi$. □

Making use of Theorem 10.1.1 we obtain formulas to calculate the optimal estimate $\widehat{A}_{NT}\xi$ of the functional $A_{NT}\xi$ based on observations of the process $\xi(t) + \eta(t)$ at points $t \leq 0$. Let $\vec{a}_j = \vec{0}, j > N$. Define a vector

$$\mathbf{a}_N = ((\vec{a}_0)^\top, (\vec{a}_1)^\top, \ldots, (\vec{a}_N)^\top, \vec{0}^\top, \ldots)^\top$$

and a vector

$$\mathbf{a}_{\tau,N} = ((\vec{a}_0^{\tau,N})^\top, (\vec{a}_1^{\tau,N})^\top, \ldots, (\vec{a}_{N+\tau d}^{\tau,N})^\top, \vec{0}^\top, \ldots)^\top$$

with the entries

$$\vec{a}_j^{\tau,N} = \vec{a}_{j-\tau d}^{-\tau,N}, \quad 0 \le j \le N + \tau d,$$

where

$$\vec{a}_m^{-\tau,N} = \sum_{l=\max\{\lceil -\frac{m}{\tau} \rceil, 0\}}^{\min\{\lfloor \frac{N-m}{\tau} \rfloor, d\}} (-1)^l \binom{d}{l} \vec{a}(m + \tau l), \quad -\tau d \le m \le N. \tag{10.9}$$

Define a linear operator $\mathbf{S}_{\tau,N}$ in the space ℓ_2 determined by the matrix with the infinite-dimensional matrix entries $(\mathbf{S}_{\tau,N})_{l,m} = S^\tau_{l+1,m-\tau d}$ for $l \ge 0$, $0 \le m \le N + \tau d$ and $(\mathbf{S}_{\tau,N})_{l,m} = \mathcal{O}$ for $l \ge 0$, $m > N + \tau d$ and a linear operator $\mathbf{Q}_N$ in the space ℓ_2 determined by the matrix with the infinite-dimensional matrix entries $(\mathbf{Q}_N)_{l,j} = Q_{l,j}$, $0 \le l, j \le N$, and $(\mathbf{Q}_N)_{l,j} = \mathcal{O}$ otherwise.

The following corollary holds.

Corollary 10.1.1. *Consider two uncorrelated processes: a stochastic process $\xi(t)$, $t \in \mathbb{R}$, with periodically stationary increments and a periodically stationary stochastic process $\eta(t)$, $t \in \mathbb{R}$. Let the conditions of Theorem* 10.1.1 *be satisfied. The optimal linear estimate $\widehat{A}_{NT}\xi$ of the functional $A_{NT}\xi$ which depends on the unobserved values $\xi(t)$, $t \le 0$, from observations of the process $\xi(t) + \eta(t)$ at points $t \le 0$ is calculated by the formula*

$$\widehat{A}_{NT}\xi = \int_0^{(N+1)T} a(t)(\xi(-t) + \eta(-t))dt - \int_{-\pi}^{\pi} (\vec{h}_{\tau,N}(\lambda))^\top d\vec{Z}_{\xi^{(d)}+\eta^{(d)}}(\lambda), \tag{10.10}$$

where

$$(\vec{h}_{\tau,N}(\lambda))^\top = \left(\sum_{j=0}^{N} \vec{a}_j e^{-i\lambda j} \right)^\top g(\lambda)(-i\lambda)^d (f(\lambda) + \lambda^{2d} g(\lambda))^{-1}$$
$$- \frac{(-i\lambda)^d}{(1 - e^{i\lambda\tau})^d} \left(\sum_{j=0}^{\infty} (\mathbf{P}_\tau^{-1} \mathbf{S}_{\tau,N} \mathbf{a}_{\tau,N})_j e^{i\lambda(j+1)} \right)^\top (f(\lambda) + \lambda^{2d} g(\lambda))^{-1}. \tag{10.11}$$

The MSE of the estimate $\widehat{A}_{NT}\xi$ is calculated by the formula

$$\Delta(f, g; \widehat{A}_{NT}\xi) = \Delta(f, g; \widehat{A}_N\vec{\eta}) = \mathsf{E}|A_N\vec{\eta} - \widehat{A}_N\vec{\eta}|^2$$

$$= \frac{1}{2\pi}\int_{-\pi}^{\pi} \frac{\lambda^{2d}}{|1-e^{i\lambda\tau}|^{2d}}((\vec{A}_{\tau,N}(e^{-i\lambda}))^{\top} g(\lambda) + (\vec{C}_{\tau,N}(e^{i\lambda}))^{\top})$$
$$\times (f(\lambda) + \lambda^{2d} g(\lambda))^{-1} f(\lambda)\,(f(\lambda) + \lambda^{2d} g(\lambda))^{-1}$$
$$\times (g(\lambda)\overline{\vec{A}_{\tau,N}(e^{-i\lambda})} + \overline{\vec{C}_{\tau,N}(e^{i\lambda})})d\lambda$$
$$+ \frac{1}{2\pi}\int_{-\pi}^{\pi} \frac{1}{|1-e^{i\lambda\tau}|^{2d}}((\vec{A}_{\tau,N}(e^{-i\lambda}))^{\top} f(\lambda) - \lambda^{2d}(\vec{C}_{\tau,N}(e^{i\lambda}))^{\top})$$
$$\times (f(\lambda) + \lambda^{2d} g(\lambda))^{-1} g(\lambda)\,(f(\lambda) + \lambda^{2d} g(\lambda))^{-1}$$
$$\times (f(\lambda)\overline{\vec{A}_{\tau,N}(e^{-i\lambda})} - \lambda^{2d}\overline{\vec{C}_{\tau,N}(e^{i\lambda})})d\lambda$$
$$= \langle \mathbf{S}_{\tau,N}\mathbf{a}_{\tau,N}, \mathbf{P}_{\tau}^{-1}\mathbf{S}_{\tau,N}\mathbf{a}_{\tau,N}\rangle + \langle \mathbf{Q}_N\mathbf{a}_N, \mathbf{a}_N\rangle, \tag{10.12}$$

where

$$\vec{A}_{\tau,N}(e^{i\lambda}) = (1-e^{i\lambda\tau})^d \sum_{j=0}^{N} \vec{a}_j e^{-i\lambda j} = \sum_{j=-\tau d}^{N} \vec{a}_j^{\,-\tau,N} e^{-i\lambda j}$$

and

$$\vec{C}_{\tau,N}(e^{i\lambda}) = \sum_{j=0}^{\infty} (\mathbf{P}_{\tau}^{-1}\mathbf{S}_{\tau,N}\mathbf{a}_{\tau,N})_j e^{i\lambda(j+1)}.$$

10.2 Minimax (robust) method of filtering

10.2.1 Least favorable spectral densities and minimax-robust spectral characteristic

Consider the estimation problem for the functional $A\xi$ based on the observations $\xi(t) + \eta(t)$ at points $t \leq 0$ when the spectral densities of the generated sequences are not exactly known while a set $\mathcal{D} = \mathcal{D}_f \times \mathcal{D}_g$ of admissible spectral densities is defined. We apply the minimax (robust) approach to estimation which is formalized by the following two definitions.

Definition 10.2.1. For a given class of spectral densities $\mathcal{D} = \mathcal{D}_f \times \mathcal{D}_g$, the spectral densities $f^0(\lambda) \in \mathcal{D}_f$ and $g^0(\lambda) \in \mathcal{D}_g$ are called least favorable in the class $\mathcal{D}$ for the optimal linear filtering of the functional $A\xi$ if the following relation holds:

$$\Delta(f^0, g^0) = \Delta(h(f^0, g^0); f^0, g^0) = \max_{(f,g)\in\mathcal{D}_f\times\mathcal{D}_g} \Delta(h(f,g); f, g).$$

Definition 10.2.2. For a given class of spectral densities $\mathcal{D} = \mathcal{D}_f \times \mathcal{D}_g$, the spectral characteristic $h^0(\lambda)$ of the optimal linear estimate of the functional $A\xi$ is called minimax-robust if the following conditions are satisfied:

$$h^0(\lambda) \in H_{\mathcal{D}} = \bigcap_{(f,g)\in\mathcal{D}_f\times\mathcal{D}_g} L_2^0(f(\lambda) + \lambda^{2d} g(\lambda)),$$

$$\min_{h\in H_{\mathcal{D}}} \max_{(f,g)\in\mathcal{D}_f\times\mathcal{D}_g} \Delta(h;f,g) = \max_{(f,g)\in\mathcal{D}_f\times\mathcal{D}_g} \Delta(h^0;f,g).$$

Taking into account the introduced definitions and the derived relations we can verify that the following lemmas hold.

Lemma 10.2.1. *Spectral densities $f^0 \in \mathcal{D}_f$ and $g^0 \in \mathcal{D}_g$ which satisfy condition* (10.1) *are least favorable in the class $\mathcal{D} = \mathcal{D}_f \times \mathcal{D}_g$ for the optimal linear filtering of the functional $A\xi$ if the operators $\mathbf{S}_\tau^0$, $\mathbf{P}_\tau^0$, $\mathbf{Q}^0$ determined by the Fourier coefficients of the functions*

$$\frac{\lambda^{2d}}{|1-e^{i\lambda\tau}|^{2d}} (g^0(\lambda)(f^0(\lambda) + \lambda^{2d} g^0(\lambda))^{-1})^\top,$$

$$\frac{\lambda^{2d}}{|1-e^{i\lambda\tau}|^{2d}} ((f^0(\lambda) + \lambda^{2d} g^0(\lambda))^{-1})^\top,$$

$$(f^0(\lambda)(f^0(\lambda) + \lambda^{2d} g^0(\lambda))^{-1} g^0(\lambda))^\top$$

determine a solution of the constrained optimization problem

$$\max_{(f,g)\in\mathcal{D}_f\times\mathcal{D}_g} (\langle \mathbf{S}_\tau \mathbf{a}^\tau, \mathbf{P}_\tau^{-1} \mathbf{S}_\tau \mathbf{a}^\tau\rangle + \langle \mathbf{Q}\mathbf{a}, \mathbf{a}\rangle) = \langle \mathbf{S}_\tau^0 \mathbf{a}^\tau, (\mathbf{P}_\tau^0)^{-1} \mathbf{S}_\tau^0 \mathbf{a}^\tau\rangle + \langle \mathbf{Q}^0 \mathbf{a}, \mathbf{a}\rangle. \tag{10.13}$$

The minimax spectral characteristic $h^0 = \vec{h}_\tau(f^0, g^0)$ is calculated by formula (10.6) *if $\vec{h}_\tau(f^0, g^0) \in H_{\mathcal{D}}$.*

The minimax spectral characteristic h^0 and the least favorable spectral densities (f^0, g^0) form a saddle point of the function $\Delta(h;f,g)$ on the set $H_{\mathcal{D}} \times \mathcal{D}$.

The saddle point inequalities

$$\Delta(h;f^0,g^0) \geq \Delta(h^0;f^0,g^0) \geq \Delta(h^0;f,g) \quad \forall f \in \mathcal{D}_f,\ \forall g \in \mathcal{D}_g,\ \forall h \in H_{\mathcal{D}}$$

hold if $h^0 = \vec{h}_\tau(f^0,g^0)$ and $\vec{h}_\tau(f^0,g^0) \in H_{\mathcal{D}}$, where (f^0,g^0) is a solution of the constrained optimization problem

$$\widetilde{\Delta}(f,g) = -\Delta(\vec{h}_\tau(f^0,g^0);f,g) \to \inf, \quad (f,g) \in \mathcal{D}, \tag{10.14}$$

where the functional $\Delta(\vec{h}_\tau(f^0,g^0);f,g)$ is calculated by the formula

$$\begin{aligned}
&\Delta(\vec{h}_\tau(f^0,g^0);f,g) \\
&= \frac{1}{2\pi} \int_{-\pi}^{\pi} \frac{\lambda^{2d}}{|1-e^{i\lambda\tau}|^{2d}} ((1-e^{i\lambda\tau})^d (\vec{A}(e^{-i\lambda}))^\top g^0(\lambda) + (\vec{C}_\tau^0(e^{i\lambda}))^\top) \\
&\quad \times (f^0(\lambda) + \lambda^{2d} g^0(\lambda))^{-1} f(\lambda)\, (f^0(\lambda) + \lambda^{2d} g^0(\lambda))^{-1}
\end{aligned}$$

$$\times ((1-e^{-i\lambda\tau})^d g^0(\lambda)\overline{\vec{A}(e^{-i\lambda})} + \overline{\vec{C}^0_\tau(e^{i\lambda})})d\lambda +$$

$$+\frac{1}{2\pi}\int_{-\pi}^{\pi}\frac{\lambda^{4d}}{|1-e^{i\lambda\tau}|^{4d}}\left(\frac{|1-e^{i\lambda\tau}|^{2d}}{\lambda^{2d}}(\vec{A}(e^{-i\lambda}))^\top f^0(\lambda) - (1-e^{-i\lambda\tau})^d(\vec{C}^0_\tau(e^{i\lambda}))^\top\right)$$

$$\times (f^0(\lambda)+\lambda^{2d}g^0(\lambda))^{-1}g(\lambda)(f^0(\lambda)+\lambda^{2d}g^0(\lambda))^{-1}$$

$$\times \left(\frac{|1-e^{i\lambda\tau}|^{2d}}{\lambda^{2d}}f^0(\lambda)\overline{\vec{A}(e^{-i\lambda})} - (1-e^{i\lambda\tau})^d\overline{\vec{C}^0_\tau(e^{i\lambda})}\right)d\lambda,$$

where

$$\vec{C}^0_\tau(e^{i\lambda}) = \sum_{j=0}^{\infty}((\mathbf{P}^0_\tau)^{-1}\mathbf{S}^0_\tau\mathbf{a}^\tau)_j e^{i\lambda(j+1)}.$$

The constrained optimization problem (10.14) is equivalent to the unconstrained optimization problem

$$\Delta_{\mathcal{D}}(f,g) = \widetilde{\Delta}(f,g) + \delta(f,g|\mathcal{D}_f\times\mathcal{D}_g) \to \inf, \tag{10.15}$$

where $\delta(f,g|\mathcal{D}_f\times\mathcal{D}_g)$ is the indicator function of the set $\mathcal{D} = \mathcal{D}_f\times\mathcal{D}_g$. Solution (f^0,g^0) to this unconstrained optimization problem is characterized by the condition $0\in\partial\Delta_{\mathcal{D}}(f^0,g^0)$, where $\partial\Delta_{\mathcal{D}}(f^0,g^0)$ is the subdifferential of the functional $\Delta_{\mathcal{D}}(f,g)$ at point $(f^0,g^0)\in\mathcal{D}=\mathcal{D}_f\times\mathcal{D}_g$. This condition makes it possible to find the least favorable spectral densities in some special classes of spectral densities $\mathcal{D}=\mathcal{D}_f\times\mathcal{D}_g$.

The form of the functional $\Delta(\vec{h}_\tau(f^0,g^0);f,g)$ is convenient for application of the Lagrange method of indefinite multipliers for finding a solution to the problem (10.15). Making use of the method of Lagrange multipliers and the form of subdifferentials of the indicator functions $\delta(f,g|\mathcal{D}_f\times\mathcal{D}_g)$ of the set $\mathcal{D}_f\times\mathcal{D}_g$ of spectral densities we describe relations that determine least favorable spectral densities in some special classes of spectral densities (see [115, 146] for additional details).

10.2.2 Least favorable spectral densities in classes $\mathcal{D}_0 \times \mathcal{D}_V^U$

Consider the minimax (robust) filtering problem for the functional $A\xi$ under the condition that the sets of admissible spectral densities $\mathcal{D}_0^k$, $\mathcal{D}_V^{Uk}$, $k = 1,2,3,4$, are defined as follows:

$$\mathcal{D}_0^1 = \left\{f(\lambda)\left|\frac{1}{2\pi}\int_{-\pi}^{\pi}\frac{|1-e^{i\lambda\tau}|^{2d}}{\lambda^{2d}}f(\lambda)d\lambda = P\right.\right\},$$

$$\mathcal{D}_0^2 = \left\{f(\lambda)\left|\frac{1}{2\pi}\int_{-\pi}^{\pi}\frac{|1-e^{i\lambda\tau}|^{2d}}{\lambda^{2d}}\,\mathrm{Tr}[f(\lambda)]d\lambda = p\right.\right\},$$

$$\mathcal{D}_0^3 = \left\{ f(\lambda) \left| \frac{1}{2\pi} \int_{-\pi}^{\pi} \frac{|1-e^{i\lambda\tau}|^{2d}}{\lambda^{2d}} f_{kk}(\lambda) d\lambda = p_k, k = \overline{1,\infty} \right.\right\},$$

$$\mathcal{D}_0^4 = \left\{ f(\lambda) \left| \frac{1}{2\pi} \int_{-\pi}^{\pi} \frac{|1-e^{i\lambda\tau}|^{2d}}{\lambda^{2d}} \langle B_1, f(\lambda)\rangle d\lambda = p \right.\right\}$$

and

$$\mathcal{D}_V^{U1} = \left\{ g(\lambda) \left| V(\lambda) \le g(\lambda) \le U(\lambda), \frac{1}{2\pi} \int_{-\pi}^{\pi} g(\lambda) d\lambda = Q \right.\right\},$$

$$\mathcal{D}_V^{U2} = \left\{ g(\lambda) \left| \mathrm{Tr}[V(\lambda)] \le \mathrm{Tr}[g(\lambda)] \le \mathrm{Tr}[U(\lambda)], \frac{1}{2\pi} \int_{-\pi}^{\pi} \mathrm{Tr}[g(\lambda)] d\lambda = q \right.\right\},$$

$$\mathcal{D}_V^{U3} = \left\{ g(\lambda) \left| v_{kk}(\lambda) \le g_{kk}(\lambda) \le u_{kk}(\lambda), \frac{1}{2\pi} \int_{-\pi}^{\pi} g_{kk}(\lambda) d\lambda = q_k, k = \overline{1,\infty} \right.\right\},$$

$$\mathcal{D}_V^{U4} = \left\{ g(\lambda) \left| \langle B_2, V(\lambda)\rangle \le \langle B_2, g(\lambda)\rangle \le \langle B_2, U(\lambda)\rangle, \frac{1}{2\pi} \int_{-\pi}^{\pi} \langle B_2, g(\lambda)\rangle d\lambda = q \right.\right\}.$$

The spectral densities $V(\lambda)$ and $U(\lambda)$ from the definitions of the sets $\mathcal{D}_0^k, \mathcal{D}_V^{Uk}$, $k = 1,2,3,4$, are known and fixed, p, p_k, q, q_k, $k = \overline{1,\infty}$, are given numbers, and P, B_1, Q, B_2 are given positive-definite Hermitian matrices.

The condition $0 \in \partial\Delta_{\mathcal{D}}(f^0, g^0)$ implies the following equations determining the least favorable spectral densities for the given sets of admissible spectral densities.

For the first set of admissible spectral densities $\mathcal{D}_0^1 \times \mathcal{D}_V^{U1}$, we have

$$\begin{aligned}
&\left((1-e^{i\lambda\tau})^d \overline{g^0(\lambda)} \vec{A}(e^{-i\lambda}) + \sum_{j=0}^{\infty} ((\mathbf{P}_\tau^0)^{-1} \mathbf{S}_\tau^0 \mathbf{a}^\tau)_j e^{i\lambda(j+1)} \right) \\
&\quad \times \left((1-e^{i\lambda\tau})^d \overline{g^0(\lambda)} \vec{A}(e^{-i\lambda}) + \sum_{j=0}^{\infty} ((\mathbf{P}_\tau^0)^{-1} \mathbf{S}_\tau^0 \mathbf{a}^\tau)_j e^{i\lambda(j-1)} \right)^* \\
&= \left(\frac{|1-e^{i\lambda\tau}|^{2d}}{\lambda^{2d}} (f^0(\lambda) + \lambda^{2d} g^0(\lambda)) \right) \\
&\quad \times \vec{a} \cdot \vec{a}^* \left(\frac{|1-e^{i\lambda\tau}|^{2d}}{\lambda^{2d}} (f^0(\lambda) + \lambda^{2d} g^0(\lambda)) \right),
\end{aligned} \tag{10.16}$$

$$\begin{aligned}
&\left(\frac{|1-e^{i\lambda\tau}|^{2d}}{\lambda^{2d}} \overline{f^0(\lambda)} \vec{A}(e^{-i\lambda}) - (1-e^{-i\lambda\tau})^d \sum_{j=0}^{\infty} ((\mathbf{P}_\tau^0)^{-1} \mathbf{S}_\tau^0 \mathbf{a}^\tau)_j e^{i\lambda(j+1)} \right) \\
&\quad \times \left(\frac{|1-e^{i\lambda\tau}|^{2d}}{\lambda^{2d}} \overline{f^0(\lambda)} \vec{A}(e^{-i\lambda}) - (1-e^{-i\lambda\tau})^d \sum_{j=0}^{\infty} ((\mathbf{P}_\tau^0)^{-1} \mathbf{S}_\tau^0 \mathbf{a}^\tau)_j e^{i\lambda(j+1)} \right)^*
\end{aligned}$$

$$= \left(\frac{|1-e^{i\lambda\tau}|^{2d}}{\lambda^{2d}}(f^0(\lambda)+\lambda^{2d}g^0(\lambda))\right)(\vec{\beta}\cdot\vec{\beta}^* + \Gamma_1(\lambda)+\Gamma_2(\lambda))$$
$$\times\left(\frac{|1-e^{i\lambda\tau}|^{2d}}{\lambda^{2d}}(f^0(\lambda)+\lambda^{2d}g^0(\lambda))\right), \tag{10.17}$$

where $\vec{\alpha}$ and $\vec{\beta}$ are vectors of Lagrange multipliers, the matrix $\Gamma_1(\lambda) \leq 0$ and $\Gamma_1(\lambda) = 0$ if $g^0(\lambda) > V(\lambda)$, and the matrix $\Gamma_2(\lambda) \geq 0$ and $\Gamma_2(\lambda) = 0$ if $g^0(\lambda) < U(\lambda)$.

For the second set of admissible spectral densities $\mathcal{D}_0^2 \times \mathcal{D}_V^{U2}$, we have

$$\left((1-e^{i\lambda\tau})^d\overline{g^0(\lambda)}\vec{A}(e^{-i\lambda}) + \sum_{j=0}^{\infty}((\mathbf{P}_\tau^0)^{-1}\mathbf{S}_\tau^0\mathbf{a}^\tau)_j e^{i\lambda(j+1)}\right)$$
$$\times\left((1-e^{i\lambda\tau})^d\overline{g^0(\lambda)}\vec{A}(e^{-i\lambda}) + \sum_{j=0}^{\infty}((\mathbf{P}_\tau^0)^{-1}\mathbf{S}_\tau^0\mathbf{a}^\tau)_j e^{i\lambda(j+1)}\right)^*$$
$$= \alpha^2\left(\frac{|1-e^{i\lambda\tau}|^{2d}}{\lambda^{2d}}(f^0(\lambda)+\lambda^{2d}g^0(\lambda))\right)^2, \tag{10.18}$$

$$\left(\frac{|1-e^{i\lambda\tau}|^{2d}}{\lambda^{2d}}\overline{f^0(\lambda)}\vec{A}(e^{-i\lambda}) - (1-e^{-i\lambda\tau})^d\sum_{j=0}^{\infty}((\mathbf{P}_\tau^0)^{-1}\mathbf{S}_\tau^0\mathbf{a}^\tau)_j e^{i\lambda(j+1)}\right)$$
$$\times\left(\frac{|1-e^{i\lambda\tau}|^{2d}}{\lambda^{2d}}\overline{f^0(\lambda)}\vec{A}(e^{-i\lambda}) - (1-e^{-i\lambda\tau})^d\sum_{j=0}^{\infty}((\mathbf{P}_\tau^0)^{-1}\mathbf{S}_\tau^0\mathbf{a}^\tau)_j e^{i\lambda(j+1)}\right)^*$$
$$= (\beta^2+\gamma_1(\lambda)+\gamma_2(\lambda))\left(\frac{|1-e^{i\lambda\tau}|^{2d}}{\lambda^{2d}}(f^0(\lambda)+\lambda^{2d}g^0(\lambda))\right)^2, \tag{10.19}$$

where α^2, β^2 are Lagrange multipliers, the function $\gamma_1(\lambda) \leq 0$ and $\gamma_1(\lambda) = 0$ if $\mathrm{Tr}[f^0(\lambda)] > \mathrm{Tr}[V(\lambda)]$, and the function $\gamma_2(\lambda) \geq 0$ and $\gamma_2(\lambda) = 0$ if $\mathrm{Tr}[f^0(\lambda)] < \mathrm{Tr}[U(\lambda)]$.

For the third set of admissible spectral densities $\mathcal{D}_0^3 \times \mathcal{D}_V^{U3}$, we have

$$\left((1-e^{i\lambda\tau})^d\overline{g^0(\lambda)}\vec{A}(e^{-i\lambda}) + \sum_{j=0}^{\infty}((\mathbf{P}_\tau^0)^{-1}\mathbf{S}_\tau^0\mathbf{a}^\tau)_j e^{i\lambda(j+1)}\right)$$
$$\times\left((1-e^{i\lambda\tau})^d\overline{g^0(\lambda)}\vec{A}(e^{-i\lambda}) + \sum_{j=0}^{\infty}((\mathbf{P}_\tau^0)^{-1}\mathbf{S}_\tau^0\mathbf{a}^\tau)_j e^{i\lambda(j+1)}\right)^*$$
$$= \left(\frac{|1-e^{i\lambda\tau}|^{2d}}{\lambda^{2d}}(f^0(\lambda)+\lambda^{2d}g^0(\lambda))\right)\{\alpha_k^2\delta_{kl}\}_{k,l=1}^{\infty}$$
$$\times\left(\frac{|1-e^{i\lambda\tau}|^{2d}}{\lambda^{2d}}(f^0(\lambda)+\lambda^{2d}g^0(\lambda))\right), \tag{10.20}$$

$$\left(\frac{|1-e^{i\lambda\tau}|^{2d}}{\lambda^{2d}}\overline{f^0(\lambda)\vec{A}(e^{-i\lambda})} - (1-e^{-i\lambda\tau})^d\sum_{j=0}^{\infty}((\mathbf{P}_\tau^0)^{-1}\mathbf{S}_\tau^0\mathbf{a}^\tau)_j e^{i\lambda(j+1)}\right)$$
$$\times\left(\frac{|1-e^{i\lambda\tau}|^{2d}}{\lambda^{2d}}\overline{f^0(\lambda)\vec{A}(e^{-i\lambda})} - (1-e^{-i\lambda\tau})^d\sum_{j=0}^{\infty}((\mathbf{P}_\tau^0)^{-1}\mathbf{S}_\tau^0\mathbf{a}^\tau)_j e^{i\lambda(j+1)}\right)^*$$
$$=\left(\frac{|1-e^{i\lambda\tau}|^{2d}}{\lambda^{2d}}(f^0(\lambda)+\lambda^{2d}g^0(\lambda))\right)\{(\beta_k^2+\gamma_{1k}(\lambda)+\gamma_{2k}(\lambda))\delta_{kl}\}_{k,l=1}^{\infty}$$
$$\times\left(\frac{|1-e^{i\lambda\tau}|^{2d}}{\lambda^{2d}}(f^0(\lambda)+\lambda^{2d}g^0(\lambda))\right), \tag{10.21}$$

where α_k^2, β_k^2 are Lagrange multipliers, δ_{kl} are Kronecker symbols, functions $\gamma_{1k}(\lambda)\le 0$ and $\gamma_{1k}(\lambda)=0$ if $g_{kk}^0(\lambda)>v_{kk}(\lambda)$, and functions $\gamma_{2k}(\lambda)\ge 0$ and $\gamma_{2k}(\lambda)=0$ if $g_{kk}^0(\lambda)<u_{kk}(\lambda)$.

For the fourth set of admissible spectral densities $\mathcal{D}_0^4\times\mathcal{D}_V^{U4}$, we have

$$\left((1-e^{i\lambda\tau})^d\overline{g^0(\lambda)\vec{A}(e^{-i\lambda})} + \sum_{j=0}^{\infty}((\mathbf{P}_\tau^0)^{-1}\mathbf{S}_\tau^0\mathbf{a}^\tau)_j e^{i\lambda(j+1)}\right)$$
$$\times\left((1-e^{i\lambda\tau})^d\overline{g^0(\lambda)\vec{A}(e^{-i\lambda})} + \sum_{j=0}^{\infty}((\mathbf{P}_\tau^0)^{-1}\mathbf{S}_\tau^0\mathbf{a}^\tau)_j e^{i\lambda(j+1)}\right)^*$$
$$=\alpha^2\left(\frac{|1-e^{i\lambda\tau}|^{2d}}{\lambda^{2d}}(f^0(\lambda)+\lambda^{2d}g^0(\lambda))\right)B_1^\top$$
$$\times\left(\frac{|1-e^{i\lambda\tau}|^{2d}}{\lambda^{2d}}(f^0(\lambda)+\lambda^{2d}g^0(\lambda))\right), \tag{10.22}$$

$$\left(\frac{|1-e^{i\lambda\tau}|^{2d}}{\lambda^{2d}}\overline{f^0(\lambda)\vec{A}(e^{-i\lambda})} - (1-e^{-i\lambda\tau})^d\sum_{j=0}^{\infty}((\mathbf{P}_\tau^0)^{-1}\mathbf{S}_\tau^0\mathbf{a}^\tau)_j e^{i\lambda(j+1)}\right)$$
$$\times\left(\frac{|1-e^{i\lambda\tau}|^{2d}}{\lambda^{2d}}\overline{f^0(\lambda)\vec{A}(e^{-i\lambda})} - (1-e^{-i\lambda\tau})^d\sum_{j=0}^{\infty}((\mathbf{P}_\tau^0)^{-1}\mathbf{S}_\tau^0\mathbf{a}^\tau)_j e^{i\lambda(j+1)}\right)^*$$
$$=(\beta^2+\gamma_1'(\lambda)+\gamma_2'(\lambda))\left(\frac{|1-e^{i\lambda\tau}|^{2d}}{\lambda^{2d}}(f^0(\lambda)+\lambda^{2d}g^0(\lambda))\right)B_2^\top$$
$$\times\left(\frac{|1-e^{i\lambda\tau}|^{2d}}{\lambda^{2d}}(f^0(\lambda)+\lambda^{2d}g^0(\lambda))\right), \tag{10.23}$$

where α^2, β^2 are Lagrange multipliers, functions $\gamma_1'(\lambda)\le 0$ and $\gamma_1'(\lambda)=0$ if $\langle B_2, g^0(\lambda)\rangle > \langle B_2, V(\lambda)\rangle$, and functions $\gamma_2'(\lambda)\ge 0$ and $\gamma_2'(\lambda)=0$ if $\langle B_2, g^0(\lambda)\rangle < \langle B_2, U(\lambda)\rangle$.

The following theorem holds.

Theorem 10.2.1. *The least favorable spectral densities $f^0(\lambda)$ and $g^0(\lambda)$ in the classes $\mathcal{D}_0^k\times\mathcal{D}_V^{Uk}$, $k=1,2,3,4$, for the optimal linear filtering of the functional $A\xi$ from observations of the process $\xi(t)+\eta(t)$ at points $t\le 0$ are determined by equations* (10.13)–(10.17),

(10.18)–(10.19), (10.20)–(10.21), *and* (10.22)–(10.23), *respectively, the minimality condition* (10.1), *the constrained optimization problem* (10.13), *and the restrictions on the densities from the classes* $\mathcal{D}_0^k$, $\mathcal{D}_V^{Uk}$, $k = 1,2,3,4$, *respectively. The minimax-robust spectral characteristic of the optimal estimate* $\widehat{A}\xi$ *is determined by formula* (10.6).

10.2.3 Least favorable spectral densities in classes $\mathcal{D}_{1\delta} \times \mathcal{D}_\varepsilon$

Consider the minimax (robust) filtering problem for the functional $A\xi$ under the condition that the sets of admissible spectral densities $\mathcal{D}_{1\delta}^k$, $\mathcal{D}_\varepsilon^k$, $k = 1,2,3,4$, are defined as follows:

$$\mathcal{D}_{1\delta}^1 = \left\{ f(\lambda) \left| \frac{1}{2\pi} \int_{-\pi}^{\pi} \frac{|1-e^{i\lambda\tau}|^{2d}}{\lambda^{2d}} \left|\mathrm{Tr}(f(\lambda) - f_1(\lambda))\right| d\lambda \le \delta \right.\right\},$$

$$\mathcal{D}_{1\delta}^2 = \left\{ f(\lambda) \left| \frac{1}{2\pi} \int_{-\pi}^{\pi} \frac{|1-e^{i\lambda\tau}|^{2d}}{\lambda^{2d}} |f_{kk}(\lambda) - f_{kk}^1(\lambda)| d\lambda \le \delta_k, k = \overline{1,\infty} \right.\right\},$$

$$\mathcal{D}_{1\delta}^3 = \left\{ f(\lambda) \left| \frac{1}{2\pi} \int_{-\pi}^{\pi} \frac{|1-e^{i\lambda\tau}|^{2d}}{\lambda^{2d}} \left|\langle B_1, f(\lambda) - f_1(\lambda)\rangle\right| d\lambda \le \delta \right.\right\},$$

$$\mathcal{D}_{1\delta}^4 = \left\{ f(\lambda) \left| \frac{1}{2\pi} \int_{-\pi}^{\pi} \frac{|1-e^{i\lambda\tau}|^{2d}}{\lambda^{2d}} |f_{ij}(\lambda) - f_{ij}^1(\lambda)| d\lambda \le \delta_i^j, i,j = \overline{1,\infty} \right.\right\}$$

and

$$\mathcal{D}_\varepsilon^1 = \left\{ g(\lambda) \left| \mathrm{Tr}[g(\lambda)] = (1-\varepsilon)\,\mathrm{Tr}[g_1(\lambda)] + \varepsilon\,\mathrm{Tr}[W(\lambda)], \frac{1}{2\pi} \int_{-\pi}^{\pi} \mathrm{Tr}[g(\lambda)] d\lambda = p \right.\right\},$$

$$\mathcal{D}_\varepsilon^2 = \left\{ g(\lambda) \left| g_{kk}(\lambda) = (1-\varepsilon) g_{kk}^1(\lambda) + \varepsilon w_{kk}(\lambda), \frac{1}{2\pi} \int_{-\pi}^{\pi} g_{kk}(\lambda) d\lambda = p_k, k = \overline{1,\infty} \right.\right\},$$

$$\mathcal{D}_\varepsilon^3 = \left\{ g(\lambda) \left| \langle B_2, g(\lambda)\rangle = (1-\varepsilon)\langle B_2, g_1(\lambda)\rangle + \varepsilon\langle B_2, W(\lambda)\rangle, \frac{1}{2\pi} \int_{-\pi}^{\pi} \langle B_2, g(\lambda)\rangle d\lambda = p \right.\right\},$$

$$\mathcal{D}_\varepsilon^4 = \left\{ g(\lambda) \left| g(\lambda) = (1-\varepsilon) g_1(\lambda) + \varepsilon W(\lambda), \frac{1}{2\pi} \int_{-\pi}^{\pi} g(\lambda) d\lambda = P \right.\right\}.$$

Here $f_1(\lambda)$ and $g_1(\lambda)$ are known fixed spectral densities, $W(\lambda)$ is an unknown spectral density, p, p_k, $k = \overline{1,\infty}$, are given numbers, P is a given positive-definite Hermitian matrix, and δ, δ_k, $k = \overline{1,\infty}$, and δ_i^j, $i,j = \overline{1,\infty}$, are given numbers.

Condition $0 \in \partial\Delta_{\mathcal{D}}(f^0, g^0)$ implies the following equations determining the least favorable spectral densities for these given sets of admissible spectral densities.

For the first set of admissible spectral densities $\mathcal{D}^1_{1\delta} \times \mathcal{D}^1_{\varepsilon}$, we have

$$\left((1-e^{i\lambda\tau})^d\overline{g^0(\lambda)\vec{A}(e^{-i\lambda})}+\sum_{j=0}^{\infty}((\mathbf{P}^0_\tau)^{-1}\mathbf{S}^0_\tau\mathbf{a}^\tau)_j e^{i\lambda(j+1)}\right)$$

$$\times\left((1-e^{i\lambda\tau})^d\overline{g^0(\lambda)\vec{A}(e^{-i\lambda})}+\sum_{j=0}^{\infty}((\mathbf{P}^0_\tau)^{-1}\mathbf{S}^0_\tau\mathbf{a}^\tau)_j e^{i\lambda(j+1)}\right)^*$$

$$=\alpha^2\gamma_2(\lambda)\left(\frac{|1-e^{i\lambda\tau}|^{2d}}{\lambda^{2d}}(f^0(\lambda)+\lambda^{2d}g^0(\lambda))\right)^2, \tag{10.24}$$

$$\left(\frac{|1-e^{i\lambda\tau}|^{2d}}{\lambda^{2d}}\overline{f^0(\lambda)\vec{A}(e^{-i\lambda})}-(1-e^{-i\lambda\tau})^d\sum_{j=0}^{\infty}((\mathbf{P}^0_\tau)^{-1}\mathbf{S}^0_\tau\mathbf{a}^\tau)_j e^{i\lambda(j+1)}\right)$$

$$\times\left(\frac{|1-e^{i\lambda\tau}|^{2d}}{\lambda^{2d}}\overline{f^0(\lambda)\vec{A}(e^{-i\lambda})}-(1-e^{-i\lambda\tau})^d\sum_{j=0}^{\infty}((\mathbf{P}^0_\tau)^{-1}\mathbf{S}^0_\tau\mathbf{a}^\tau)_j e^{i\lambda(j+1)}\right)^*$$

$$=(\beta^2+\gamma_1(\lambda))\left(\frac{|1-e^{i\lambda\tau}|^{2d}}{\lambda^{2d}}(f^0(\lambda)+\lambda^{2d}g^0(\lambda))\right)^2, \tag{10.25}$$

$$\frac{1}{2\pi}\int_{-\pi}^{\pi}\frac{|1-e^{i\lambda\tau}|^{2d}}{\lambda^{2d}}|\mathrm{Tr}(f^0(\lambda)-f_1(\lambda))|d\lambda=\delta, \tag{10.26}$$

where α^2, β^2 are Lagrange multipliers, the function $\gamma_1(\lambda) \le 0$ and $\gamma_1(\lambda) = 0$ if $\mathrm{Tr}[g^0(\lambda)] > (1-\varepsilon)\,\mathrm{Tr}[g_1(\lambda)]$, the function $|\gamma_2(\lambda)| \le 1$, and

$$\gamma_2(\lambda)=\mathrm{sign}(\mathrm{Tr}(f^0(\lambda)-f_1(\lambda))): \ \mathrm{Tr}(f^0(\lambda)-f_1(\lambda))\neq 0.$$

For the second set of admissible spectral densities $\mathcal{D}^2_{1\delta} \times \mathcal{D}^2_{\varepsilon}$, we have

$$\left((1-e^{i\lambda\tau})^d\overline{g^0(\lambda)\vec{A}(e^{-i\lambda})}+\sum_{j=0}^{\infty}((\mathbf{P}^0_\tau)^{-1}\mathbf{S}^0_\tau\mathbf{a}^\tau)_j e^{i\lambda(j+1)}\right)$$

$$\times\left((1-e^{i\lambda\tau})^d\overline{g^0(\lambda)\vec{A}(e^{-i\lambda})}+\sum_{j=0}^{\infty}((\mathbf{P}^0_\tau)^{-1}\mathbf{S}^0_\tau\mathbf{a}^\tau)_j e^{i\lambda(j+1)}\right)^*$$

$$=\left(\frac{|1-e^{i\lambda\tau}|^{2d}}{\lambda^{2d}}(f^0(\lambda)+\lambda^{2d}g^0(\lambda))\right)\{\alpha_k^2\gamma_k^2(\lambda)\delta_{kl}\}_{k,l=1}^{\infty}$$

$$\times\left(\frac{|1-e^{i\lambda\tau}|^{2d}}{\lambda^{2d}}(f^0(\lambda)+\lambda^{2d}g^0(\lambda))\right), \tag{10.27}$$

$$\left(\frac{|1-e^{i\lambda\tau}|^{2d}}{\lambda^{2d}}\overline{f^0(\lambda)\vec{A}(e^{-i\lambda})}-(1-e^{-i\lambda\tau})^d\sum_{j=0}^{\infty}((\mathbf{P}^0_\tau)^{-1}\mathbf{S}^0_\tau\mathbf{a}^\tau)_j e^{i\lambda(j+1)}\right)$$

$$\times\left(\frac{|1-e^{i\lambda\tau}|^{2d}}{\lambda^{2d}}\overline{f^0(\lambda)\vec{A}(e^{-i\lambda})}-(1-e^{-i\lambda\tau})^d\sum_{j=0}^{\infty}((\mathbf{P}^0_\tau)^{-1}\mathbf{S}^0_\tau\mathbf{a}^\tau)_j e^{i\lambda(j+1)}\right)^*$$

$$= \left(\frac{|1 - e^{i\lambda\tau}|^{2d}}{\lambda^{2d}} (f^0(\lambda) + \lambda^{2d} g^0(\lambda)) \right) \{(\beta_k^2 + \gamma_k^1(\lambda))\delta_{kl}\}_{k,l=1}^{\infty}$$
$$\times \left(\frac{|1 - e^{i\lambda\tau}|^{2d}}{\lambda^{2d}} (f^0(\lambda) + \lambda^{2d} g^0(\lambda)) \right), \tag{10.28}$$

$$\frac{1}{2\pi} \int_{-\pi}^{\pi} \frac{|1 - e^{i\lambda\tau}|^{2d}}{\lambda^{2d}} |f_{kk}^0(\lambda) - f_{kk}^1(\lambda)| d\lambda = \delta_k, \tag{10.29}$$

where α_k^2, β_k^2 are Lagrange multipliers, functions $\gamma_k^1(\lambda) \le 0$ and $\gamma_k^1(\lambda) = 0$ if $g_{kk}^0(\lambda) > (1-\varepsilon) g_{kk}^1(\lambda)$, functions $|\gamma_k^2(\lambda)| \le 1$, and

$$\gamma_k^2(\lambda) = \operatorname{sign}(f_{kk}^0(\lambda) - f_{kk}^1(\lambda)) : f_{kk}^0(\lambda) - f_{kk}^1(\lambda) \ne 0, \quad k = \overline{1, \infty}.$$

For the third set of admissible spectral densities $\mathcal{D}_{1\delta}^3 \times \mathcal{D}_\varepsilon^3$, we have

$$\left((1 - e^{i\lambda\tau})^d \overline{g^0(\lambda)} \vec{A}(e^{-i\lambda}) + \sum_{j=0}^{\infty} ((\mathbf{P}_\tau^0)^{-1} \mathbf{S}_\tau^0 \mathbf{a}^\tau)_j e^{i\lambda(j+1)} \right)$$
$$\times \left((1 - e^{i\lambda\tau})^d \overline{g^0(\lambda)} \vec{A}(e^{-i\lambda}) + \sum_{j=0}^{\infty} ((\mathbf{P}_\tau^0)^{-1} \mathbf{S}_\tau^0 \mathbf{a}^\tau)_j e^{i\lambda(j+1)} \right)^*$$
$$= \alpha^2 \gamma_2'(\lambda) \left(\frac{|1 - e^{i\lambda\tau}|^{2d}}{\lambda^{2d}} (f^0(\lambda) + \lambda^{2d} g^0(\lambda)) \right) B_1^\top$$
$$\times \left(\frac{|1 - e^{i\lambda\tau}|^{2d}}{\lambda^{2d}} (f^0(\lambda) + \lambda^{2d} g^0(\lambda)) \right), \tag{10.30}$$

$$\left(\frac{|1 - e^{i\lambda\tau}|^{2d}}{\lambda^{2d}} \overline{f^0(\lambda)} \vec{A}(e^{-i\lambda}) - (1 - e^{-i\lambda\tau})^d \sum_{j=0}^{\infty} ((\mathbf{P}_\tau^0)^{-1} \mathbf{S}_\tau^0 \mathbf{a}^\tau)_j e^{i\lambda(j+1)} \right)$$
$$\times \left(\frac{|1 - e^{i\lambda\tau}|^{2d}}{\lambda^{2d}} \overline{f^0(\lambda)} \vec{A}(e^{-i\lambda}) - (1 - e^{-i\lambda\tau})^d \sum_{j=0}^{\infty} ((\mathbf{P}_\tau^0)^{-1} \mathbf{S}_\tau^0 \mathbf{a}^\tau)_j e^{i\lambda(j+1)} \right)^*$$
$$= (\beta^2 + \gamma_1'(\lambda)) \left(\frac{|1 - e^{i\lambda\tau}|^{2d}}{\lambda^{2d}} (f^0(\lambda) + \lambda^{2d} g^0(\lambda)) \right) B_2^\top$$
$$\times \left(\frac{|1 - e^{i\lambda\tau}|^{2d}}{\lambda^{2d}} (f^0(\lambda) + \lambda^{2d} g^0(\lambda)) \right), \tag{10.31}$$

$$\frac{1}{2\pi} \int_{-\pi}^{\pi} \frac{|1 - e^{i\lambda\tau}|^{2d}}{\lambda^{2d}} |\langle B_1, f^0(\lambda) - f_1(\lambda) \rangle| d\lambda = \delta, \tag{10.32}$$

where α^2, β^2 are Lagrange multipliers, function $\gamma_1'(\lambda) \le 0$ and $\gamma_1'(\lambda) = 0$ if $\langle B_2, g^0(\lambda) \rangle > (1-\varepsilon) \langle B_2, g_1(\lambda) \rangle$, function $|\gamma_2'(\lambda)| \le 1$, and

$$\gamma_2'(\lambda) = \operatorname{sign}\langle B_1, f^0(\lambda) - f_1(\lambda) \rangle : \langle B_1, f^0(\lambda) - f_1(\lambda) \rangle \ne 0.$$

For the fourth set of admissible spectral densities $\mathcal{D}_{1\delta}^4 \times \mathcal{D}_\varepsilon^4$, we have

$$\left((1-e^{i\lambda\tau})^d \overline{g^0(\lambda)\vec{A}(e^{-i\lambda})} + \sum_{j=0}^{\infty}((\mathbf{P}_\tau^0)^{-1}\mathbf{S}_\tau^0\mathbf{a}^\tau)_j e^{i\lambda(j+1)}\right)$$
$$\times\left((1-e^{i\lambda\tau})^d \overline{g^0(\lambda)\vec{A}(e^{-i\lambda})} + \sum_{j=0}^{\infty}((\mathbf{P}_\tau^0)^{-1}\mathbf{S}_\tau^0\mathbf{a}^\tau)_j e^{i\lambda(j+1)}\right)^*$$
$$= \left(\frac{|1-e^{i\lambda\tau}|^{2d}}{\lambda^{2d}}(f^0(\lambda)+\lambda^{2d}g^0(\lambda))\right)\{\alpha_{ij}(\lambda)\gamma_{ij}(\lambda)\}_{i,j=1}^{\infty}$$
$$\times\left(\frac{|1-e^{i\lambda\tau}|^{2d}}{\lambda^{2d}}(f^0(\lambda)+\lambda^{2d}g^0(\lambda))\right), \tag{10.33}$$

$$\left(\frac{|1-e^{i\lambda\tau}|^{2d}}{\lambda^{2d}}\overline{f^0(\lambda)\vec{A}(e^{-i\lambda})} - (1-e^{-i\lambda\tau})^d \sum_{j=0}^{\infty}((\mathbf{P}_\tau^0)^{-1}\mathbf{S}_\tau^0\mathbf{a}^\tau)_j e^{i\lambda(j+1)}\right)$$
$$\times\left(\frac{|1-e^{i\lambda\tau}|^{2d}}{\lambda^{2d}}\overline{f^0(\lambda)\vec{A}(e^{-i\lambda})} - (1-e^{-i\lambda\tau})^d \sum_{j=0}^{\infty}((\mathbf{P}_\tau^0)^{-1}\mathbf{S}_\tau^0\mathbf{a}^\tau)_j e^{i\lambda(j+1)}\right)^*$$
$$= \left(\frac{|1-e^{i\lambda\tau}|^{2d}}{\lambda^{2d}}(f^0(\lambda)+\lambda^{2d}g^0(\lambda))\right)(\vec{\beta}\cdot\vec{\beta}^* + \Gamma(\lambda))$$
$$\times\left(\frac{|1-e^{i\lambda\tau}|^{2d}}{\lambda^{2d}}(f^0(\lambda)+\lambda^{2d}g^0(\lambda))\right), \tag{10.34}$$

$$\frac{1}{2\pi}\int_{-\pi}^{\pi}\frac{|1-e^{i\lambda\tau}|^{2d}}{\lambda^{2d}}|f_{ij}^0(\lambda)-f_{ij}^1(\lambda)|d\lambda = \delta_i^j, \tag{10.35}$$

where $\vec{\alpha}, \beta_{ij}$ are Lagrange multipliers, function $\Gamma(\lambda) \le 0$ and $\Gamma(\lambda) = 0$ if $g^0(\lambda) > (1-\varepsilon)g_1(\lambda)$, functions $|\gamma_{ij}(\lambda)| \le 1$, and

$$\gamma_{ij}(\lambda) = \frac{f_{ij}^0(\lambda)-f_{ij}^1(\lambda)}{|f_{ij}^0(\lambda)-f_{ij}^1(\lambda)|} : f_{ij}^0(\lambda)-f_{ij}^1(\lambda) \neq 0, \quad i,j=\overline{1,\infty}.$$

The following theorem holds.

Theorem 10.2.2. *The least favorable spectral densities $f^0(\lambda)$ and $g^0(\lambda)$ in classes $\mathcal{D}_{1\delta}^k\times\mathcal{D}_\varepsilon^k$, $k = 1,2,3,4$, for the optimal linear filtering of the functional $A\xi$ from observations of the process $\xi(t)+\eta(t)$ at points $t \le 0$ are determined by equations* (10.24)–(10.26), (10.27)–(10.29), (10.30)–(10.32), *and* (10.33)–(10.35), *respectively, minimality condition* (10.1), *the constrained optimization problem* (10.13), *and restrictions on densities from the corresponding classes $\mathcal{D}_\varepsilon^k$, $\mathcal{D}_{1\delta}^k$, $k = 1,2,3,4$. The minimax-robust spectral characteristic of the optimal estimate $\widehat{A}\xi$ is determined by formula* (10.6).

Conclusions

In this chapter, we describe methods of solution of the filtering problem for linear functionals which are constructed from unobserved values of stochastic processes with periodically stationary increments. Estimates of functionals are based on observations of the considered process $\xi(t)$ with a periodically stationary stochastic noise process $\eta(t)$, uncorrelated with the process $\xi(t)$. The problem is solved by representing these processes as infinite-dimensional sequences with stationary increments.

The Hilbert space projection method is applied to derive formulas for calculating the spectral characteristics and the MSEs of the optimal estimates of the functionals in the case of spectral certainty, where the spectral densities of the generated sequences are exactly known.

In the case of spectral uncertainty, where the spectral densities are not exactly known while, instead, a set of admissible spectral densities is specified, the minimax-robust method is applied. We propose a representation of the MSE in the form of a linear functional in L_1 space with respect to spectral densities, which allows us to solve the corresponding constrained optimization problem and describe the minimax (robust) estimates of the functionals.

This method is illustrated with the help of concrete sets $\mathcal{D}_0 \times \mathcal{D}_V^U$ and $\mathcal{D}_{1\delta} \times \mathcal{D}_\varepsilon$ of admissible spectral densities. Here D_0 are classes of spectral densities with the moment restrictions, D_V^U are classes describing the "strip" models of spectral densities, $D_{1\delta}$ are classes describing the "δ-neighborhood" models in the space L_1 of a fixed bounded spectral density, and D_ε are classes describing the "ε-contaminated" models of a fixed bounded spectral density.

The results presented in this chapter rely on and develop prior studies of the minimax filtering problem by Moklyachuk and Masyutka [150] for multidimensional stationary processes, Golichenko (Dubovets'ka) and Moklyachuk [40] for periodically correlated processes, and Luz and Moklyachuk [108] for processes with stationary increments.

11 Filtering problem when signal and noise have periodically correlated increments

In this chapter, we present the results of our investigation of the filtering problem in the case where both the signal and the noise processes are processes with periodically stationary dth increments. Recall that solution of the filtering problem with periodically stationary noise is described in Chapter 10. In the present chapter we will show that additional conditions are required for the function $a(t)$, $t \geq 0$, determining the functional $A\xi$. In particular, the filtering problem for the functional $A_{NT}\xi$ defined on the bounded interval $[0,(N+1)T]$ cannot be solved with the proposed approach in the general case.

11.1 Hilbert space projection method of filtering

11.1.1 Filtering based on Fourier transformations of spectral densities

Let the uncorrelated increment processes $\{\xi^{(d)}(t,\tau T) : t \in \mathbb{R}\}$ and $\{\eta^{(d)}(t,\tau T) : t \in \mathbb{R}\}$ generate infinite-dimensional vector stationary increment sequences $\{\vec{\xi}_j^{(d)}, j \in \mathbb{Z}\}$ and $\{\vec{\eta}_j^{(d)}, j \in \mathbb{Z}\}$ (6.7) with the spectral density matrices $f(\lambda) = \{f_{ij}(\lambda)\}_{i,j=1}^{\infty}$ and $g(\lambda) = \{g_{ij}(\lambda)\}_{i,j=1}^{\infty}$.

As in the previous chapters, assume that the mean values of the sequences $\vec{\xi}_j^{(d)}$ and $\vec{\eta}_j^{(d)}$ equal $\vec{0}$ and consider the increment step $\tau > 0$.

Recall that by the classical *filtering problem* we understand the problem of the mean square optimal linear estimation of the functional

$$A\xi = \int_0^{\infty} a(t)\xi(-t)dt,$$

which depends on the unknown values of the stochastic process $\xi(t)$ with periodically correlated dth increments. Estimates are based on observations of the process $\zeta(t) = \xi(t) + \eta(t)$ at points $t \leq 0$.

To apply the Hilbert pace projection method of estimation, we represent the functional $A\xi$ in terms of the generated increment sequence $\vec{\xi}_j^{(d)}$.

Lemma 11.1.1. *Any linear functional*

$$A\xi = \int_0^{\infty} a(t)\xi(-t)dt$$

allows the representation

$$A\xi = B\xi,$$

https://doi.org/10.1515/9783111325620-011

where

$$B\xi = \int_0^\infty b^\tau(t)\xi^{(d)}(-t,\tau T)dt,$$

where

$$b^\tau(t) = \sum_{l=0}^{[\frac{t}{\tau T}]} a(t-\tau Tl)d(l) = D^{\tau T}\mathbf{a}(t), \quad t \geq 0, \tag{11.1}$$

where $[x]$ *denotes the integer part of* x, *coefficients* $\{d(l) : l \geq 0\}$ *are determined by the relation*

$$\sum_{l=0}^{\infty} d(l)x^l = \left(\sum_{j=0}^{\infty} x^j\right)^d,$$

and $D^{\tau T}$ *is the linear transformation acting on an arbitrary function* $x(t)$, $t \geq 0$, *as follows:*

$$D^{\tau T}\mathbf{x}(t) = \sum_{l=0}^{[\frac{t}{\tau T}]} x(t-\tau Tl)d(l).$$

Proof. From the definition of the increment process we obtain the formal equality

$$\xi(-t) = \frac{1}{(1-B_{\tau T})^d}\xi^{(d)}(-t,\tau T) = \sum_{j=0}^{\infty} d(j)\xi^{(d)}(-t-\tau Tj,\tau T). \tag{11.2}$$

Relation (11.2) implies that

$$\int_0^\infty a(t)\xi(-t)dt = \int_0^\infty \xi^{(d)}(-t,\tau T)\sum_{j=0}^{[\frac{t}{\tau T}]} a(t-\tau Tj)d(j)dt,$$

which proves the statement of the lemma. □

The functional $A\xi$ allows a representation in terms of the increment sequence $\vec{\xi}_j^{(d)} = (\xi_{kj}^{(d)}, k = 1,2,\dots)^\top, j \in \mathbb{Z}$, which is described in the following lemma.

Lemma 11.1.2. *The linear functional* $A\xi$ *allows a representation*

$$A\xi = B\xi = \sum_{j=0}^{\infty} (\vec{b}_j^\tau)^\top \vec{\xi}_{-j}^{(d)} =: B\vec{\xi},$$

where the infinite-dimensional vectors $\vec{\xi}_j^{(d)}$ *and* $\vec{b}_j^\tau$ *are defined as follows:*

$$\vec{\xi}_j^{(d)} = (\xi_{kj}^{(d)}, k = 1, 2, \dots)^\top,$$
$$\vec{b}_j^\tau = (b_{kj}^\tau, k = 1, 2, \dots)^\top = (b_{1j}^\tau, b_{3j}^\tau, b_{2j}^\tau, \dots, b_{2k+1,j}^\tau, b_{2k,j}^\tau, \dots)^\top.$$

Here

$$b_{kj}^\tau = \langle b_j^\tau, \widetilde{e}_k \rangle = \frac{1}{\sqrt{T}} \int_0^T b_j^\tau(v) e^{-2\pi i\{(-1)^k [\frac{k}{2}]\}v/T} dv,$$

$$b_j^\tau(u) = \sum_{l=0}^{j} a(u + jT - \tau Tl) d_\tau(l) = D^\tau a_j(u), \quad u \in [0, T),\ j = 0, 1, \dots,$$

and

$$b_{k,j}^\tau = \sum_{l=0}^{j} a_{k,j-\tau l} d_\tau(l), \quad j = 0, 1, \dots,\ k = 1, 2 \dots,$$

$$a_{kj} = \langle a_j, \widetilde{e}_k \rangle = \frac{1}{\sqrt{T}} \int_0^T a_j(v) e^{-2\pi i\{(-1)^k [\frac{k}{2}]\}v/T} dv,$$

$$\vec{b}_j^\tau = \sum_{m=0}^{j} \mathrm{diag}_\infty(d_\tau(j - m)) \vec{a}_m = (D^\tau \mathbf{a})_j, \quad j = 0, 1, 2, \dots, \tag{11.3}$$

where

$$\mathbf{a} = ((\vec{a}_0)^\top, (\vec{a}_1)^\top, (\vec{a}_2)^\top, \dots)^\top,$$
$$\vec{a}_j = (a_{kj}, k = 1, 2, \dots)^\top = (a_{1j}, a_{3j}, a_{2j}, \dots, a_{2k+1,j}, a_{2k,j}, \dots)^\top,$$

the coefficients $\{d_\tau(k) : k \geq 0\}$ *are determined by the relationship*

$$\sum_{k=0}^{\infty} d_\tau(k) x^k = \left(\sum_{j=0}^{\infty} x^{\tau j} \right)^d,$$

and D^τ *is a linear transformation determined by a matrix with infinite-dimensional matrix entries* $D^\tau(k, j)$, $k, j = 0, 1, 2, \dots$, *such that* $D^\tau(k, j) = \mathrm{diag}_\infty(d_\tau(k - j))$ *if* $0 \leq j \leq k$ *and* $D^\tau(k, j) = \mathcal{O}$ *if* $0 \leq k < j$; $\mathrm{diag}_\infty(x)$ *denotes an infinite-dimensional diagonal matrix with the entry* x *on its diagonal.*

Proof. The representation of the functional $A\vec{\xi}$ comes from Lemma 8.1.2. □

Suppose that the vectors $\vec{b}_j^\tau$, $j \geq 0$, generated by the function $a(t)$, $t \geq 0$, satisfy the conditions

$$\sum_{j=0}^{\infty} \|\vec{b}_j^\tau\| < \infty, \quad \sum_{j=0}^{\infty} (j + 1) \|\vec{b}_j^\tau\|^2 < \infty, \quad \|\vec{b}_j^\tau\|^2 = \sum_{k=1}^{\infty} |b_{kj}^\tau|^2. \tag{11.4}$$

Under such conditions the functional $B\vec{\xi}$ belongs to the Hilbert space $H = L_2(\Omega, \mathcal{F}, P)$.

Remark 11.1.1. Restrictions (11.4) on coefficients $\vec{b}_j^{\tau}, j \geq 0$, are strong. Let us show that they are not satisfied if $\vec{a}_j = \vec{0}$ for $j \geq j_0$. Consider the simple case when $\|\vec{b}_0^{\tau}\|^2 = \|\vec{a}_0\|^2 = a$ and $\vec{a}_j = \vec{0}$ for $j \geq 1$. Then the vectors $\vec{b}_j = d_\tau(j)\vec{a}_0$, $k \geq 0$, do not satisfy conditions (11.4).

Remark 11.1.2. An example when conditions (11.4) are satisfied for the functional

$$A_{NT}\xi = \int_0^{(N+1)T} a(t)\xi(-k)dt$$

is a class of functionals allowing representation as a functional

$$A_{NT}\xi = B_M\vec{\xi} = \sum_{j=0}^{M} (\vec{b}_j^{\tau})^\top \vec{\xi}_{-j}^{(d)},$$

which depends on the increments $\vec{\xi}_{-j}^{(d)}$ generated by the process, namely, $A_{NT}\xi = B_M\vec{\xi}$. In terms of the vectors $\vec{a}_j$, we have the following condition: there exist vectors $\{\vec{b}_j^{\tau} : j = 0, 1, 2, \dots, M\}$ such that

$$\vec{a}_j = \sum_{l=\max\{0;\lceil \frac{j-M}{\tau} \rceil\}}^{\min\{d;[\frac{j}{\tau}]\}} (-1)^l \binom{n}{l} \vec{b}_{j-\tau l}^{\tau}$$

for $j = 0, 1, 2, \dots, N$, $N = M + \tau d$. By $\lceil x \rceil$ we denote the least integer greater than or equal to x.

Denote by $\Delta(f, g; \widehat{A}\xi) = \mathsf{E}|A\xi - \widehat{A}\xi|^2$ and $\Delta(f, g; \widehat{B}\vec{\xi}) := \mathsf{E}|B\vec{\xi} - \widehat{B}\vec{\xi}|^2$ the MSEs of the optimal estimates $\widehat{A}\xi$ and $\widehat{B}\vec{\xi}$ of the functionals $A\xi$ and $B\vec{\xi}$. Since $A\xi = B\vec{\xi}$, we have the following relations:

$$\widehat{A}\xi = \widehat{B}\vec{\xi} \tag{11.5}$$

and

$$\Delta(f, g; \widehat{A}\xi) = \mathsf{E}|A\xi - \widehat{A}\xi|^2 = \mathsf{E}|B\vec{\xi} - \widehat{B}\vec{\xi}|^2 = \Delta(f, g; \widehat{B}\vec{\xi}).$$

Let the spectral densities $f(\lambda)$ and $g(\lambda)$ satisfy the minimality condition

$$\int_{-\pi}^{\pi} \frac{\lambda^{2d}}{|1 - e^{i\lambda\tau}|^{2d}} (f(\lambda) + g(\lambda))^{-1} d\lambda < \infty. \tag{11.6}$$

The estimate of the functional $B\vec{\xi}$ can be found applying the method of orthogonal projection in the Hilbert space.

Define a subspace

$$H^0(\vec{\xi}^{(d)} + \vec{\eta}^{(d)}) = \overline{\text{span}}\{\xi_k^{(d)}(j,\tau) + \eta_k^{(d)}(j,\tau) : k = 1, \dots, \infty; j \le 0\}$$

of $H = L_2(\Omega, \mathcal{F}, \mathsf{P})$ generated by the observations $\{\zeta(t) : t \le 0\}$ and a subspace

$$L_2^0(f+g) = \overline{\text{span}}\left\{e^{i\lambda j}(1 - e^{-i\lambda\tau})^d \frac{1}{(i\lambda)^d}\vec{\delta}_k : k = 1, \dots, \infty; j \le 0\right\}$$

of the Hilbert space $L_2(f+g)$ of vector-valued functions endowed with the inner product $\langle g_1; g_2 \rangle = \int_{-\pi}^{\pi}(g_1(\lambda))^\top(f(\lambda) + g(\lambda))\overline{g_2(\lambda)}d\lambda$. Here $\vec{\delta}_k = \{\delta_{kl}\}_{l=1}^{\infty}$, where δ_{kl} are Kronecker symbols. Note that $f(\lambda) + g(\lambda)$ is a spectral density matrix of the increment sequence $\vec{\xi}_j^{(d)} + \vec{\eta}_j^{(d)}$.

Remark 11.1.3. Representation (6.9) yields a map between a function

$$e^{i\lambda j}(1 - e^{-i\lambda\tau})^d(i\lambda)^{-d}\vec{\delta}_k$$

from the space $L_2^0(f+g)$ and an increment $\xi_k^{(d)}(j,\tau) + \eta_k^{(d)}(j,\tau)$ from the space $H^0(\vec{\xi}^{(d)} + \vec{\eta}^{(d)})$.

The linear estimates $\widehat{A}\xi$ and $\widehat{B}\vec{\xi}$ have the following representations:

$$\widehat{A}\xi = \widehat{B}\vec{\xi} = \int_{-\pi}^{\pi} (\vec{h}_\tau(\lambda))^\top d\vec{Z}_{\xi^{(d)}+\eta^{(d)}}(\lambda), \tag{11.7}$$

where $\vec{h}_\tau(\lambda)$ is a spectral characteristic of the optimal estimate $\widehat{B}\vec{\xi}$. The estimate $\widehat{B}\xi$ is an orthogonal projection of the element $B\xi$ of the Hilbert space $H = L_2(\Omega, \mathcal{F}, P)$ on the subspace $H^0(\vec{\xi}^{(d)} + \vec{\eta}^{(d)})$:

$$\widehat{A}\xi = \widehat{B}\vec{\xi} = \text{Proj}_{H^0(\vec{\xi}^{(d)}+\vec{\eta}^{(d)})}B\vec{\xi}.$$

We define the following Fourier coefficients for all $l, j \ge 0$:

$$R_{l,j} = \frac{1}{2\pi}\int_{-\pi}^{\pi} e^{-i\lambda(l+j)}(f(\lambda)(f(\lambda) + g(\lambda))^{-1})^\top d\lambda,$$

$$P_{l,j}^\tau = \frac{1}{2\pi}\int_{-\pi}^{\pi} e^{-i\lambda(l-j)}\frac{\lambda^{2d}}{|1 - e^{i\lambda\tau}|^{2d}}((f(\lambda) + g(\lambda))^{-1})^\top d\lambda,$$

$$Q_{l,j}^\tau = \frac{1}{2\pi}\int_{-\pi}^{\pi} e^{-i\lambda(l-j)}\frac{|1 - e^{i\lambda\tau}|^{2d}}{\lambda^{2d}}(f(\lambda)(f(\lambda) + g(\lambda))^{-1}g(\lambda))^\top d\lambda.$$

We define a vector

$$\mathbf{b}^\tau = ((\vec{b}_0^\tau)^\top, (\vec{b}_1^\tau)^\top, (\vec{b}_2^\tau)^\top, \dots)^\top$$

and linear operators $\mathbf{P}_\tau$, $\mathbf{R}$, and $\mathbf{Q}_\tau$ in the space ℓ_2 determined by matrices with the infinite-dimensional entries $(\mathbf{P}_\tau)_{l,j} = P_{l,j}^\tau$, $(\mathbf{R})_{l,j} = R_{l+1,j}$, and $(\mathbf{Q}_\tau)_{l,j} = Q_{l,j}^\tau$, $l, j \geq 0$.

The following theorem provides a solution of the classical filtering problem.

Theorem 11.1.1. *Consider two uncorrelated stochastic processes $\xi(t)$ and $\eta(t)$, $t \in \mathbb{R}$, with periodically stationary increments, which determine generated stationary dth increment sequences $\vec{\xi}_j^{(d)}$ and $\vec{\eta}_j^{(d)}$ with the spectral density matrices $f(\lambda) = \{f_{kn}(\lambda)\}_{k,n=1}^\infty$ and $g(\lambda) = \{g_{kn}(\lambda)\}_{k,n=1}^\infty$, respectively. Let the coefficients $\vec{b}_j^\tau$, $j = 0, 1, \dots$, generated by the function $a(t)$, $t \geq 0$, satisfy conditions* (11.4). *Let the minimality condition* (11.6) *be satisfied. The optimal linear estimate $\widehat{A}\xi$ of the functional $A\xi$ which depends on the unknown values $\xi(t)$ for $t \leq 0$ based on observations of the process $\xi(t) + \eta(t)$ at points $t \leq 0$ is determined by formula* (11.7). *The spectral characteristic $\vec{h}_\tau(\lambda)$ of the optimal estimate is calculated by the formula*

$$\begin{aligned}(\vec{h}_\tau(\lambda))^\top &= (\vec{B}_\tau(e^{-i\lambda}))^\top \frac{(1 - e^{-i\lambda\tau})^d}{(i\lambda)^d} f(\lambda)(f(\lambda) + g(\lambda))^{-1} \\ &\quad - \frac{(-i\lambda)^d}{(1 - e^{i\lambda\tau})^d} (\vec{C}_\tau(e^{i\lambda}))^\top (f(\lambda) + g(\lambda))^{-1}, \end{aligned} \tag{11.8}$$

where

$$\vec{B}_\tau(e^{-i\lambda}) = \sum_{j=0}^\infty \vec{b}_j^\tau e^{-i\lambda j} = \sum_{j=0}^\infty (D^\tau \mathbf{a})_j e^{-i\lambda j},$$

$$\vec{C}_\tau(e^{i\lambda}) = \sum_{j=0}^\infty (\mathbf{P}_\tau^{-1} \mathbf{R} D^\tau \mathbf{a})_j e^{i\lambda(j+1)}.$$

The MSE of the estimate $\widehat{A}\xi$ is calculated by the formula

$$\begin{aligned}\Delta(f, g; \widehat{A}\xi) &= \Delta(f, g; \widehat{B}\vec{\xi}) = \mathsf{E}|B\vec{\xi} - \widehat{B}\vec{\xi}|^2 \\ &= \frac{1}{2\pi} \int_{-\pi}^{\pi} \frac{\lambda^{2d}}{|1 - e^{i\lambda\tau}|^{2d}} \left((\vec{B}_\tau(e^{-i\lambda}))^\top \frac{|1 - e^{i\lambda\tau}|^{2d}}{\lambda^{2d}} g(\lambda) + (\vec{C}_\tau(e^{i\lambda}))^\top \right) \\ &\quad \times (f(\lambda) + g(\lambda))^{-1} f(\lambda) (f(\lambda) + g(\lambda))^{-1} \\ &\quad \times \left(g(\lambda) \overline{\vec{B}_\tau(e^{-i\lambda})} \frac{|1 - e^{i\lambda\tau}|^{2d}}{\lambda^{2d}} + \overline{\vec{C}_\tau(e^{i\lambda})} \right) d\lambda \\ &\quad + \frac{1}{2\pi} \int_{-\pi}^{\pi} \frac{\lambda^{2d}}{|1 - e^{i\lambda\tau}|^{2d}} \left((\vec{B}_\tau(e^{-i\lambda}))^\top \frac{|1 - e^{i\lambda\tau}|^{2d}}{\lambda^{2d}} f(\lambda) - (\vec{C}_\tau(e^{i\lambda}))^\top \right) \\ &\quad \times (f(\lambda) + g(\lambda))^{-1} g(\lambda) (f(\lambda) + g(\lambda))^{-1}\end{aligned}$$

$$\times\left(f(\lambda)\overline{\vec{B}_\tau(e^{-i\lambda})}\frac{|1-e^{i\lambda\tau}|^{2d}}{\lambda^{2d}}-\overline{\vec{C}_\tau(e^{i\lambda})}\right)d\lambda$$
$$=\langle \mathbf{R}D^\tau\mathbf{a},\mathbf{P}_\tau^{-1}\mathbf{R}D^\tau\mathbf{a}\rangle+\langle \mathbf{Q}_\tau D^\tau\mathbf{a},D^\tau\mathbf{a}\rangle. \quad (11.9)$$

Proof. The projection $\widehat{B}\vec{\xi}$ is characterized by the following conditions:
(1) $\widehat{B}\vec{\xi}\in H^0(\vec{\xi}^{(d)}+\vec{\eta}^{(d)})$;
(2) $(B\vec{\xi}-\widehat{B}\vec{\xi})\perp H^0(\vec{\xi}^{(d)}+\vec{\eta}^{(d)})$.

Condition (2) implies that for all $j\le 0$ and $k\ge 1$ the function $\vec{h}_\tau(\lambda)$ satisfies the relation

$$\mathsf{E}(B\vec{\xi}-\widehat{B}\vec{\xi})^\top\overline{(\xi_k^{(d)}(j,\tau)+\eta_k^{(d)}(j,\tau))}$$
$$=\int_{-\pi}^{\pi}\left((\vec{B}_\tau(e^{-i\lambda}))^\top\frac{(1-e^{-i\lambda\tau})^d}{(i\lambda)^d}f(\lambda)-(\vec{h}_\tau(\lambda))^\top(f(\lambda)+g(\lambda))\right)$$
$$\times\frac{(1-e^{i\lambda\tau})^d}{(-i\lambda)^d}e^{-i\lambda j}\vec{\delta}_k d\lambda=0.$$

Thus, the spectral characteristic $\vec{h}_\tau(\lambda)$ allows a representation

$$(\vec{h}_\tau(\lambda))^\top=(\vec{B}_\tau(e^{-i\lambda}))^\top\frac{(1-e^{-i\lambda\tau})^d}{(i\lambda)^d}f(\lambda)(f(\lambda)+g(\lambda))^{-1}$$
$$-\frac{(-i\lambda)^d}{(1-e^{i\lambda\tau})^d}(\vec{C}_\tau(e^{i\lambda}))^\top(f(\lambda)+g(\lambda))^{-1},$$

where

$$\vec{C}_\tau(e^{i\lambda})=\sum_{j=0}^{\infty}\vec{c}_j^{\,\tau}e^{i\lambda(j+1)},$$

where $\vec{c}_j^{\,\tau}=\{c_{kj}^\tau\}_{k=1}^\infty$, $j\ge 0$, are unknown vectors to be found.

Condition (1) implies that the function $\vec{h}_\tau(\lambda)$ is of the form

$$\vec{h}_\tau(\lambda)=\vec{h}(\lambda)(1-e^{-i\lambda\tau})^d\frac{1}{(i\lambda)^d},\quad \vec{h}(\lambda)=\sum_{j=0}^{\infty}\vec{s}_je^{-i\lambda j}.$$

Thus, for all $l\ge 1$

$$\int_{-\pi}^{\pi}\left((\vec{B}_\tau(e^{-i\lambda}))^\top f(\lambda)-\frac{\lambda^{2d}}{|1-e^{i\lambda\tau}|^{2d}}(\vec{C}_\tau(e^{i\lambda}))^\top\right)(f(\lambda)+g(\lambda))^{-1}e^{-i\lambda l}d\lambda=0. \quad (11.10)$$

One can conclude that relation (11.10) is equivalent to the following system of linear equations:

$$\sum_{m=0}^{\infty} R_{l+1,m}\vec{b}_m^{\tau} = \sum_{j=0}^{\infty} P_{l+1,j+1}^{\tau}\vec{c}_j^{\,\tau}, \quad l \geq 1,$$

which can be written in matrix form as

$$\mathbf{R}\mathbf{b}^{\tau} = \mathbf{P}_{\tau}\mathbf{c}_{\tau},$$

where

$$\mathbf{c}^{\tau} = ((c_1^{\tau})^{\top}, (c_2^{\tau})^{\top}, (c_3^{\tau})^{\top}, \ldots)^{\top}.$$

Therefore, the coefficients $\vec{c}_j^{\,\tau}, j \geq 0$, are calculated by the formula

$$\vec{c}_j^{\,\tau} = (\mathbf{P}_{\tau}^{-1}\mathbf{R}\mathbf{b}^{\tau})_j,$$

where $(\mathbf{P}_{\tau}^{-1}\mathbf{R}\mathbf{b}^{\tau})_j$ is the jth entry of the vector $\mathbf{P}_{\tau}^{-1}\mathbf{R}\mathbf{b}^{\tau}$.

The derived expressions justify the use of formulas (11.8) and (11.9) for calculating the spectral characteristic $\vec{h}_{\tau}(\lambda)$ and the MSE $\Delta(f,g;\widehat{A}\xi)$ of the estimate $\widehat{A}\xi$ of the functional $A\xi$. □

Making use of Theorem 11.1.1 we derive the optimal estimate of the unobserved value $\vec{b}^{\top}\vec{\xi}_{-p}^{(d)}, p \geq 0$, based on observations of the increment sequence $\xi_j^{(d)} + \eta_j^{(d)}$ at points $j = 0, -1, -2, \ldots$, under the assumption that $\|\vec{b}\| < \infty$. Consider the vector $\mathbf{b}^{\tau}$ with the entry $\vec{b}$ on the position p and with the entries $\vec{0}$ on the other positions. The following corollary holds.

Corollary 11.1.1. *Let $\vec{b}$ be an infinite-dimensional vector such that $\|\vec{b}\| < \infty$. The optimal linear estimate $\widehat{\vec{b}^{\top}\vec{\xi}_{-p}^{(d)}}$ of the value $\vec{b}^{\top}\vec{\xi}_{-p}^{(d)}$, $p \geq 0$, based on observations of the increment sequence $\xi_j^{(d)} + \eta_j^{(d)}$ at points $j = 0, -1, -2, \ldots$ is determined by the formula*

$$\widehat{\vec{b}^{\top}\vec{\xi}_{-p}^{(d)}} = \int_{-\pi}^{\pi} (\vec{h}_{\tau,-p}(\lambda))^{\top} d\vec{Z}_{\xi^{(d)}+\eta^{(d)}}(\lambda).$$

The spectral characteristic $\vec{h}_{\tau,-p}(\lambda)$ of the optimal estimate $\widehat{\vec{b}^{\top}\vec{\xi}_{-p}^{(d)}}$ is calculated by the formula

$$\begin{aligned}(\vec{h}_{\tau,-p}(\lambda))^{\top} &= \frac{e^{-i\lambda p}(1-e^{-i\lambda\tau})^d}{(i\lambda)^d}\vec{b}^{\top}f(\lambda)(f(\lambda)+g(\lambda))^{-1} \\ &\quad - \frac{(-i\lambda)^d}{(1-e^{i\lambda\tau})^d}(\vec{C}_{\tau,-p}(e^{i\lambda}))^{\top}(f(\lambda)+g(\lambda))^{-1},\end{aligned}$$

where

$$\vec{C}_{\tau,-p}(e^{i\lambda}) = \sum_{k=0}^{\infty} (\mathbf{P}_\tau^{-1}\mathbf{r}_p\vec{b})_k e^{i\lambda(k+1)},$$

where $\mathbf{r}_p = ((R_{1,p})^\top, (R_{2,p})^\top, \ldots)^\top$. *The MSE is calculated by the formula*

$$\begin{aligned}
\Delta(f,g;\vec{b}^\top\widehat{\vec{\xi}_{-p}^{(d)}}) &= \frac{1}{2\pi}\int_{-\pi}^{\pi} \frac{\lambda^{2d}}{|1-e^{i\lambda\tau}|^{2d}}\left(e^{-i\lambda p}\frac{|1-e^{i\lambda\tau}|^{2d}}{\lambda^{2d}}\vec{b}^\top g(\lambda) + (\vec{C}_{\tau,-p}(e^{i\lambda}))^\top\right) \\
&\quad \times (f(\lambda)+g(\lambda))^{-1} f(\lambda)(f(\lambda)+g(\lambda))^{-1} \\
&\quad \times \left(e^{i\lambda p}\frac{|1-e^{i\lambda\tau}|^{2d}}{\lambda^{2d}} g(\lambda)\overline{\vec{b}} + \overline{\vec{C}_{\tau,-p}(e^{i\lambda})}\right) d\lambda \\
&\quad + \frac{1}{2\pi}\int_{-\pi}^{\pi} \frac{\lambda^{2d}}{|1-e^{i\lambda\tau}|^{2d}}\left(e^{-i\lambda p}\frac{|1-e^{i\lambda\tau}|^{2d}}{\lambda^{2d}}\vec{b}^\top f(\lambda) - (\vec{C}_{\tau,-p}(e^{i\lambda}))^\top\right) \\
&\quad \times (f(\lambda)+g(\lambda))^{-1} g(\lambda)(f(\lambda)+g(\lambda))^{-1} \\
&\quad \times \left(e^{i\lambda p}\frac{|1-e^{i\lambda\tau}|^{2d}}{\lambda^{2d}} f(\lambda)\overline{\vec{b}} - \overline{\vec{C}_{\tau,-p}(e^{i\lambda})}\right) d\lambda \\
&= \langle \mathbf{r}_p\vec{b}, \mathbf{P}_\tau^{-1}\mathbf{r}_p\vec{b}\rangle + \langle Q_{p,p}^\tau\vec{b}, \vec{b}\rangle.
\end{aligned}$$

11.1.2 Filtering based on factorizations of the spectral densities

Formulas (11.8) and (11.9), which provide a solution of the filtering problem for the functional $A\xi$, are derived using the Fourier transformations of the functions which involve the spectral densities $f(\lambda)$ and $g(\lambda)$. In this section, we derive the formulas for the spectral characteristic and the MSE in terms of coefficients of canonical factorizations of these spectral densities.

Suppose that the spectral density $f(\lambda) + g(\lambda)$ admits the canonical factorization

$$\frac{|1-e^{i\lambda\tau}|^{2d}}{\lambda^{2d}}(f(\lambda)+g(\lambda)) = \Theta_\tau(e^{-i\lambda})\Theta_\tau^*(e^{-i\lambda}), \quad \Theta_\tau(e^{-i\lambda}) = \sum_{k=0}^{\infty}\theta_\tau(k)e^{-i\lambda k} \tag{11.11}$$

Define the matrix-valued function $\Psi_\tau(e^{-i\lambda}) = \{\Psi_{ij}(e^{-i\lambda})\}_{i=\overline{1,M}}^{j=\overline{1,\infty}}$ by the equation

$$\Psi_\tau(e^{-i\lambda})\Theta_\tau(e^{-i\lambda}) = E_M,$$

where E_M is an identity $M \times M$ matrix. Then the following factorization takes place:

$$\frac{\lambda^{2d}}{|1-e^{i\lambda\tau}|^{2d}}(f(\lambda)+g(\lambda))^{-1} = \Psi_\tau^*(e^{-i\lambda})\Psi_\tau(e^{-i\lambda}), \quad \Psi_\tau(e^{-i\lambda}) = \sum_{k=0}^{\infty}\psi_\tau(k)e^{-i\lambda k}. \tag{11.12}$$

Define the linear operators Ψ_τ and Θ_τ in the space ℓ_2 by the matrices with the matrix entries $(\Psi_\tau)_{k,j} = \psi_\tau(k-j)$, $(\Theta_\tau)_{k,j} = \theta_\tau(k-j)$ for $0 \le j \le k$ and $(\Psi_\tau)_{k,j} = \mathcal{O}$, $(\Theta_\tau)_{k,j} = \mathcal{O}$ for $0 \le k < j$. Making use of Lemma 8.1.3 for factorizations (11.11)–(11.12) one can obtain the factorizations

$$\mathbf{P}_\tau = (\Psi_\tau)^\top \overline{\Psi}_\tau, \quad (\mathbf{P}_\tau)^{-1} = \overline{\Theta}_\tau(\Theta_\tau)^\top.$$

Remark 11.1.4. The entries of the matrix of the operator $(\mathbf{P}_\tau)^{-1}$ are calculated by the formula

$$((\mathbf{P}_\tau)^{-1})_{l,j} = \sum_{p=0}^{\min(l,j)} \overline{\theta}_\tau(j-p)(\theta_\tau(l-p))^\top, \quad l,j \ge 0.$$

Let the canonical factorizations

$$\frac{|1-e^{i\lambda\tau}|^{2d}}{\lambda^{2d}} f(\lambda) = \sum_{k=-\infty}^{\infty} f_\tau(k) e^{i\lambda k} = \varphi_\tau(e^{-i\lambda})\varphi_\tau^*(e^{-i\lambda}), \quad \varphi_\tau(e^{-i\lambda}) = \sum_{k=0}^{\infty} \varphi_\tau(k) e^{-i\lambda k} \tag{11.13}$$

and

$$\frac{|1-e^{i\lambda\tau}|^{2d}}{\lambda^{2d}} g(\lambda) = \sum_{k=-\infty}^{\infty} g_\tau(k) e^{i\lambda k} = \Phi_\tau(e^{-i\lambda})\Phi_\tau^*(e^{-i\lambda}), \quad \Phi_\tau(e^{-i\lambda}) = \sum_{k=0}^{\infty} \phi_\tau(k) e^{-i\lambda k} \tag{11.14}$$

take place.

Lemma 11.1.3. *Let factorizations* (11.11), (11.13), *and* (11.14) *take place. Define by* $\vec{e}_m^{\,\tau} = (\Theta_\tau^\top \mathbf{R}\mathbf{b}^\tau)_m$, $m \ge 0$, *the mth entry of the vector* $\widetilde{\mathbf{e}}_\tau = \Theta_\tau^\top \mathbf{R}\mathbf{b}^\tau$. *Then*

$$\vec{e}_m^{\,\tau} = \sum_{j=0}^{\infty} Z_\tau(m+j+1)\vec{b}_j^{\,\tau} = -\sum_{j=0}^{\infty} X_\tau(m+j+1)\vec{b}_j^{\,\tau},$$

where $Z_\tau(j)$ *and* $X_\tau(j)$ *are defined as*

$$Z_\tau(j) = \sum_{l=0}^{\infty} \overline{\psi}_\tau(l)\overline{f}_\tau(l-j), \quad j \in \mathbb{Z},$$

$$f_\tau(k) = \sum_{m=\max\{0,-k\}}^{\infty} \varphi_\tau(m)\varphi_\tau^*(k+m), \quad k \in \mathbb{Z},$$

and

$$X_\tau(j) = \sum_{l=0}^{\infty} \overline{\psi}_\tau(l)\overline{g}_\tau(l-j), \quad j \in \mathbb{Z},$$

$$g_\tau(j) = \sum_{m=\max\{0,-j\}}^{\infty} \phi_\tau(m)\phi_\tau^*(j+m), \quad j \in \mathbb{Z}.$$

Proof. Factorizations (11.13) and (11.12) and Remark 8.1.3 imply

$$\begin{aligned}
\sum_{j\in\mathbb{Z}} \vec{s}(j)e^{i\lambda j} &= \left(f(\lambda)(f(\lambda)+g(\lambda))^{-1}\right)^\top \\
&= (\Psi_\tau(e^{-i\lambda}))^\top \overline{\Psi}_\tau(e^{-i\lambda})\frac{|1-e^{i\lambda\tau}|^{2d}}{\lambda^{2d}}\overline{f}(\lambda) \\
&= \sum_{l=0}^{\infty} \psi_\tau^\top(l)e^{-i\lambda l}\sum_{j\in\mathbb{Z}} Z_\tau(j)e^{i\lambda j} \\
&= \sum_{j\in\mathbb{Z}}\sum_{l=0}^{\infty} \psi_\tau^\top(l)Z_\tau(l+j)e^{i\lambda j}.
\end{aligned}$$

Then

$$\begin{aligned}
(\Theta_\tau^\top \mathbf{R}\mathbf{b}_\tau)_m &= \sum_{j=0}^{\infty}\sum_{p=m}^{\infty} \theta_\tau^\top(p-m)R(p+j+1)\vec{b}_j^\tau \\
&= \sum_{j=0}^{\infty}\sum_{p=m}^{\infty}\sum_{l=0}^{\infty} \theta_\tau^\top(p-m)\psi_\tau^\top(l)Z_\tau(l+p+j+1)\vec{b}_j^\tau \\
&= \sum_{j=0}^{\infty}\sum_{p=m}^{\infty}\sum_{l=p}^{\infty} (\psi_\tau(l-p)\theta_\tau(p-m))^\top Z_\tau(l+j+1)\vec{b}_j^\tau \\
&= \sum_{j=0}^{\infty}\sum_{l=m}^{\infty} \operatorname{diag}_M(\delta_{l,m})Z_\tau(l+j+1)\vec{b}_j^\tau \\
&= \sum_{j=0}^{\infty} Z_\tau(m+j+1)\vec{b}_j^\tau.
\end{aligned}$$

A representation for $Z_\tau(j)$ follows from

$$\sum_{j\in\mathbb{Z}} Z_\tau(j)e^{i\lambda j} = \overline{\Psi}_\tau(e^{-i\lambda})\frac{|1-e^{i\lambda\tau}|^{2d}}{\lambda^{2d}}\overline{f}(\lambda) = \sum_{j\in\mathbb{Z}}\sum_{l=0}^{\infty} \overline{\Psi}_\tau(l)\overline{f}_\tau(l-j)e^{i\lambda j}.$$

On the other hand,

$$\begin{aligned}
\sum_{j\in\mathbb{Z}} \vec{s}(j)e^{i\lambda j} &= \left(f(\lambda)(f(\lambda)+g(\lambda))^{-1}\right)^\top \\
&= \left((f(\lambda)+g(\lambda)-g(\lambda))(f(\lambda)+g(\lambda))^{-1}\right)^\top \\
&= \left(E_\infty - g(\lambda)(f(\lambda)+g(\lambda))^{-1}\right)^\top \\
&= E_\infty - (\Psi_\tau(e^{-i\lambda}))^\top \overline{\Psi}_\tau(e^{-i\lambda})\frac{|1-e^{i\lambda\tau}|^{2d}}{\lambda^{2d}}\overline{g}(\lambda) \\
&= E_\infty - \sum_{l=0}^{\infty} \psi_\tau^\top(l)e^{-i\lambda l}\sum_{j\in\mathbb{Z}} X_\tau(j)e^{i\lambda j}
\end{aligned}$$

$$= E_\infty - \sum_{j\in\mathbb{Z}}\sum_{l=0}^{\infty}\psi_\tau^\top(l)X_\tau(l+j)e^{i\lambda j}$$
$$= \sum_{j=0}^{\infty}Z_\tau(m+j+1)\vec{b}_j^\tau,$$

where

$$\sum_{j\in\mathbb{Z}}X_\tau(j)e^{i\lambda j} = \overline{\Psi}_\tau(e^{-i\lambda})\frac{|1-e^{i\lambda\tau}|^{2d}}{\lambda^{2d}}\overline{g}(\lambda) = \sum_{k\in\mathbb{Z}}\sum_{l=0}^{\infty}\overline{\Psi}_\tau(l)\overline{g}_\tau(l-k)e^{i\lambda k}.$$

Thus, coefficients $Z_\tau(j)$ and $X_\tau(j)$ are related as

$$\begin{cases} E_\infty - \sum_{l=0}^{\infty}\psi_\tau^\top(l)X_\tau(l) = \sum_{l=0}^{\infty}\psi_\tau^\top(l)Z_\tau(l), \\ -\sum_{l=0}^{\infty}\psi_\tau^\top(l)X_\tau(l+k) = \sum_{l=0}^{\infty}\psi_\tau^\top(l)Z_\tau(l+k), \quad k\neq 0, \end{cases}$$

which implies

$$(\Theta_\tau^\top\mathbf{R}\mathbf{b}_\tau)_m = \sum_{j=0}^{\infty}\sum_{p=m}^{\infty}\sum_{l=0}^{\infty}\theta_\tau^\top(p-m)\psi_\tau^\top(l)Z_\tau(l+p+j+1)\vec{b}_j^\tau$$
$$= -\sum_{j=0}^{\infty}\sum_{p=m}^{\infty}\sum_{l=0}^{\infty}\theta_\tau^\top(p-m)\psi_\tau^\top(l)X_\tau(l+p+j+1)\vec{b}_j^\tau$$
$$= -\sum_{j=0}^{\infty}X_\tau(m+j+1)\vec{b}_j^\tau. \qquad \square$$

Define the linear operators $\mathbf{G}_\tau^f$ and $\mathbf{G}_\tau^g$ in the space ℓ_2 by matrices with the matrix entries $(\mathbf{G}_\tau^f)_{l,j} = \overline{f}_\tau(l-j)$ and $(\mathbf{G}_\tau^g)_{l,j} = \overline{g}_\tau(l-j)$, $l,j\geq 0$, and the linear operators $\widetilde{\varphi}_\tau$ and $\widetilde{\Phi}_\tau$ in the space ℓ_2 determined by a matrix with the matrix entries $(\widetilde{\varphi}_\tau)_{l,j} = \varphi_\tau^\top(l-j)$ and $(\widetilde{\Phi}_\tau)_{l,j} = \phi_\tau^\top(l-j)$ for $0\leq j\leq l$ and $(\widetilde{\varphi}_\tau)_{l,j} = \mathcal{O}$ and $(\widetilde{\Phi}_\tau)_{l,j} = \mathcal{O}$ for $0\leq l<j$.

Theorem 11.1.2. *Consider two stochastic processes with periodically stationary increments, $\xi(t)$ and $\eta(t)$, generating stationary increment sequences $\vec{\xi}_j^{(d)}$ and $\vec{\eta}_j^{(d)}$ with spectral densities $f(\lambda)$ and $g(\lambda)$. Assume that condition* (11.4) *is satisfied. The solution $\widehat{A}\xi$ to the filtering problem for the linear functional $A\xi$ is calculated by formula* (11.7). *If factorizations* (11.11)–(11.12) *take place, then the spectral characteristic $\vec{h}_\tau(\lambda)$ of the estimate $\widehat{A}\xi$ can be calculated by the formula*

$$\vec{h}_\tau(\lambda) = \frac{(1-e^{-i\lambda\tau})^d}{(i\lambda)^d}\left(\sum_{k=0}^{\infty}\psi_\tau^\top(k)e^{-i\lambda k}\right)\sum_{m=0}^{\infty}(\overline{\psi}_\tau\mathbf{C}_\tau^f)_m e^{-i\lambda m}, \tag{11.15}$$

where $\overline{\psi}_\tau = (\overline{\psi}_\tau(0), \overline{\psi}_\tau(1), \overline{\psi}_\tau(2), \dots)$ and

$$(\overline{\psi}_\tau \mathbf{c}_\tau^f)_m = \sum_{j=0}^{\infty} \overline{\psi}_\tau(j)\mathbf{c}_\tau^f(j+m),$$

where

$$\mathbf{c}_\tau^f(m) = \sum_{l=0}^{\infty} \overline{\varphi}_\tau(l)(\widetilde{\boldsymbol{\varphi}}_\tau D^\tau \mathbf{a})_{l+m} = (\widetilde{\boldsymbol{\varphi}}_\tau^* \widetilde{\boldsymbol{\varphi}}_\tau D^\tau \mathbf{a})_m,$$

and the MSE is calculated by the formula

$$\Delta(f,g;\widehat{A}\xi) = \|\widetilde{\boldsymbol{\varphi}}_\tau D^\tau \mathbf{a}\|^2 - \|\overline{\psi}_\tau \mathbf{c}_\tau^f\|^2. \tag{11.16}$$

If factorizations (11.11) *and* (11.14) *take place, then the spectral characteristic $\vec{h}_\tau(\lambda)$ can be calculated by the formula*

$$\vec{h}_\tau(\lambda) = \frac{(1-e^{-i\lambda\tau})^d}{(i\lambda)^d}\left(\vec{B}_\tau(e^{-i\lambda}) - \left(\sum_{k=0}^{\infty}\psi_\tau^\top(k)e^{-i\lambda k}\right)\sum_{m=0}^{\infty}(\overline{\psi}_\tau \mathbf{c}_\tau^g)_m e^{-i\lambda m}\right), \tag{11.17}$$

where

$$(\overline{\psi}_\tau \mathbf{c}_\tau^g)_m = \sum_{k=0}^{\infty}\overline{\psi}_\tau(k)\mathbf{c}_\tau^g(k+m),$$

$$\mathbf{c}_\tau^g(m) = \sum_{l=0}^{\infty}\overline{\phi}_\tau(l)(\widetilde{\Phi}_\tau D^\tau \mathbf{a})_{l+m} = (\widetilde{\Phi}_\tau^* \widetilde{\Phi}_\tau D^\tau \mathbf{a})_m,$$

and the MSE is calculated by the formula

$$\Delta(f,g;\widehat{A}\xi) = \|\widetilde{\Phi}_\tau D^\tau \mathbf{a}\|^2 - \|\overline{\psi}_\tau \mathbf{c}_\tau^g\|^2. \tag{11.18}$$

Proof. Making use of Lemma 11.1.3 we rewrite formulas (11.8) and (11.9). We make the following transformations:

$$\begin{aligned}
&\frac{\lambda^{2d}}{|1-e^{i\lambda\tau}|^{2d}}\left((f(\lambda)+g(\lambda))^{-1}\right)^\top\left(\sum_{k=0}^{\infty}(\mathbf{P}_\tau^{-1}\mathbf{R}\mathbf{b}^\tau)_k e^{i\lambda(k+1)}\right)\\
&=\left(\sum_{k=0}^{\infty}\psi_\tau^\top(k)e^{-i\lambda k}\right)\sum_{j=0}^{\infty}\sum_{k=0}^{\infty}\overline{\psi}_\tau(j)(\overline{\Theta}_\tau \vec{\mathbf{e}}_\tau)_k e^{i\lambda(k+j+1)}\\
&=\left(\sum_{k=0}^{\infty}\psi_\tau^\top(k)e^{-i\lambda k}\right)\sum_{m=0}^{\infty}\sum_{p=0}^{m}\sum_{k=p}^{m}\overline{\psi}_\tau(m-k)\overline{\theta}_\tau(k-p)\widetilde{\vec{e}}_\tau(p)e^{i\lambda(m-1)}\\
&=\left(\sum_{k=0}^{\infty}\psi_\tau^\top(k)e^{-i\lambda k}\right)\sum_{m=0}^{\infty}\sum_{p=0}^{m}\operatorname{diag}_M(\delta_{m,p})\widetilde{\vec{e}}_\tau(m)e^{i\lambda(m+1)}\\
&=\left(\sum_{k=0}^{\infty}\psi_\tau^\top(k)e^{-i\lambda k}\right)\sum_{m=0}^{\infty}\widetilde{\vec{e}}_\tau(m)e^{i\lambda(m+1)}
\end{aligned}$$

and

$$
\begin{aligned}
&((f(\lambda)+g(\lambda))^{-1})^{\top}(f(\lambda))^{\top}\vec{B}_{\tau}(e^{-i\lambda})\\
&=\Psi_{\tau}^{\top}(e^{-i\lambda})\overline{\Psi_{\tau}(e^{-i\lambda})\frac{|1-e^{i\lambda\tau}|^{2d}}{\lambda^{2d}}f(\lambda)B_{\tau}(e^{-i\lambda})}\\
&=\left(\sum_{k=0}^{\infty}\psi_{\tau}^{\top}(k)e^{-i\lambda k}\right)\sum_{m\in\mathbb{Z}}\sum_{j=0}^{\infty}Z_{\tau}(m+j)\vec{b}_{j}^{\tau}e^{i\lambda m}.
\end{aligned}
$$

Then we obtain

$$
\begin{aligned}
\vec{h}_{\tau}(\lambda)&=\frac{(1-e^{-i\lambda\tau})^{d}}{(i\lambda)^{d}}\left(\sum_{k=0}^{\infty}\psi_{\tau}^{\top}(k)e^{-i\lambda k}\right)\sum_{m=0}^{\infty}\sum_{j=0}^{\infty}Z_{\tau}(j-m)\vec{b}_{j}^{\tau}e^{-i\lambda m}\\
&=\frac{(1-e^{-i\lambda\tau})^{d}}{(i\lambda)^{d}}\left(\sum_{k=0}^{\infty}\psi_{\tau}^{\top}(k)e^{-i\lambda k}\right)\sum_{m=0}^{\infty}\sum_{j=0}^{\infty}\sum_{p=m}^{\infty}\overline{\psi}_{\tau}(p-m)\overline{f}_{\tau}(p-j)\vec{b}_{j}^{\tau}e^{-i\lambda m}\\
&=\frac{(1-e^{-i\lambda\tau})^{d}}{(i\lambda)^{d}}\left(\sum_{k=0}^{\infty}\psi_{\tau}^{\top}(k)e^{-i\lambda k}\right)\sum_{m=0}^{\infty}((\widetilde{\Psi}_{\tau})^{*}\mathbf{G}_{\tau}^{f}\mathbf{b}^{\tau})_{m}e^{-i\lambda m}\\
&=\frac{(1-e^{-i\lambda\tau})^{d}}{(i\lambda)^{d}}\left(\sum_{k=0}^{\infty}\psi_{\tau}^{\top}(k)e^{-i\lambda k}\right)\sum_{m=0}^{\infty}(\overline{\psi}_{\tau}\mathbf{c}_{\tau,}^{f})_{m}e^{-i\lambda m}.
\end{aligned}
$$

Here

$$
\begin{aligned}
(\overline{\psi}_{\tau}\mathbf{c}_{\tau}^{f})_{m}&=\sum_{k=0}^{\infty}\overline{\psi}_{\tau}(k)\mathbf{c}_{\tau}^{f}(k+m),\\
\mathbf{c}_{\tau}^{f}(m)&=\sum_{k=0}^{\infty}\overline{f}_{\tau}(m-k)\vec{b}_{k}^{\tau}=\sum_{l=0}^{\infty}\overline{\varphi}_{\tau}(l)\sum_{k=0}^{l+m}\varphi_{\tau}^{\top}(l+m-k)\vec{b}_{k}^{\tau}\\
&=\sum_{l=0}^{\infty}\overline{\varphi}_{\tau}(l)(\widetilde{\varphi}_{\tau}\mathbf{b}^{\tau})_{l+m}=(\widetilde{\varphi}_{\tau}^{*}\overline{\widetilde{\varphi}}_{\tau}\mathbf{b}^{\tau})_{m}.
\end{aligned}
$$

The MSE $\Delta(f,g;\widehat{A}\xi)$ is calculated by the formula

$$
\begin{aligned}
\Delta(f,g;\widehat{A}\xi)&=\Delta(f,g;\widehat{B}\vec{\xi})=\mathsf{E}|B\vec{\xi}-\widehat{B}\vec{\xi}|^{2}\\
&=\frac{1}{2\pi}\int_{-\pi}^{\pi}(\vec{B}_{\tau}(e^{-i\lambda}))^{\top}\frac{|1-e^{i\lambda\tau}|^{2d}}{\lambda^{2d}}f(\lambda)\overline{\vec{B}_{\tau}(e^{-i\lambda})}d\lambda\\
&\quad+\frac{1}{2\pi}\int_{-\pi}^{\pi}(\vec{h}_{\tau}(\lambda))^{\top}(f(\lambda)+g(\lambda))\overline{\vec{h}_{\tau}(\lambda)}d\lambda\\
&\quad-\frac{1}{2\pi}\int_{-\pi}^{\pi}(\vec{h}_{\tau}(\lambda))^{\top}\frac{(1-e^{i\lambda\tau})^{d}}{(-i\lambda)^{d}}\overline{\vec{B}_{\tau}(e^{-i\lambda})}d\lambda
\end{aligned}
$$

$$- \frac{1}{2\pi} \int_{-\pi}^{\pi} (\vec{B}_\tau(e^{-i\lambda}))^\top \frac{(1-e^{-i\lambda\tau})^d}{(i\lambda)^d} f(\lambda) \overline{\vec{h}_\tau(\lambda)} d\lambda$$
$$= \|\widetilde{\varphi}_\tau \mathbf{b}^\tau\|^2 - \|\overline{\psi}_\tau \mathbf{c}_\tau^f\|^2.$$

Suppose that factorizations (11.11) and (11.14) take place. Then from Lemma 11.1.3 we obtain

$$\frac{\lambda^{2d}}{|1-e^{i\lambda\tau}|^{2d}} ((f(\lambda)+g(\lambda))^{-1})^\top \vec{C}_\tau(e^{i\lambda})$$
$$= -\left(\sum_{k=0}^{\infty} \psi_\tau^\top(k) e^{-i\lambda k}\right) \sum_{m=0}^{\infty}\sum_{j=0}^{\infty} X_\tau(m+j+1)\vec{b}_j^\tau e^{i\lambda(m+1)}.$$

Next,

$$((f(\lambda)+g(\lambda))^{-1})^\top (f(\lambda))^\top \vec{B}_\tau(e^{-i\lambda})$$
$$= \vec{B}_\tau(e^{-i\lambda}) - ((f(\lambda)+g(\lambda))^{-1})^\top (g(\lambda))^\top \vec{B}_\tau(e^{-i\lambda})$$
$$= \vec{B}_\tau(e^{-i\lambda}) - \Psi_\tau^\top(e^{-i\lambda})\overline{\Psi_\tau(e^{-i\lambda})} \frac{|1-e^{i\lambda\tau}|^{2d}}{\lambda^{2d}} \overline{g(\lambda)\vec{B}_\tau(e^{-i\lambda})}$$
$$= \vec{B}_\tau(e^{-i\lambda}) - \left(\sum_{k=0}^{\infty} \psi_\tau^\top(k) e^{-i\lambda k}\right) \sum_{m\in\mathbb{Z}}\sum_{j=0}^{\infty} X_\tau(m+j)\vec{b}_j^\tau e^{i\lambda m}.$$

Finally,

$$\vec{h}_\tau(\lambda) = \frac{(1-e^{-i\lambda\tau})^d}{(i\lambda)^d}\left(\vec{B}_\tau(e^{-i\lambda}) - \Psi_\tau^\top(e^{-i\lambda}) \sum_{m=0}^{\infty}\sum_{j=0}^{\infty} X_\tau(j-m)\vec{b}_j^\tau e^{-i\lambda m}\right)$$
$$= \frac{(1-e^{-i\lambda\tau})^d}{(i\lambda)^d}\left(\vec{B}_\tau(e^{-i\lambda}) - \Psi_\tau^\top(e^{-i\lambda}) \sum_{m=0}^{\infty} (\overline{\Psi}_\tau^* \mathbf{G}_\tau^g \mathbf{b}^\tau)_m e^{-i\lambda m}\right)$$
$$= \frac{(1-e^{-i\lambda\tau})^d}{(i\lambda)^d}\left(\vec{B}_\tau(e^{-i\lambda}) - \Psi_\tau^\top(e^{-i\lambda}) \sum_{m=0}^{\infty} (\overline{\psi}_\tau \mathbf{c}_\tau^g)_m e^{-i\lambda m}\right),$$

where

$$(\overline{\psi}_\tau \mathbf{c}_\tau^g)_m = \sum_{k=0}^{\infty} \overline{\psi}_\tau(k) \mathbf{c}_\tau^g(k+m),$$
$$\mathbf{c}_\tau^g(m) = \sum_{k=0}^{\infty} \overline{g}_\tau(m-k)\vec{b}_k^\tau = \sum_{l=0}^{\infty} \overline{\phi}_\tau(l) \sum_{k=0}^{l+m} \phi_\tau^\top(l+m-k)\vec{b}_k^\tau$$
$$= \sum_{l=0}^{\infty} \overline{\phi}_\tau(l)(\widetilde{\Phi}_\tau \mathbf{b}^\tau)_{l+m} = (\widetilde{\Phi}_\tau^* \widetilde{\Phi}_\tau \mathbf{b}^\tau)_m.$$

The MSE is calculated by the formula

$$\Delta(f,g;\widehat{A}\xi) = \|\widetilde{\Phi}_\tau \mathbf{b}^\tau\|^2 - \|\overline{\psi}_\tau \mathbf{C}^g_\tau\|^2.$$ □

Remark 11.1.5. Let

$$\frac{|\chi^{(d)}_\tau(e^{-i\lambda})|^2}{|\beta^{(d)}(i\lambda)|^2} = \frac{|1-e^{i\lambda\tau}|^{2d}}{\lambda^{2d}}$$

in Corollary 2.1.2. Then

$$\frac{|1-e^{i\lambda\tau}|^{2d}}{\lambda^{2d}} = |w_\tau(e^{-i\lambda})|^2 = \left|\sum_{k=0}^{\infty} w_\tau(k)e^{-i\lambda k}\right|^2.$$

Suppose, for example, that factorization (11.12) takes place. Then the spectral density $f(\lambda)+g(\lambda)$ admits a factorization

$$f(\lambda)+g(\lambda) = \Theta(e^{-i\lambda})\Theta^*(e^{-i\lambda}), \quad \Theta(e^{-i\lambda}) = \sum_{k=0}^{\infty}\theta(k)e^{-i\lambda k} \tag{11.19}$$

and the following relations hold:

$$\begin{aligned}\Theta_\tau(e^{-i\lambda}) &= w_\tau(e^{-i\lambda})\Theta(e^{-i\lambda}),\\ \theta_\tau(k) &= \sum_{m=0}^{k} w_\tau(k-m)\theta(m), \quad k=0,1,\dots,\end{aligned} \tag{11.20}$$

or

$$\theta_{\tau,ij}(k) = \sum_{m=0}^{k} w_\tau(k-m)\theta_{ij}(m), \quad i=1,\dots,\infty,\ j=1,\dots,M,\ k=0,1,\dots.$$

Define the matrix-valued function $\Psi(e^{-i\lambda}) = \{\Psi_{ij}(e^{-i\lambda})\}_{i=\overline{1,M}}^{j=\overline{1,\infty}}$ by the equation

$$\Psi(e^{-i\lambda})\Theta(e^{-i\lambda}) = E_M,$$

where E_M is an identity $M\times M$ matrix. Then the following factorization takes place:

$$(f(\lambda)+g(\lambda))^{-1} = \Psi^*(e^{-i\lambda})\Psi(e^{-i\lambda}), \quad \Psi(e^{-i\lambda}) = \sum_{k=0}^{\infty}\psi(k)e^{-i\lambda k},$$

and

$$\Psi(e^{-i\lambda}) = \Psi_\tau(e^{-i\lambda})w_\tau(e^{-i\lambda}). \tag{11.21}$$

Define a linear operator $\mathbf{W}_\tau$ in the space ℓ_2 by a matrix with the entries $(\mathbf{W}_\tau)_{k,j} = w_\tau(k-j)$ for $0 \leq j \leq k$ and $(\mathbf{W}_\tau)_{k,j} = \mathcal{O}$ for $0 \leq k < j$. Relation (11.20) can be represented in the form

$$\boldsymbol{\theta}_\tau = \mathbf{W}^\tau \boldsymbol{\theta}, \tag{11.22}$$

where

$$\boldsymbol{\theta}_\tau = ((\theta_\tau(0))^\top, (\theta_\tau(1))^\top, (\theta_\tau(2))^\top, \dots)^\top,$$
$$\boldsymbol{\theta} = ((\theta(0))^\top, (\theta(1))^\top, (\theta(2))^\top, \dots)^\top$$

are vectors with the matrix entries $\theta_\tau(k) = \{\theta_{\tau,ij}(k)\}_{i=\overline{1,\infty}}^{j=\overline{1,M}}$, $k = 0,1,2,\dots$, and $\theta(k) = \{\theta_{ij}(k)\}_{i=\overline{1,\infty}}^{j=\overline{1,M}}$, $k = 0,1,2,\dots$.

11.2 Minimax filtering based on observations with periodically stationary increment noise

11.2.1 Least favorable spectral densities and minimax-robust spectral characteristic

Consider the estimation problem for the functional $A\xi$ based on the observations $\xi(t) + \eta(t)$ at points $t \leq 0$ when the spectral densities of the generated stationary increment sequences are not exactly known while a set $\mathcal{D} = \mathcal{D}_f \times \mathcal{D}_g$ of admissible spectral densities is defined. Here $\xi(t)$ and $\eta(t)$ are two uncorrelated processes with periodically stationary increments. We apply the minimax (robust) approach of estimation, which is formalized by the following two definitions.

Definition 11.2.1. For a given class of spectral densities $\mathcal{D} = \mathcal{D}_f \times \mathcal{D}_g$, the spectral densities $f^0(\lambda) \in \mathcal{D}_f$ and $g^0(\lambda) \in \mathcal{D}_g$ are called the least favorable spectral densities in the class $\mathcal{D}$ for the optimal linear filtering of the functional $A\xi$ if

$$\Delta(f^0, g^0) = \Delta(h(f^0, g^0); f^0, g^0) = \max_{(f,g)\in\mathcal{D}_f\times\mathcal{D}_g} \Delta(h(f,g); f, g).$$

Definition 11.2.2. For a given class of spectral densities $\mathcal{D} = \mathcal{D}_f \times \mathcal{D}_g$, the spectral characteristic $h^0(e^{i\lambda})$ of the optimal estimate of the functional $A\xi$ is called minimax-robust if

$$h^0(e^{i\lambda}) \in H_{\mathcal{D}} = \bigcap_{(f,g)\in\mathcal{D}_f\times\mathcal{D}_g} L_2^0(f+g),$$

$$\min_{h\in H_{\mathcal{D}}} \max_{(f,g)\in\mathcal{D}_f\times\mathcal{D}_g} \Delta(h; f, g) = \sup_{(f,g)\in\mathcal{D}_f\times\mathcal{D}_g} \Delta(h^0; f, g).$$

Taking into account the introduced definitions and the relations derived in this chapter, as well as Remark 11.1.5 for the spectral densities $f(\lambda)$, $g(\lambda)$, and $f(\lambda) + g(\lambda)$, we can verify that the following lemmas hold.

Lemma 11.2.1. *The spectral densities $f^0 \in \mathcal{D}_f$ and $g^0 \in \mathcal{D}_g$ which satisfy condition* (11.6) *are the least favorable densities in the class $\mathcal{D}$ for the optimal linear filtering of the functional $A\xi$ if operators $\mathbf{P}_\tau^0$, $\mathbf{R}^0$, and $\mathbf{Q}_\tau^0$ defined by the Fourier coefficients of the functions*

$$\frac{\lambda^{2d}}{|1-e^{i\lambda\tau}|^{2d}}\left((f^0(\lambda)+g^0(\lambda))^{-1}\right)^\top,$$

$$\left(g^0(\lambda)(f^0(\lambda)+g^0(\lambda))^{-1}\right)^\top,$$

$$\frac{|1-e^{i\lambda\tau}|^{2d}}{\lambda^{2d}}\left(f^0(\lambda)(f^0(\lambda)+g^0(\lambda))^{-1}g^0(\lambda)\right)^\top$$

determine a solution of the constrained optimization problem

$$\begin{aligned}&\max_{f\in\mathcal{D}}\left(\langle \mathbf{R}D^\tau\mathbf{a}, \mathbf{P}_\tau^{-1}\mathbf{R}D^\tau\mathbf{a}\rangle + \langle \mathbf{Q}_\tau D^\tau\mathbf{a}, D^\tau\mathbf{a}\rangle\right)\\ &\quad= \langle \mathbf{R}^0D^\tau\mathbf{a}, (\mathbf{P}_\tau^0)^{-1}\mathbf{R}^0D^\tau\mathbf{a}\rangle + \langle \mathbf{Q}_\tau^0D^\tau\mathbf{a}, D^\tau\mathbf{a}\rangle.\end{aligned} \tag{11.23}$$

Lemma 11.2.2. *The spectral densities $f^0 \in \mathcal{D}_f$ and $g^0 \in \mathcal{D}_g$ which admit the canonical factorizations* (11.11) *and* (11.13) *or canonical factorizations* (11.11) *and* (11.14) *are the least favorable densities in the class $\mathcal{D}$ for the optimal linear filtering of the functional $A\xi$ based on observations of the process $\xi(t)+\eta(t)$ at points $t \le 0$ if matrix coefficients $\{\psi^0(k), \varphi^0(k), \phi^0(k) : k \ge 0\}$ of the canonical factorizations*

$$(f^0(\lambda)+g^0(\lambda))^{-1} = \left(\sum_{k=0}^{\infty}\psi^0(k)e^{-i\lambda k}\right)^*\left(\sum_{k=0}^{\infty}\psi^0(k)e^{-i\lambda k}\right), \tag{11.24}$$

$$f^0(\lambda) = \left(\sum_{k=0}^{\infty}\varphi^0(k)e^{-i\lambda k}\right)\left(\sum_{k=0}^{\infty}\varphi^0(k)e^{-i\lambda k}\right)^*, \tag{11.25}$$

$$g^0(\lambda) = \left(\sum_{k=0}^{\infty}\phi^0(k)e^{-i\lambda k}\right)\left(\sum_{k=0}^{\infty}\phi^0(k)e^{-i\lambda k}\right)^* \tag{11.26}$$

determine a solution of the constrained optimization problem

$$\begin{aligned}&\langle \widetilde{\varphi}_\tau D^\tau\mathbf{a}^\tau, \widetilde{\varphi}_\tau D^\tau\mathbf{a}^\tau\rangle - \langle \overline{\psi}_\tau\mathbf{C}_\tau^f, \overline{\psi}_\tau\mathbf{C}_\tau^f\rangle \to \sup,\\ &f(\lambda) = \varphi(e^{-i\lambda})\varphi^*(e^{-i\lambda}) \in \mathcal{D}_f,\\ &g(\lambda) = \Theta(e^{-i\lambda})\Theta^*(e^{-i\lambda}) - \varphi(e^{-i\lambda})\varphi^*(e^{-i\lambda}) \in \mathcal{D}_g\end{aligned} \tag{11.27}$$

or of the constrained optimization problem

$$\langle \Phi_\tau D^\tau\mathbf{a}_\tau, \Phi_\tau D^\tau\mathbf{a}_\tau\rangle - \langle \overline{\psi}_\tau\mathbf{C}_\tau^g, \overline{\psi}_\tau\mathbf{C}_\tau^g\rangle \to \sup, \tag{11.28}$$

$$f(\lambda) = \Theta(e^{-i\lambda})\Theta^*(e^{-i\lambda}) - \Phi(e^{-i\lambda})\Phi^*(e^{-i\lambda}) \in \mathcal{D}_f,$$
$$g(\lambda) = \Phi(e^{-i\lambda})\Phi^*(e^{-i\lambda}) \in \mathcal{D}_g,$$

where $\Psi(e^{-i\lambda})\Theta(e^{-i\lambda}) = E_M$. *The minimax spectral characteristic* $h^0 = \vec{h}_\tau(f^0, g^0)$ *is calculated by formula* (11.15) *or* (11.17) *if* $\vec{h}_\tau(f^0, g^0) \in H_\mathcal{D}$.

Lemma 11.2.3. *The spectral density* $f^0(\lambda) \in \mathcal{D}_f$ *which admits the canonical factorizations* (11.11) *and* (11.13) *with a known regular spectral density* $g(\lambda)$ *is the least favorable spectral density in the class* $\mathcal{D}_f$ *for the optimal linear filtering of the functional* $A\xi$ *based on observations of the process* $\xi(t) + \eta(t)$ *at points* $t \le 0$ *if the matrix coefficients* $\{\psi^0(k) : k \ge 0\}$ *from the canonical factorization*

$$(f^0(\lambda) + g(\lambda))^{-1} = \left(\sum_{k=0}^{\infty} \psi^0(k)e^{-i\lambda k}\right)^* \left(\sum_{k=0}^{\infty} \psi^0(k)e^{-i\lambda k}\right) \tag{11.29}$$

determine a solution of the constrained optimization problem

$$\langle \overline{\psi}_\tau \mathbf{C}_\tau^g, \overline{\psi}_\tau \mathbf{C}_\tau^g \rangle \to \inf, \tag{11.30}$$
$$f(\lambda) = \Theta(e^{-i\lambda})\Theta^*(e^{-i\lambda}) - g(\lambda) \in \mathcal{D}_f,$$

where $\Psi(e^{-i\lambda})\Theta(e^{-i\lambda}) = E_M$. *The minimax spectral characteristic* $h^0 = \vec{h}_\tau(f^0, g)$ *is calculated by formula* (11.17) *if* $\vec{h}_\tau(f^0, g) \in H_\mathcal{D}$.

Lemma 11.2.4. *The spectral density* $g^0 \in \mathcal{D}_g$ *which admits the canonical factorizations* (11.12) *and* (11.14) *with a known regular spectral density* $f(\lambda)$ *is the least favorable spectral density in the class* $\mathcal{D}_g$ *for the optimal linear filtering of the functional* $A\xi$ *based on observations of the process* $\xi(t) + \eta(t)$ *at points* $t \le 0$ *if the matrix coefficients* $\{\psi^0(k) : k \ge 0\}$ *from the canonical factorization*

$$(f(\lambda) + g^0(\lambda))^{-1} = \left(\sum_{k=0}^{\infty} \psi^0(k)e^{-i\lambda k}\right)^* \left(\sum_{k=0}^{\infty} \psi^0(k)e^{-i\lambda k}\right) \tag{11.31}$$

determine a solution of the constrained optimization problem

$$\langle \overline{\psi}_\tau \mathbf{C}_\tau^f, \overline{\psi}_\tau \mathbf{C}_\tau^f \rangle \to \inf, \tag{11.32}$$
$$g(\lambda) = \Theta(e^{-i\lambda})\Theta^*(e^{-i\lambda}) - f(\lambda) \in \mathcal{D}_g,$$

where $\Psi(e^{-i\lambda})\Theta(e^{-i\lambda}) = E_M$. *The minimax spectral characteristic* $h^0 = \vec{h}_\tau(f, g^0)$ *is calculated by formula* (11.15) *if* $\vec{h}_\tau(f, g^0) \in H_\mathcal{D}$.

The least favorable spectral densities (f^0, g^0) and the function h^0 form a saddle point of the function $\Delta(h; f, g)$ on the set $H_\mathcal{D} \times \mathcal{D}$. The saddle point inequalities

$$\Delta(h;f^0,g^0) \geq \Delta(h^0;f^0,g^0) \geq \Delta(h^0;f,g) \quad \forall f \in \mathcal{D}_f,\ \forall g \in \mathcal{D}_g,\ \forall h \in H_{\mathcal{D}}$$

hold if $h^0 = \vec{h}_\tau(f^0,g^0)$ and $\vec{h}_\tau(f^0,g^0) \in H_{\mathcal{D}}$, where (f^0,g^0) is a solution of the constrained optimization problem

$$\widetilde{\Delta}(f,g) = -\Delta(\vec{h}_\tau(f^0,g^0);f,g) \to \inf, \quad (f,g) \in \mathcal{D}. \tag{11.33}$$

Here the functional $\Delta(\vec{h}_\tau(f^0,g^0);f,g)$ is calculated by the formula

$$\begin{aligned}\Delta(\vec{h}_\tau(f^0,g^0);f,g) &= \frac{1}{2\pi}\int_{-\pi}^{\pi} \frac{\lambda^{2d}}{|1-e^{i\lambda\tau}|^{2d}}(\mathbf{C}_\tau^{f0}(e^{i\lambda}))^\top (f^0(\lambda)+g^0(\lambda))^{-1} f(\lambda)\\ &\times (f^0(\lambda)+g^0(\lambda))^{-1}\overline{\mathbf{C}_\tau^{f0}(e^{i\lambda})}d\lambda\\ &+ \frac{1}{2\pi}\int_{-\pi}^{\pi} \frac{\lambda^{2d}}{|1-e^{i\lambda\tau}|^{2d}}(\mathbf{C}_\tau^{g0}(e^{i\lambda}))^\top (f^0(\lambda)+g^0(\lambda))^{-1} g(\lambda)\\ &\times (f^0(\lambda)+g^0(\lambda))^{-1}\overline{\mathbf{C}_\tau^{g0}(e^{i\lambda})}d\lambda,\end{aligned}$$

where

$$\begin{aligned}(\mathbf{C}_\tau^{f0}(e^{i\lambda}))^\top &= (\vec{B}_\tau(e^{-i\lambda}))^\top \frac{|1-e^{i\lambda\tau}|^{2d}}{\lambda^{2d}} g^0(\lambda) + (\vec{C}_\tau^0(e^{i\lambda}))^\top,\\ (\mathbf{C}_\tau^{g0}(e^{i\lambda}))^\top &= (\vec{B}_\tau(e^{-i\lambda}))^\top \frac{|1-e^{i\lambda\tau}|^{2d}}{\lambda^{2d}} f^0(\lambda) - (\vec{C}_\tau^0(e^{i\lambda}))^\top,\end{aligned}$$

where

$$\vec{C}_\tau^0(e^{i\lambda}) = \sum_{j=0}^{\infty}((\mathbf{P}_\tau^0)^{-1}\mathbf{R}^0 D^\tau \mathbf{a})_j e^{i\lambda(j+1)},$$

or, taking into account (11.21), it is calculated by the formula

$$\begin{aligned}\Delta(\vec{h}_\tau(f^0,g^0);f,g) &= \frac{1}{2\pi}\int_{-\pi}^{\pi} (\mathbf{h}_{\tau,g}^0(e^{-i\lambda}))^\top \Psi^0(e^{-i\lambda}) f(\lambda)(\Psi^0(e^{-i\lambda}))^* \overline{\mathbf{h}_{\tau,g}^0(e^{-i\lambda})}d\lambda\\ &+ \frac{1}{2\pi}\int_{-\pi}^{\pi} (\mathbf{h}_{\tau,f}^0(e^{-i\lambda}))^\top \Psi^0(e^{-i\lambda}) g(\lambda)(\Psi^0(e^{-i\lambda}))^* \overline{\mathbf{h}_{\tau,f}^0(e^{-i\lambda})}d\lambda,\end{aligned}$$

where

$$\mathbf{h}_{\tau,g}^0(e^{-i\lambda}) = \sum_{j=0}^{\infty}(\overline{\psi}_\tau^0(\mathbf{C}_\tau^g)^0)_j e^{-i\lambda j}, \quad \mathbf{h}_{\tau,f}^0(e^{-i\lambda}) = \sum_{j=0}^{\infty}(\overline{\psi}_\tau^0(\mathbf{C}_\tau^f)^0)_j e^{-i\lambda j}.$$

This constrained optimization problem is equivalent to the unconstrained optimization problem

$$\Delta_{\mathcal{D}}(f,g) = \widetilde{\Delta}(f,g) + \delta(f,g|\mathcal{D}_f \times D_g) \to \inf .$$

Solution (f^0, g^0) to this unconstrained optimization problem is characterized by the condition $0 \in \partial\Delta_{\mathcal{D}}(f^0, g^0)$, where $\partial\Delta_{\mathcal{D}}(f^0, g^0)$ is the subdifferential of the functional $\Delta_{\mathcal{D}}(f,g)$ at point $(f^0, g^0) \in \mathcal{D} = \mathcal{D}_f \times \mathcal{D}_g$.

11.2.2 Least favorable spectral densities in classes $D_0 \times \mathcal{D}_\varepsilon$

Consider the minimax (robust) filtering problem for the functional $A\xi$ in case of two uncorrelated stochastic processes $\xi(t)$ and $\eta(t)$ with periodically stationary increments under the condition that the spectral densities belong to the sets of admissible spectral densities $\mathcal{D} = \mathcal{D}_0^k \times \mathcal{D}_\varepsilon^k$, where $\mathcal{D}_0^k$, $k = 1, 2, 3, 4$, are

$$\mathcal{D}_0^1 = \left\{ f(\lambda) \,\middle|\, \frac{1}{2\pi} \int_{-\pi}^{\pi} f(\lambda) d\lambda = P \right\},$$

$$\mathcal{D}_0^2 = \left\{ f(\lambda) \,\middle|\, \frac{1}{2\pi} \int_{-\pi}^{\pi} \mathrm{Tr}[f(\lambda)] d\lambda = p \right\},$$

$$\mathcal{D}_0^3 = \left\{ f(\lambda) \,\middle|\, \frac{1}{2\pi} \int_{-\pi}^{\pi} f_{kk}(\lambda) d\lambda = p_k, k = \overline{1,\infty} \right\},$$

$$\mathcal{D}_0^4 = \left\{ f(\lambda) \,\middle|\, \frac{1}{2\pi} \int_{-\pi}^{\pi} \langle B_1, f(\lambda)\rangle d\lambda = p \right\},$$

where p, p_k, $k = \overline{1,\infty}$, are given numbers, P, B_1 are given positive-definite Hermitian matrices, and $\mathcal{D}_\varepsilon^k$, $k = 1, 2, 3, 4$, are

$$\mathcal{D}_\varepsilon^1 = \left\{ g(\lambda) \,\middle|\, \mathrm{Tr}[g(\lambda)] = (1-\varepsilon)\,\mathrm{Tr}[g_1(\lambda)] + \varepsilon\,\mathrm{Tr}[W(\lambda)], \frac{1}{2\pi} \int_{-\pi}^{\pi} \mathrm{Tr}[g(\lambda)] d\lambda = q \right\},$$

$$\mathcal{D}_\varepsilon^2 = \left\{ g(\lambda) \,\middle|\, g_{kk}(\lambda) = (1-\varepsilon) g_{kk}^1(\lambda) + \varepsilon w_{kk}(\lambda), \frac{1}{2\pi} \int_{-\pi}^{\pi} g_{kk}(\lambda) d\lambda = q_k, k = \overline{1,\infty} \right\},$$

$$\mathcal{D}_\varepsilon^3 = \left\{ g(\lambda) \,\middle|\, \langle B_2, g(\lambda)\rangle = (1-\varepsilon)\langle B_2, g_1(\lambda)\rangle + \varepsilon\langle B_2, W(\lambda)\rangle, \frac{1}{2\pi} \int_{-\pi}^{\pi} \langle B_2, g(\lambda)\rangle d\lambda = q \right\},$$

$$\mathcal{D}_\varepsilon^4 = \left\{ g(\lambda) \,\middle|\, g(\lambda) = (1-\varepsilon) g_1(\lambda) + \varepsilon W(\lambda), \frac{1}{2\pi} \int_{-\pi}^{\pi} g(\lambda) d\lambda = Q \right\},$$

where $g_1(\lambda)$ is a fixed spectral density, $W(\lambda)$ is an unknown spectral density, q, q_k, $k = \overline{1,\infty}$, are given numbers, and Q is a given positive-definite Hermitian matrix.

The condition $0 \in \partial\Delta_{\mathcal{D}}(f^0, g^0)$ implies that the least favorable spectral densities $f^0(\lambda) \in \mathcal{D}_f^0$ and $g^0(\lambda) \in \mathcal{D}_g^0$ satisfy the equations described below for each pair of sets of admissible spectral densities.

For the first pair of sets of admissible spectral densities $\mathcal{D}_0^1 \times \mathcal{D}_\varepsilon^1$, we have

$$\frac{\lambda^{2d}}{|1-e^{i\lambda\tau}|^{2d}}(\mathbf{C}_\tau^{f0}(e^{i\lambda}))(\mathbf{C}_\tau^{f0}(e^{i\lambda}))^* = (f^0(\lambda)+g^0(\lambda))\vec{\alpha}\cdot\vec{\alpha}^*(f^0(\lambda)+g^0(\lambda)), \tag{11.34}$$

$$\frac{\lambda^{2d}}{|1-e^{i\lambda\tau}|^{2d}}(\mathbf{C}_\tau^{g0}(e^{i\lambda}))(\mathbf{C}_\tau^{g0}(e^{i\lambda}))^* = (\alpha^2+\gamma_1(\lambda))(f^0(\lambda)+g^0(\lambda))^2, \tag{11.35}$$

where α^2, $\vec{\alpha}$ are Lagrange multipliers and the function $\gamma_1(\lambda) \le 0$ and $\gamma_1(\lambda) = 0$ if $\mathrm{Tr}[g^0(\lambda)] > (1-\varepsilon)\,\mathrm{Tr}[g_1(\lambda)]$.

For the second pair of sets of admissible spectral densities $\mathcal{D}_{f0}^2 \times \mathcal{D}_\varepsilon^2$, we have

$$\frac{\lambda^{2d}}{|1-e^{i\lambda\tau}|^{2d}}(\mathbf{C}_\tau^{f0}(e^{i\lambda}))(\mathbf{C}_\tau^{f0}(e^{i\lambda}))^* = \alpha^2(f^0(\lambda)+g^0(\lambda))^2, \tag{11.36}$$

$$\frac{\lambda^{2d}}{|1-e^{i\lambda\tau}|^{2d}}(\mathbf{C}_\tau^{g0}(e^{i\lambda}))(\mathbf{C}_\tau^{g0}(e^{i\lambda}))^* = (f^0(\lambda)+g^0(\lambda))\{(\alpha_k^2+\gamma_k^1(\lambda))\delta_{kl}\}_{k,l=1}^{\infty}(f^0(\lambda)+g^0(\lambda)), \tag{11.37}$$

where α^2, α_k^2 are Lagrange multipliers and functions $\gamma_k^1(\lambda) \le 0$ and $\gamma_k^1(\lambda) = 0$ if $g_{kk}^0(\lambda) > (1-\varepsilon)g_{kk}^1(\lambda)$.

For the third pair of sets of admissible spectral densities $\mathcal{D}_{f0}^3 \times \mathcal{D}_\varepsilon^3$, we have

$$\frac{\lambda^{2d}}{|1-e^{i\lambda\tau}|^{2d}}(\mathbf{C}_\tau^{f0}(e^{i\lambda}))(\mathbf{C}_\tau^{f0}(e^{i\lambda}))^* = (f^0(\lambda)+g^0(\lambda))\{\alpha_k^2\delta_{kl}\}_{k,l=1}^{\infty}(f^0(\lambda)+g^0(\lambda)), \tag{11.38}$$

$$\frac{\lambda^{2d}}{|1-e^{i\lambda\tau}|^{2d}}(\mathbf{C}_\tau^{g0}(e^{i\lambda}))(\mathbf{C}_\tau^{g0}(e^{i\lambda}))^* = (\alpha^2+\gamma_1'(\lambda))(f^0(\lambda)+g^0(\lambda))B_2^{\top}(f^0(\lambda)+g^0(\lambda)), \tag{11.39}$$

where α_k^2, α^2 are Lagrange multipliers, function $\gamma_1'(\lambda) \le 0$ and $\gamma_1'(\lambda) = 0$ if $\langle B_2, g^0(\lambda)\rangle > (1-\varepsilon)\langle B_2, g_1(\lambda)\rangle$, and δ_{kl} are Kronecker symbols.

For the fourth pair of sets of admissible spectral densities $\mathcal{D}_{f0}^4 \times \mathcal{D}_\varepsilon^4$, we have

$$\frac{\lambda^{2d}}{|1-e^{i\lambda\tau}|^{2d}}(\mathbf{C}_\tau^{f0}(e^{i\lambda}))(\mathbf{C}_\tau^{f0}(e^{i\lambda}))^* = \alpha^2(f^0(\lambda)+g^0(\lambda))B_1^{\top}(f^0(\lambda)+g^0(\lambda)), \tag{11.40}$$

$$\frac{\lambda^{2d}}{|1-e^{i\lambda\tau}|^{2d}}(\mathbf{C}_\tau^{g0}(e^{i\lambda}))(\mathbf{C}_\tau^{g0}(e^{i\lambda}))^* = (f^0(\lambda)+g^0(\lambda))(\vec{\alpha}\cdot\vec{\alpha}^*+\Gamma(\lambda))(f^0(\lambda)+g^0(\lambda)), \tag{11.41}$$

where α^2, $\vec{\alpha}$ are Lagrange multipliers and function $\Gamma(\lambda) \le 0$ and $\Gamma(\lambda) = 0$ if $g^0(\lambda) > (1-\varepsilon)g_1(\lambda)$.

The following theorem holds.

Theorem 11.2.1. *The matrix functions $f^0(\lambda)$ and $g^0(\lambda)$ determined by the pairs of equations* (11.34)–(11.35), (11.36)–(11.37), (11.38)–(11.39), *and* (11.40)–(11.41) *and restrictions on the spectral density matrices from the corresponding classes $\mathcal{D}_0^k \times \mathcal{D}_\varepsilon^k$, $k = 1, 2, 3, 4$, are the least favorable in the class $\mathcal{D}_0^k \times \mathcal{D}_\varepsilon^k$, $k = 1, 2, 3, 4$, for the optimal linear filtering of the functional $A\xi$ if they satisfy minimality condition* (11.6) *and determine a solution of the constrained optimization problem* (11.23). *The minimax-robust spectral characteristic $\vec{h}_\tau(f^0, g^0)$ of the optimal estimate of the functional $A\xi$ is determined by formula* (11.8).

Let the spectral densities $f(\lambda)$ and $g(\lambda)$ admit the canonical factorizations (11.11), (11.13), and (11.14). Then equations (11.34)–(11.41) for the least favorable spectral densities can be presented in terms of the coefficients of these canonical factorizations.

For the set $\mathcal{D}_0^1 \times \mathcal{D}_\varepsilon^1$ of admissible spectral densities, we have

$$(\mathbf{h}_{\tau,g}^0(e^{-i\lambda}))(\mathbf{h}_{\tau,g}^0(e^{-i\lambda}))^* = (\Theta^0(e^{-i\lambda}))^\top \vec{\alpha} \cdot \vec{\alpha}^* \overline{\Theta^0(e^{-i\lambda})}, \tag{11.42}$$

$$(\mathbf{h}_{\tau,f}^0(e^{-i\lambda}))(\mathbf{h}_{\tau,f}^0(e^{-i\lambda}))^* = (\alpha^2 + \gamma_1(\lambda))(\Theta^0(e^{-i\lambda}))^\top \overline{\Theta^0(e^{-i\lambda})}, \tag{11.43}$$

for the set $\mathcal{D}_0^2 \times \mathcal{D}_\varepsilon^2$ of admissible spectral densities, we have

$$(\mathbf{h}_{\tau,g}^0(e^{-i\lambda}))(\mathbf{h}_{\tau,g}^0(e^{-i\lambda}))^* = \alpha^2(\Theta^0(e^{-i\lambda}))^\top \overline{\Theta^0(e^{-i\lambda})}, \tag{11.44}$$

$$(\mathbf{h}_{\tau,f}^0(e^{-i\lambda}))(\mathbf{h}_{\tau,f}^0(e^{-i\lambda}))^* = (\Theta^0(e^{-i\lambda}))^\top \{(\alpha_k^2 + \gamma_k^1(\lambda))\delta_{kl}\}_{k,l=1}^\infty \overline{\Theta^0(e^{-i\lambda})}, \tag{11.45}$$

for the set $\mathcal{D}_0^3 \times \mathcal{D}_\varepsilon^3$ of admissible spectral densities, we have

$$(\mathbf{h}_{\tau,g}^0(e^{-i\lambda}))(\mathbf{h}_{\tau,g}^0(e^{-i\lambda}))^* = (\Theta^0(e^{-i\lambda}))^\top \{\alpha_k^2 \delta_{kl}\}_{k,l=1}^\infty \overline{\Theta^0(e^{-i\lambda})}, \tag{11.46}$$

$$(\mathbf{h}_{\tau,f}^0(e^{-i\lambda}))(\mathbf{h}_{\tau,f}^0(e^{-i\lambda}))^* = (\alpha^2 + \gamma_1'(\lambda))(\Theta^0(e^{-i\lambda}))^\top B_2^\top \overline{\Theta^0(e^{-i\lambda})}, \tag{11.47}$$

and for the set $\mathcal{D}_0^4 \times \mathcal{D}_\varepsilon^4$ of admissible spectral densities, we have

$$(\mathbf{h}_{\tau,g}^0(e^{-i\lambda}))(\mathbf{h}_{\tau,g}^0(e^{-i\lambda}))^* = \alpha^2(\Theta^0(e^{-i\lambda}))^\top B_1^\top \overline{\Theta^0(e^{-i\lambda})}, \tag{11.48}$$

$$(\mathbf{h}_{\tau,f}^0(e^{-i\lambda}))(\mathbf{h}_{\tau,f}^0(e^{-i\lambda}))^* = (\Theta^0(e^{-i\lambda}))^\top (\vec{\alpha} \cdot \vec{\alpha}^* + \Gamma(\lambda)) \overline{\Theta^0(e^{-i\lambda})}, \tag{11.49}$$

where the Lagrange multipliers α^2, $\vec{\alpha}$, α_k^2 and the functions $\gamma_1(\lambda)$, $\gamma_k^1(\lambda)$, $\gamma_1'(\lambda)$, and $\Gamma(\lambda)$ are defined in the same way as in equations (11.34)–(11.41); δ_{kl} are Kronecker symbols.

The following theorem holds.

Theorem 11.2.2. *The matrix functions $f^0(\lambda)$ and $g^0(\lambda)$ determined by canonical factorizations* (11.24), (11.25), *and* (11.26), *pairs of equations* (11.42)–(11.43), (11.44)–(11.45), (11.46)–(11.47), *and* (11.48)–(11.49), *and restrictions on the spectral density matrices from the corresponding classes $\mathcal{D}_0^k \times \mathcal{D}_\varepsilon^k$, $k = 1, 2, 3, 4$, are the least favorable in the class $\mathcal{D}_0^k \times \mathcal{D}_\varepsilon^k$, $k = 1, 2, 3, 4$, for the optimal linear filtering of the functional $A\xi$ if they satisfy minimality* *condition* (11.6) *and determine a solution of the constrained optimization problem* (11.27) *(or* (11.28)*). The minimax-robust spectral characteristic $\vec{h}_\tau(f^0, g^0)$ of the optimal estimate of the functional $A\xi$ is determined by formula* (11.15) *(or* (11.17)*).*

11.2.3 Semi-uncertain filtering problem in classes $\mathcal{D}_0$

Consider the minimax (robust) filtering problem for the functional $A\xi$ when the spectral density $g(\lambda)$ is known and the spectral density $f(\lambda)$ belongs to the sets of admissible spectral densities $\mathcal{D}_0^k$, $k = 1, 2, 3, 4$, defined in Section 11.2.2.

Define

$$\vec{C}_\tau^{f0,\mathrm{semi}}(e^{i\lambda}) = \sum_{j=0}^{\infty}((\mathbf{P}_\tau^{f0,\mathrm{semi}})^{-1}\mathbf{R}^{f0,\mathrm{semi}}D^\tau\mathbf{a})_j e^{i\lambda(j+1)},$$

where the matrices $\mathbf{P}_\tau^{f0,\mathrm{semi}}$ and $\mathbf{R}^{f0,\mathrm{semi}}$ are defined by the Fourier coefficients of the functions

$$\frac{\lambda^{2d}}{|1-e^{i\lambda\tau}|^{2d}}((f^0(\lambda)+g(\lambda))^{-1})^{\top},\ (f^0(\lambda)(f^0(\lambda)+g(\lambda))^{-1})^{\top}.$$

For the sets of admissible spectral densities $\mathcal{D}_0^k$, $k = 1, 2, 3, 4$, we have the following equations, respectively:

$$\begin{aligned}
&\frac{\lambda^{2d}}{|1-e^{i\lambda\tau}|^{2d}}\left(\frac{|1-e^{i\lambda\tau}|^{2d}}{\lambda^{2d}}\overline{g(\lambda)\vec{B}_\tau(e^{-i\lambda})}+\vec{C}_\tau^{f0,\mathrm{semi}}(e^{i\lambda})\right)\\
&\quad\times\left(\frac{|1-e^{i\lambda\tau}|^{2d}}{\lambda^{2d}}\overline{g(\lambda)\vec{B}_\tau(e^{-i\lambda})}+\vec{C}_\tau^{f0,\mathrm{semi}}(e^{i\lambda})\right)^*\\
&\quad=(f^0(\lambda)+g(\lambda))\vec{\alpha}\cdot\vec{\alpha}^*(f^0(\lambda)+g(\lambda)), &(11.50)\\
&\frac{\lambda^{2d}}{|1-e^{i\lambda\tau}|^{2d}}\left(\frac{|1-e^{i\lambda\tau}|^{2d}}{\lambda^{2d}}\overline{g(\lambda)\vec{B}_\tau(e^{-i\lambda})}+\vec{C}_\tau^{f0,\mathrm{semi}}(e^{i\lambda})\right)\\
&\quad\times\left(\frac{|1-e^{i\lambda\tau}|^{2d}}{\lambda^{2d}}\overline{g(\lambda)\vec{B}_\tau(e^{-i\lambda})}+\vec{C}_\tau^{f0,\mathrm{semi}}(e^{i\lambda})\right)^*\\
&\quad=\alpha^2(f^0(\lambda)+g(\lambda))^2, &(11.51)\\
&\frac{\lambda^{2d}}{|1-e^{i\lambda\tau}|^{2d}}\left(\frac{|1-e^{i\lambda\tau}|^{2d}}{\lambda^{2d}}\overline{g(\lambda)\vec{B}_\tau(e^{-i\lambda})}+\vec{C}_\tau^{f0,\mathrm{semi}}(e^{i\lambda})\right)\\
&\quad\times\left(\frac{|1-e^{i\lambda\tau}|^{2d}}{\lambda^{2d}}\overline{g(\lambda)\vec{B}_\tau(e^{-i\lambda})}+\vec{C}_\tau^{f0,\mathrm{semi}}(e^{i\lambda})\right)^*\\
&\quad=(f^0(\lambda)+g(\lambda))\{\alpha_k^2\delta_{kl}\}_{k,l=1}^{\infty}(f^0(\lambda)+g(\lambda)), &(11.52)
\end{aligned}$$

$$
\begin{aligned}
&\frac{\lambda^{2d}}{|1-e^{i\lambda\tau}|^{2d}}\left(\frac{|1-e^{i\lambda\tau}|^{2d}}{\lambda^{2d}}\overline{g(\lambda)\vec{B}_\tau(e^{-i\lambda})} + \vec{C}_\tau^{f0,\mathrm{semi}}(e^{i\lambda})\right)\\
&\quad\times\left(\frac{|1-e^{i\lambda\tau}|^{2d}}{\lambda^{2d}}\overline{g(\lambda)\vec{B}_\tau(e^{-i\lambda})} + \vec{C}_\tau^{f0,\mathrm{semi}}(e^{i\lambda})\right)^*\\
&\quad= \alpha^2(f^0(\lambda)+g(\lambda))B_1^\top(f^0(\lambda)+g(\lambda)),
\end{aligned} \tag{11.53}
$$

where α^2, $\vec{\alpha}$, α_k^2, are Lagrange multipliers.

Theorem 11.2.3. *The matrix function $f^0(\lambda)$ determined by equations* (11.50), (11.51), (11.52), *or* (11.53) *and restrictions on the spectral density matrix from the corresponding classes $\mathcal{D}_0^k$, $k=1,2,3,4$, is the least favorable in the class $\mathcal{D}_0^k$, $k=1,2,3,4$, for the optimal linear filtering of the functional $A\xi$ if the known spectral density $g(\lambda)$ and the spectral density $f^0(\lambda)$ satisfy minimality condition* (11.6) *and $f^0(\lambda)$ determines a solution of the constrained optimization problem* (11.27). *The minimax-robust spectral characteristic $\vec{h}_\tau(f^0,g)$ of the optimal estimate of the functional $A\xi$ is determined by formula* (11.8).

Let the spectral density $g(\lambda)$ be known and let the spectral densities $f^0(\lambda)$ and $g(\lambda)$ admit the canonical factorizations (11.11), (11.13), and (11.14). In this case we obtain the following equations for the least favorable spectral densities in terms of the coefficients of these canonical factorizations:

$$
\begin{aligned}
&\left(\sum_{j=0}^{\infty}(\overline{\psi}_\tau^0\mathbf{c}_\tau^g)_j e^{-i\lambda j}\right)\left(\sum_{j=0}^{\infty}(\overline{\psi}_\tau^0\mathbf{c}_\tau^g)_j e^{-i\lambda j}\right)^*\\
&\quad=\left(\sum_{k=0}^{\infty}\theta^0(k)e^{-i\lambda k}\right)^\top \vec{\alpha}\cdot\vec{\alpha}^*\overline{\left(\sum_{k=0}^{\infty}\theta^0(k)e^{-i\lambda k}\right)},
\end{aligned} \tag{11.54}
$$

$$
\begin{aligned}
&\left(\sum_{j=0}^{\infty}(\overline{\psi}_\tau^0\mathbf{c}_\tau^g)_j e^{-i\lambda j}\right)\left(\sum_{j=0}^{\infty}(\overline{\psi}_\tau^0\mathbf{c}_\tau^g)_j e^{-i\lambda j}\right)^*\\
&\quad=\alpha^2\left(\sum_{k=0}^{\infty}\theta^0(k)e^{-i\lambda k}\right)^\top \overline{\left(\sum_{k=0}^{\infty}\theta^0(k)e^{-i\lambda k}\right)},
\end{aligned} \tag{11.55}
$$

$$
\begin{aligned}
&\left(\sum_{j=0}^{\infty}(\overline{\psi}_\tau^0\mathbf{c}_\tau^g)_j e^{-i\lambda j}\right)\left(\sum_{j=0}^{\infty}(\overline{\psi}_\tau^0\mathbf{c}_\tau^g)_j e^{-i\lambda j}\right)^*\\
&\quad=\left(\sum_{k=0}^{\infty}\theta^0(k)e^{-i\lambda k}\right)^\top \{\alpha_k^2\delta_{kl}\}_{k,l=1}^{\infty}\overline{\left(\sum_{k=0}^{\infty}\theta^0(k)e^{-i\lambda k}\right)},
\end{aligned} \tag{11.56}
$$

$$
\begin{aligned}
&\left(\sum_{j=0}^{\infty}(\overline{\psi}_\tau^0\mathbf{c}_\tau^g)_j e^{-i\lambda j}\right)\left(\sum_{j=0}^{\infty}(\overline{\psi}_\tau^0\mathbf{c}_\tau^g)_j e^{-i\lambda j}\right)^*\\
&\quad=\alpha^2\left(\sum_{k=0}^{\infty}\theta^0(k)e^{-i\lambda k}\right)^\top B_1^\top\overline{\left(\sum_{k=0}^{\infty}\theta^0(k)e^{-i\lambda k}\right)},
\end{aligned} \tag{11.57}
$$

where $\alpha^2, \vec{\alpha}, \alpha_k^2$ are Lagrange multipliers and the matrix $\mathbf{C}_\tau^g$ is known and defined by the coefficients of the canonical factorization of the spectral density $g(\lambda)$.

The following theorem holds.

Theorem 11.2.4. *Suppose that the spectral density $g(\lambda)$ is known and allow the canonical factorization* (11.14). *The matrix function $f^0(\lambda)$ determined by canonical factorizations* (11.24), (11.25), *or* (11.26), *equations* (11.54), (11.55), (11.56), *and* (11.57), *and restrictions on the spectral density matrix from the corresponding classes $\mathcal{D}_0^k$, $k = 1, 2, 3, 4$, is the least favorable in the class $\mathcal{D}_0^k$, $k = 1, 2, 3, 4$, for the optimal linear filtering of the functional $A\xi$ if the function $f^0(\lambda) + g(\lambda)$ satisfies minimality condition* (11.6) *and determines a solution of the constrained optimization problem* (11.30). *Coefficients $\{\psi_\tau^0(k) : k \geq 0\}$ and $\{\theta^0(k) : k \geq 0\}$ are connected by relations $\Psi^0(e^{-i\lambda})\Theta^0(e^{-i\lambda}) = E_M$ and* (11.21). *The minimax-robust spectral characteristic $\vec{h}_\tau(f^0, g)$ of the optimal estimate of the functional $A\xi$ is determined by formula* (11.17).

Conclusions

In this chapter, we describe methods of solution of the filtering problem for linear functionals in the case where both the signal and the noise processes are processes with periodically stationary dth increments. Estimates of functionals are based on observations of the process $\zeta(t) = \xi(t) + \eta(t)$ at points $t \leq 0$, where $\xi(t)$ and $\eta(t)$ are uncorrelated processes with periodically stationary dth increments. The problem was solved by representing these processes as infinite-dimensional sequences with stationary increments.

The Hilbert space projection method is applied to derive formulas for calculating the spectral characteristics and the MSEs of the optimal estimates of the functionals in the case of spectral certainty, where the spectral densities of the generated sequences are exactly known.

In the case of spectral uncertainty, where the spectral densities are not exactly known while, instead, a set of admissible spectral densities is specified, the minimax-robust method is applied. This method is illustrated with the help of concrete sets $\mathcal{D}_0 \times \mathcal{D}_\varepsilon$ of admissible spectral densities, where D_0 are classes of spectral densities with the moment restrictions and D_ε are classes describing the "ε-contaminated" models of a fixed bounded spectral density.

The results described in this chapter are published in [125]. These results rely on and develop prior studies of the minimax filtering problem by Moklyachuk and Masyutka [150] for multidimensional stationary processes, Golichenko (Dubovets'ka) and Moklyachuk [40] for periodically correlated processes, Luz and Moklyachuk [108] for processes with stationary increments, and Luz and Moklyachuk [105] for sequences with stationary increments when the noise is a sequence with stationary increments as well.

Problems

Let ϵ_m be a sequence of i. i. d. random variables with mean value 0 and variance σ^2.

1. Consider a one-dimensional stochastic sequence ξ_m, $m \in \mathbb{Z}$, defined by the equations
   a. $\xi_m = 2\xi_{m-1} - \xi_{m-2} + \epsilon_m + a\epsilon_{m-1}$;
   b. $\xi_m = (\alpha + 1)\xi_{m-1} - \alpha\xi_{m-2} + \epsilon_m + a\epsilon_{m-1}$, $|\alpha| < 1$.
   I. Check that the stochastic sequence ξ_m allows an ARIMA(p, d, q) representation and find the parameters p, d, q.
   II. Find the spectral function $F(\lambda)$ and the spectral density $f(\lambda)$ of the sequence ξ_m.
2. Find the structural function $D^{(n)}(m; \mu_1, \mu_2)$ of a one-dimensional sequence ξ_m, $m \in \mathbb{Z}$, at points $\mu_1 = 1, 2$, $\mu_2 = 1, 2$:
   a. $\xi_m = 2\xi_{m-1} - \xi_{m-2} + \epsilon_m + a\epsilon_{m-1}$;
   b. $\xi_m = (\alpha + 1)\xi_{m-1} - \alpha\xi_{m-2} + \epsilon_m + a\epsilon_{m-1}$, $|\alpha| < 1$.
3. Consider a one-dimensional stochastic sequence

$$\xi_m = \xi_{m-1} + \epsilon_m + a\epsilon_{m-1}, \quad m \in \mathbb{Z}.$$

   Find the spectral density of the stochastic sequence $\eta_m := \xi_m - \xi_{m-\mu}$, $\mu > 1$.
4. Consider a one-dimensional stochastic sequence ξ_m, $m \in \mathbb{Z}$, defined by the equation

$$\xi_m = \xi_{m-2} + \epsilon_m + a\epsilon_{m-1}.$$

   Using results of the previous exercise, find the spectral function $F(\lambda)$ and the structural function $D_s^{(n)}(m; \mu_1, \mu_2)$ of the sequence ξ_m.
5. Consider the ARIMA sequences
   a. $\xi_m = \xi_{m-1} + \epsilon_m + a\epsilon_{m-1}$,
   b. $\xi_m = 2\xi_{m-1} - \xi_{m-2} + \epsilon_m + a\epsilon_{m-1}$.
   Find the optimal linear estimate and the MSE of the estimate of value of the functional

$$A_2\xi = a_0\xi_0 + a_1\xi_1 + a_2\xi_2$$

   based on observations of the sequence ξ_m
   I. at points $m < 0$ (extrapolation problem);
   II. at points $m < 0$ and $m > 2$ (interpolation problem).
6. Consider a one-dimensional seasonal time series

$$\xi_m = \phi\xi_{m-1} + \xi_{m-2} - \phi\xi_{m-3} + \epsilon_m$$

   with a seasonal parameter $s = 2$. Represent the sequence in terms of a GM increment sequence. Find the optimal linear estimate and MSE of the estimate of the functional

https://doi.org/10.1515/9783111325620-012

$$A_1\xi = \alpha\xi_0 + (1-\alpha)\xi_1$$

based on observations of the sequence ξ_m at points

I. $m < 0$ (extrapolation problem),

II. $m \in \mathbb{Z} \setminus \{0,1\}$ (interpolation problem).

7. Consider an integrated PAR model

$$(1-B^T)(X_{mT+\nu} - \phi(\nu)X_{mT+\nu-1} - \rho(\nu)X_{mT+\nu-2}) = \varepsilon_{mT+\nu}, \tag{58}$$

where $B^T X_{mT+\nu} = X_{(m-1)T+\nu}$, $\varepsilon_{mT+\nu}$ is an uncorrelated with $X_{mT+\nu}$ white noise sequence with $\mathsf{E}\varepsilon_{mT+\nu} = 0$ and $\mathsf{E}|\varepsilon_{mT+\nu}|^2 = 1$, $1 \le \nu \le T$.

I. Derive the conditions on $\phi(\nu)$ and $\rho(\nu)$ under which model (58) allows a vector ARIMA representation.

II. Find the VARIMA representation $\mathbf{X}_m$ of the model (58).

III. Find the spectral density of the one-step increment sequence $\Delta\mathbf{X}_m$.

IV. Find an optimal linear estimate of a functional

$$A_N X = \sum_{m=0}^{N}\sum_{\nu=1}^{T} X_{mT+\nu} = \sum_{m=0}^{N} \mathbf{a}_m^\top \mathbf{X}_m,$$

where $\mathbf{a}_m^\top = (1,1,\dots,1)$, based on the observations of $X_{mT+\nu}$ at points $m < 0$, $1 \le \nu \le T$, and the MSE of the estimate.

8. Consider an integrated PAR model

$$(1-B^2)(X_{2m+\nu} - \phi(\nu)X_{2m+\nu-1}) = \varepsilon_{2m+\nu}, \quad \nu = 1,2, \tag{59}$$

where $\phi(1) = 1$, $\phi(2) = 1/2$, and $B^2 X_{2m+\nu} = X_{2(m-1)+\nu}$, $\varepsilon_{2m+\nu}$ is an uncorrelated with $X_{2m+\nu}$ white noise sequence with $\mathsf{E}\varepsilon_{2m+\nu} = 0$ and $\mathsf{E}|\varepsilon_{2m+\nu}|^2 = 1$.

I. Find the vector ARIMA representation $\mathbf{X}_m$ of the model (59).

II. Find the spectral density of the one-step increment sequence $\Delta\mathbf{X}_m$.

III. Find an optimal linear estimate of a functional

$$A_N X = \sum_{m=0}^{1}\sum_{\nu=1}^{2} X_{2m+\nu} = \mathbf{a}_0^\top \mathbf{X}_0 + \mathbf{a}_1^\top \mathbf{X}_1,$$

where $\mathbf{a}_m^\top = (1,1)$, based on observations of $X_{2m+\nu}$ at the points $m \in \mathbb{X} \setminus \{0,1\}$, $\nu = 1,2$, and the MSE of the estimate.

9. Consider a time series ξ_m, $m \in \mathbb{Z}$, with two seasonal patterns and periodic parameters

$$(1-B^s)^2(1-B^{us})\xi_m = \varepsilon_m - a_{i_m}\varepsilon_{m-1} - b_{i_m}\varepsilon_{m-s}, \quad i_m = (m \mod s) + 1,$$

where ε_m, $m \in \mathbb{Z}$, are i. i. d. random variables with $\mathsf{E}\varepsilon_m = 0$ and $\mathsf{E}|\varepsilon_m|^2 = 1$.

I. Find the vector SARIMA representation of ξ_m.
II. Find the spectral density of the one-step GM increment sequence.
III. Find an optimal linear estimate of the functional

$$A_{2s}\xi = \alpha\left(\frac{1}{s}\sum_{k=0}^{s-1}\xi_k\right) + (1-\alpha)\left(\frac{1}{s}\sum_{k=s}^{2s-1}\xi_k\right)$$

based on the observations of ξ_m at points $m < 0$ and the MSE of the estimate.

10. Consider a time series $\xi_m, m \in \mathbb{Z}$, with two fractional seasonal patterns and periodic parameters

$$(1-B^s)^{1+D_0}(1-B^{us})^{1+D_2}\xi_m = \varepsilon_m - a_{i_m}\varepsilon_{m-1} - b_{i_m}\varepsilon_{m-s}, \quad i_m = (m \mod s) + 1,$$

where $\varepsilon_m, m \in \mathbb{Z}$, are i. i. d. random variables with $\mathsf{E}\varepsilon_m = 0$ and $\mathsf{E}|\varepsilon_m|^2 = 1$, $|D_0 + D_1| < 1/2$, and $|D_1| < 1/2$.
I. Find the vector fractional SARIMA representation of ξ_m.
II. Find the spectral density of the one-step GM increment sequence.
III. Find an optimal linear estimate of the functional

$$A_{3s}\xi = \sum_{k=0}^{3s-1}\rho^k\xi_k$$

based on the observations of ξ_m at points $m < 0$ and the MSE of the estimate.
I. Find the vector SARIMA representation of ξ_m.
II. Find the spectral density of the one-step GM increment sequence.
III. Find an optimal linear estimate of a functional

$$A_{2s}\xi = \alpha\left(\frac{1}{s}\sum_{k=0}^{s-1}\xi_k\right) + (1-\alpha)\left(\frac{1}{s}\sum_{k=s}^{2s-1}\xi_k\right)$$

based on the observations of ξ_m at the points $m < 0$ and the MSE of the estimate.

11. Find the minimax estimate of the functional

$$A_2\xi = a_0\xi_0 + a_1\xi_1 + a_2\xi_2$$

and the least favorable spectral density in the class

$$\mathcal{D}_f^0 = \left\{f(\lambda)\,\middle|\,\frac{1}{2\pi}\int_{-\pi}^{\pi}\frac{|\chi_{\overline{\mu}}^{(d)}(e^{-i\lambda})|^2}{|\beta^{(d)}(i\lambda)|^2}f(\lambda)d\lambda \le P\right\}$$

for the extrapolation problem without noise, where $\xi_m, m \in \mathbb{Z}$, is a one-dimensional sequence with stationary GM increments.

12. Find an expression for the least favorable spectral density in the class

$$\mathcal{D}_{1\varepsilon} = \left\{ f(\lambda) \,\middle|\, \frac{1}{2\pi} \int_{-\pi}^{\pi} \frac{|\chi_{\overline{\mu}}^{(d)}(e^{-i\lambda})|^2}{|\beta^{(d)}(i\lambda)|^2} |f(\lambda) - \nu(\lambda)| d\lambda \leq \varepsilon \right\}$$

for the mean square optimal extrapolation problem for the functional $A\xi$, where ξ_m, $m \in \mathbb{Z}$, is a one-dimensional sequence with stationary GM increments.

13. Find the minimax-robust estimate and the least favorable spectral density in the class

$$\mathcal{D}_0^- = \left\{ f(\lambda) \,\middle|\, \frac{1}{2\pi} \int_{-\pi}^{\pi} \frac{|\beta^{(d)}(i\lambda)|^2}{|\chi_{\overline{\mu}}^{(d)}(e^{-i\lambda})|^2}, f^{-1}(\lambda) d\lambda = P \right\},$$

where P is a given positive-definite Hermitian matrix, for the mean square optimal interpolation problem for the functional $A_N\xi$, where $\xi(m)$, $m \in \mathbb{Z}$, is a sequence with periodically correlated GM increments.

14. Consider a cointegrated stochastic sequence $(\xi(m), \zeta(m))$, $m \in \mathbb{Z}$, defined by relations

$$\xi(m) = 2\xi(m-1) - \xi(m-2) + \epsilon_1(m) + a\epsilon_1(m-1),$$
$$\zeta(m) = \xi(m) + \epsilon_2(m),$$

where $\epsilon_1(m)$ and $\epsilon_2(m)$, $m \in \mathbb{Z}$, are two uncorrelated white noise sequences with mean value 0 and variance 1.

Find the optimal linear estimate and the MSE of the estimate of the functional

$$A_1\xi = \xi(0) + 2\xi(1)$$

based on observations of the sequence $\zeta(m)$ at points

I. $m < 0$ (extrapolation problem),

II. $m \in \mathbb{Z} \setminus \{0, 1\}$ (interpolation problem).

15. Consider a cointegrated stochastic sequence $(\xi(m), \zeta(m))$, $m \in \mathbb{Z}$, defined by relations

$$\xi(m) = 2\xi(m-1) - \xi(m-2) + \epsilon_1(m) + 0.5\epsilon_1(m-1),$$
$$\zeta(m) = \xi(m) + \epsilon_2(m) - 0.3\epsilon_2(m-1),$$

where $\epsilon_1(m)$ and $\epsilon_2(m)$, $m \in \mathbb{Z}$, are two uncorrelated white noise sequences with mean value 0 and variance 1.

Find the optimal linear estimate and the MSE of the estimate of the functional

$$A_1\xi = \xi(0) + 3\xi(-1)$$

based on observations of the sequence $\zeta(m)$ at points $m \leq 0$ (filtering problem).

16. Find an expression for the least favorable spectral densities $f^0(\lambda)$ and $g^0(\lambda)$ in the class $\mathcal{D} = \mathcal{D}_\varepsilon \times D_{1\varepsilon}$, where

$$\mathcal{D}_\varepsilon = \left\{ f(\lambda) \,\middle|\, f(\lambda) = (1-\varepsilon)w(\lambda) + \varepsilon u(\lambda), \frac{1}{2\pi}\int_{-\pi}^{\pi} \frac{|\chi_{\overline{\mu}}^{(d)}(e^{-i\lambda})|^2}{|\beta^{(d)}(i\lambda)|^2} f(\lambda)d\lambda = p \right\},$$

$$\mathcal{D}_{1\varepsilon} = \left\{ g(\lambda) \,\middle|\, \frac{1}{2\pi}\int_{-\pi}^{\pi} |g(\lambda) - v(\lambda)|d\lambda \le \varepsilon \right\},$$

for the mean square optimal
I. extrapolation problem;
II. interpolation problem;
III. filtering problem
for the functional $A_N\xi$ based on observations of the stochastic sequence $\xi(m)+\eta(m)$. Here $\xi(m)$, $m \in \mathbb{Z}$, is a one-dimensional sequence with stationary GM increments and $\eta(m)$, $m \in \mathbb{Z}$, is a one-dimensional stationary sequence uncorrelated with $\xi(m)$.

17. Find a representation for the least favorable spectral density $f^0(\lambda)$ in the class

a. $$\mathcal{D}_\varepsilon = \left\{ f(\lambda) \,\middle|\, f(\lambda) = (1-\varepsilon)w(\lambda) + \varepsilon u(\lambda), \frac{1}{2\pi}\int_{-\pi}^{\pi} \frac{|\chi_{\overline{\mu}}^{(d)}(e^{-i\lambda})|^2}{|\beta^{(d)}(i\lambda)|^2} f(\lambda)d\lambda = p \right\},$$

b. $$\mathcal{D}_{1\varepsilon} = \left\{ f(\lambda) \,\middle|\, \frac{1}{2\pi}\int_{-\pi}^{\pi} \frac{|\chi_{\overline{\mu}}^{(d)}(e^{-i\lambda})|^2}{|\beta^{(d)}(i\lambda)|^2} |f(\lambda) - v(\lambda)|d\lambda \le \varepsilon \right\},$$

c. $$\mathcal{D}_{2\varepsilon} = \left\{ f(\lambda) \,\middle|\, \frac{1}{2\pi}\int_{-\pi}^{\pi} \frac{|\chi_{\overline{\mu}}^{(d)}(e^{-i\lambda})|^2}{|\beta^{(d)}(i\lambda)|^2} |f(\lambda) - v(\lambda)|^2 d\lambda \le \varepsilon \right\}$$

for the mean square optimal
I. extrapolation problem,
II. interpolation problem,
III. filtering problem
for the functional $A_N\xi$ based on observation of the stochastic sequence $\xi(m)+\eta(m)$. Here $\xi(m)$, $m \in \mathbb{Z}$, is a one-dimensional sequence with stationary GM increments and $\eta(m)$, $m \in \mathbb{Z}$, is
(i) a known white noise stationary sequence,
(ii) an known MA(1) stationary sequence.

18. Derive the optimal estimate $\widehat{\xi}(0)$ and the MSE $\Delta(f,g;\widehat{\xi}(0))$ of the missed value $\xi(0)$ of the stochastic sequence $\xi(m)$ with the periodically correlated GM increments based on observations of the sequences $\xi(m)$ or $\xi(m)+\eta(m)$ at points $m \in \mathbb{Z}\setminus\{0\}$, where the noise sequence $\eta(m)$ is a periodically correlated sequence, uncorrelated with $\xi(m)$.

19. As a conclusion from Theorem 5.1.2, derive a solution to the filtering problem for the functional

$$A_{M,N}\vec{\xi} = \sum_{k=M}^{N} (\vec{a}(k))^{\top}\vec{\xi}(-k)$$

for $M \geq n(\gamma)$ in the case when the spectral densities of the vector sequences $\vec{\xi}(m)$ and $\vec{\eta}(m)$ admit canonical factorizations.

20. Prove the factorizations

$$\mathbf{G}^{-} = \widetilde{\Phi}^{*}\widetilde{\Phi}, \quad \mathbf{G}^{+} = \widetilde{\Phi}^{*}\widetilde{\Phi}^{+}$$

from Remark 5.1.2.

21. Derive representations of the mean value $c^{(d)}(\overline{\mu})$ and the structural function $D^{(d)}(m,\overline{\mu}_1,\overline{\mu}_2)$ of the stationary dth increment sequence $\vec{\xi}^{(d)}(m,\overline{\mu}) := (1 - B^{\mu_1}) \cdot \cdots \cdot (1 - B^{\mu_r})\vec{\xi}(m)$ without seasonal patterns.

22. Let $a(t) \equiv 1$, $t \in [0;T]$. For the functionals

$$A_0\xi = \int_0^T \xi(t)dt, \quad A_0^{-}\xi = \int_0^T \xi(-t)dt,$$

find the spectral characteristics and the MSE of the optimal estimate

I. $\widehat{A}_0\xi$ based on observations of the process $\xi(t)+\eta(t)$ at points $t < 0$ (extrapolation problem),

II. $\widehat{A}_0\xi$ based on observations of the process $\xi(t) + \eta(t)$ at points $t \in \mathbb{R} \setminus [0;T]$ (interpolation problem),

III. $\widehat{A}_0^{-}\xi$ based on observations of the process $\xi(t) + \eta(t)$ at points $t \leq 0$ (filtering problem),

where $\xi(t)$ is a process with T-periodically correlated increments and $\eta(t)$ is a T-periodically correlated process, uncorrelated with $\xi(t)$.

23. Find an expression for the least favorable spectral densities $f^0(\lambda)$ and $g^0(\lambda)$ in the class $\mathcal{D} = \mathcal{D}_{1\varepsilon} \times D_{2\varepsilon}$, where

$$\mathcal{D}_{1\delta} = \left\{ f(\lambda) \,\middle|\, \frac{1}{2\pi}\int_{-\pi}^{\pi} \frac{|1 - e^{i\lambda\tau}|^{2d}}{\lambda^{2d}} |\mathrm{Tr}(f(\lambda) - v_1(\lambda))| d\lambda \leq \delta_1 \right\},$$

$$\mathcal{D}_{2\delta} = \left\{ g(\lambda) \,\middle|\, \frac{1}{2\pi}\int_{-\pi}^{\pi} |\mathrm{Tr}(g(\lambda) - v_2(\lambda))|^2 d\lambda \leq \delta_2 \right\},$$

for the mean square optimal

I. extrapolation problem;

II. interpolation problem;

III. filtering problem

for the functional $A_{NT}\xi$ based on observations of the stochastic sequence $\xi(t)+\eta(t)$. Here $\xi(t)$, $t \in \mathbb{R}$, is a process with periodically stationary increments and $\eta(t)$, $t \in \mathbb{R}$, is a periodically stationary process uncorrelated with $\xi(t)$.

24. Consider the minimax interpolation problem for the functional $A_{NT}\xi$ depending on the unobserved values of the stochastic process $\xi(t)$ with periodically stationary increments from observations without noise for the set of admissible spectral densities

$$\mathcal{D}_0^- = \left\{ f(\lambda) \,\middle|\, \frac{1}{2\pi} \int_{-\pi}^{\pi} \frac{\lambda^{2d}}{|1-e^{i\lambda\tau}|^{2d}} f^{-1}(\lambda)d\lambda = P \right\},$$

where P is a given positive-definite Hermitian matrix. The Fourier coefficients $F_{l,j}^{0\tau}$ of the function $\lambda^{2d}|1-e^{i\lambda\tau}|^{-2d}((f^0(\lambda))^{-1})^\top$ satisfy an equation

$$(\mathbf{F}_N^\tau)^0 \vec{a}^{j^*,\tau} = [D_N^\tau \mathbf{a}_N]_{+\tau d},$$

as derived in Section 9.2.3. Here

$$\vec{a}^{j^*,\tau} = (\vec{0}^\top, \ldots, \vec{0}^\top, \vec{a}^\top, \vec{0}^\top, \ldots, \vec{0}^\top)^\top$$

is an $(N+\tau d+1)$-dimensional vector with infinite-dimensional entries, where the entry $\vec{a}^\top$ occupies the j^*th place. Find the least favorable spectral density and the minimax spectral characteristic choosing $j^* = N$.

A Some models of non-stationary time series

A wide range of different time series models are described in the classical book by Box and Jenkins [21]. See also the book by Luz and Moklyachuk [115].

Autoregressive moving average. The ARMA(p,q) sequence x_t, $t \in \mathbb{Z}$, with parameters (p,q) is a common model defined as

$$\psi(B)x_t = \theta(B)\varepsilon_t,$$

where ε_t, $t \in \mathbb{Z}$, is a sequence of i. i. d. random variables, $B = B_1$ is the one-step shift operator, and $\psi(z)$ and $\theta(z)$ are polynomials of degrees p and q, respectively. The stochastic sequence x_t is stationary when the roots of the *autoregressive polynomial* $\psi(z) = 0$ lie outside the unit circle.

Integrated sequence. If the autoregressive polynomial has one or more unit roots, we obtain an integrated ARMA (ARIMA(p,d,q)) model with parameters (p,d,q) defined as

$$\psi(B)(1-B)^d x_t = \theta(B)\varepsilon_t. \tag{60}$$

In terms of sequences with stationary increments, we can define an increment $w_t = (1-B)^d x_t$ of order d with the step $\mu = 1$. The model (60) is an integrated sequence of order d, with notation $I(d)$.

Seasonal time series. When the unit roots of the autoregressive polynomial satisfy the equation $1 - z^s = 0$, we obtain a model

$$\psi(B)(1-B^s)x_t = \theta(B)\varepsilon_t, \tag{61}$$

which is the simplest seasonal time series, or SARMA. An increment $w_t = (1-B^s)x_t$ of the first order with the step $\mu = s$ corresponds to this model.

General multiplicative model. A deeper investigation of (61) shows that it is reasonable to describe seasonal behavior by an equation

$$\Psi(B^s)(1-B^s)^D x_t = \Theta(B^s)\eta_t, \tag{62}$$

with two polynomials $\Psi(z)$ and $\Theta(z)$ of degrees P and Q, respectively, with the roots lying outside the unit circle. Additionally, the sequence η_t can be chosen as an integrated sequence

$$\psi(B)(1-B)^d \eta_t = \theta(B)\varepsilon_t$$

so that we obtain a general multiplicative model

$$\Psi(B^s)\psi(B)(1-B)^d(1-B^s)^D x_t = \Theta(B^s)\theta(B)\varepsilon_t \tag{63}$$

https://doi.org/10.1515/9783111325620-013

with parameters $(p,d,q) \times (P,D,Q)$. An increment $w_t = (1-B)^d(1-B^s)^D x_t$ has an order $D+d$ and two different steps $\mu_1 = 1$ and $\mu_2 = s$.

Seasonal fractional integrated autoregressive moving average model. The fractional increment orders d and D in (63) imply an example of a *seasonal fractional integrated autoregressive moving average* (SARFIMA) model.

GARMA model. Another model of fractional integration is a GARMA process described by the equation (see Gray, Cheng, and Woodward [63])

$$(1-2uB+B^2)^d x_t = \varepsilon_t, \quad |u| \le 1. \tag{64}$$

Cointegrated time series. Consider a multidimensional time series $\mathbf{x}_t = (x_t^{(1)}, x_t^{(2)}, \dots, x_t^{(m)})^\top$, where $x_t^{(i)}$, $i = \overline{1,m}$, are integrated sequences $I(d)$. The entries of the vector $\mathbf{x}_t$ are called *cointegrated of order d, b* ($\mathrm{CI}(d,b)$) if there exists a vector $\alpha \neq \overline{0}$ such that a time series $z_t = \alpha'\mathbf{x}_t$ is an integrated sequence of order $d-b$, $b > 0$, i. e., $z_t \sim I(d-b)$. The vector α is a *cointegrating* vector. For the m-dimensional cointegrated time series $\mathbf{x}_t$, more than one cointegrating vector may exist. Let there be exactly r linearly independent cointegrating vectors, $r \le m-1$. Then one can define an $m \times r$-dimensional matrix α composed of cointegrating vectors. The rank r of the matrix α is called a *cointegrated rank* of $\mathbf{x}_t$. The following statement connects cointegrated sequences and the error-correction model

$$\mathbf{A}(B)(1-B)\mathbf{x}_t = -\gamma\mathbf{z}_{t-1} + \varepsilon_t, \tag{65}$$

where ε_t is a stationary sequence, $\mathbf{z}_t = \alpha'\mathbf{x}_t$, $\gamma \neq 0$, $A(0) = I$, and all elements of the vector $\mathbf{A}(1)$ are finite.

Granger representation [62]. A stochastic vector sequence $\mathbf{x}_t \sim I(1)$ is a cointegrated sequence if and only if it allows representation (65).

Bibliography

[1] L. Aggoun and R. J. Elliott. *Measure theory and filtering: introduction and applications*. CUP, 2004.
[2] N. Ahlgren and J. Nyblom. Tests against stationary and explosive alternatives in vector autoregressive models. *J. Time Ser. Anal.*, 29(3):421–443, 2008.
[3] N. I. Akhiezer and I. M. Glasmann. *Theorie der linearen Operatoren im Hilbert-Raum*. Akademie-Verlag, Berlin, 1968.
[4] S. Albeverio and S. Kawasaki. On a localization property of wavelet coefficients for processes with stationary increments, and applications. I. Localization with respect to shift. *Adv. Appl. Probab.*, 37:938–962, 2005.
[5] S. Albeverio and S. Kawasaki. On localization property of wavelet coefficients for processes with stationary increments, and applications. II. Localization with respect to scale. *Osaka J. Math.*, 51:1–37, 2014.
[6] D. Z. Arov and H. Dym. *Multivariate prediction, de Branges spaces, and related extension and inverse problems*. Birkhäuser, Cham, 2018.
[7] S. Arvanitis and T. Magdalinos. Mildly explosive autoregression under stationary conditional heteroskedasticity. *J. Time Ser. Anal.*, 39(6):892–908, 2018.
[8] M. Avarucci and D. Marinucci. Polynomial cointegration between stationary processes with long memory. *J. Time Ser. Anal.*, 28(6):923–942, 2007.
[9] C. Baek, R. A. Davis, and V. Pipiras. Periodic dynamic factor models: estimation approaches and applications. *Electron. J. Stat.*, 12(2):4377–4411, 2018.
[10] R. T. Baillie, C. Kongcharoen, and G. Kapetanios. Prediction from ARFIMA models: comparisons between MLE and semiparametric estimation procedures. *Int. J. Forecast.*, 28:46–53, 2012.
[11] A. Banerjee, J. J. Dolado, J. W. Galbraith, and D. F. Hendry. *Co-integration, error correction and the econometric analysis of non-stationary data*. Oxford University Press, 1993.
[12] I. V. Basawa, R. Lund, and Q. Shao. First-order seasonal autoregressive processes with periodically varying parameters. *Stat. Probab. Lett.*, 67(4):299–306, 2004.
[13] I. V. Basawa and B. L. S. Prakasa Rao. *Statistical inference for stochastic processes*. Academic Press, London, 1980.
[14] W. Bell. Signal extraction for nonstationary time series. *Ann. Stat.*, 12(2):646–664, 1984.
[15] W. Bell. On RegComponent time series models and their applications. In Andrew C. Harvey, Siem Jan Koopman, and Neil Shephard, editors, *State space and unobserved component models. Theory and applications*. Cambridge University Press, Cambridge, UK, 2004.
[16] W. Bell and D. Martin. Computation of asymmetric signal extraction filters and meansquared error for ARIMA component models. *J. Time Ser. Anal.*, 25:603–625, 2004.
[17] D. Bertsimas, X. V. Doan, K. Natarajan, and C. P. Teo. Models for minimax stochastic linear optimization problems with risk aversion. *Math. Oper. Res.*, 35(3):580–602, 2010.
[18] P. S. Bhattacharya and D. D. Thomakos. Robust model rankings of forecasting performance. *J. Forecast.*, 37(6):676–690, 2018.
[19] P. Bondon. Prediction with incomplete past of a stationary process. *Stoch. Process. Appl.*, 98:67–76, 2002.
[20] P. Bondon. Influence of missing values on the prediction of a stationary time series. *J. Time Ser. Anal.*, 26(4):519–525, 2005.
[21] G. E. P. Box, G. M. Jenkins, G. C. Reinsel, and G. M. Ljung. *Time series analysis. forecasting and control. 5rd ed.* Wiley, 2016.
[22] L. Breiman. A note of minimax filtering. *Ann. Probab.*, 1:175–179, 1973.
[23] W. M. Brelsford and R. H. Jones. Time series with periodic structure. *Biometrica*, 54(3–4):403–408, 1967.
[24] J. Campbell and R. J. Shiller. Cointegration and tests of present value models. *J. Polit. Econ.*, 95:1062–1088, 1987.

https://doi.org/10.1515/9783111325620-014

[25] E. Carrizosaa, A. Olivares-Nadal, and P. Ramirez-Cobob. Robust newsvendor problem with autoregressive demand. *Comput. Oper. Res.*, 68:123–133, 2016.

[26] F. Chaari, J. Leskow, A. Napolitano, and A. Sanchez-Ramirez, editors. *Cyclostationarity: theory and methods*. Springer Cham, 2014.

[27] F. Chaari, J. Leskow, A. Napolitano, R. Zimroz, and A. Wylomanska, editors. *Cyclostationarity: theory and methods III. Contributions to the 9th workshop on cyclostationary systems and their applications, Grodek, Poland, 2016*. Springer Cham, 2017.

[28] F. Chaari, J. Leskow, A. Napolitano, R. Zimroz, A. Wylomanska, and A. Dudek, editors. *Cyclostationarity: theory and methods – II. Contributions to the 7th workshop on cyclostationary systems and their applications, Grodek, Poland, 2014*. Springer Cham, 2015.

[29] F. Chaari, J. Leskow, R. Zimroz, A. Wylomanska, and A. Dudek, editors. *Cyclostationarity: theory and methods – IV. Contributions to the 10th workshop on cyclostationary systems and their applications, February 2017, Grodek, Poland*. Springer Cham, 2020.

[30] M. J. Chambers. The estimation of continuous time models with mixed frequency data. *J. Econom.*, 193:390–404, 2016.

[31] M. J. Chambers and M. A. Thornton. Discrete time representation of continuous time ARMA processes. *Econom. Theory*, 28:219–238, 2012.

[32] H. Chigira and T. Yamamoto. Forecasting in large cointegrated processes. *J. Forecast.*, 28(7):631–650, 2009.

[33] P. Christoffersen and F. X. Diebold. Cointegration and long-horizon forecasting. *J. Bus. Econ. Stat.*, 16:450–458, 1998.

[34] M. P. Clements and D. F. Hendry. Forecasting in cointegrated systems. *J. Appl. Econom.*, 10:127–146, 1995.

[35] S. Cohen and R. J. Elliott. *Stochastic calculus and applications*. Birkhauser Basel, 2015.

[36] H. Cramér. On the theory of stationary stochastic processes. *Ann. Math.*, 41:215–230, 1940.

[37] L. De Branges. *Hilbert spaces of entire functions*. Prentice-Hall, Englewood Cliffs, New York, 1968.

[38] E. Delage and Y. Ye. Distributionally robust optimization under moment uncertainty with application to data-driven problems. *Oper. Res.*, 58(3):596–612, 2010.

[39] I. I. Dubovets'ka, O. Yu. Masyutka, and M. P. Moklyachuk. Interpolation of periodically correlated stochastic sequences. *Theory Probab. Math. Stat.*, 84:43–56, 2012.

[40] I. I. Dubovets'ka and M. P. Moklyachuk. Filtering of periodically correlated processes. *Prykl. Stat., Aktuarna Finans. Mat.*, 2:149–158, 2012.

[41] I. I. Dubovets'ka and M. P. Moklyachuk. Filtration of linear functionals of periodically correlated sequences. *Theory Probab. Math. Stat.*, 86:51–64, 2013.

[42] I. I. Dubovets'ka and M. P. Moklyachuk. Minimax estimation problem for periodically correlated stochastic processes. *J. Math. Syst. Sci.*, 3(1):26–30, 2013.

[43] I. I. Dubovets'ka and M. P. Moklyachuk. Extrapolation of periodically correlated processes from observations with noise. *Theory Probab. Math. Stat.*, 88:67–83, 2014.

[44] I. I. Dubovets'ka and M. P. Moklyachuk. On minimax estimation problems for periodically correlated stochastic processes. *Contemp. Math. Stat.*, 2(1):123–150, 2014.

[45] A. Dudek, H. Hurd, and W. Wojtowicz. Parma methods based on Fourier representation of periodic coefficients. *Wiley Interdiscip. Rev.: Comput. Stat.*, 8(3):130–149, 2016.

[46] G. Dudek. Forecasting time series with multiple seasonal cycles using neural networks with local learning. In L. Rutkowski, M. Korytkowski, R. Scherer, R. Tadeusiewicz, L. A. Zadeh, and J. M. Zurada, editors, *Artificial intelligence and soft computing. ICAISC 2013*, volume 7894 of *Lecture Notes in Computer Science*, pages 52–63. Springer, Berlin, Heidelberg, 2013.

[47] H. Dym and H. P. McKean. Extrapolation and interpolation of stationary Gaussian processes. *Ann. Math. Stat.*, 41:1817–1844, 1970.

[48] H. Dym and H. P. McKean. *Gaussian processes, function theory, and the inverse spectral problem*. Academic Press, New York, 1976.

[49] R. F. Engle and C. W. J. Granger. Co-integration and error correction: representation, estimation and testing. *Econometrica*, 55:251–276, 1987.

[50] D. Findley and D. Martin. Frequency domain analyses of SEATS and X-11/12-ARIMA seasonal adjustment filters for short and moderate-length time series. *J. Off. Stat.*, 22:1–34, 2006.

[51] J. Franke. On the robust prediction and interpolation of time series in the presence of correlated noise. *J. Time Ser. Anal.*, 5(4):227–244, 1984.

[52] J. Franke. Minimax robust prediction of discrete time series. *Z. Wahrscheinlichkeitstheor. Verw. Geb.*, 68:337–364, 1985.

[53] J. Franke and H. V. Poor. Minimax-robust filtering and finite-length robust predictors. In *Robust and nonlinear time series analysis*, volume 26 of *Lecture Notes in Statistics*, pages 87–126. Springer-Verlag, 1984.

[54] W. A. Gardner, editor. *Cyclostationarity in communications and signal processing*. IEEE, New York, NY, 1994.

[55] W. A. Gardner, A. Napolitano, and L. Paura. Cyclostationarity: half a century of research. *Signal Process.*, 86(4):639–697, 2006.

[56] A. Gholipour, J. A. Estroff, and S. K. Warfield. Robust super-resolution volume reconstruction from slice acquisitions: application to fetal brain MRI. *IEEE Trans. Med. Imaging*, 29(10):1739–1758, 2010.

[57] L. Giraitis and R. Leipus. A generalized fractionally differencing approach in long-memory modeling. *Lith. Math. J.*, 35:53–65, 1995.

[58] E. G. Gladyshev. Periodically correlated random sequences. *Sov. Math. Dokl.*, 2:385–388, 1961.

[59] E. G. Gladyshev. Periodically and almost-periodically correlated random processes with continuous time parameter. *Theory Probab. Appl.*, 8:173–177, 1963.

[60] I. Golichenko and M. Moklyachuk. Estimation problems for periodically correlated stochastic sequences with missed observations. In M. Moklyachuk, editor, *Stochastic processes: fundamentals and emerging applications*, pages 111–162. Nova Science Publishers, New York, 2023.

[61] P. G. Gould, A. B. Koehler, J. K. Ord, R. D. Snyder, R. J. Hyndman, and F. Vahid-Araghi. Forecasting time-series with multiple seasonal patterns. *Eur. J. Oper. Res.*, 191:207–222, 2008.

[62] C. W. J. Granger. Cointegrated variables and error correction models. *UCSD Discussion paper 83-13a*, 1983.

[63] H. Gray, N.-F. Zhang, and W. Woodward. On generalized fractional processes. *J. Time Ser. Anal.*, 10(3):233–257, 1989.

[64] S. Gregoir. Fully modified estimation of seasonally cointegrated processes. *Econom. Theory*, 25(5):1491–1528, 2010.

[65] U. Grenander. A prediction problem in game theory. *Ark. Mat.*, 3:371–379, 1957.

[66] M. S. Grewal and A. P. Andrews. *Kalman filtering. Theory and practice with MATLAB*. John Wiley & Sons, Hoboken, NJ, 2015.

[67] E. J. Hannan. Measurement of a wandering signal amid noise. *J. Appl. Probab.*, 4:90–102, 1967.

[68] E. J. Hannan. *Multiple time series. 2nd rev. ed.* John Wiley & Sons, New York, 2009.

[69] L. Hansen and T. Sargent. Robust control and model uncertainty. *Am. Econ. Rev.*, 91(2):60–66, 2001.

[70] A. Harvey and J. Stock. Estimation, smoothing, interpolation, and distribution for structural time-series models in continuous time. In P. C. B. Phillips (ed.). *Models Methods Appl. Econom.*, 1:55–70, 1993.

[71] U. Hassler and M. O. Pohle. Forecasting under long memory. *J. Financ. Econom.*, 21(3):742–778, 2023.

[72] Y. Hosoya. Robust linear extrapolations of second-order stationary processes. *Ann. Probab.*, 6(4):574–584, 1967.

[73] P. J. Huber. Robust estimation of a location parameter. *Ann. Math. Stat.*, 35(1):73–101, 1964.

[74] P. J. Huber. A robust version of the probability ratio test. *Ann. Math. Stat.*, 36(6):1753–1758, 1965.

[75] H. Hurd and V. Pipiras. Modeling periodic autoregressive time series with multiple periodic effects. In F. Chaari, J. Leskow, R. Zimroz, A. Wylomanska, and A. Dudek, editors, *Cyclostationarity: theory and methods – IV. CSTA 2017*, volume 16 of *Applied Condition Monitoring*, pages 1–18. Springer, Cham, 2020.

[76] H. L. Hurd and A. Miamee. *Periodically correlated random sequences. Spectral theory and practice*. Wiley Series in Probability and Statistics. John Wiley & Sons, Hoboken, NJ, 2007.
[77] J. Ishihara, M. Terra, and J. Campos. Robust Kalman filter for descriptor systems. *IEEE Trans. Autom. Control*, 51(8):1354–1358, 2006.
[78] T. Jiang, C. Huang, and L Guo. Control of uncertain nonlinear systems based on observers and estimators. *Automatica*, 59:35–47, 2015.
[79] S. Johansen. Estimation and hypothesis testing of cointegration vectors in Gaussian vector autoregressive models. *Econometrica*, 59(6):1551–1580, 1991.
[80] S. Johansen. Representation of cointegrated autoregressive processes with application to fractional processes. *Econom. Rev.*, 28:121–145, 2009.
[81] K. Juselius and R. MacDonald. The international parities between USA and Japan. *Jpn. World Econ.*, 16:17–34, 2004.
[82] G. Kallianpur. *Stochastic filtering theory*. Springer-Verlag, New York, Heidelberg, Berlin, 1980.
[83] G. Kallianpur and V. Mandrekar. Multiplicity and representation theory of purely non-deterministic processes. *Teor. Veroâtn. Primen.*, 10(4):614–644, 1965.
[84] G. Kallianpur and V. Mandrekar. Spectral theory of stationary h-valued processes. *J. Multivar. Anal.*, 1:1–16, 1971.
[85] K. Karhunen. Uber lineare Methoden in der Wahrscheinlichkeitsrechnung. *Ann. Acad. Sci. Fenn., I*, 37, 1947.
[86] S. A. Kassam. Robust hypothesis testing for bounded classes of probability densities. *IEEE Trans. Inf. Theory*, 27:242–247, 1981.
[87] S. A. Kassam. Robust hypothesis testing and robust time series interpolation and regression. *J. Time Ser. Anal.*, 3(3):185–194, 1982.
[88] S. A. Kassam and H. V. Poor. Robust techniques for signal processing: a survey. *Proc. IEEE*, 73(3):433–481, 1985.
[89] Yu. Kharin. *Robustness in statistical forecasting*. Springer International Publishing, 2013.
[90] A. N. Kolmogorov. *Selected works by A. N. Kolmogorov. Vol. II: probability theory and mathematical statistics. Ed. by A. N. Shiryayev*. volume 26 of *Mathematics and its Applications. Soviet Series*. Kluwer, Dordrecht, 1992.
[91] S. Koopman and A. Harvey. Computing observation weights for signal extraction and filtering. *J. Econ. Dyn. Control*, 27:1317–1333, 2003.
[92] D. Koroliouk. Stationary statistical experiments and the optimal estimator for a predictable component. *J. Math. Sci.*, 214(2):220–228, 2016.
[93] D. V. Koroliouk and V. S. Koroliuk. Filtration of stationary Gaussian statistical experiments. *J. Math. Sci.*, 229(1):30–35, 2018.
[94] P. Kozak, M. Luz, and M. Moklyachuk. Minimax prediction of sequences with periodically stationary increments. *Carpath. Math. Publ.*, 13(2):352–376, 2021.
[95] P. S. Kozak and M. P. Moklyachuk. Estimates of functionals constructed from random sequences with periodically stationary increments. *Theory Probab. Math. Stat.*, 97:85–98, 2018.
[96] M. G. Krein. On a fundamental approximation problem in the theory of extrapolation and filtration of stationary random processes. *Sel. Transl. Math. Stat. Probab.*, 4:127–132, 1963.
[97] H. Krim and J.-C. Pesquet. Multiresolution analysis of a class of nonstationary processes. *IEEE Trans. Inf. Theory*, 41(4):1010–1020, 1995.
[98] Y. A. Kutoyants. *Statistical inference for ergodic diffusion processes*. Springer, London, 2004.
[99] Yu. A. Kutoyants. *Parameter estimation for stochastic processes. Translated and edited by B. L. S. Prakasa Rao*. Heldermann, Berlin, 1984.
[100] Yu. A. Kutoyants. *Statistical inference for spatial Poisson processes*. Springer, New York, 1998.
[101] A. Lazaridis. A novel method of estimation under co-integration. pages 23–42. Hackensack, NJ: World Scientific, 2009.

[102] Y. Liu, Yu. Xue, and M. Taniguchi. Robust linear interpolation and extrapolation of stationary time series in Lp. *J. Time Ser. Anal.*, 41(2):229–248, 2020.

[103] R. Lund. Choosing seasonal autocovariance structures: PARMA or SARMA. In W. R. Bell, S. H. Holan, and T. S. McElroy, editors, *Economic time series: modelling and seasonality*, pages 63–80. Chapman and Hall, London, 2011.

[104] M. Luz and M. Moklyachuk. Interpolation of functional of stochastic sequences with stationary increments. *Theory Probab. Math. Stat.*, 87:117–133, 2013.

[105] M. Luz and M. Moklyachuk. Minimax-robust filtering problem for stochastic sequences with stationary increments. *Theory Probab. Math. Stat.*, 89:127–142, 2014.

[106] M. Luz and M. Moklyachuk. Minimax-robust filtering problem for stochastic sequences with stationary increments and cointegrated sequences. *Stat. Optim. Inf. Comput.*, 2(3):176–199, 2014.

[107] M. Luz and M. Moklyachuk. Robust extrapolation problem for stochastic processes with stationary increments. *Math. Stat.*, 2(2):78–88, 2014.

[108] M. Luz and M. Moklyachuk. Filtering problem for stochastic processes with stationary increments. *Contemp. Math. Stat.*, 3(1):8–27, 2015.

[109] M. Luz and M. Moklyachuk. Minimax interpolation problem for stochastic processes with stationary increments. *Stat. Optim. Inf. Comput.*, 3(1):30–41, 2015.

[110] M. Luz and M. Moklyachuk. Minimax-robust prediction problem for stochastic sequences with stationary increments and cointegrated sequences. *Stat. Optim. Inf. Comput.*, 3(2):160–188, 2015.

[111] M. Luz and M. Moklyachuk. Minimax interpolation of sequences with stationary increments and cointegrated sequences. *Mod. Stoch. Theory Appl.*, 3(1):59–87, 2016.

[112] M. Luz and M. Moklyachuk. Minimax prediction of stochastic processes with stationary increments from observations with stationary noise. *Cogent Math.*, 3:1–17, 2016.

[113] M. Luz and M. Moklyachuk. Minimax-robust filtering problem for stochastic sequences with stationary increments and cointegrated sequences. *Cogent Math.*, 3:1–21, 2016.

[114] M. Luz and M. Moklyachuk. Minimax interpolation of stochastic processes with stationary increments from observations with noise. *Theory Probab. Math. Stat.*, 94:121–135, 2017.

[115] M. Luz and M. Moklyachuk. *Estimation of stochastic processes with stationary increments and cointegrated sequences*. John Wiley & Sons, London: ISTE; Hoboken, NJ 2019.

[116] M. Luz and M. Moklyachuk. Minimax-robust estimation problems for sequences with periodically stationary increments observed with noise. *Bull. Taras Shevchenko Natl. Univ. Kyiv, Phys. Math.*, 2020(3):68–83, 2020.

[117] M. Luz and M. Moklyachuk. Minimax-robust forecasting of sequences with periodically stationary long memory multiple seasonal increments. *Stat. Optim. Inf. Comput.*, 8(3):684–721, 2020.

[118] M. Luz and M. Moklyachuk. Robust filtering of sequences with periodically stationary multiplicative seasonal increments. *Stat. Optim. Inf. Comput.*, 9(4):1010–1030, 2021.

[119] M. Luz and M. Moklyachuk. Minimax filtering of sequences with periodically stationary increments. *Cybern. Syst. Anal.*, 58(1):126–143, 2022.

[120] M. Luz and M. Moklyachuk. Robust forecasting of sequences with periodically stationary long memory multiplicative seasonal increments observed with noise and cointegrated sequences. *Stat. Optim. Inf. Comput.*, 10(2):295–338, 2022.

[121] M. Luz and M. Moklyachuk. Robust interpolation of sequences with periodically stationary multiplicative seasonal increments. *Carpath. Math. Publ.*, 14(1):105–126, 2022.

[122] M. Luz and M. Moklyachuk. Estimation problem for continuous time stochastic processes with periodically correlated increments. *Stat. Optim. Inf. Comput.*, 11(4):811–828, 2023.

[123] M. Luz and M. Moklyachuk. Minimax interpolation of continuous time stochastic processes with periodically correlated increments observed with noise. *arXiv:2307.02676*, 2023.

[124] M. Luz and M. Moklyachuk. Minimax prediction of sequences with periodically stationary increments observed with noise and cointegrated sequences. In M. Moklyachuk, editor, *Stochastic processes: fundamentals and emerging applications*, pages 189–247. Nova Science Publishers, New York, 2023.

[125] M. Luz and M. Moklyachuk. Filtering of stochastic processes having periodically correlated increments. *arXiv:2402.06396*, 2024.

[126] M. Luz and M. Moklyachuk. Filtering problem for sequences with periodically stationary multi-seasonal increments with spectral densities allowing canonical factorizations. *Stat. Optim. Inf. Comput.*, 12(2):343–363, 2024.

[127] M. Luz and M. Moklyachuk. Prediction problem for continuous time stochastic processes with periodically correlated increments observed with noise. *arXiv:2401.08642*, 2023.

[128] S. Mallat. A theory for multiresolution signal decomposition: the wavelet representation. *IEEE Trans. Pattern Anal. Mach. Intell.*, 11:14–38, 1989.

[129] E. Masry. The wavelet transform of stochastic processes with stationary increments and its application to fractional Brownian motion. *IEEE Trans. Inf. Theory*, 39:260–264, 1993.

[130] O. Yu. Masyutka, M. P. Moklyachuk, and M. I. Sidei. Extrapolation problem for multidimensional stationary sequences with missing observations. *Stat. Optim. Inf. Comput.*, 7(1):97–117, 2019.

[131] O. Yu. Masyutka, M. P. Moklyachuk, and M. I. Sidei. Interpolation problem for multidimensional stationary processes with missing observations. *Stat. Optim. Inf. Comput.*, 7(1):118–132, 2019.

[132] T. McElroy. Signal extraction for non-stationary multivariate time series with illustrations for trend inflation. *J. Time Ser. Anal.*, 36(2):209–227, 2015.

[133] T. McElroy and A. Sutcliffe. An iterated parametric approach to nonstationary signal extraction. *Comput. Stat. Data Anal.*, 50:2206–2231, 2006.

[134] M. Moklyachuk. *Convex optimization. Introductory course*. John Wiley & Sons, London: ISTE; Hoboken, NJ, 2020.

[135] M. Moklyachuk, M. Sidei, and O. Masyutka. *Estimation of stochastic processes with missing observations. Mathematics Research Developments*. Nova Science Publishers, New York, 2019.

[136] M. P. Moklyachuk. Estimation of linear functionals of stationary stochastic processes and a two-person zero-sum game. *Stanford University Technical Report*, 169:1–87, 1981.

[137] M. P. Moklyachuk. On a problem of game theory and the extrapolation of stochastic processes with values in a Hilbert space. *Theory Probab. Math. Stat.*, 24:107–114, 1981.

[138] M. P. Moklyachuk. On an antagonistic game and prediction of stationary random sequences in a Hilbert space. *Theory Probab. Math. Stat.*, 25:107–113, 1982.

[139] M. P. Moklyachuk. Minimax extrapolation and autoregression-moving average processes. *Theory Probab. Math. Stat.*, 41:77–84, 1990.

[140] M. P. Moklyachuk. Minimax filtration of linear transformations of stationary sequences. *Ukr. Math. J.*, 43:92–99, 1991.

[141] M. P. Moklyachuk. On a filtering problem for vector-valued sequences. *Theory Probab. Math. Stat.*, 47:107–118, 1993.

[142] M. P. Moklyachuk. Estimation of stochastic processes from observations with noise. *Theory Stoch. Process.*, 3(3–4):330–338, 1997.

[143] M. P. Moklyachuk. Robust procedures in time series analysis. *Theory Stoch. Process.*, 6(3–4):127–147, 2000.

[144] M. P. Moklyachuk. Game theory and convex optimization methods in robust estimation problems. *Theory Stoch. Process.*, 7(1–2):253–264, 2001.

[145] M. P. Moklyachuk. *Robust estimates for functionals of stochastic processes*. Kyiv University, Kyiv, 2008.

[146] M. P. Moklyachuk. Minimax-robust estimation problems for stationary stochastic sequences. *Stat. Optim. Inf. Comput.*, 3(4):348–419, 2015.

[147] M. P. Moklyachuk and I. I. Golichenko. *Periodically correlated processes estimates*. LAP Lambert Academic Publishing, 2016.

[148] M. P. Moklyachuk and O. Yu. Masyutka. Robust estimation problems for stochastic processes. *Theory Stoch. Process.*, 12(3–4):88–113, 2006.

[149] M. P. Moklyachuk and O. Yu. Masyutka. On the problem of filtration of vector stationary sequences. *Theory Probab. Math. Stat.*, 75:109–119, 2007.

[150] M. P. Moklyachuk and O. Yu. Masyutka. Robust filtering of stochastic processes. *Theory Stoch. Process.*, 13(1–2):16–181, 2008.

[151] M. P. Moklyachuk and O. Yu. Masyutka. *Minimax-robust estimation technique for stationary stochastic processes*. LAP Lambert Academic Publishing, 2012.

[152] M. P. Moklyachuk and O. Yu. Masyutka. Interpolation of multidimensional stationary sequences. *Theory Probab. Math. Stat.*, 73:125–133, 2006.

[153] M. P. Moklyachuk and O. Yu. Masyutka. Minimax prediction problem for multidimensional stationary stochastic sequences. *Theory Stoch. Process.*, 14(3–4):89–103, 2008.

[154] M. P. Moklyachuk and O. Yu. Masyutka. Minimax prediction problem for multidimensional stationary stochastic processes. *Commun. Stat., Theory Methods*, 40(19–20):3700–3710, 2011.

[155] M. P. Moklyachuk and V. I. Ostapenko. Minimax filtering of harmonizable stochastic sequences. *Nauk. Visn. Uzhgorod. Univ., Ser. Mat.*, (1):80–89, 2016.

[156] M. P. Moklyachuk and V. I. Ostapenko. Minimax interpolation of harmonizable sequences. *Theory Probab. Math. Stat.*, 92:135–146, 2016.

[157] M. P. Moklyachuk and V. I. Ostapenko. Minimax extrapolation problem for harmonizable stable sequences with noise observations. *Stoch. Model. Appl.*, 21(1):1–22, 2017.

[158] T. L. Molloy and J. J. Ford. Asymptotic minimax robust quickest change detection for dependent stochastic processes with parametric uncertainty. *IEEE Trans. Inf. Theory*, 62(11):6594–6608, 2016.

[159] A. Napolitano. *Generalizations of cyclostationary signal processing. Spectral analysis and applications.* Hoboken, NJ: John Wiley & Sons, 2012.

[160] A. Napolitano. Cyclostationarity: limits and generalizations. *Signal Process.*, 120:323–347, 2016.

[161] A. Napolitano. Cyclostationarity: new trends and applications. *Signal Process.*, 120:385–408, 2016.

[162] A. Napolitano. *Cyclostationary processes and time series: theory, applications, and generalizations.* Academic Press, 2019.

[163] N. Nguyen and P. Milanfar. A wavelet-based interpolation-restoration method for super resolution. *Circuits Syst. Signal Process. Mag.*, 19(4):321–338, 2000.

[164] D. Osborn. The implications of periodically varying coefficients for seasonal time-series processes. *J. Econom.*, 48(3):373–384, 1991.

[165] W. Palma and P. Bondon. On the eigenstructure of generalized fractional processes. *Stat. Probab. Lett.*, 65(93–101), 2003.

[166] P. C. B. Phillips. Optimal inference in cointegrated systems. *Econometrica*, 58:283–306, 1991.

[167] M. S. Pinsker. The theory of curves with *n*th stationary increments in Hilbert spaces. *Izv. Akad. Nauk SSSR, Ser. Mat.*, 19:319–344, 1955.

[168] M. S. Pinsker and A. M. Yaglom. On linear extrapolation of random processes with *n*th stationary increments. *Dokl. Akad. Nauk SSSR*, 94:385–388, 1954.

[169] H. Poor. On robust wiener filtering. *IEEE Trans. Autom. Control*, 25:531–536, 1980.

[170] H. Poor. Robust matched filters. *IEEE Trans. Inf. Theory*, 29(5):677–687, 1983.

[171] S. Porter-Hudak. An application of the seasonal fractionally differenced model to the monetary aggregates. *J. Am. Stat. Assoc.*, 85(410):338–344, 1990.

[172] B. L. S. Prakasa Rao. *Statistical inference for fractional diffusion processes*. John Wiley & Sons, Chichester, 2010.

[173] B. L. S. Prakasa Rao. *Associated sequences, demimartingales and nonparametric inference*. Birkhäuser, Basel, 2012.

[174] T. Proietti, M. Marczak, and G Mazzi. A class of periodic trend models for seasonal time series. *J. Forecast.*, 38(2):106–121, 2018.

[175] B. N. Pshenichnyj. *Necessary conditions of an extremum*, volume 4 of *Pure and Applied mathematics*. Marcel Dekker, New York, 1971.

[176] M. B. Rajarshi. *Statistical inference for discrete time stochastic processes*. Springer India, New Dehli, 2012.

[177] V. A. Reisen, E. Z. Monte, G. C. Franco, A. M. Sgrancio, F. A. F. Molinares, P. Bondond, F. A. Ziegelmann, and B. Abraham. Robust estimation of fractional seasonal processes: modeling and forecasting daily average SO2 concentrations. *Math. Comput. Simul.*, 146:27–43, 2018.
[178] V. A. Reisen, B. Zamprogno, W. Palma, and J. Arteche. A semiparametric approach to estimate two seasonal fractional parameters in the SARFIMA model. *Math. Comput. Simul.*, 98:1–17, 2014.
[179] F. Riesz and B. Sz.-Nagy. *Lectures in functional analysis. VIII*. Academie des Sciences de Hongrie, Budapest, 1952.
[180] R. T. Rockafellar. *Convex analysis*. Princeton University Press, Princeton, NJ, 1997.
[181] F. Roueff and A. Sanchez-Perez. Prediction of weakly locally stationary processes by auto-regression. *ALEA Lat. Am. J. Probab. Math. Stat.*, 15(2):1215–1239, 2018.
[182] Yu. A. Rozanov. Spectral theory of multidimensional stationary stochastic processes with discrete time. *Usp. Mat. Nauk*, 13(2):93–142, 1958.
[183] Yu. A. Rozanov. *Stationary stochastic processes*. Holden-Day, San Francisco–Cambridge–London–Amsterdam, 1967.
[184] B. L. Rozovsky and S. V. Lototsky. *Stochastic evolution systems. Linear theory and applications to non-linear filtering*. Springer, Cham, 2018.
[185] P. Ruckdeschel, B. Spangl, and D. Pupashenko. Robust Kalman tracking and smoothing with propagating and non-propagating outliers. *Stat. Pap.*, 55(1):93–123, 2014.
[186] B. Seong, S. K. Ahn, and P. A. Zadrozny. Estimation of vector error correction models with mixed-frequency data. *J. Time Ser. Anal.*, 34:194–205, 2013.
[187] E. Serpedin, F. Panduru, I. Sarı, and G. B. Giannakis. Bibliography on cyclostationarity. *Signal Process.*, 85(12):2233–2303, 2005.
[188] E. Sobel. *Extraction of slowly changing seasonals and of sums of accumulating sequences. PhD thesis*. Johns Hopkins University, 1967.
[189] E. Sobel. Prediction of a noise-distorted, multivariate, non-stationary signal. *J. Appl. Probab.*, 4:330–342, 1967.
[190] E. Sobel. *Prediction of a sum of accumulating sequences. Technical report. Dept. of Statistics, No. 49*. Johns Hopkins University, 1967.
[191] C. C. Solci, V. A. Reisen, Sarnaglia A. J. Q., and P. Bondon. Empirical study of robust estimation methods for par models with application to the air quality area. *Commun. Stat., Theory Methods*, 48(1):152–168, 2020.
[192] A. R. Soltani and M. Mohammadpour. Moving average representations for multivariate stationary processes. *J. Time Ser. Anal.*, 27(6):831–841, 2006.
[193] S. Song, Y. Gong, Y. Zhang, G. Huang, and G.-B. Huang. Dimension reduction by minimum error minimax probability machine. *IEEE Trans. Syst. Man Cybern. Syst.*, 47(1):58–69, 2017.
[194] B. Spangl and R. Dutter. Robustness in time series: robust frequency domain analysis. In C. Becker, R. Fried, S. Kuhnt, editors, *Robustness and complex data structures*. Springer, Berlin, Heidelberg, 2013.
[195] M. H. Stone. Linear transformations in Hilbert space. *AMS Colloq. Publ.*, 15, 1932.
[196] R. Stresing, D. Lindenberger, and R. Kummel. Cointegration of output, capital, labor, and energy. *Eur. Phys. J. B, Condens. Matter Complex Syst.*, 66(2):279–287, 2008.
[197] M. Taniguchi. Robust regression and interpolation for time series. *J. Time Ser. Anal.*, 2(1):53–62, 1981.
[198] H. Tsai, H. Rachinger, and E. M. H. Lin. Inference of seasonal long-memory time series with measurement error. *Scand. J. Stat.*, 42(1):137–154, 2015.
[199] S. Van Bellegem, P. Fryzlewicz, and R. Von Sachs. Forecasting non-stationary time series by wavelet process modeling. *Ann. Inst. Stat. Math.*, 55(4):737–764, 2003.
[200] S. K. Vastola and H. V. Poor. An analysis of the effects of spectral uncertainty on Wiener filtering. *Automatica*, 19(3):289–293, 1983.
[201] S. K. Vastola and H. V. Poor. Robust Wiener–Kolmogorov theory. *IEEE Trans. Inf. Theory*, 30(2):316–327, 1984.

[202] S. Verdu and H. V. Poor. On minimax robustness: a general approach and applications. *IEEE Trans. Inf. Theory*, 30(2):328–340, 1984.

[203] J. S. White. The limiting distribution of the serial correlation coefficient in the explosive case. *Ann. Math. Stat.*, 29:1188–1197, 1958.

[204] N. Wiener. *Extrapolation, interpolation and smoothing of stationary time series: with engineering applications*. The M. I. T. Press, Massachusetts Institute of Technology, Cambridge, Mass., 1966.

[205] N. Wiener and P. Masani. The prediction theory of multivariate stochastic processes. Part 1: the regularity condition. *Acta Math.*, 98:111–150, 1957.

[206] N. Wiener and P. Masani. The prediction theory of multivariate stochastic processes. Part 2: the linear predictor. *Acta Math.*, 99:93–137, 1957.

[207] M. Wildi. Real-time signal extraction: a shift of perspective. *Estud. Econ. Apl.*, 28(3):497–518, 2010.

[208] H. Wold. On prediction in stationary time series. *Ann. Math. Stat.*, 19(4):558–567, 1948.

[209] H. Wold. *A study in the analysis of stationary time series. 2nd ed.* Almqvist & Wiksell, Uppsala, 1954.

[210] W. Woodward, Q. Cheng, and H. Gray. A k-factor GARMA long memory model. *J. Time Ser. Anal.*, 19(4):485–504, 1998.

[211] M. I. Yadrenko. *Spectral theory of random fields. Translation Series in Mathematics and Engineering*. Optimization Software, Inc.; Springer-Verlag, New York; New York–Heidelberg–Berlin, 1983.

[212] A. M. Yaglom. Some classes of random fields in n-dimensional space related to stationary random processes. *Theory Probab. Appl.*, 2:27–320, 1957.

[213] A. M. Yaglom. Correlation theory of stationary and related random processes with stationary nth increments. *Transl. Am. Math. Soc. (2)*, 8:87–141, 1958.

[214] A. M. Yaglom. Effective solutions of linear approximation problems for multivariate stationary processes with a rational spectrum. *Theory Probab. Appl.*, 5:239–264, 1960.

[215] A. M. Yaglom. *Correlation theory of stationary and related random functions. Vol. 1: basic results. Springer Series in Statistics*. Springer-Verlag, New York etc., 1987.

[216] A. M. Yaglom. *Correlation theory of stationary and related random functions. Vol. 2: supplementary notes and references. Springer Series in Statistics*. Springer-Verlag, New York etc., 1987.

[217] A. M. Yaglom and M. S. Pinsker. Random processes with stationary increments of the nth order. *Dokl. Akad. Nauk SSSR, n. Ser.*, 90:731–734, 1953.

[218] V. N. Zasuhin. On the theory of multidimensional stationary random processes. *C. R. (Dokl.) Acad. Sci URSS, n. Ser.*, 33:435–437, 1941.

[219] M. Zhong, Y. Song, and S. Ding. Parity spacebased fault detection for linear discrete time-varying systems with unknown input. *Automatica*, 59:120–126, 2015.

Index

https://doi.org/10.1515/9783111325620-015

www.ingramcontent.com/pod-product-compliance
Lightning Source LLC
Chambersburg PA
CBHW061201240926
51846CB00006B/1079
* 9 7 8 3 1 1 1 3 2 5 3 3 0 *